Diatom Ecology

Scrivener Publishing
100 Cummings Center, Suite 541J
Beverly, MA 01915-6106

Diatoms: Biology and Applications

Series Editors: Richard Gordon and Joseph Seckbach

There are about 100,000 species and as many papers written about them since their discovery over three hundred years ago. The literature on diatoms shows about 60,000 papers during the period 2006-2023. In this context, it is timely to review the progress to date, highlight cutting-edge discoveries, and discuss exciting future perspectives. To fulfill this objective, this Diatom Series has been launched under the leadership of two experts in diatoms and related disciplines. The aim is to provide a comprehensive and reliable source of information on diatom biology and applications and enhance interdisciplinary collaborations required to advance knowledge and applications of diatoms.

Publishers at Scrivener
Martin Scrivener (martin@scrivenerpublishing.com)
Phillip Carmical (pcarmical@scrivenerpublishing.com)

Diatom Ecology

From Molecules to Metacommunities

Edited by

Nora I. Maidana

Universidad de Buenos Aires-CONICET, Argentina

Magdalena Licursi

Instituto Nacional de Limnología (CONICET-UNL), Argentina

and

Eduardo Morales†

Scrivener Publishing

WILEY

This edition first published 2024 by John Wiley & Sons, Inc., 111 River Street, Hoboken, NJ 07030, USA and Scrivener Publishing LLC, 100 Cummings Center, Suite 541J, Beverly, MA 01915, USA

For more information about Scrivener publications please visit www.scrivenerpublishing.com.

Wiley Global Headquarters
111 River Street, Hoboken, NJ 07030, USA

For details of our global editorial offices, customer services, and more information about Wiley products visit us at www.wiley.com.

Library of Congress Cataloging-in-Publication Data

ISBN 978-1-394-17445-4

Cover image: The Manzores Stream is a tributary of the Uruguay River. Photograph by Dr. Martin D. Novoa, Universidad Nacional de Entre Rios, Argentina
Cover design by Russell Richardson

Set in size of 11pt and Minion Pro by Manila Typesetting Company, Makati, Philippines

Printed in the USA

10 9 8 7 6 5 4 3 2 1

Dedicated to Eduardo A. Morales
In Memoriam 1968-2023

Eduardo Antonio Morales Luizaga (November 20, 1968 – May 19, 2023) was a distinguished biologist and educator, born in Argentina and proudly holding Bolivian citizenship. His academic journey was characterized by an unwavering commitment to advancing the field of ecology and fostering knowledge both locally and internationally.

Eduardo's educational achievements included a bachelor's degree in Biology from the University Mayor de San Simón in 1993. Seeking global perspectives, he pursued higher education in the United States, earning a Master of Science in Ecology from the University of Connecticut in 1997 and a PhD in Ecology in 2000. Eduardo's dedication to learning was further demonstrated by his Diplomat in Higher Education from the Universidad Católica Boliviana in 2008.

Throughout his career, Eduardo held various positions that showcased his expertise and passion for ecology. He served as a lecturer at the Universidad Católica Boliviana, Cochabamba, Bolivia, and took on the role of curator at the Cryptogamic Herbarium within the same institution. Additionally, he contributed significantly as an associate researcher at the Phycology Section of the Patrick Center for Environmental Research, Academy of Natural Sciences of Drexel University in Philadelphia, Pennsylvania, USA.

Eduardo's commitment to international collaboration was evident in his participation in numerous graduate courses and workshops spanning several countries. Notable among these were workshops on taxonomy, limnology, freshwater phytoplankton, and the harmonization of algal taxonomy in the United States and Latin America.

With a wealth of experience, Eduardo held key roles in academia and research. He played a pivotal part in the National Water Quality Assessment Program, serving as the taxonomy coordinator and analyst. As a research scientist at the Diatom Laboratory in the Phycology Section, Patrick Center for Environmental Research, he led workshops and contributed significantly to projects on river and lake diatoms in the continental United States.

Eduardo's dedication to education extended beyond borders. He served as an invited lecturer and associate researcher in Bolivia, impacting the scientific community and inspiring future generations. His roles as a lecturer in various biology and ecology courses at the Universidad Católica Boliviana highlighted his commitment to knowledge dissemination.

In recognition of his contributions, Eduardo was an active member of several professional societies, including the Northeast Algal Society, Bolivian Limnological Society, Bolivian Scientific Society, Latin American and Caribbean Phycological Society, and Peruvian Phycological Society.

Eduardo received numerous honors, awards, and distinctions, such as the Fulbright Scholarship, Francis Rice Trainor Scholarship Award, and multiple staff education and development grants from the Academy of Natural Sciences of Philadelphia. His legacy continues through ongoing grants and research projects, where he focused on assessing diatom diversity in Bolivian cloud forests and studying algal biodiversity in various regions of Bolivia.

His impact on the scientific community, his commitment to education, and his contributions to environmental research leave an enduring legacy that will inspire future generations in Bolivia and beyond.

Bibliography of Eduardo Morales

[1.1] Cadima, M. and Morales, E.A. (1992) Estudio preliminar cualitativo de la comunidad ficológica en tres zonas del río Rocha, Cochabamba, Bolivia [Preliminary qualitative study of the algal community in three zones of the Rocha River, Cochabamba, Bolivia]. *Acta Limnol. Brasil.* 4, 359-370.

[1.2] Cadima, M. and Morales, E.A. (1994) Estudio cualitativo del fitoplancton de 6 estanques de la Estación Pscícola Pirahiba-Valle del Sajta.

Cochabamba, Bolivia [Qualitative study of the phytoplankton from 6 tanks in the Pirahiba-Valle del Sajta Fisheries, Cochabamba, Bolivia]. *Pub. Fac. Cs. Tecnol. UMSS. Ser. Científ.* 3(2), 4-16.

[1.3] Cadima, M., Morales, E.A. and Dalence, M. (1994) Aporte al estudio de la ficoflora de Tablas Monte. Cochabamba, Bolivia [Contribution to the study of the phycoflora from Tablas Monte, Cochabamba, Bolivia]. *Pub.. Fac. Cs. Tecnol. UMSS. Ser. Científ.* 3(2), 17-37.

[1.4] Morales, E.A. and Trainor, F.R. (1996) Phytoplankton community of Alalay Pond, Cochabamba, Bolivia. *Algae* 11(2), 207.

[1.5] Morales, E.A. and Trainor, F.R. (1997) Algal phenotypic plasticity: Its importance in developing new concepts. *Algae* 12, 147-157.

[1.6] Trainor, F.R. and Morales, E.A. (1997) Implications of microalgal phenotypic plasticity [abstract]. *Phycologia* 36(4), 112. WOS:000202958000428

[1.7] Morales, E.A. (2000) *Limnological studies in Connecticut lakes: The use of periphytic diatoms in the assessment of lake trophic status.* University of Connecticut, Storrs, Connecticut, USA.

[1.8] Morales, E.A. and Potapova, M. (2000) *Third NAWQA Diatom Taxonomy Harmonization Workshop. Patrick Center for Environmental Research [Report No. 00-8].* The Academy of Natural Sciences of Philadelphia, Philadelphia, PA, USA.

[1.9] Morales, E.A. (2001) Morphological studies in selected fragilarioid diatoms (Bacillariophyceae) from Connecticut waters (USA). *Proceedings of the Academy of Natural Sciences of Philadelphia* 151, 105-120. WOS:000173000500009

[1.10] Morales, E.A. (2001) *Fifth NAWQA Taxonomy Workshop on Harmonization of Algal Taxonomy June 2001.*

[1.11] Morales, E.A., Siver, P.A. and Trainor, F.R. (2001) Identification of diatoms (Bacillariophyceae) during ecological assessments: Comparison between light microscopy and scanning electron microscopy techniques. *Proceedings of the Academy of Natural Sciences of Philadelphia* 151, 95-103. WOS:000173000500008

[1.12] Clason, T., Acker, F., Morales, E. and Marr, L. (2002) *Analysis of Diatoms on Microscope Slides Prepared From USGS NAWQA Program Algae Samples. Protocols for the analysis of algal samples collected as part of the US Geological Survey National Water-Quality Assessment Program.* 61.

[1.13] Morales, E.A. (2002) Studies in selected fragilarioid diatoms of potential indicator value from Florida (USA) with notes on the genus *Opephora* Petit (Bacillariophyceae). *Limnologica* 32(2), 102-113. WOS:000178729900002

[1.14] Morales, E.A. (2002) *Eighth NAWQA Taxonomy Workshop on Harmonization of Algal Taxonomy October 25-27, 2002.*

[1.15] Morales, E.A. and Charles, D.F. (2002) *Tenth NAWQA Taxonomy Workshop on Harmonization of Algal Taxonomy* June 13-15, 2002.

[1.16] Morales, E.A. and Hamilton, P.B. (2002) *Seventh NAWQA Taxonomy Workshop on Harmonization of Algal Taxonomy*, Philadelphia, PA, USA.

[1.17] Manoylov, K.M., Morales, E.A. and Stoermer, E.F. (2003) *Staurosira stevensonii* sp. nov. (Bacillariophyta), a new taxon from Florida, USA. *European Journal of Phycology* 38(1), 65-71. WOS:000184702800007

[1.18] Morales, E.A. (2003) *Fragilaria pennsylvanica*, a new diatom (Bacillariophyceae) species from North America, with comments on the taxonomy of the genus *Synedra* Ehrenberg. *Proceedings of the Academy of Natural Sciences of Philadelphia* 153, 155-166. WOS:000187500900011

[1.19] Morales, E.A. (2003) On the taxonomic status of the genera *Belonastrum* and *Synedrella* proposed by Round and Maidana (2001). *Cryptogamie Algologie* 24(3), 277-288. WOS:000185932400007

[1.20] Morales, E.A. and Edlund, M.B. (2003) Studies in selected fragilarioid diatoms (Bacillariophyceae) from Lake Hovsgol, Mongolia. *Phycological Research* 51(4), 225-239.

[1.21] Morales, E.A. (2005) Observations of the morphology of some known and new fragilariold diatoms (Bacillariophyceae) from rivers in the USA. *Phycological Research* 53(2), 113-133. WOS:000231180800004

[1.22] Morales, E.A., Bahls, L.L. and Cody, W.R. (2005) Morphological studies of *Distrionella incognita* (Reichardt) Williams (Bacillariophyceae) from North America with comments on the taxonomy of *Distrionella* Williams. *Diatom Research* 20(1), 115-135. WOS:000228697800004

[1.23] Morales, E.A. and Le, M. (2005) A new species of the diatom genus *Adlafia* (Bacillariophyceae) from the United States. *Proceedings of the Academy of Natural Sciences of Philadelphia* 154, 149-154. WOS:000230829500011

[1.24] Morales, E.A. and Le, M. (2005) The taxonomy of the diatom *Lacunicula sardiniensis* Lange-Bertalot, Cavacini, Tagliaventi et Alfinito and its relationship with the genus *Craticula* Grunow (Bacillariophyceae). *Proceedings of the Academy of Natural Sciences of Philadelphia* 154, 155-161. WOS:000230829500012

[1.25] Lowe, R.L., Morales, E. and Kilroy, C. (2006) *Frankophila biggsii* (Bacillariophyceae), a new diatom species from New Zealand. *New Zealand Journal of Botany* 44(1), 41-46. WOS:000237690900004

[1.26] Morales, E.A. (2006) Small *Planothidium* Round et Bukhtiyarova (Bacillariophyceae) taxa related to *P. daui* (Foged) Lange Bertalot from the United States. *Diatom Research* 21(2), 325-342. WOS:000241822400007

[1.27] Morales, E.A. and Manoylov, K.M. (2006) Morphological studies on selected taxa in the genus *Staurosirella* Williams et Round (Bacillariophyceae) from rivers in North America. *Diatom Research* 21(2), 343-364. WOS:000241822400008

[1.28] Siver, P.A., Morales, E.A., Van de Vijver, B., Smits, M., Hamilton, P.B., Lange-Bertalot, H. and Hains, J.J. (2006) Observations on *Fragilaria longifusiformis* comb. nov. et nom. nov (Bacillariophyceae), a widespread planktic diatom documented from North America and Europe. *Phycological Research* 54(3), 183-192. WOS:000240492400002

[1.29] Morales, E.A. and Vis, M.L. (2007) Epilithic diatoms (Bacillariophyceae) from cloud forest and alpine streams in Bolivia, South America. *Proceedings of the Academy of Natural Sciences of Philadelphia* 156, 123-155. WOS:000261854400010

[1.30] Morales, E.A., Vis, M.L., Fernández, E. and Kociolek, J.P. (2007) Epilithic diatoms (Bacillariophyta) from cloud forest and alpine streams in Bolivia, South America II: A preliminary report on the diatoms from Sorata, Department of La Paz. *Acta Nova* 3(4), 680-696.

[1.31] Ramirez-Herrera, M.A., Teresa, M., Cundy, A., Kostoglodov, V., Carranza-Edwards, A., Morales, E. and Metcalfe, S. (2007) Sedimentary record of late-Holocene relative sea-level change and tectonic deformation from the Guerrero Seismic Gap, Mexican Pacific Coast. *The Holocene* 17(8), 1211-1220.

[1.32] Siver, P.A., Hamilton, P.B. and Morales, E.A. (2007) Notes on the genus *Nupela* (Bacillariophyceae) including the description of a new species, *Nupela scissura* sp. nov. and an expanded description of *Nupela paludigena*. *Phycological Research* 55(2), 125-134. WOS:000245745600005

[1.33] Morales, E.A. Fernandez, E., Fernandez, C.E., Lizarro, D. and Alcoreza, P. (2008) Algal studies in Bolivia: A compilation and preliminary analysis of existing phycological literature. *Gayana Botanica* 65(1), 93-108. WOS:000258598700009

[1.34] Siver, P.A., Hamilton, P.B. and Morales, E.A. (2008) *Brevilinea pocosinensis* Siver Hamilton & Morales gen. et sp nov., a new diatom (Bacillariophyceae) genus from North Carolina, USA. *Phycological Research* 56(3), 141-148. WOS:000258597800001

[1.35] Acs, É., Morales, E.A., Kiss, K.T., Bolla, B., Plenkovic-Moraj, A., Reskóné, M.N. and Ector, L. (2009) *Staurosira grigorszkyi* nom. nov. (Bacillariophyceae) an araphid diatom from Lake Balaton, Hungary, with notes on *Fragilaria hungarica* Pantocsek. *Nova Hedwigia* 89(3-4), 469-483. WOS:000272160700013

[1.36] Morales, E.A., Fernández, E. and Kociolek, P.J. (2009) Epilithic diatoms (Bacillariophyta) from cloud forest and alpine streams in Bolivia, South America 3: diatoms from Sehuencas, Carrasco National Park, Department of Cochabamba. *Acta Botanica Croatica* 68(2), 263-283. WOS:000271400100009

[1.37] Morales, E.A. and Manoylov, K.M. (2009) *Mayamaea cahabaensis* sp. nov. (Bacillariophyceae), a new freshwater diatom from streams in the southern United States. *Proceedings of the Academy of Natural Sciences of Philadelphia* 158(1), 49-59.

[1.38] Morales, E.A., Edlund, M.B. and Spaulding, S.A. (2010) Description and ultrastructure of araphid diatom species (Bacillariophyceae) morphologically similar to *Pseudostaurosira elliptica* (Schumann) Edlund *et al.* *Phycological Research* 58(2), 97-107. WOS:000276036300003

[1.39] Morales, E.A., Lücking, R. & Anze, R. 2009. Una introducción al estudio de los líquenes de Bolivia [An introduction to the study of the lichens from Bolivia]. *Universidad Católica Boliviana San Pablo. Serie Ecología No. 1*. Cochabamba, Bolivia. 58 pp. ISSN 2076-779X.

[1.39] Morales, E.A., Fernández, E. and Chávez, V.S. (2010) Diatomeas (Bacillariophyta): ¿Por qué debemos incorporarlas en estudios de la biodiversidad boliviana [Diatoms (Bacillariophyta): Why should we incorporate them into Bolivian biodiversity studies?]. In: *Biodiversidad y Ecología en Bolivia, Simposio XXX Aniversario Instituto de Ecología Universidad Mayor de San Andrés*. Plural Editores, La Paz, Bolivia.

[1.40] Morales, E.A. and Manoylov, K.M. (2010) Three new araphid diatoms (Bacillariophyta) from rivers in North America. *Proceedings of the Academy of Natural Sciences of Philadelphia* 160(1), 29-46. WOS:000208413100005

[1.41] Morales, E.A., Wetzel, C.E. and Ector, L. (2010) Two short-striated species of *Staurosirella* (Bacillariophyceae) from Indonesia and the United States. *Polish Botanical Journal* 55(1), 107-117.

[1.42] Williams, D.M. and Morales, E.A. (2010) *Pseudostaurosira medliniae*, a new name for *Pseudostaurosira elliptica* (Gasse) Jung et Medlin. *Diatom Research* 25(1), 225-226. WOS:000277752200019

[1.43] Acs, E., Ector, L., Kiss, K.T., Cserháti, C., Morales, E.A. and Levkov, Z. (2011) Morphological observations and emended description of *Amphora micrometra* from the Bolivian Altiplano, South America. *Diatom Research* 26(1-2), 199-212. WOS:000298270200021

[1.44] Cejudo-Figueiras, C., Morales, E.A., Wetzel, C.E., Blanco, S., Hoffmann, L. and Ector, L. (2011) Analysis of the type of *Fragilaria construens* var. *subsalina* (Bacillariophyceae) and description of two morphologically related taxa from Europe and the United States. *Phycologia* 50(1), 67-77. WOS:000286357100005

[1.45] Morales, E.A., Ector, L., Fernández, E., Novais, M.H., Hlúbiková, D., Hamilton, P.B. and Kociolek, J.P. (2011) The genus *Achnanthidium* Kütz. (Achnanthales, Bacillariophyceae) in Bolivian streams: A report of taxa found in recent investigations. *Algological Studies* 136(1), 89.

[1.46] Taylor, J.C., Morales, E.A. and Ector, L. (2011) *Achnanthidium standeri* (Cholnoky) comb. nov. and *Achnanthidium taiaense* (J. R. Carter et Denny) comb. nov., two new combinations of morphologically similar *Achnanthidium* species from Africa. *Algological Studies* 136/137, 151–166.

[1.47] Wojtal, A.Z., Ector, L., Van de Vijver, B., Morales, E.A., Blanco, S., Piatek, J. and Smieja, A. (2011) The *Achnanthidium minutissimum* complex (Bacillariophyceae) in southern Poland. *Algological Studies*, 211-238.

[1.48] Morales, E.A., Manoylov, K.M. and Bahls, L.L. (2012) *Fragilariforma horstii* sp. nov. (Bacillariophyceae) a new araphid species from the northern United States of America. *Nova Hedwigia*, 141-153. WOS:000301990400010

[1.49] Morales, E.A., Novais, M.H., Chávez, G., Hoffmann, L. and Ector, L. (2012) Diatoms (Bacillariophyceae) from the Bolivian Altiplano: Three new araphid species from the Desaguadero River draining Lake Titicaca. *Fottea* 12(1), 41-58. WOS:000302453000004

[1.50] Morales, E.A., Rivera, S.F., Pedrazas, D. and Azero, M. (2012) A cinco años de la apertura del Herbario Criptogámico UCB [Five years after the opening of the UCB Cryptogamic Herbarium]. *Acta Nova* 5(3), 411-414.

[1.51] Morales, E.A., Rivera, S.F., Veizaga, A. and Fiorini, R. (2012) *Didymosphenia geminata* (Lyngbye) M. Schmidt (Bacillariophyta), una especie invasora y potencial amenaza para ecosistemas acuáticos bolivianos [*Didymosphenia geminata* (Lyngbye) M. Schmidt (Bacillariophyta), an invasive species and potential threat to Bolivian aquatic ecosystems]. *Acta Nova* 5(3), 327-343.

[1.52] Morales, E.A., Wetzel, C.E., Van de Vijver, B., Novais, M.H., Hoffmann, L. and Ector, L. (2012) Examination of type material of araphid diatoms (Bacillariophyta) nomenclaturally associated to *Fragilaria*, *Staurosira* or *Pseudostaurosira*: A long history of species concept drift. In: *Twenty-second International Diatom Symposium, Aula Academica, Ghent, 26–31 August 2012, Abstracts. Special Publication*. Vol. 58.

[1.53] Solak, C.N., Ector, L., Wojtal, A.Z., Acs, É. and Morales, E.A. (2012) A review of investigations on diatoms (Bacillariophyta) in Turkish inland waters. *Nova Hedwigia*, 431-461. WOS:000301990400025

[1.54] Wetzel, C.E., Lange-Bertalot, H., Morales, E.A., Bicudo, D.D., Hoffmann, L. and Ector, L. (2012) *Bicudoa amazonica* gen. nov. et sp. nov. (Bacillariophyta)-a new freshwater diatom from the Amazon basin with a complete raphe loss in the Eunotioid lineage. *Phytotaxa* 75, 1-18. WOS:000312188400001

[1.55] Lücking, R., Dal-Forno, M., Lawrey, J.D., Bungartz, F., Rojas, M.E.H., Hernandez, J.E., Marcelli, M.P., Moncada, B., Morales, E.A., Nelsen, M.P., Paz, E., Salcedo, L., Spielmann, A.A., Wilk, K., Will-Wolf, S. and Yánez-Ayabaca, A. (2013) Ten new species of lichenized Basidiomycota in the genera *Dictyonema* and *Cora* (Agaricales: Hygrophoraceae), with a key to all accepted genera and species in the *Dictyonema* clade. *Phytotaxa* 139(1), 1-38. WOS:000325990400001

[1.56] Morales, E.A., Guerrero, J.M., Wetzel, C.E., Sala, S. and Ector, L. (2013) Unraveling the identity of *Fragilaria pinnata* Ehrenberg and *Staurosira pinnata* Ehrenberg: Research in progress on a convoluted story. *Cryptogamie Algologie* 34(2), 89-102. WOS:000338714800003

[1.57] Morales, E.A. and Rivera, S.F. (2013) Macrófitas poco frecuentes o desconocidas de la laguna Alalay, Cochabamba, Bolivia [Rare or unknown macrophytes from Alalay Lagoon, Cochabamba, Bolivia]. *Acta Nova* 6(1-2), 36-52.

[1.58] Morales, E.A., Rivera, S.F., Ayala, R. and Acosta, F. (2013) Ocurrencia de *Cyclotella meduanae* Germain (Bacillariophyta) en la laguna Alalay, un cuerpo de agua salobre y eutrófico (Cochabamba, Bolivia) [Occurrence of *Cyclotella meduanae* Germain (Bacillariophyta) in Alalay Lagoon, a brackish and eutrophic body of water (Cochabamba, Bolivia)]. *Acta Nova* 6(1-2), 82-93.

[1.59] Rivera, S.F. and Morales, E.A. (2013) Nuevos taxa de diatomeas (Bacillariophyta) descritos en el Herbario Criptogámico de la Universidad Católica Boliviana San Pablo (HCUCB) [New taxa of diatoms (Bacillariophyta) described in the Cryptogamic Herbarium of the Universidad Católica Boliviana San Pablo (HCUCB)]. *Acta Nova* 6(1-2), 142-152.

[1.60] Rivera, S.F. and Morales, E.A. (2013) Publicaciones científicas del Herbario Criptogámico de la Universidad Católica Boliviana San Pablo (HCUCB) [Scientific publications of the Cryptogamic Herbarium of the Universidad Católica Boliviana San Pablo (HCUCB)]. *Acta Nova* 6(1-2), 122-141.

[1.61] Wetzel, C.E., Morales, E.A., Blanco, S. and Ector, L. (2013) *Pseudostaurosira cataractarum* comb. nov. (Bacillariophyta): type analysis, ecology and world distribution of a former "centric" diatom. *Acta Nova* 6(1-2), 53-63.

[1.62] Wetzel, C.E., Morales, E.A., Hinz, F., Bicudo, D.C. and Ector, L. (2013) *Fragilariforma javanica* comb. nov.: Analysis of type material of a widely reported species with a tropical distribution. *Diatom Research* 28(4), 373-379. WOS:000326562100004

[1.63] Bolaños Angulo, A., Azero, M. and Morales, E.A. (2014) Evaluación de la influencia de tres especies: Tunal (*Opuntia ficus-indica* L.), chacatea (Dodonea viscosa Jacq.) y molle (*Schinus molle* L.) sobre las propiedades edáficas de un sistema agroforestal sucesional en Combuyo-Vinto [Evaluation of the influence of three species: Tunal (*Opuntia ficus-indica* L.), *chacatea* (*Dodonea viscosa* Jacq.) and molle (*Schinus molle* L.) on the edaphic properties of a successional agroforestry system in Combuyo-Vinto]. *Acta Nova* 6(4), 523-524.

[1.64] Buczkó, K., Ognjanova-Rumenova, N., Magyari, E.K., Morales, E.A., Wojtal, A.Z., Magyar, I. and Ector, L. (2014) Structure, biostratigraphy and distribution of *Navicula haueri* Grunow, a Miocene diatom (Bacillariophyta) from the Carpathian Realm. *Nova Hedwigia*, 369-379. WOS:000339535600019

[1.65] Enache, M.D., Potapova, M. and Morales, E.A. (2014) *Platessa strelnikovae* (Bacillariophyta), a new species from Maine and Vermont lakes, USA. *Nova Hedwigia*, 239-244. WOS:000339535600012

[1.66] Khan-Bureau, D.A., Beauchene, M.S., Ector, L., Morales, E.A. and Lewis, L.A. (2014) Observations of two nuisance stalk-forming diatoms (Bacillariophyta) from a river in Connecticut, Northeastern

USA: *Didymosphenia* sp. and *Cymbella janischii* (A. Schmidt) De Toni. *Bioinvasions Records* 3(3), 139-149. WOS:000372414700002

[1.67] Morales, E.A., Buczkó, K., Wetzel, C.E., Novais, M.H., Ognjanova-Rumenova, N., Hoffmann, L. and Ector, L. (2014) Transfer of *Staurosira grunowii* to *Staurosirella*. *Diatom Research* 29(2), 105-110. WOS:000331336600001

[1.68] Morales, E.A., Rivera, S.F., Wetzel, C.E., Novais, M.H., Hamilton, P.B., Hoffmann, L. and Ector, L. (2014) New epiphytic araphid diatoms in the genus *Ulnaria* (Bacillariophyta) from Lake Titicaca, Bolivia. *Diatom Research* 29(1), 41-54. WOS:000330933100006

[1.69] Morales, E.A., Wetzel, C.E., Rivera, S.F., Novais, M.H., Hoffmann, L. and Ector, L. (2014) *Craticula strelnikoviana* sp. nov. and *Craticula guaykuruorum* sp. nov. (Bacillariophyta) from South American saline lakes. *Nova Hedwigia*, 223-237. WOS:000339535600011

[1.70] Morales, E.A., Wetzel, C.E., Rivera, S.F., Van de Vijver, B. and Ector, L. (2014) Current taxonomic studies on the diatom flora (Bacillariophyceae) of the Bolivian Altiplano, South America, with possible consequences for palaeoecological assessments. *Journal of Micropalaeontology* 33, 121-129. WOS:000343130100003

[1.71] Rioual, P., Morales, E.A., Chu, G.Q., Han, J.T., Li, D., Liu, J.Q., Liu, Q., Mingram, J. and Ector, L. (2014) *Staurosira longwanensis* sp. nov., a new araphid diatom (Bacillariophyta) from Northeast China. *Fottea* 14(1), 91-100. WOS:000334587400007

[1.72] Talgatti, D., Wetzel, C.E., Morales, E.A., Ector, L. and Torgan, L.C. (2014) Transfer of *Fragilaria atomus* Hust. to the genus *Stauroforma* (Bacillariophyta) based on observation of type and newly collected material. *Phytotaxa* 158(1), 43-56. WOS:000330510600003

[1.73] van de Vijver, B., Morales, E.A. and Kopalová, K. (2014) Three new araphid diatoms (Bacillariophyta) from the Maritime Antarctic Region. *Phytotaxa* 167(3), 256-266. WOS:000335889800004

[1.74] Almeida, P.D., Wetzel, C.E., Morales, E.A., Ector, L. and Bicudo, D.C. (2015) *Staurosirella acidophila* sp. nov., a new araphid diatom (Bacillariophyta) from southeastern Brazil: ultrastructure, distribution and autecology. *Cryptogamie Algologie* 36(3), 255-270. WOS:000360577500003

[1.75] Grana, L., Morales, E.A., Besta, T., Echazu, D., Wetzel, C.E., Novais, M.H., Rajdlova, H., Ector, L. and Maidana, N.I. (2015) On the geographical distribution and ecology of *Pseudostaurosira cataractarum* (Bacillariophyceae): New findings in the Palearctic and Neotropic ecozones. *Brazilian Journal of Botany* 38(4), 809-821. WOS:000368593500013

[1.76] Morales, E.A. (2015) On the synonymy of *Nitzschia frustulum* var. *subsalina*, *N. inconspicua* and *N. boliviana* sensu Trobajo *et al.* 2013. *Acta Nova* 8(2), 114-121.

[1.77] Morales, E.A. and Rivera, S.F. (2015) Ocurrencia de *Batrachospermum macrosporum* y su fase Chantransia (= *Audouinella macrospora*) en la cuenca del río Yacuma (Beni, Bolivia) [Occurrence of *Batrachospermum macrosporum* and its Chantransia phase (= *Audouinella macrospora*) in the Yacuma river basin (Beni, Bolivia)]. *Acta Nova* 7(1), 47-60.

[1.78] Morales, E.A., Rivera, S.F., Lozano, R. and Bicudo, C.E.D.M. (2015) Primer reporte confirmado de *Hydrodictyon reticulatum* (Chlorophyta, Hydrodictyaceae) para Bolivia [First confirmed report of *Hydrodictyon reticulatum* (Chlorophyta, Hydrodictyaceae) for Bolivia]. *Acta Nova* 7(2), 109-113.

[1.79] Morales, E.A., Rivera, S.F., Rubin, S.D.C., Vis, M.L. and Houk, V. (2015) *Aulacoseira kruegeriana* (Diatomeae, Coscinodiscophyceae): A new centric diatom from high-elevation Andean rivers and streams of Bolivia. *Diatom Research* 30(3), 269-277. WOS:000362502400003

[1.80] Morales, E.A., Rivera, S.F., Wetzel, C.E., Hamilton, P.B., Bicudo, D.C., Pibernat, R.A. and Ector, L. (2015) Hipótesis: La agrupación *Microcystis aeruginosa* Kütz.-*Nitzschia palea* (Kütz.) W. Sm.-bacterias en la laguna Alalay, Cochabamba, Bolivia es de tipo simbiótico [Hypothesis: The group *Microcystis aeruginosa* Kütz.-*Nitzschia palea* (Kütz.) W. Sm.-bacteria in the Alalay lagoon, Cochabamba, Bolivia is of symbiotic type]. *Acta Nova* 7(2), 122-142.

[1.81] Morales, E.A., Wetzel, C.E., Van de Vijver, B. and Ector, L. (2015) Morphological studies on type material of widely cited araphid diatoms (Bacillariophyta). *Phycologia* 54(5), 455-470. WOS:000365398100004

[1.82] Almeida, P.D., Morales, E.A., Wetzel, C.E., Ector, L. and Bicudo, D.C. (2016) Two new diatoms in the genus *Fragilaria* Lyngbye (Fragilariophyceae) from tropical reservoirs in Brazil and comparison with type material of *F. tenera*. *Phytotaxa* 246(3), 163-183. WOS:000370224200001

[1.83] Bicudo, D.C., Tremarin, P.I., Almeida, P.D., Zorzal-Almeida, S., Wengrat, S., Faustino, S.B., Costa, L.F., Bartozek, E.C.R., Rocha, A.C.R., Bicudo, C.E.M. and Morales, E.A. (2016) Ecology and distribution of *Aulacoseira* species (Bacillariophyta) in tropical reservoirs from Brazil. *Diatom Research* 31(3), 199-215. WOS:000385847400003

[1.84] Khan-Bureau, D.A., Morales, E.A., Ector, L., Beauchene, M.S. and Lewis, L.A. (2016) Characterization of a new species in the genus *Didymosphenia* and of *Cymbella janischii* (Bacillariophyta) from Connecticut, USA. *European Journal of Phycology* 51(2), 203-216. WOS:000373496900008

[1.85] Lücking, R., Nelsen, M.P., Aptroot, A., Barillas De Klee, R., Bawingan, P.A., Benatti, M.N., Binh, N.Q., Bungartz, F., Cáceres, M.E.S., Canêz, L.D., Chaves, J.L., Ertz, D., Esquivel, R.E., Ferraro, L.I., Grijalva, A., Gueidan, C., Hernandez, J.E., Knight, A., Lumbsch, H.T., Marcelli, M.P., Mercado-Díaz, J.A., Moncada, B., Morales, E.A., Naksuwankul, K., Orozco, T., Parnmen, S., Rivas Plata, E., Salazar-Allen, N., Spielmann, A.A. and Ventura, N. (2016) A phylogenetic framework for reassessing generic concepts and species delimitation in the lichenized family

Trypetheliaceae (Ascomycota: Dothideomycetes). *Lichenologist* 48(6), 739-762. WOS:000390364600008

[1.86] Wengrat, S., Morales, E.A., Wetzel, C.E., Almeida, P.D., Ector, L. and Bicudo, D.C. (2016) Taxonomy and ecology of *Fragilaria billingsii* sp. nov. and analysis of type material of *Synedra rumpens* var. *fusa* (Fragilariaceae, Bacillariophyta) from Brazil. *Phytotaxa* 270(3), 191-202. WOS:000382037400003

[1.87] Almeida, P.D., Wetzel, C.E., Morales, E.A., Ector, L. and Bicudo, D.C. (2017) New species and combinations on Fragilariforma (Bacillariophyta) from tropical freshwater environments. *Fottea* 17(2), 277-292. WOS:000416485100011

[1.88] García, M.L., Maidana, N.I., Ector, L. and Morales, E.A. (2017) *Staurosira patagonica* sp. nov., a new diatom (Bacillariophyta) from southern Argentina, with a discussion on the genus *Staurosira* Ehrenberg. In: *Diatom Taxonomy and Ecology: From France to the Sub-Antarctic Islands.* B. VanDeVijver, L. Tudesque and L. Ector, (eds.) J Cramer, Stuttgart. 146: 103-123. WOS:000473347500009

[1.89] Khan-Bureau, D.A., Ector, L., Morales, E.A., Wade, E.J. and Lewis, L.A. (2017) Contrasting morphological and DNA barcoding methods for diatom (Bacillariophyta) identification from environmental samples in the Eightmile River in Connecticut, USA. In: *Diatom Taxonomy and Ecology: From France to the Sub-Antarctic Islands.* B. VanDeVijver, L. Tudesque and L. Ector, (eds.) J Cramer, Stuttgart. 146: 279-302. WOS:000473347500020

[1.90] Lücking, R., Dal Forno, M., Moncada, B., Coca, L.F., Vargas-Mendoza, L.Y., Aptroot, A., Arias, L.J., Besal, B., Bungartz, F., Cabrera-Amaya, D.M., Cáceres, M.E.S., Chaves, J.L., Eliasaro, S., Gutiérrez, M.C., Marin, J.E.H., Herrera-Campos, M., Holgado-Rojas, M.E., Jonitz, H., Kukwa, M., Lucheta, F., Madriñán, S., Marcelli, M.P., Martins, S.M.D., Mercado-Díaz, J.A., Molina, J.A., Morales, E.A., Nelson, P.R., Nugra, F., Ortega, F., Paredes, T., Patiño, A.L., Peláez-Pulido, R.N., Pérez, R.E.P., Perlmutter, G.B., Rivas-Plata, E., Robayo, J., Rodríguez, C., Simijaca, D.F., Soto-Medina, E., Spielmann, A.A., Suárez-Corredor, A., Torres, J.M., Vargas, C.A., Yánez-Ayabaca, A., Weerakoon, G., Wilk, K., Pacheco, M.C., Diazgranados, M., Brokamp, G., Borsch, T., Gillevet, P.M., Sikaroodi, M. and Lawrey, J.D. (2017) Turbo-taxonomy to assemble a megadiverse lichen genus: Seventy new species of Cora (Basidiomycota: Agaricales: Hygrophoraceae), honouring David Leslie Hawksworth's seventieth birthday. *Fungal Diversity* 84(1), 139-207. WOS:000403785900007

[1.91] Maidana, N.I., Aponte, G.A., Fey, M., Schäbitz, F. and Morales, E.A. (2017) *Cyclostephanos salsae* and *Placoneis patagonica*, two new diatoms (Bacillariophyta) from Laguna Chaltel in southern Patagonia, Argentina. In: *Diatom Taxonomy and Ecology: From France to the Sub-Antarctic*

Islands. B. VanDeVijver, L. Tudesque and L. Ector, (eds.) J Cramer, Stuttgart. 146: 89-102. WOS:000473347500008

[1.92] Maidana, N.I., Morales, E.A., Bradbury, J.P., Schäbitz, F. and Houk, V. (2017) A new order and family of diatoms: Arcanodiscales, Arcanodiscaceae (Bacillariophyta) to accommodate *Arcanodiscus platti* gen. nov. et sp. nov. from the Argentinian Patagonia. In: *Diatom Taxonomy and Ecology: From France to the Sub-Antarctic Islands [Nova Hedwigia Beiheft 146]*. B. VanDeVijver, L. Tudesque and L. Ector, (eds.) J. Cramer, Stuttgart, Germany. 146: 63-72. WOS:000473347500006

[1.93] Mogul, R., Vaishampayan, P., Bashir, M., McKay, C.P., Schubert, K., Bornaccorsi, R., Gomez, E., Tharayil, S., Payton, G., Capra, J., Andaya, J., Bacon, L., Bargoma, E., Black, D., Boos, K., Brant, M., Chabot, M., Chau, D., Cisneros, J., Chu, G., Curnutt, J., DiMizio, J., Engelbrecht, C., Gott, C., Harnoto, R., Hovanesian, R., Johnson, S., Lavergne, B., Martinez, G., Mans, P., Morales, E., Oei, A., Peplow, G., Piaget, R., Ponce, N., Renteria, E., Rodriguez, V., Rodriguez, J., Santander, M., Sarmiento, K., Scheppelmann, A., Schroter, G., Sexton, D., Stephenson, J., Symer, K., Russo-Tait, T., Weigel, B. and Wilhelm, M.B. (2017) Microbial community and biochemical dynamics of biological soil crusts across a gradient of surface coverage in the central Mojave Desert. *Frontiers in Microbiology* 8, #1974. WOS:000413413700001

[1.94] Morales, E.A., Achá, M. and Villarroel, M.L. (2017) En busca de alternativas de alimentación frente al cambio climático: estudio preliminar del impacto nutricional del *Nostoc* en estudiantes de primaria de la comunidad de Putucuni, Cordillera del Tunari, Quillacollo, Cochabamba-Bolivia [In search of food alternatives in the face of climate change: Preliminary study of the nutritional impact of *Nostoc* on primary school students in the community of Putucuni, Cordillera del Tunari, Quillacollo, Cochabamba-Bolivia]. *Acta Nova* 8(2), 205-215.

[1.95] Morales, E.A., Rivera, S.F., Vildozo, L.H. and Pol, A. (2017) Floración algal nociva (FAN) producida por cianobacterias en la laguna Alalay, Cochabamba, Bolivia [Harmful algal bloom (HAB) produced by cyanobacteria in Alalay Lagoon, Cochabamba, Bolivia]. *Acta Nova* 8(1), 50-75.

[1.96] Morales, E.A., Rivera, S.F., Vildozo, L.H. and Pol, A. (2017) Harmful algal bloom (HAB) produced by cyanobacteria in Alalay Shallow Lake, Cochabamba, Bolivia. *Acta Nova* 8(1), 50-75.

[1.97] Morales, E.A., Rivera, S.F., Wetzel, C.E., Hamilton, P.B., Houk, V. and Ector, L. (2017) A new centric diatom belonging to the *Cyclotella meneghiniana* Kutz. species complex (Bacillariophyta): *Cyclotella longirimoportulata* from Alalay Pond, Cochabamba, Bolivia. In: *Diatom Taxonomy and Ecology: From France to the Sub-Antarctic Islands*. B. VanDeVijver, L. Tudesque and L. Ector, (eds.) J Cramer, Stuttgart. 146: 73-87. WOS:000473347500007

[1.98] Rubin, S.D.C., Marín, I., Gómez, M.J., Morales, E.A., Zekker, I., San Martín-Uriz, P., Rodríguez, N. and Amils, R. (2017) Prokaryotic diversity and community composition in the Salar de Uyuni, a large scale, chaotropic salt flat. *Environmental Microbiology* 19(9), 3745-3754. WOS:000410660900027

[1.99] Grana, L., Morales, E.A., Maidana, N.I. and Ector, L. (2018) Two new species of *Staurosira* and *Pseudostaurosira* (Bacillariophyta) from the highlands of Argentina (south-central Andes) and two new nomenclatural combinations. *Phytotaxa* 365(1), 60-72. WOS:000442280600002

[1.100] Rusanov, A.G., Ector, L., Morales, E.A., Kiss, K.T. and Acs, É. (2018) Morphometric analyses of *Staurosira inflata* comb. nov. (Bacillariophyceae) and the morphologically related *Staurosira tabellaria* from north-western Russia. *European Journal of Phycology* 53(3), 336-349. WOS:000441948900008

[1.101] Seeligmann, C.T., Maidana, N.I. and Morales, E.A. (2018) Fragilariaceae (Bacillariophyta) in high altitude wetlands of Catamarca (Argentina). *Boletin de la Sociedad Argentina de Botanica* 53(4), 507-519. WOS:000459326400001

[1.102] García, M.L., Morales, E.A., Guerrero, J.M., Tremarin, P.I. and Maidana, N.I. (2019) New *Aulacoseira* species (Bacillarliphyta) from the Argentinean Patagonia and re-examination of type material of *Melosira perpusilla* Frenguelii. *Phytotaxa* 408(3), 161-177. WOS:000473254300002

[1.103] Guerrero, J.M., García, M.L. and Morales, E.A. (2019) *Staurosirella andino-patagunica* sp. nov. (Bacillariophyta) from lake sediments in Patagonia, Argentina. *Phytotaxa* 402(3), 131-144. WOS:000466483100001

[1.104] Lehmkuhl, E.A., Morales, E.A., Tremarin, P.I., Bartozek, E.C.R., Zorzal-Almeida, S., Ludwig, T.A.V. and Bicudo, C.E.D. (2019) Two new species of *Nitzschia* (Bacillariaceae, Bacillariophyta) from tropical reservoirs of southeastern Brazil. *Phytotaxa* 399(1), 83-99. WOS:000462041000009

[1.105] Morales, E.A., Wetzel, C.E., Haworth, E.Y. and Ector, L. (2019) Ending a 175-year taxonomic uncertainty: Description of *Staurosirella neopinnata* sp. nov. (Bacillariophyta) to accommodate *Fragilaria pinnata*, a highly misconstrued taxon with a purported worldwide distribution. *Phytotaxa* 402(2), 75-87. WOS:000465412500001

[1.106] Morales, E.A., Wetzel, C.E., Novais, M.H., Buczkó, K., Morais, M.M. and Ector, L. (2019) Morphological reconsideration of the araphid genus *Pseudostaurosira* (Bacillariophyceae), a revision of *Gedaniella*, *Popovskayella* and *Serratifera*, and a description of a new *Nanofrustulum* species. *Plant Ecology and Evolution* 152(2), 262-284. WOS:000475963600013

[1.107] Novais, M.H., Penha, A.M., Morales, E.A., Potes, M., Salgado, R. and Morais, M. (2019) Vertical distribution of benthic diatoms in a large reservoir (Alqueva, Southern Portugal) during thermal stratification. *Science of the Total Environment* 659, 1242-1255. WOS:000457293700120

[1.108] Barbosa, L.G., Amorim, C.A., Parra, G., Portinho, J.L., Morais, M., Morales, E.A. and Menezes, R.F. (2020) Advances in limnological research in Earth's drylands. *Inland Waters* 10(4), 429-437. WOS:000581678000001

[1.109] García, M.L., Morales, E.A., Mann, D.G. and Maidana, N.I. (2020) *Sellaphora mayrii* (Bacillariophyceae), a new diatom from the Argentinean Patagonia. *Phytotaxa* 437(3), 135-146. WOS:000524757700002

[1.110] Morales, E.A. (2020) *Frustulia tunariensis* sp. nov. (Bacillariophyceae) from the Andes of Bolivia, South America. *Diversity-Basel* 12(9), #362. WOS:000581209900001

[1.111] Morales, E.A., Wetzel, C.E., Novais, M.H., Morais, M.M. and Ector, L. (2020) *Nitzschia transtagensis* sp. nov. (Bacillariophyceae) from a spring in Southern Portugal. *Botany Letters* 167(1), 32-41. WOS:000496786800001

[1.112] Novais, M.H., Morales, E.A., Penha, A.M., Potes, M., Bouchez, A., Barthès, A., Costa, M.J., Salgado, R., Santos, J. and Morais, M. (2020) Benthic diatom community dynamics in Mediterranean intermittent streams: Effects of water availability and their potential as indicators of dry-phase ecological status. *Science of the Total Environment* 719, 19, #137462. WOS:000521936300021

[1.113] Pinto, P., Oliveira, J.M., Leitao, F., Morais, M.M., Chícharo, L., Vaz, P., Delgado, S.M.A., Voreadou, C., Morales, E.A. and Teodósio, M.A. (2020) Development of a metric of aquatic invertebrates for volunteers (MAIV): A simple and friendly biotic metric to assess ecological quality of streams. *Water* 12(3), #654. WOS:000529249500039

[1.114] Zorzal-Almeida, S., Bartozek, E.C.R., Morales, E.A. and Bicudo, D.C. (2020) *Brachysira aristidesii* sp. nov. (Bacillariophyceae, Brachysiraceae): A new species from oligotrophic and mesotrophic tropical reservoirs in southeastern Brazil. *Phytotaxa* 456(1), 105-113. WOS:000563782900008

[1.115] García, M.L., Bustos, S., Villacís, L.A., Laprida, C., Mayr, C., Moreno, P.I., Maidana, N.I. and Morales, E.A. (2021) New araphid species of the genus *Pseudostaurosira* (Bacillariophyceae) from southern Patagonia. *European Journal of Phycology* 56(3), 255-272. WOS:000606927400001

[1.116] Morais, M., Penha, A.M., Novais, M.H., Landim, L., Victória, S.S., Morales, E.A. and Barbosa, L.G. (2021) Some observations on phytoplankton community structure, dynamics and their relationship to water quality in five Santiago Island Reservoirs, Cape Verde. *Water* 13(20), #2888. WOS:000716267000001

[1.117] Morales, E., Novais, M.H., García, M.L., Maidana, N.I. and Morais, M.M. (2021) *Planothidium audax* sp. nov. (Bacillariophyta, Achnanthidiaceae), a new diatom from temporary streams in southern Portugal. *Phytotaxa* 510(3), 288-294. WOS:000677644400003

[1.118] Morales, E., Wetzel, C. and Ector, L. (2021) New and poorly known "araphid" diatom species (Bacillariophyta) from regions near Lake Titicaca, South America and a discussion on the continued use of morphological characters in "araphid" diatom taxonomy. *PhytoKeys* 187, 23-70.

[1.119] Morales, E.A. (2021) Eduardo A. Morales. https://diatoms.org/contributor/EduardoAMorales

[1.120] Morales, E.A. (2021) *Punctastriata altera* sp. nov. (Bacillariophyta, Fragilariophyceae), a diatom from the United States of America. *Phytotaxa* 516(3), 296-300. WOS:000690737800003

[1.121] Morales, E.A., Novais, M.H., García, M.L., Maidana, N.I. and Morais, M.M. (2021) A new species of *Punctastriata* (Bacillariophyta, Fragilariophyceae) from temporary streams in southern Portugal. *Phytotaxa* 507(3), 261-265. WOS:000659907900002

[1.122] Morales, E.A., Wetzel, C.E., Ector, L. and Van de Vijver, B. (2021) Analysis of type material of *Opephora pacifica* and emendation of the genus *Opephora* (Bacillariophyta). *Botanica Marina* 64(1), 55-69. WOS:000621737800006

[1.123] Harper, M.A., Morales, E.A. and Van De Vijver, B. (2022) An unusual freshwater diatom with bilobate ends from the Mid-Miocene of East Antarctica: *Staurosirella tigris* sp. nov. (Fragilariaceae, Bacillariophyta). *Phytotaxa* 541(2), 201-208. WOS:000778751600004

[1.124] Morales, E.A. (2022) Two new species of *Pseudostaurosiropsis* (Bacillariophyta, Fragilario-phyceae) from the United States of America, with taxonomic comments on the genus. *Algae* 37(1), 33-47. WOS:000789108500003

[1.125] Morales, E.A., Novais, M.H., Wetzel, C.E., Ector, L. and Morais, M.M. (2022) *Fragilaria odeloucaensis* sp. nov. (Bacillariophyta, Fragilariaceae) a diatom from temporary streams in southern Portugal. *Phytotaxa* 555(3), 217-230. WOS:000837197200001

[1.126] Van de Vijver, B., Ector, L., Wetzel, C.E. and Morales, E.A. (2022) Lectotypification of four former *Odontidium* species described by William Smith (Bacillariophyta, Fragilariales). *Notulae Algarum* 226, 1-3.

[1.127] Van de Vijver, B., Morales, E.A., Schuster, T.M., Wetzel, C.E. and Ector, L. (2022) Typification and morphology of *Staurosirella lapponica* (Grunow) DM Williams & Round and *Staurosirella pinnata* var. *intercedens* (Grunow) PB Hamilton (Staurosiraceae, Bacillariophyta). *Nova Hedwigia* 115(1-2), 31-45. WOS:000814206100001

[1.128] Vouilloud, A.A., Guerrero, J.M. and Morales, E.A. (2022) *Fragilaria pinnata* var. *subcapitata* Frenguelli, typification and current taxonomic status. *Phytotaxa* 566(1), 133-139. WOS:000865727300005

[1.129] Khan-Bureau, D.A., Blanco, S., Morales, E.A., Ector, L., Beauchene, M.S. and Lewis, L.A. (2023) Using geometric morphometrics to disentangle *Didymosphenia hullii* and *D. geminata* (Bacillariophyceae) from Connecticut, USA, and two congeneric relatives. *European Journal of Phycology* 58(1), 99-109. WOS:000808165100001

[1.130] Morales, E.A. and Manoylov, K.M. (2023) *Achnanthidium lucectorii* sp. nov. and *A. enigmaticum* sp. nov., novel diatoms (Bacillariophyta,

Achnanthidiaceae) from Texas, USA. *Phytotaxa* 592(1), 1-20. WOS:000980410500001

[1.131] Morales, E.A., Novais, M.E., García, M.L., Maidana, N.I. and Morais, M.M. (2023) Two new species of *Nanofrustulum* (Bacillariophyta) from temporary rivers in the Alentejo region, southern Portugal. *Nova Hedwigia* 117(1-4), 71-92.

[1.132] Maidana, N.I., Licursi, M. and Morales, E.A. (eds.) (2024) *Diatom Ecology: from Molecules to Metacommunities [DIEC, Volume in the series: Diatoms: Biology & Applications, series editors: Richard Gordon & Joseph Seckbach, in press, THIS VOLUME].* Wiley-Scrivener, Beverly, MA, USA.

Contents

Preface **xxvii**

1 Using Diatoms to Guide Successful Ecological Restoration **1**
Evelyn E. Gaiser
1.1 Introduction 2
1.2 Components of Successful Ecosystem Restoration 3
1.3 Incorporating Diatoms in Ecological Restoration 8
1.3.1 Incorporating Diatom Indicators in Aquatic Ecosystem Restoration Planning 8
1.3.2 Using Diatoms to Establish Restoration Goals and Targets 10
1.3.3 Diatom Metrics of Restoration Success 18
1.3.4 Adaptively Managing Ecological Condition 22
1.4 Communicating Diatom-Based Indicators of Restoration Progress 24
1.4.1 Communicating Restoration Progress 24
1.4.2 Public Education and Engagement 25
1.5 Summary and Opportunities 27
Acknowledgments 29
References 29

2 Abundance-Occupancy Relationships in Freshwaters: The Case of Pond Diatoms **41**
Saúl Blanco, María Borrego-Ramos and Óscar Fernández-Ramos
2.1 Introduction 42
2.2 Materials and Methods 43
2.2.1 Study Area 43
2.2.2 Sample Collection and Processing 45
2.2.3 Statistical Analysis 46
2.3 Results 47

2.4 Discussion 49
Acknowledgments 53
References 53

3 The Ecology of Diatom Reproduction **59**
David G. Mann and Mark B. Edlund
3.1 Introduction 60
3.2 Reports of Auxosporulation in Natural Populations 63
3.3 Ecological Significance of Auxosporulation 71
3.4 Future Prospects for Phenological Studies 74
Acknowledgments 76
References 77

4 Deterministic and Stochastic Effects on Freshwater Diatom Biodiversity and Community Composition **85**
Xavier Benito, Sophia I. Passy, Annika Vilmi, Aurélien Jamoneau, Juliette Tison-Rosebery, Maria Kahlert, Chad A. Larson, Joseph L. Mruzek, Janne Soininen and Andrew Bramburger
4.1 Introduction 86
4.2 Deterministic Effects on Diatom Metacommunities 94
4.3 Stochastic Effects on Diatom Metacommunities 98
4.4 Relative Importance of Deterministic and Stochastic Effects on Diatom Metacommunities 100
4.5 Assembly Processes Over Time 102
4.6 Assembly Processes Over Space 104
4.7 Concluding Remarks and Future Directions 105
References 106

5 Recent Insights into Diatom Distributions and the Contributions of Molecular Approaches **123**
Jana Kulichová and Rosa Trobajo
5.1 Introduction 124
5.2 A Metacommunity Framework 142
5.2.1 Abiotic Factors 143
5.2.2 Biotic Interactions 145
5.2.3 Dispersal 145
5.3 Metabarcoding 146
5.4 The Problem of Undersampling 149
5.5 Conclusions 150
Acknowledgment 150
References 150

6 Diatom Taxonomy: Disentangled from Ecology? 163
Eileen J. Cox
6.1 Introduction 164
6.2 Diatom Taxonomy 165
6.2.1 Overview of Species Concepts and Changing Emphases 165
6.2.2 Naming and the Type Method 167
6.2.3 Impact of Molecular Studies 168
6.2.4 Morphological Variation Within Clones 170
6.2.5 Phenotypic Plasticity – Implications 171
6.2.6 Use of Infraspecific Categories for Diatom Taxa 173
6.3 Diatom Ecology 174
6.3.1 History of Diatom Ecology 174
6.3.1.1 Phytoplankton 174
6.3.1.2 Benthic Ecology 175
6.3.1.3 Paleolimnology 176
6.3.2 Diatoms as Bioindicators 176
6.3.3 Problems with Establishing Species 'Preferences' 177
6.3.4 Implications of Experimental Work 179
6.3.5 What Do Molecular Studies Tell Us? 180
6.4 Where Are the Challenges? 181
6.5 Integrating Information 182
6.6 Conclusions 183
Acknowledgments 183
References 183

7 Diatom Biodiversity in Urban Freshwater Ecosystems: Opportunities and Challenges for Conservation 203
Majoi de Novaes Nascimento and Luciane Fontana
7.1 Contextualizing Urban and Novel Ecosystems 204
7.2 Diatom Diversity in Novel Freshwater Ecosystems: Insights and Future Directions from a Systematic Literature Review 208
7.3 Mapping Gaps and Opportunities 213
7.4 Ecological Importance of Diatoms in Monitoring and Managing Urban Freshwater Ecosystems 216
7.5 Climate Change and Conservation of Diatom Biodiversity in Urban Systems 217
7.6 New Approaches and Perspectives for the Future 219
References 220

8 Guilds for Diatoms: History and Future Prospects **233**
Csilla Stenger-Kovács and Viktória B-Béres
8.1 The Origin and History of the Guild Concept 234
8.2 Passy's Idea: Use of the Guild Concept in Diatom Research 236
8.3 Relationship Between Guilds and Environmental Variables 242
8.3.1 Nutrients 242
8.3.2 Physical Disturbances 244
8.3.3 Light 247
8.3.4 Alkalinity and pH 248
8.3.5 Extreme Environments 248
8.4 Effect of Global Threatening Processes on Guilds 249
8.4.1 Direct Anthropogenic Pressures: Land Use, Urbanization, Mining and Salinization 249
8.4.2 Climatic Extremes, Intermittence, Floods 251
8.5 Diatom Guilds in Plankton and Their Ecological Role 252
8.6 Guild Dispersal, Guild-Based Metacommunity Analysis and Β-Diversity 253
8.6.1 Dispersal 254
8.6.2 Metacommunity Studies 254
8.6.3 Beta Diversity 257
8.7 Guilds in Conservation Biology Studies 258
8.8 Guilds in the Past, Present and Future 260
8.8.1 Guilds in Paleolimnological Studies 260
8.8.2 Recent Trends of Guild Distributions 260
8.8.3 Future Distribution of Diatom Guilds 262
8.9 Conclusion and Future Directions of the Guild Concept in Diatom Community Ecology 263
Acknowledgment 264
References 265

9 Soil Diatoms and Their Use in Bioindication **281**
Jasper Foets and Eveline Pinseel
9.1 Introduction 282
9.2 Taxonomy 284
9.2.1 Methodological Protocols 288
9.3 Ecology 291
9.3.1 Adaptations to Non-Aquatic Conditions 291
9.3.2 Functionality on Soils 293
9.3.3 The Influence of Environmental Factors 294
9.3.3.1 Vitality and Absolute Abundances 294
9.3.3.2 Community Composition 295

9.3.3.3 The Impact of Herbicides and Oil Pollution on Soil Diatoms 296
9.4 Diatoms as Environmental Markers 297
9.4.1 Soil Quality Biomonitoring with Diatoms 297
9.4.2 Tracing Hydrological Connectivity with Soil Diatoms 298
9.4.3 Soil Diatoms as Environmental Markers: Major Challenges and Solutions 300
9.5 Soil Diatoms in Forensic Sciences 301
9.6 Future Perspectives 302
Appendix 304
References 315

10 Lacustrine Diatoms as Paleoclimate Proxies and Their Use in Climate Change Research 329
Gabriela S. Hassan
10.1 Introduction 330
10.2 Lacustrine Diatoms as Archives of Climatic Signals 332
10.3 Diatoms, Limnology and Climate 335
10.3.1 Numerical Methods 335
10.3.2 Salinity and Conductivity 335
10.3.3 Water-Level Fluctuations 337
10.3.4 Temperature 340
10.3.5 pH 343
10.3.6 Nutrients 345
10.3.7 Ice Cover 348
10.4 Conclusions 351
Acknowledgments 353
References 353

11 Diatoms in Temporary Rivers: Importance in a Global Climate Change Context 361
Maria Helena Novais and Manuela Morais
11.1 Introduction 361
11.2 Diatoms Adaptation Mechanisms in Temporary Rivers 366
11.3 Flow Reduction/Lentification 370
11.4 Desiccation Stress 372
11.5 Thermal Stress 374
11.6 UV Radiation Exposure 376
11.7 Community Dynamics in Temporary Rivers 380
11.8 Diatoms as Bioindicators in Temporary Rivers 382

11.9 Final Remarks 382
References 383

12 The Ecology of Diatoms in Peatlands: Communities from Tierra del Fuego Peat Bogs as a Study Case 393
Valeria Casa, María V. Quiroga, Bart Van de Vijver and Gabriela Mataloni
12.1 Introduction 394
12.1.1 Wetlands and Peatlands: General Characteristics and Global Significance 394
12.1.2 Peatland Classification and Environmental Characterization 395
12.2 Singularity of Peatland Diatoms 396
12.2.1 Main Genera and Species in Peatlands 396
12.2.2 Physical, Chemical and Landscape-Scale Drivers of Diatom Communities 398
12.3 Diatoms in Tierra del Fuego Peat Bogs 400
12.3.1 Prior Knowledge 400
12.3.2 Study Area, Sampling and Statistical Analysis 401
12.3.3 Changes in Diatom Composition Between Peat Bogs and Their Surrounding Environments 403
12.3.4 Diatom Communities Inside the Peat Bogs 404
12.4 Closing Remarks and Future Perspectives 409
References 410

Index 421

Preface

In recent years, there has been a considerable increase in the amount of research devoted to diatom ecology, with a wide spectrum of approaches. This large amount of information, published in many different journals and books, makes it very difficult for both trained researchers and students to keep up-to-date. Therefore, the idea of bringing together the most recent information on some of the topics related to diatom ecology was enthusiastically promoted by our dear colleague and friend, Eduardo A. Morales.

In order to carry out this project, Eduardo posed a series of questions on relevant topics in diatom ecology and, with these in mind, invited prestigious colleagues from different parts of the world to be part of this adventure, some as authors of chapters and others as reviewers of manuscripts.

The chapters finally included in this book attempt to answer some of the questions raised in the original proposal:

- Are diatoms suitable tools for ecological restoration?
- What would be the features that make them reliable in this context?
- What makes diatoms (so) ecologically successful?
- In an ecological sense, why is there such variability in diatom reproductive strategies and why are these worth considering?
- What do new approaches in ecological synthesis provide to diatom ecology, biogeography and metacommunities?
- Are all diatoms widely distributed and is each species uniquely characterized by its own, unaltered phenotype?
- Can we really make any ecological consideration without knowing (with a high degree of certainty) the identity of taxa?
- Are urban ecosystems important repositories of biodiversity?
- What benefits have been realized due to the progress made in the ecological guilds of diatoms?

- Why, how, and when are soil diatoms used in bioindication and what are the benefits of such an approach?
- Are diatoms bona fide indicators of climate change?
- Are diatom communities in temporary rivers important for these lotic ecosystems as they are subjected to the effects of climate change?
- Do diatoms in peatlands behave differently from their terrestrial and aquatic (rivers, lakes, others) counterparts?

Eduardo died in May 2023, and we decided to complete the task and turn it into a tribute to his memory. We hope we have succeeded. For the colleagues (authors and reviewers) who helped us make this possible, we have nothing but words of thanks.

Nora I. Maidana
Magdalena Licursi
July 2024

1

Using Diatoms to Guide Successful Ecological Restoration

Evelyn E. Gaiser

Institute of Environment and Department of Biological Sciences, Florida International University, Miami, Florida, USA

Abstract

The number of aquatic ecosystems severely affected by human activities is rapidly increasing across the globe. Recognition of the extent of degradation of aquatic ecosystems and their services has catalyzed efforts to reverse these trends. To be successful, projects to restore or rehabilitate degraded aquatic ecosystems and their services require explicit achievable goals and regular assessments of progress toward those goals to guide adaptive management. Restoration targets are often established from knowledge of pre-disturbance or reference ecosystem states. Assessment monitoring can then include measurements of chemical and ecological criteria associated with that target. Diatoms have long been employed in assessing ecosystem degradation, particularly water quality deterioration, but there are fewer examples of their use in guiding the goals, assessment, and adaptive management of ecosystem restoration and resilience. This chapter evaluates whether diatoms provide an equally powerful tool for adaptively managing ecosystem restoration and resilience as they do for signaling ecosystem degradation. Examples of successful diatom-based restoration guidance are provided from the Florida Everglades, which is currently undergoing the largest-scale ecosystem restoration program in the world. The chapter also discusses novel approaches for building public literacy about diatoms to encourage more effective and widespread application in successfully rehabilitating and sustaining the planet's threatened aquatic resources.

Keywords**:** Diatom indicators, ecological restoration, ecological resilience, Everglades restoration, adaptive management, ecosystem assessment

---

Email: gaisere@fiu.edu

Nora I. Maidana, Magdalena Licursi and Eduardo Morales (eds.) Diatom Ecology: From Molecules to Metacommunities, (1–40) © 2024 Scrivener Publishing LLC

1.1 Introduction

Aquatic ecosystems are experiencing unprecedented degradation due to human activities, including climate change, widespread damming and diversion, agricultural development, urbanization, overfishing, introduction of non-native species, and pollution from chemicals, erosion, plastics, and solid waste [1.1]. Because of their ubiquity and sensitivity to environmental conditions, diatoms are frequently used to signal the effects of these human activities on aquatic ecosystems, including changes in pH, conductivity, transparency, hydrology, trophic state, oxygen, and nutrients, among other factors [1.2]. Diatoms are considered ecosystem engineers because they drive food webs, form soils, regulate biogeochemical cycles, and modify flow regimes. Because of their fast and reliable response to changes, diatoms also provide important indications of impending ecosystem-scale cascades [1.3]. Early detection and prevention of these cascades is important because they can lead to ecological state shifts that are difficult or almost impossible to reverse [1.4]. In recognition that state shifts often entail undesirable losses in aquatic ecosystem services, many countries have employed diatom-based water quality monitoring programs to identify degrading ecosystems. Targeting degrading ecosystems before they have reached these irreversible states is invaluable. Examples of diatom-based monitoring programs exist throughout North America [1.5] [1.6], Europe [1.7] [1.8] [1.9], and the tropics [1.10] [1.11] [1.12]. Diatoms and other organisms are sounding alarms that we are close to crossing planetary boundaries [1.13], warning of further degradation if widespread actions are not taken to reverse losses. However, comparatively little research has been conducted on (1) how efficiently these diatom signals are acted upon to limit degradation while ecosystems are still in a recoverable state, or (2) whether diatoms can serve as equally effective indicators of ecosystem recovery. This chapter reviews the literature related to these topics, addressing the question: what features make diatoms reliable indicators of ecological recovery, and how can we best use them to guide restoration and management of aquatic ecosystems for resilience in a rapidly changing world?

Determining whether and how diatoms can be used to guide more effective aquatic ecosystem management depends on how well they meet the needs of restoration managers. Those needs include: (1) reliable ecological condition targets based on pre-degradation, reference, and/or predicted (and achievable) desired conditions, (2) accurate indication of whether restoration actions are nearing or achieving target conditions, and also

(3) clear and easily understood presentation of results to resource managers and the public, who finance and depend on the results [1.14]. This chapter addresses the key components of successful restoration programs and the desired features of diatom bioindicators that best help guide the restoration process.

While incorporation of diatoms into bioassessment of ecosystem recovery has received much less attention than their use in assessing risk of degradation [1.15], multiple studies have incorporated diatoms in establishing targets and assessing and communicating restoration progress. These studies include bioassessments of lakes, rivers, and wetlands undergoing restoration, which are often coupled with paleoecological, experimental, and modeling studies that have been used to establish targets and metrics. This chapter reviews these studies with a focus on the Florida Everglades, which is currently subject to the largest-scale global ecosystem restoration project ever attempted. The qualities of diatoms that best serve the needs of restoration managers are evaluated and then strategies are recommended for incorporating these qualities into restoration planning, assessments of ecosystem recovery and resilience, science communication, and engagement with the public.

1.2 Components of Successful Ecosystem Restoration

Ecological restoration is defined as an "intentional activity that initiates or accelerates the recovery of an ecosystem with respect to its health, integrity and sustainability" [1.16]. The process of restoring ecosystems typically entails phases of planning and design, implementation, monitoring and assessment, and evaluation to adaptively manage progress [1.17] [1.18]. The planning and design phase establishes objectives to scientifically evaluate the causes of degradation and propose actions necessary to reverse degradation to achieve desired ecosystem conditions [1.19]. Ideally, restoration plans are co-produced by scientists and stakeholders working together to generate targets that account for scientific and societal constraints and benefits [1.20] (Figure 1.1). Restoration targets are commonly based on biological conditions indicative of healthy ecosystem status, as determined from long-term data, historical knowledge, paleoecological information, or reference/undisturbed neighboring ecosystems. However, targets based on pre-disturbance baselines are not always achievable because most ecosystems are complex adaptive systems where more than one key driver is changing, including drivers that are irreversible at the scale of the restoration project [1.21]. For example, global climate change and non-native species introductions can continue to alter ecosystems even as local-scale drivers are being restored. In addition, most degraded ecosystems are so highly disturbed that they have lost critical features that are unrecoverable in realistic restoration timeframes, prompting use of more realistic terms like "rehabilitation" or "recovery" in the restoration science community [1.22]. Reasonable targets of restored conditions, therefore, are informed by multidimensional data and often analytical and statistical modeling efforts that can project outcomes of scenarios, which account for ongoing changes in other drivers as well as disturbance legacies and hysteretic ecosystem behavior. Societal factors also influence what might be achievable through restoration, including legal, regulatory, and funding constraints, as well as the competing needs within a landscape for sustaining other societal roles [1.20]. As the multiple effects of rapid land-use and climate change intersect to influence ecosystems of the future, restoration planning must also entail managing ecosystem resilience to future disturbances and stressors.

The restoration planning process includes designing appropriate monitoring programs, evaluating uncertainties, and establishing alternative plans if condition targets are not met. Once designed, monitoring ideally occurs throughout all phases to capture variability in pre-restoration conditions [1.23]. This design process ideally engages scientists and managers

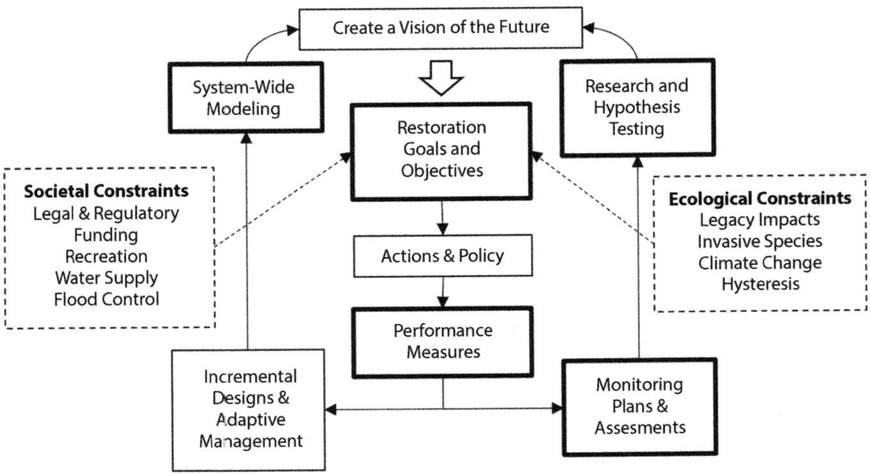

Figure 1.1 A conceptual framework developed for the Comprehensive Everglades Restoration Program to describe how society and science are linked to create an adaptive vision for the future of the Everglades. The left side is the Design Flow Path; the right side is the Assessment Flow Path. Boxes with thick lines represent components that can be informed by diatoms. (Adapted from Sklar *et al.* [1.20].)

in co-production of falsifiable hypotheses about anticipated restoration outcomes [1.24]. Depicting hypotheses using conceptual ecological models that motivate and guide monitoring and modeling can be helpful (Figure 1.1). A key outcome of the planning process is creation of performance measures—key attributes (species, traits, functions) of ecosystems with established, reliable relationships to environmental drivers over desirable spatiotemporal scales. Doren *et al.* [1.25] conducted a co-production process to establish criteria for selecting 12 indicators to be used to guide the Everglades adaptive restoration process (Table 1.1). These criteria were used to select a suite of indicators to guide the restoration process in conjunction with a set of landscape, trophic, biodiversity, and physicochemical features that characterize reference or target conditions.

A final important component of the restoration process is effectively communication with policymakers and the public—a process as important as the performance of the indicators themselves [1.26]. The communication process needs to be transparent, unambiguous, and able to portray key uncertainties.

Table 1.1 Twelve criteria used to evaluate indicators of ecosystem restoration [1.25] and supporting or limiting attributes of diatoms that pertain to each criterion.

Criterion	Supporting attributes of diatoms	Limiting attributes of diatoms
(1) Is the indicator relevant to the ecosystem?	Ubiquitous and ecologically important in aquatic ecosystems	Limited application in terrestrial ecosystems
(2) Does it respond to variability at a scale that makes it applicable to the entire system?	Wide dispersal and environmental selection can promote synchronous variability at large spatial scales	Isolated populations may vary in their response to environmental variability; diatoms will also reflect small-scale patchiness in heterogenous habitats

(*Continued*)

Table 1.1 Twelve criteria used to evaluate indicators of ecosystem restoration [1.25] and supporting or limiting attributes of diatoms that pertain to each criterion. (*Continued*)

Criterion	Supporting attributes of diatoms	Limiting attributes of diatoms
(3) Is the indicator feasible to implement and is it measurable?	Field collection protocols are well-established, easy, and rapid relative to other organisms	Microscopic evaluations are time-consuming and expensive; molecular tools are expensive and underdeveloped
(4) Is the indicator sensitive to system drivers and is it predictable?	Highly sensitive, predictable responses to most environmental drivers	N/A
(5) Is the indicator interpretable in a common language?	Individual, population, and community attributes can be described using familiar terms	Some education may be required to effectively communicate diatom-based indicators to the public
(6) Are there situations where an optimistic trend with regard to an indicator might suggest a pessimistic restoration trend?	N/A	Increased oligotrophic diatoms may indicate lack of restored water flows
(7) Are there situations where a pessimistic trend with regard to an indicator may be unrelated to restoration activities?	Negative trends in response to large-scale drivers like climate change or invasion by a non-native diatom species may overwhelm positive restoration signals	N/A

(*Continued*)

Table 1.1 Twelve criteria used to evaluate indicators of ecosystem restoration [1.25] and supporting or limiting attributes of diatoms that pertain to each criterion. (*Continued*)

Criterion	Supporting attributes of diatoms	Limiting attributes of diatoms
(8) Is the indicator scientifically defensible?	Significant literature providing guidance for applying diatoms as indicators of ecosystem health	Regionality in species pools and environmental conditions often requires place-based studies of species, traits, and environmental responses prior to defensible application in restoration
(9) Can clear, measurable targets be established for the indicator to allow for assessments of success?	Diatom-based restoration targets can be based on reference system conditions, experiments, models, and long-term or paleoecological data sources	Reference systems may not exist; valves may be poorly preserved or absent in some sediments
(10) Does the indicator have specificity to be able to result in corrective action?	Narrow optima and tolerances of species and traits can guide restoration actions	Some drivers are correlated with each other and have correlated influences on indicator attributes
(11) What level of ecosystem process or structure does the indicator address?	Base of food web, soil formation, biogeochemical regulation	N/A
(12) Does the indicator provide early warning signs of ecological change?	Responds very quickly and is an early signal of ecosystem response cascades	N/A

1.3 Incorporating Diatoms in Ecological Restoration

1.3.1 Incorporating Diatom Indicators in Aquatic Ecosystem Restoration Planning

Determining reliable metrics of ecological integrity for gauging the success of restoration is an important early step in the restoration planning process. The process of selecting indicators of success can help refine restoration goals. Ideally, selected indicators should meet many of the desirable criteria (Table 1.1) and collectively cover the different characteristics of the restored ecosystem, including trophic interactions and other ecosystem functions that occur across a variety of spatiotemporal scales [1.25]. Diatoms have been used for more than a century to assess the condition of water bodies due to their ubiquity, sensitivity, and reliability as indicators of a variety of water quality variables (see Stevenson and Smol [1.27] for a thorough review). Stevenson [1.28] provides exhaustive guidance on how to best design algae (particularly diatom)-based environmental assessments to improve water resource management. Charles *et al.* [1.29] review key approaches of diatom-based indicator application in the expansive monitoring programs established in the U.S. and Europe. This chapter highlights some of the key strategies suggested in these reviews, particularly strategies that improve diatom-based restoration guidance.

Diatoms check many of the boxes in Table 1.1 for aquatic ecosystems where they thrive, although they shouldn't be ignored as possible indicators of restoration of terrestrial habitats—or even deserts—where diatomaceous crusts or arboreal or subaerial assemblages can occur. Diatoms react to environmental variability at the spatial scale of their cells (sub-millimeter)—prompting the phrase "think like a diatom" by diatomists [1.30]. Diatom assemblages reflect the small-scale patchiness of heterogeneous benthic environments that require spatially intensive monitoring to capture [1.31]. Such spatially explicit studies have demonstrated that diatom assemblage composition, when sampled appropriately, provides a strong signal of regional-scale environmental variance [1.32], due to (1) concordance of small and large-scale environmental variance, and (2) the few dispersal barriers within most aquatic ecosystems. Similarly, diatom assemblages respond temporally to environmental changes over weeks to months, integrating finer-scale temporal variability in water chemistry while providing an early warning signal of environmental change relative to other organisms that have slower turnover rates. Much like with environmental heterogeneity, this rapid response to environmental disturbance requires temporally intensive sampling to detect signals of restoration occurring over longer timescales and over the higher-frequency fluctuations. For these reasons, diatoms and other microbial environmental indicators are best used alongside other indicators that operate on larger temporal and spatial scales, such as macroinvertebrate, fish, and plant communities [1.25] (Figure 1.2).

Selecting indicators of restoration success can be guided by conceptual ecological models that depict linkages among multiple drivers, stressors, ecological effects, and attributes that characterize the ecosystem and are expected to change in response to restoration [1.33]. Development of conceptual models can guide restoration by identifying a minimum set of driver characteristics and response variables that must be achieved for restoration to be considered successful. For instance, the goals of restoration of the Florida Everglades are captured by a suite of conceptual ecological models, one of which captures interrelationships between sheet flow, water depth patterns, landscape patterns, and oligotrophic nutrient status as indicated by diatoms [1.34]. Diatoms are an important component of the periphyton (benthic microbial) indicator, providing a particularly good indication of ecosystem trophic status while also interacting with other drivers and ecosystem responses. The process of mapping linkages within

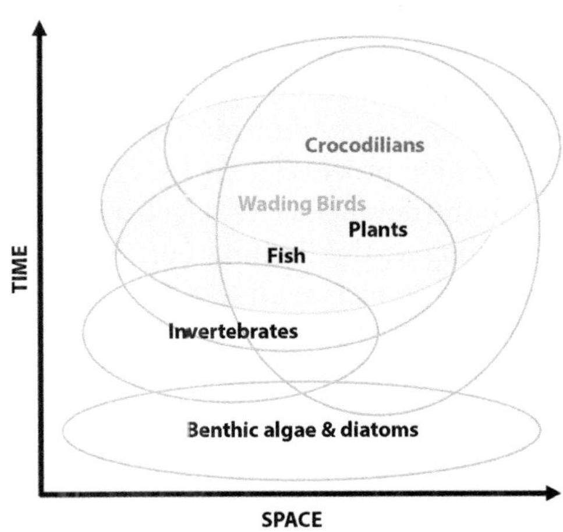

Figure 1.2 Graphical depiction of the spatial and temporal scales represented by different biological indicators of the restoration of the Everglades. For example, benthic algae and diatoms are broadly distributed and react very quickly to changes in nutrients and hydrology while crocodilians respond more slowly and at large spatial scales. This figure is not meant to capture literal aspects of spatial and temporal interactions with exactness, but rather illustrates the importance of including diatoms alongside other indicators to capture the full spatial-temporal context of a restoration project. (Adapted from Doren *et al.* [1.25].)

a conceptual model clarifies the importance of using multiple indicators to track restoration success. In the Everglades example, benthic diatoms provide a strong indication of ecosystem departure from oligotrophy while plant community and soil attributes provide a better indication of changes in landscape patterning and sheet flow, which are equally important restoration goals [1.25]. Conceptual ecological models can be used to formulate hypotheses about the anticipated effects of changes in drivers on key indicators while also revealing uncertainties about effects of correlations among drivers and indirect interactions on those indicators.

Box 1.1. Example of diatom-guided restoration planning

In the Everglades, an expansive Comprehensive Everglades Restoration Plan is intended to restore the ecological benefits of the Everglades ecosystem by recovering wetland hydrology. Among the desired outcomes is the recovery of large populations of wading birds which were once plentiful and whose decline was viewed as emblematic of systematic failure of ecosystem function. The Everglades restoration planning process included the development of conceptual ecological models [1.33]. One model guided a trophic hypothesis that predicts wading bird recovery will follow the re-establishment of prey populations that are dependent upon benthic algae, which provide habitat and a food source [1.35] [1.36]. Diatoms are a critical component of the trophic hypothesis because they are a primary energy source to the food web, and have a strong and predictable response to the primary drivers of nutrient status and hydrology, both of which are managed through the restoration process [1.35] [1.37]. Hydrologic restoration is expected to increase the dominance of oligotrophic diatoms in benthic algal communities as long as the limiting nutrient, phosphorus, is kept at concentrations below a protective limit of 10 ppb [1.35]. As hydrology recovers, diatoms are expected to increase relative to filamentous cyanobacteria in benthic algal communities, which will provide a more palatable food source to aquatic consumers and increase the prey base for desired wading bird communities and recreational fisheries [1.36] (Figure 1.3).

1.3.2 Using Diatoms to Establish Restoration Goals and Targets

Establishing clear criteria for the fundamental features that characterize a restored ecosystem is essential for successful restoration [1.38]. These criteria may extend across multiple indicator organisms that respond predictably to changes in critical drivers across different spatiotemporal scales.

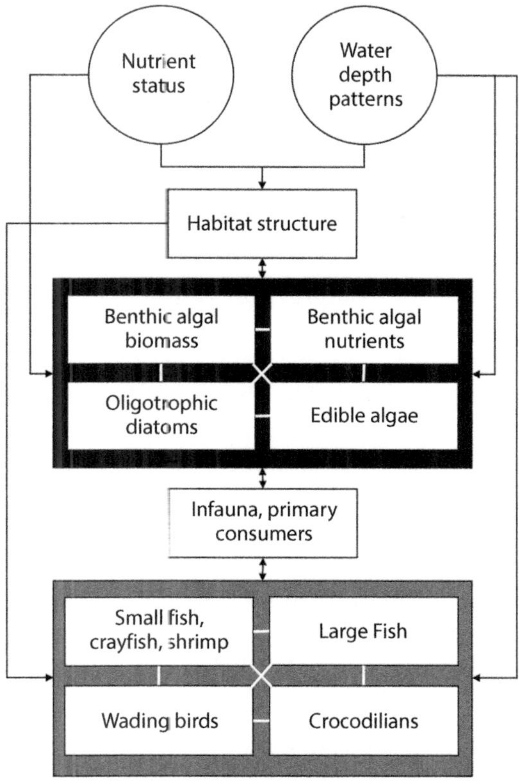

Figure 1.3 A conceptual ecological model used to guide the restoration of the Florida Everglades. Drivers being restored are in the top circles and the performance measures, including benthic algal and diatom indicators and dependent consumers (including infauna that live within the periphyton and primary consumers that eat it), are indicated in the black and gray boxes, respectively. Linkages portray direct and indirect interactions and feedbacks expressed in the trophic hypothesis guiding restoration. (Adapted from Gaiser [1.35].)

Establishing performance measures for an indicator requires a quantifiable expectation of their status in the restored state. Long-term data can directly provide targets based on pre-disturbance conditions. In the best cases, these data are quantitative, but even qualitative data can help guide interpretations of pre-disturbance condition [1.39]. In the absence of long-term data, reference conditions can be determined from paleoecological studies, spatial surveys, expert judgement, or experiments. Most of these approaches entail modeling of attribute responses to the stressor and other potentially interacting variables. Diatoms are desirable indicators because they are amenable to each of these approaches used to establish reliable restoration criteria. The use of diatoms in bioassessment extends back 150+ years, which provides a rich literature of diatom- and other algae-based metrics of ecological status (including reviews by Lowe and Pan [1.40] and Stevenson and Smol [1.27]). Much of this literature is geared toward bioassessment of environmental health rather than explicitly restoration, but both require establishing a baseline condition where biological integrity is defined by the difference between current conditions from this expected status. Rather than revisiting that well-reviewed literature, this review illuminates the strengths and weaknesses of common approaches while highlighting some newer techniques that have come from recent literature.

Paleolimnological approaches: In the absence of long-term data, paleolimnological analysis of diatoms in chronologically dated sediment records can be used to identify restoration targets by hindcasting pre-disturbance limnological conditions [1.41] [1.42] [1.43]. Paleolimnological records are particularly useful when historical limnological data are unavailable [1.44] or require validation [1.45]. For instance, paleolimnological data have been useful for pinpointing the timing of acidification and eutrophication impacting lake biota [1.46]. The magnitude of change in diatom assemblage composition between pre-industrial and modern sediments can then be used to prioritize ecosystems requiring restoration [1.47] [1.48]. In this way, diatoms preserved in pre-disturbance sediments have provided quantitative targets for both physicochemical conditions and diatom assemblage criteria to assess recovery of nutrient status [1.47] [1.48], acidity [1.46], organic carbon [1.49], hydrology [1.50] [1.51], and trace metals [1.52]. Alternatively, contemporary assemblages can be examined across ecosystems that encompass the range of predicted past ecological conditions. This process results in a training set of data on diatom assemblages (i.e., relative species abundances, trait characteristics) and variables of interest (i.e., pH, total phosphorus). A variety of modeling approaches can then be used to develop transfer functions from the training set to quantitatively infer the variable in question from pre-disturbances assemblages.

Paleolimnological approaches to diatom-based target setting are not without pitfalls [1.41]. While many lakes contain chronologically resolved sediment records with well-preserved diatoms, some records suffer from high levels of perturbation or poor diatom preservation. Both drawbacks are common in shallow lakes and wetlands where sediments can be highly physically and biologically disturbed or have poor preservation of siliceous microfossils for unknown reasons—possibly frequent drying or porewater chemistry extremes [1.53]. There are also statistical and ecological constraints involved in setting targets based on paleolimnological reconstructions. Typically, models for inferring water chemistry or other variables from diatoms in sediments are based on transfer functions developed from relationships between diatoms and those variables in the modern environment in training sets. Aside from the space-for-time assumption, other weaknesses of paleoenvironmental calibration using transfer functions include: (1) the environmental variable targeted for management is often correlated with other environmental variables, (2) many species may not have strong responses to the target variable but may be changing for other reasons, and (3) calibration models have poor replicability across similar ecosystems [1.54], partly due to measurable dispersal limitation [1.55]. In the case of trophic state reconstructions, for instance, pre-disturbance inferences often do not align with conditions anticipated from other sources of knowledge [1.55].

Finally, in their comprehensive review of paleolimnological approaches to restoration target setting, Bennion *et al.* [1.48] identify some broader considerations. First, there may be no modern analogue to the historical state: no ecosystem today is pristine, so it is unreasonable to expect that an ecosystem can return to that state [1.56]. Climate change and widespread land-use alterations alter regional diatom species pools and shift baseline physicochemical conditions of ecosystems, which creates a different setting for recovery than was present prior to disturbance. For instance, lakes that are warming or drying due to climate change exhibit changes similar to those caused by nutrient pollution [1.57], so climate change prevents a return to the pre-disturbance state even when excessive nutrient loads are abated. In addition, damaged ecosystems are unlikely to recover in the same trajectory as they declined, as hysteretic ecosystem behavior makes returning to a pre-disturbance state or assemblage unlikely despite restoration of a key driving variable (Figure 1.4). For example, internal loading of legacy phosphorus accumulated in sediments may lengthen or prevent restoration to a pre-loading state in a lake where external loading has been reduced to pre-disturbance rates [1.59]. Nevertheless, paleoecological data can provide a range of past conditions that can be considered by managers and local stakeholders to guide realistic restoration goals.

Paleolimnological studies have overcome the problem of multiple interacting drivers by using more than one proxy and the weakness of having no modern analogue by reconstructing past conditions in both historically disturbed and undisturbed ecosystem. The undisturbed system can serve as a control where changes due to other sources of variability, such as climate change or regional land use shifts, can be measured. For instance, Sayer and Roberts [1.60] established restoration goals for polluted Attenborough ponds in the English Midlands by comparing long-term monitoring and sediment diatom data from isolated (less disturbed)

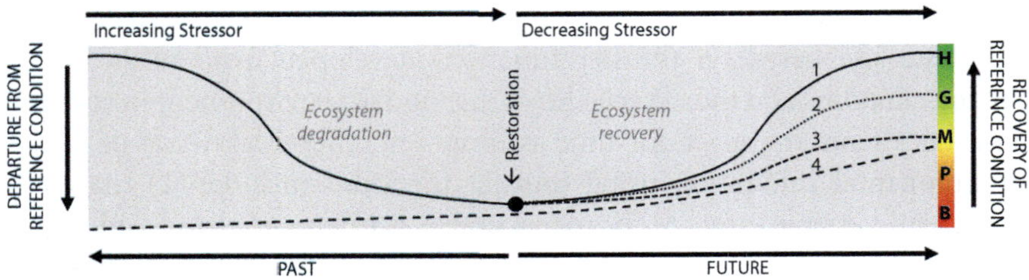

Figure 1.4 Modification of diagram developed to guide the European Water Framework Directive [1.58] to show the impact of climate change (or other press) on the ability of an ecosystem to recover reference conditions upon driver restoration. Possible trajectories include: 1 – return to reference conditions defined by paleoecological or historical/long-term data; 2 – alternative trajectory typical of hysteretic behavior leading to a "good" ecological status upon recovery; 3 – alternative trajectory driven by interactions with an ongoing press; 4 – trajectory of underlying press. The green to red gradient at the right represents the transitions between high (H), good (G), moderate (M), poor (P), and bad (B) condition categories.

ponds to the diatom record in a short sediment core in a disturbed pond. Similarities between the pre-disturbance flora in the disturbed pond and the past and modern communities in the undisturbed setting supported use of the pre-disturbance community as a modern analogue for expected habitat structure in the disturbed pond undergoing restoration. Bennion and Simpson [1.47] used the technique of "modern analogue matching" to determine ecological and chemical reference conditions for nine European lakes. By comparing pre-disturbance flora in the nine lakes to the flora of 347 lakes in a training set, they were able to identify the best lakes to use as modern analogues to guide restoration. Advancements in quantitative methods also now allow the identification of different confounding factors driving paleolimnological change [1.61], improving the ability to forecast independent effects of change in particular drivers of interest. When re-assembling species assemblages becomes impossible due to irreversible change, the pre-disturbance paleolimnological assemblage becomes a useful benchmark rather than a target. There is likely untapped potential to focus restoration on the return or establishment of particular habitats or ecosystem functions characterized by particular types of diatom assemblages, rather than a fixed specific species assemblage [1.48].

Survey approaches: When paleoecological data are not available, such as in most flowing water and many coastal ecosystems [1.62], restoration targets can be developed from surveys that evaluate diatom attributes across multiple ecosystems that vary in environmental condition. Sampling should be designed in a manner that allows for hypotheses to be tested in a way that includes estimation of error or uncertainty. Repeated surveys of sites that are randomly or strategically distributed in space can provide an estimate of means and precision. A stratified random sampling design may be beneficial for capturing variance in regions with significant environmental heterogeneity, while a more deliberate spacing, such as transect approaches, can be useful in regions with known patterns in underlying gradients. Neighboring "reference" ecosystems that aren't undergoing restoration but that are subject to similar stresses may be assessed contemporaneously to provide estimates of baseline variability and restoration effects. Stevenson [1.28] recommends designs that provide equal representation along the anticipated condition gradient. Alternatively or additionally, sampling of the ecosystem undergoing restoration should occur for a significant duration prior to the onset of restoration to capture the variability existing along the condition gradient in the ecosystem before project implementation. Before-after-control-intervention (BACI) or Bayesian analytical designs are common in restoration ecology and are especially useful in the absence of a contemporaneous reference ecosystem or where temporal variability is high in both the reference and recovering ecosystem [1.63].

Once a sampling design is in place, strategies can follow established protocols for parameters of interest. Ample guidance is available on the selection of substrate or habitat types for sampling and quantitative, qualitative, and semi-quantitative approaches for their sampling [1.27] [1.64]. A large variety of attributes of diatom assemblages have been used to assess ecological condition, ranging from biomass (chlorophyll *a*, ash-free dry mass, cell densities and/or biovolumes), taxonomic composition (species relative abundances and/or biovolumes, functional groups), diversity (richness at different taxonomic resolutions, evenness), chemical composition (i.e., nutrient and/or chlorophyll *a* content), or aggregate functional variables such as rates of photosynthesis, respiration, growth, or nutrient uptake [1.27] [1.65]. In addition to compositional information, population-level attributes such as cell size and appearance (cell health status, lipid bodies, valve deformities) or motility can also be helpful in water quality assessment [1.66].

Diatom assessment surveys typically include evaluation of the distribution of physicochemical and/or biological data to define condition gradients and determine stressor-response relationships. Where a restoration target falls along this gradient depends on economic, technological, and ecological feasibility and may or may not equate to a baseline (i.e., natural, reference, or unimpaired) state. Ecosystems exhibiting such near-natural conditions are often scarce or difficult to identify. Therefore, frequency distributions are used to determine 'good' ecological status [1.67], where conditions present in the upper quartile of reference sites, or the lower quartile of all sites. are used to define the baseline condition. Multiple applications of this quartile approach to defining biological reference condition have been applied in diatom-based approaches to the European Water Framework Directive (WFD), which required that all aquatic ecosystems achieve at least good status by 2015. For instance, Birk and Hering [1.68] examined environmental data from 356 sampling sites in five countries to determine sites in at least good environmental status using threshold values from the literature and environmental standards. They then selected the upper or lower quartile of the distribution of different diatom metrics across these sites to establish biological benchmarks that were then used to categorize degrees of impairment of the remaining sites cross countries. In another WFD study, Bennion *et al.* [1.69] used paleolimnological data to validate environmental criteria of reference lakes determined through prior analysis. They then applied a diatom-based trophic state index to 1000+ lakes by measuring the difference between the observed and expected trophic state index values for each lake, and then assigning the upper quartile of those differences in the pre-determined reference lakes as the boundary for "good" ecological status.

Professional opinion: Professional judgement is applied in the development of most assessment frameworks in ecology and health sciences [1.70] [1.71]. A robust and validated use of expert opinion in gauging ecological response to increasing levels of human disturbance known as the Biological Condition Gradient (BCG) model, was developed by biologists in a coordinated effort by the U.S. Environmental Protection Agency [1.72]. In BCG models, experts determine attributes that characterize distinct levels of ecosystem health and signal shifts to different levels. They develop numeric rules for these levels based on a set of "learning" sites and test those rules with independent datasets to validate them. Resultant quantitative rules can then be applied to any dataset to determine ecological status. Hausman *et al.* [1.71] followed these steps to develop a BCG for diatoms in New Jersey, using threshold models to determine the changepoints in diatom assemblages along the BCG gradient. Charles *et al.* [1.6] then determined the phosphorus and nitrogen levels corresponding to those BCG categories to determine levels (criteria) associated with baseline and impaired conditions.

Experimental approaches: Indices of biological integrity that rely on comparison to a reference state characterization have been criticized for rarely diagnosing the causes for ecological impairment or deviation from that state. Stressor-response models based on observational data reveal the forms of those relationships, providing hypotheses that can be tested experimentally. Experimental validation of drivers of degradation are particularly useful in management and restoration since achieving rehabilitated states is reliant on adjusting the particular variable(s) causing degradation. Such controlled experiments have frequently identified similar changepoints to those observed in large-scale surveys of many ecosystems providing supporting lines of evidence for controlling that variable. For instance, Gaiser *et al.* [1.73] found similar responses of diatoms to phosphorus along concentration gradients and in experimental manipulations. This allowed them to successfully employ experimentally derived diatom thresholds to assess compliance to a protective phosphorus criterion. However, fewer experiments have been conducted to determine the relationship between diatom response to stressor reduction, which does not always follow the same pathway as the response to its increase, and this is an area ripe for development.

Box 1.2. Example of diatom-based restoration assessment
In diatom-based assessment of system-wide Everglades restoration, a finite-sample Generalized Recursive Tesselation Sampling (GRTS) is used to create a stratified random set of sampling locations [1.74]. GRTS is considered a highly advantageous design for ecological monitoring because it provides spatial balance or spread, that allows each site to have an independent chance of inclusion, which greatly simplifies computation of confidence intervals about estimates [1.36]. Habitat sampleable for diatoms and aquatic fauna is mapped via photo interpretation and an iterative process is used to determine sites within larger landscape units that are repeatedly sampled each wet season. At each site, diatoms are sampled from 1-m^2 quadrats alongside all of the features represented in the conceptual ecological model, which guides the trophic hypothesis (nutrient status, hydrological data, habitat characteristics, benthic algal attributes, aquatic consumers; Figure 1.3). To assess effectiveness of specific restoration projects, diatoms are also sampled annually at fixed sites placed at logarithmic distances downstream of structures and parallel to the flow of restored freshwater supplies [1.75]. In both survey and transect studies, "before" period sampling has occurred for at least a decade prior to the initiation of the restoration project for a full BACI design (Figure 1.5a).

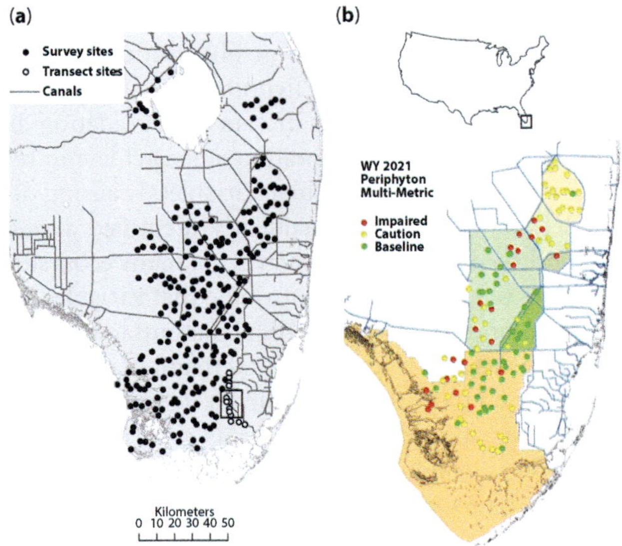

Figure 1.5 (a) Map of the Florida Everglades in the southeast U.S.A (inset) showing center points of primary assessment survey units (closed circles) within which random sampleable sites are assessed annually for diatoms, habitat characteristics, and aquatic consumers. The small box shows the location of transects providing data for Figure 1.8. (b) Status of the benthic algal multimetric for water year 2021 across the Everglades ecosystem [1.76].

1.3.3 Diatom Metrics of Restoration Success

Diatom-based restoration assessment depends on identifying the most useful metrics, which can be guided by the rich theoretical and empirical science in this field and exploration of regional datasets generated by the types of studies described above. Biological data collected alongside driver variables can be explored in a variety of ways to determine attributes that respond most reliably and predictably to the gradients of interest. Good metrics provide explicit, sensitive, and precise estimates of the stressor being restored [1.27]. Both parametric and non-parametric approaches can be used to determine relationships between metrics and stressors, frequently requiring ordination and other approaches that accommodate multidimensional data. Common biological metrics include diatom species richness and relative abundances, compositional similarity to reference conditions, percentage of diatoms with particular life histories (i.e., motile, attached, planktonic), and percentage of aberrant valves [1.28]. Indicator taxa might be selected based on general knowledge of their distributions (i.e., [1.77] [1.78]), although it is important to regionally verify preferences due to population-level variability, gradient length effects, and interactions of other local sources of variability. Prediction accuracy is often improved by combining the different strengths of individual metrics into a multimetric index after first normalizing the range of each metric to ensure equal weighting. Multimetric indices were first employed with fish and macroinvertebrate assemblages to produce an index of biological integrity [1.79] and are now commonly incorporated in diatom-based ecological assessment to provide a single metric of integrity [1.28].

Defining boundaries between "good" reference conditions and those indicative of impairment (or vice-versa) is guided by the quantitative evaluation of the relationships between the stressors and individual metrics or combined index values, especially since these relationships tend to be nonlinear. This nonlinearity can be used to establish criteria based on levels of the stressor that protects biological integrity as indicated by the metric. Sites with conditions closer to the criterion are more sustainable than sites

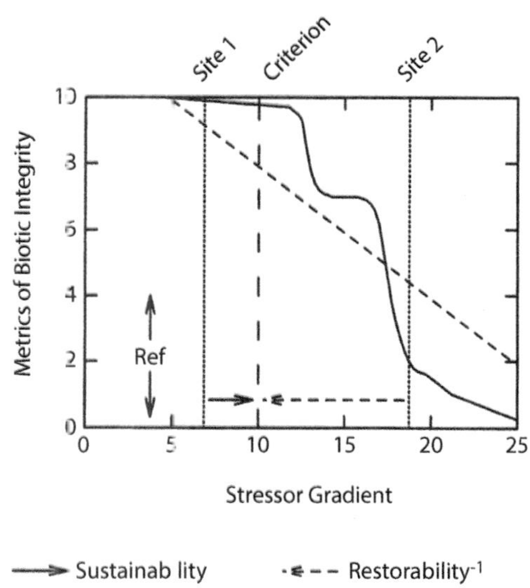

Figure 1.6 Illustration from Stevenson and Smol [1.27] showing a nonlinear relationship of a metric of biological integrity to a stressor gradient (solid line). Reference conditions (Ref) are indicated by the vertical arrows at the low end of the stressor gradient. The dashed vertical line represents the protective criterion for the stressor, while the vertical lines represent the stressor and metric condition at an unimpaired and impaired site (Sites 1 and 2, respectively). Sites are more sustainable the closer they are to the criterion, sites become less sustainable and potentially unrestorable the further they are to the right of the criterion.

with conditions that have greatly deviated from the reference condition—a distance can be used to gauge restorability [1.25] (Figure 1.6).

Nonlinear analytical approaches, such as classification and regression tree or breakpoint analyses, can help define conditions associated with departure from baseline condition and set criteria for impairment, or conversely, restoration [1.80]. Breakpoint regression can be used to identify changepoints for single stressor-response surfaces, but more complicated approaches are necessary for multiple drivers and response variables. Threshold indicator taxa analysis (TITAN) is now a frequently used approach for identifying changes in species composition along environmental degradation gradients [1.81]. TITAN assesses synchrony among species changepoints along environmental gradients as evidence of community thresholds, identifies robust indicator taxa, and estimates uncertainty around the changepoint thresholds. González-Paz *et al.* [1.82] reviewed diatom-based ecological status classification systems and found that TITAN identified lower boundaries for good/moderate status than the multimetric indices developed for the regions (streams in the NW Iberian Peninsula and Balearic Islands, Spain). Another analytical approach to determining thresholds is self-organizing maps (SOMs) that uses machine learning to represent multidimensional data in two dimensions representing clusters of similar data. Grenier *et al.* [1.5] used SOMs to identify integrity classes based on differences between diatom index values (using the approach of Lavoie *et al.* [1.83]) and reference conditions. They identified four biotypes related to meaningful thresholds that induce significant shifts in diatom community structure. Lavoie *et al.* [1.83] extended this approach to account for spatiotemporal variability in geology and nutrient conditions, creating an index that is now used to assess streams in eastern Canada. Rapid advancements in mathematical tools for data exploration and change detection will continue to improve, probably through applications of machine learning and artificial intelligence [1.84] [1.85].

Box 1.3. Example of diatom-based metric development

In Everglades restoration assessment, benthic algal samples are analyzed for a broad suite of variables known to provide reliable indication of nutrient status in aquatic ecosystems and the Everglades, including ash-free dry biomass, chlorophyll *a* and nutrient content, and diatom and soft algae taxon relative abundances and biovolumes [1.86] [1.87]. Transect studies have included artificial substrate deployment to minimize local effects of habitat structure and to provide estimates of accumulation rates [1.88]. Multidimensional non-parametric and parametric exploratory analyses were used to determine key stressors and responses [1.89] [1.90], confirming hydrology, phosphorus, and conductivity as primary drivers [1.27] [1.91]. The best performing metrics of oligotrophic nutrient status were ash-free dry biomass, phosphorus content, and relative abundance of oligotrophic diatoms [1.35]. Oligotrophic diatoms were designated as the 11 taxa whose abundances decreased with increasing availability of phosphorus, identified through an indicator species analysis [1.92]. To determine thresholds for assessment, the frequency distributions of these three biological variables were examined. Following Barbour *et al.* [1.67], the upper 75% of the relative abundances of oligotrophic diatoms were used to indicate "good" (oligotrophic) nutrient status, while samples with percent oligotrophic diatoms below one and two standard deviations from the mean were considered "caution" (changing toward undesirable state) and "impaired" states, respectively. The threshold of oligotrophic diatoms for total phosphorus (TP), defined by a benthic algal concentration of 150 ug TP g^{-1} of dry mass, was confirmed through a TITAN analysis (Figure 1.7). Benthic algal TP concentrations were used instead of water concentrations because the oligotrophic nature of this habitat causes all P to be immediately sequestered by biota. Oligotrophic diatom taxa and thresholds for all three parameters were experimentally validated in a large-scale manipulative experiment [1.4]. This experimental dataset also provided a precision estimate for each metric (75–85%), which improved by 20% after normalizing the metric distributions and combining them into a multimetric [1.90].

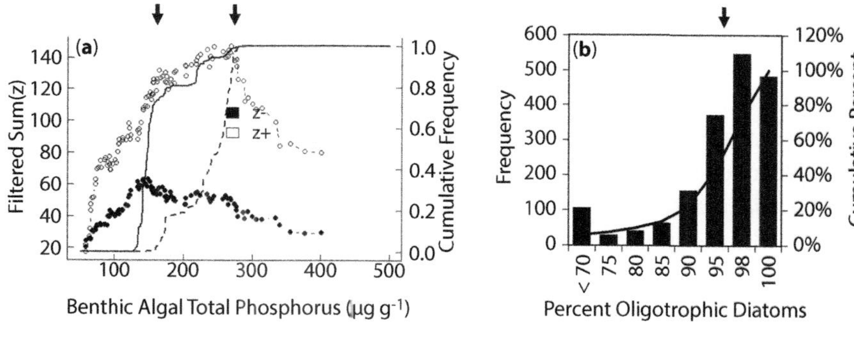

Figure 1.7 Threshold Indicator Taxa Analysis (TITAN) of benthic diatom response to benthic algal total phosphorus content (a) with arrows indicating change points where a peak in filtered sum (z-) and sum (z+) indicates an assemblage threshold where a coincident decline or increase in taxa occurs. Solid circles are negative taxa (z-); open circles are positive taxa (z+). Oligotrophic taxa were defined as negative responding taxa. Their frequency distribution in 1650 survey samples (b) suggest a protective level at the point of the arrow (>98% relative abundance) representing the upper 75th percentile of samples. (E. Gaiser, unpubl. data)

1.3.4 Adaptively Managing Ecological Condition

Once restoration targets and metrics are established, continued assessment of the metrics should ideally be used to guide the adaptive management of the stressor to achieve the target ecological condition [1.93]. Diatoms are typically used alongside other performance measures to evaluate restoration progress. In some instances, features of recovering diatom assemblages can even provide helpful insight into activities that might accelerate recovery. For example, in a stream restoration study, diatom assemblages and trait composition were studied along restored and unrestored stream reaches, providing not only a useful indication of restoration progress but also guiding channel reconfiguration to maximize floodplain connectivity and habitat complexity [1.94].

While diatoms may respond faster than many other organisms to reduction in a stressor, it is important to recognize that their responses may lag physical and chemical recovery. Zheng *et al.* [1.95] found that algal compositional change could be used to evaluate salt marsh restoration but only after changes in nutrient storage were observed. Bortolotti *et al.* [1.96] documented a 10-year recovery period for diatom assemblages in restored prairie wetlands to become indistinguishable from natural wetlands. Similarly, Diamond *et al.* [1.97] found a 15-year lag in diatom assemblage response to recovery from acidification in a mountain pond. The recovered assemblage was compositionally different than the acid-sensitive assemblage present prior to acidification. This is a good example of a biological constraint preventing restoration due to factors such as dispersal limitation, ongoing changes unrelated to the restoration, and hysteretic ecosystem behavior. Once recognized, these constraints can be used to adaptively manage the restoration process. For instance, Janousek *et al.* [1.98] found that structural heterogeneity aided driver microphytobenthic compositional recovery, allowing for return to reference conditions within a year of *Spartina* marsh restoration. Mayer and Galatowitsch [1.99] overcame dispersal limitation in the recovery of restored prairie wetlands by transplanting diatoms from reference wetlands to the restoration sites, a process of assisted migration that is common in restoration of wetland vegetation. Translocation studies can also improve expectations for the rate of diatom assemblage recovery following water quality improvement [1.100] [1.101] and guide metric selection. For example, a biofilm translocation from high to low impact urban streams in the Argentine Pampean plain showed how nuclear membrane integrity and ecological guilds responded much more reliably to improvement than motility endpoints, providing an improved suite of indicators to monitor recovery of lowland streams [1.102].

Modeling approaches can be used alongside assessment to estimate outcomes of restoration scenarios against which recovery trajectories can be calibrated. For instance, Naja *et al.* [1.103] modeled outcomes of five alternative scenarios for hydrologic restoration of the Everglades, including a benthic algal multimetric index of oligotrophic nutrient status. These outcomes not only informed selection of optimal restoration scenarios but also provided estimates of the acreage of benthic algal treatment marsh necessary to achieve biological conditions in compliance with a protective nutrient criterion. Similarly, Larras *et al.* [1.104] evaluated large-scale monitoring data to estimate the risk of stream impairment and gauge recovery of streams from a diversity of anthropogenic pressures using diatom trait and taxonomic data. They guided managers to prioritize risk-reducing actions by providing these model outcomes in user-friendly R scripts. This is a rare example of incorporating successful rapid conveyance of highly accurate diatom-based risk assessment into actions that improve management outcomes, and highlights the value of durable collaborations between scientists and decision-makers. Similar examples appear in the Everglades restoration arena where academic and agency scientists have been working closely together for decades, resulting in tools that effectively share and communicate the status and trends in key indicators of restoration status [1.24].

Box 1.4 Example using diatom indicators to adaptively manage restoration
Since the hydrologic restoration of the Everglades is not yet occurring at the landscape scale, diatom data obtained from the system-wide survey are not yet being employed in change detection, but rather to provide ideal multidecadal baseline data on pre-restoration condition. Notably, one of the most significant changes detected in the system-wide survey is related to the effects of climate change, specifically sea level rise, which is reducing oligotrophic diatom abundance in near-coastal freshwater marshes exposed to novel sources of phosphorus from seawater intrusion [1.105]. This finding is incentivizing efforts to manage climate change resilience while restoring wetland hydrology. Meanwhile, freshwater flow restoration is occurring in earnest along the boundaries of Everglades National Park. Restoration has an unintended consequence of mobilizing legacy phosphorus that accumulated prior to the establishment of a protective phosphorus criterion (Figure 1.8). By working closely with agency scientists, these results led to changes in the timing and distribution of restored flows to minimize legacy phosphorus mobilization [1.106] while maximizing potential for freshwater flows to the coast to offset the effects of rising seas.

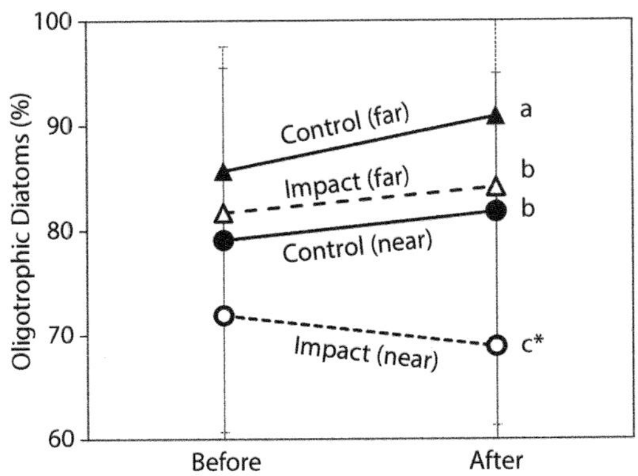

Figure 1.8 Percent oligotrophic diatoms present in benthic algal assemblages upstream and downstream of water management structures replenishing freshwater flows to impact but not control transects before (2006–2011) and after (2012–2019) implementation. Oligotrophic diatoms declined along impact transects near the legacy sources of phosphorus, a finding that is being used to guide the adaptive management of restoration to improve both water quantity and quality (symbols with asterisks denote significant differences in a before-after control-intervention design). (E. Gaiser, unpubl. data)

1.4 Communicating Diatom-Based Indicators of Restoration Progress

1.4.1 Communicating Restoration Progress

Communicating progress toward restoring ecological conditions can take many forms depending on the audience, which can range from other scientists to managers, decision-makers, stakeholders, and different sectors of the public. Across all communication platforms, it is important to provide evidence that ecological interpretations are scientifically credible, objective, and unambiguously vetted [1.25]. Expressing multimetric index values using grades, score cards, symbols, or colors can help convey take-home messages as long as these are accompanied by materials that allow the audience to "drill down" into the supporting material and/or data [1.25]. The transition to open data sharing practices can be extremely helpful in this regard, especially when developers of associated metadata are cognizant of different levels of expertise in user communities [1.107]. Public audiences also desire information about the social values of ecological recovery [1.108]. For this reason, the effectiveness of diatom-based indicators in adaptive management of restoration might be determined in part by how well the audience understands their linkage to the ecosystem services they value, including the charismatic megafauna at the top of the food webs that diatoms feed [1.109].

Box 1.5. Example of a diatom indicator communication tool
The Everglades benthic algal (periphyton) multimetric indicator is one of eleven indicators used to track and report system-wide success of restoration, the status of which is reported biennially to the U.S. Congress in a Stoplight Report Card [1.76]. The status of indicators is conveyed using a simple stoplight system where green, yellow, and red represent baseline (reference condition), caution (changing), and impaired conditions. More granular information about each indicator is provided in an accompanying report. The benthic algal stoplight is updated biennially with data from the system-wide survey, and regional-scale means are reported in time series and maps (Figure 1.5b). The 2021 report showed that the system-wide benthic algal multimetric indicator is below the restoration target (yellow stoplight), with the most impaired sites occurring in areas that have not yet recovered from a history of agricultural nutrient loading but also in areas receiving additional phosphorus due to seawater intrusion. However, recovery trajectories were observed in regions where delivery of freshwater meeting the protective nutrient standard has been restored.

1.4.2 Public Education and Engagement

Most of the public has never heard of diatoms, yet is responsible for and dependent upon diatom use in protecting the world's most valuable water resources. Everyone desires clean drinking water and access to productive and safely consumable fisheries, but the role that diatoms play in ensuring that those ecosystem services continue is almost entirely beyond the scope of public awareness. Botanists lament and debate cures for "plant blindness" (lack of knowledge of the trees in one's own backyard), but how much worse it is for microbes that are invisible to most people. How do we cure diatom blindness? How can the power of diatoms be more effectively communicated and embraced by the public? What difference would it make if a larger fraction of the public knew about diatoms? Since the invention of the first microscope, microscopists have tried to portray the incredible beauty of diatoms and other microorganisms—most visibly including the works of Ernst Haeckel, which can be purchased as coffee table books [1.110]. The development and rapid progression of very high-resolution microscopic imaging tools has revolutionized both diatom science and art (for example, artwork by British diatomist Klaus D. Kemp available online). Still, the gorgeous array of images circulating on the internet remain known to but a few. To improve public accountability to our most valuable resources, there must be more avenues for public education and engagement with diatoms. Social media provides an important outlet for sharing and learning about diatoms, including through organizations like the Diatom Appreciation Society and the International Diatom Society's "Diatom of the Month" blog. Museums and parks can also provide opportunities for the public to view diatoms through a microscope for the first time, sparking curiosity and a desire to learn more, not just about diatoms [1.111]. Experiential art can be even more effective when combined with scientific literacy projects. For instance, Edwards *et al.* [1.94] employed work-study students to evaluate diatom indicators of river restoration success near Portland, Oregon, where these same students view human-sized sculptures of diatoms by artist Fernanda D'Agostino while on public transit. In another example, Bahls [1.112] described 12 new diatom species based on samples collected by volunteers, engaging them in the value of cataloguing diatom diversity and building diatom herbaria. Such efforts to connect sectors of the public with diatoms are becoming more common as the value of citizen science projects and other activities to broaden public scientific literacy of aquatic ecosystems are increasingly appreciated [1.113]. Collaborations between scientists and artists have the added benefit of also inspiring scientists to think differently and creatively about the scientific process and the meaning of discoveries.

Box 1.6. Examples of participatory diatom art

During the course of studying diatoms of the Everglades, I've had the opportunity to engage with a variety of local artists who became excited about depicting diatoms in their art and using them to increase public literacy about Everglades restoration. My first collaboration was with participatory artist, Xavier Cortada, who spent time in my lab as an artist in residence through a program at my university. Cortada created public displays of diatoms across many media types (ceramic sculptures, paintings, engravings) and built participatory citizen science programming at public venues to improve understanding and actions to reverse the course of climate change. For instance, the diatom court at the public Pinecrest Gardens in Miami, Florida, contains three 10-foot sculptures of diatoms and a diatom-decorated viewing platform (Figure 1.9). A placard contains a QR code that once scanned, provides the viewer with a prompt to stand on the platform and ask questions about climate change to the three pillars meant to represent magistrates. Answers to a variety of questions about climate change are provided by the audio feed, given from the perspective of diatoms that are described as "incapable of telling a lie." These works by Cortada received

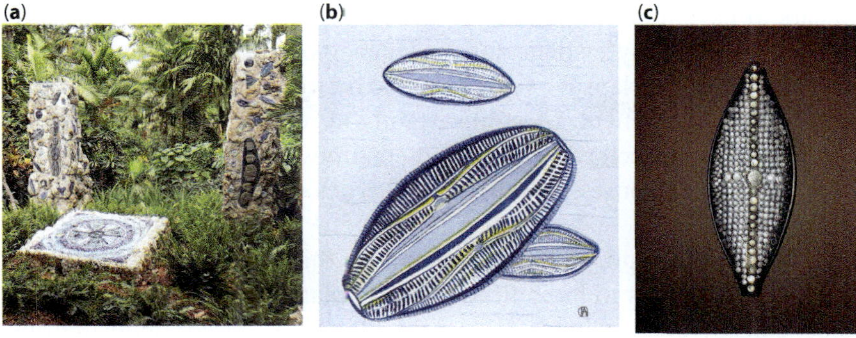

Figure 1.9 Artist depictions of Everglades diatoms, including (a) ceramic sculpture of the "diatom court" by Xavier Cortada, a permanent installation at Pinecrest Gardens, FL [1.114], (b) watercolor of *Amphora proteus* by Pauline Goldsmith of the Tropical Botanic Artists exhibited in the Atlanta-Hartsfield International Airport, Atlanta, GA [1.115], and (c) glass sculpture of *Proschkinia browderiana* by Richard Cohen shown at Dante Fascell Visitor Center, Everglades National Park, Homestead, FL [1.116].

national attention, including an invitation to speak about science-based participatory art with Dr. John Holdren at the U.S. White House when he was Chief Science Advisor to President Barack Obama. Other experiential diatom art exhibits include In Deep with Diatoms by the Tropical Botanic Artists that was featured at the Frost Art Museum (Miami, FL), throughout South Florida galleries and institutions, and at other research sites in the U.S. Long Term Ecological Research Network. Artist Pauline Goldsmith's depiction of the diatom Amphora proteus (Figure 1.9) went on to the Hartsfield-Jackson Atlanta International Airport for 1.5 years as part of the National Park Service Centennial Display. More recently, the glass beaded diatom sculptures of artist Richard Cohen (Figure 1.9) have been on public display at the Glenn Hubert Library (Miami, FL) and the Dante Fascell Visitor Center Gallery at Biscayne National Park. For the latter exhibit, microscopes were also available with slides containing live diatoms from Biscayne Bay for visitors to observe.

1.5 Summary and Opportunities

While diatoms are well-proven tools for assessing risk of aquatic ecosystem deterioration, this review shows similar promise for their application in setting targets for and tracking effectiveness of ecosystem restoration. Delays in applying established principles of diatom bioassessment to restoration appears to have more to do with societal constraints to restoring ecosystems rather than any evidence of inadequacy of diatoms as tools. There is considerable overlap in methodologies between the processes of risk and recovery assessment. Most uncertainty in recovery assessment stems from nonlinear trajectories, indirect interactions, and time lags characterizing recovery responses that apply to all types of indicators of biological condition.

Opportunities ripe for development in this field include applying genomic approaches to bioindication, particularly metabarcoding, which holds considerable promise in paleoecologically based target setting [1.117] and bioassessment [1.118]. However, these methods still require much more attention to improve reproducibility that is challenged by the abundance of unassignable sequences [1.119]. Increased reproducibility of both morphologically and genetically based taxonomic data will radically improve with expanded data availability afforded now by openly accessible data repositories but also through more widespread adoption of practices to improve diatom data quality, including taxonomic certification [1.120]. The internet is not only improving access to data but also expert-verified taxonomic identification information through expansive repositories like diatoms.org [1.121]. The advantages of internet-enabled sharing extend beyond data and taxonomic guidance into program code that facilitates development and reproducibility of multimetric diatom indices (i.e., DiaThor R package [1.122]). At the opposite end of the spatial scale, advances in remote sensing may also revolutionize the scale at which diatoms can be used to monitor the health and recovery of ecosystems [1.123].

Restoration ecologists are beginning to embrace the reality that fully recovering prior ecosystem states on a warming planet is unlikely, and that restoration planning should include scenarios of novel or designer ecosystems that may look very different than they had in the past but maximize service provisions [1.124]. Diatoms may play a helpful role in designing ecosystems that conserve biodiversity, support societally relevant functions [1.125], mitigate extinction [1.126], and enable greener infrastructure for a more sustainable future [1.127], especially in urban ecosystems [1.128]. Particular attention should be given to the application of diatoms in managing resilience in low-latitude tropical regions where warm water and extended growing seasons complicate ecosystem recovery from eutrophication [1.129] and high-latitude/altitude regions where endemic species extirpations may be more likely in a warming climate. In addition, genetic, morphometric, and breeding studies of diatoms show that species classifications are likely too coarse [1.130]. Finer-grained classification reveals a high degree of endemism in diatom assemblages, calling into question the degree to which rapid environmental change (including invasion of non-native species) may be threatening diatom diversity and the provisioning of ecosystem services. Since species with broad ecological tolerances tend to replace specialists in stressful, disturbed environments, restoration efforts must embrace conservation strategies that protect endemic and rare diatom species that contribute meaningfully to both biodiversity and bioindication.

Acknowledgments

The author's time was supported in part by the George M. Barley, Jr., Endowment at the Institute of Environment, Florida International University. Everglades examples were funded by various grants and contracts to the author from the U.S. Army Corps of Engineers, South Florida Water Management District, and the Department of Interior, National Park Service in collaboration with the Florida Coastal Everglades Long Term Ecological Research program (National Science Foundation #DEB-2025954). The author thanks Nathan Dorn, Gabriel Kamener, Jeffrey Kline, Franco Tobias, and Rafael Travieso for their contributions to the research, and Pauline Goldsmith, Samantha Hormiga, Paige Kleindl, Thomas Shannon, and two anonymous reviewers who improved the chapter through careful review. This chapter is dedicated to the memory of Dr. Eduardo A. Morales who inspired its development.

References

[1.1] Albert, J.S., G. Destouni, S.M. Duke-Sylvester, A.E. Magurran, T. Oberdorff, R.E. Reis, K.O. Winemiller, & W.J. Ripple. 2021. Scientists' warning to humanity on the freshwater biodiversity crisis. *Ambio* 50:85-94.

[1.2] Smol, J.P. & E.F. Stoermer. 2010. *The Diatoms: Applications for the Environmental and Earth Sciences*. Second edition. Cambridge University Press. Cambridge, NY.

[1.3] Dixit, S.S., J.P. Smol, J.C. Kingston, & D.F. Charles. 1992. Diatoms: powerful indicators of environmental changes. *Environ. Sci. Technol.* 26:22-33.

[1.4] Gaiser, E.E., J. Trexler, J. Richards, D. Childers, D. Lee, A. Edwards, L. Scinto, K. Jayachandran, G. Noe, & R. Jones. 2005. Cascading ecological effects of low-level phosphorus enrichment in the Florida Everglades. *J. Environ. Qual.* 34:717-723.

[1.5] Grenier, M., I. Lavoie, A.N. Rousseau, S. Campeau. 2010. Defining ecological thresholds to determine class boundaries in a bioassessment tool: The case of the Eastern Canadian Diatom Index (IDEC). *Ecol. Indic.* 10:980-989.

[1.6] Charles, D.F., A.P. Tuccillo, T.J. Belton. 2019. Use of diatoms for developing nutrient criteria for rivers and streams: A Biological Condition Gradient approach. *Ecol. Indic.* 96: 258-269.

[1.7] Kelly, M., C. Bennett, M. Coste, C. Delgado, F. Delmas, L. Denys, L. Ector, C. Fauville, M. Ferreol, M. Golub, A. Jarlman, M. Kahlert, J. Lucey, B. Ní Chatháin, I. Pardo, P. Pfister, J. Picinska-Faltynowicz, C. Schranz, J. Schaumburg, J. Tison, H. van Dam, & S. Vilbaste. 2009. A comparison of national approaches to setting ecological status boundaries in phytobenthos assessment for the European Water Framework Directive: results of an intercalibration exercise. *Hydrobiologia* 621: 169–82.

[1.8] Birk, S., W. Bonne, A. Borja, S. Brucet, A. Courrat, S. Poikane, A. Solimini, W. van de Bund, N. Zampoukas & D. Hering. 2012. Three hundred ways to assess Europe's surface waters: an almost complete overview of biological methods to implement the Water Framework Directive. *Ecol. Indic.* 18: 31–41.

[1.9] Birk, S., Chapman, D., Carvalho, L. et al. 2020. Impacts of multiple stressors on freshwater biota across spatial scales and ecosystems. *Nat. Ecol. Evol.* 4:1060–1068.

[1.10] Gómez, N. & M. Licursi. 2001. The Pampean Diatom Index (IDP) for assessment of rivers and streams in Argentina. *Aquat. Ecol.* 35:173–181.

[1.11] Salomoni, S.E., O. Rocha, V.L. Callegaro, V.L. *& E.A.* Lobo. 2006. Epilithic Diatoms as Indicators of Water Quality in the Gravataí River, Rio Grande do Sul, Brazil. *Hydrobiologia* **559**:233–246.

[1.12] Dalu, T. & P.W. Froneman. 2016. Diatom-based water quality monitoring in southern Africa: challenges and future prospects: review. *Water* SA 42.

[1.13] Rockström, J., W. Steffen, K. Noone, Å. Persson, F.S. Chapin, E. Lambin, *et al.* 2009. Planetary Boundaries: Exploring the Safe Operating Space for Humanity. *Ecol. Soc.* 14.

[1.14] Nilsson, C., A. L. Aradottir, D. Hagen, G. Halldórsson, K. Høegh, R. J. Mitchell, K. Raulund-Rasmussen, K. Svavarsdóttir, A. Tolvanen, & S.D. Wilson. 2016. Evaluating the process of ecological restoration. *Ecol. Soc.* 21:41.

[1.15] Stevenson, R.J., Y. Pan, & H. van Dam. 2010. *Assessing environmental conditions in rivers and streams with diatoms.* In Smol, J.P. & E.F. Stoermer (Eds). *The Diatoms: Applications for the Environmental and Earth Sciences.* Second edition. Cambridge University Press. Cambridge, NY.

[1.16] SER 2004. *The SER Primer on Ecological Restoration,* Version 2. Society for Ecological Restoration Science and Policy Working Group. Washington, DC.

[1.17] Hobbs, R.J. & D.A. Norton, D. A. 1996. Towards a conceptual framework for restoration ecology. *Restor. Ecol.* 4:93-110.

[1.18] Murray, C. & D. Marmorek. 2003. Adaptive management and ecological restoration. In *Ecological restoration of southwestern ponderosa pine forests.* Pp. 417-428. Washington: Island Press.

[1.19] Williams, B.K. 2011. Adaptive management of natural resources—framework and issues. *J. Environ. Manage.* 92:1346-1353.

[1.20] Sklar, F., J. Beerens, L. Brandt, C. Coronado, S.E. Davis III, T. Francovich, C. Madden, A. McLean, J. Trexler, & W. Wilcox. 2019. Back to the future: rebuilding the Everglades. Pp 202-231 In Childers, D.L., E.E. Gaiser, & L.A. Ogden (eds.) *The Coastal Everglades: The Dynamics of Social-Ecological Transformation in the South Florida Landscape.* Oxford University Press, Oxford, UK.

[1.21] Johnson, C.R., R.H. Chabot, M.P. Marzloff, & S. Wotherspoon. 2016. Knowing when (not) to attempt ecological restoration. *Restor. Ecol.* 25:140-147.

[1.22] Elliott, M., D. Burdon, K.L. Hemingway, & S.E. Apitz. 2007. Estuarine, coastal and marine ecosystem restoration: Confusing management and science – A revision of concepts. *Estuar. Coast. Shelf Sci.* 74: 349-366.

[1.23] Michener, W.K. 2008. Quantitatively evaluating restoration experiments: research design, statistical analysis, and data management considerations. *Restor. Ecol.* 5:324-337.

[1.24] Gaiser, E.E., L.A. Ogden, D.L. Childers, & C. Hopkinson. 2019. Chapter 9: Re-Imagining Ecology through an Everglades Lens. In Childers, D.L., E.E. Gaiser, & L.A. Ogden (eds.) *The Coastal Everglades: The Dynamics of Social-Ecological Transformation in the South Florida Landscape.* Oxford University Press, Oxford, UK.

[1.25] Doren, R. F., J.C. Trexler, A.D. Gottlieb, and M.C. Harwell. 2009. Ecological indicators for system-wide assessment of the greater everglades ecosystem restoration program. *Ecol. Indic.* 9:S2-S16.

[1.26] Dennison, W.C., T.R. Lookingbill, T.J.B. Carruthers, J.M. Hawkey, S.L. Carter. 2007. An eye-opening approach to developing and communicating integrated environmental assessments. *Front. Ecol. Environ.* 5:307-314.

[1.27] Stevenson, R.J. & J.P. Smol. 2015. Use of algae in environmental assessments. Pp 921-962 In Wehr, J.D. & R.G. Sheath (eds). Freshwater algae of North America: Ecology and Classification. Elsevier.

[1.28] Stevenson, R.J. 2014. Ecological assessments with algae: a review and synthesis. *J. Phycol.* 50:437-461.

[1.29] Charles, D., M. Kelly, R.J. Stevenson, & S. Poikane. 2021. Benthic algae assessments in the EU and the US: Striving for consistency in the face of great ecological diversity. *Ecol. Indic.* 121.

[1.30] Gaiser, E.E. 2012. Think like a diatom. Word and World 32: 21-25.

[1.31] Burkholder, J.M. 1996. Interactions of Benthic Algae with Their Substrata. Pp. 253-289 In Stevenson, R.J., M.L. Bothwell, and R.L. Lowe. *Algal Ecology: Freshwater Benthic Ecosystems.* Academic Press. New York.

[1.32] Soininen, J. & A. Teittinen. 2019. Fifteen important questions in the spatial ecology of diatoms. *Freshw. Biol.* 64:2071-2083.

[1.33] Ogden, J.C., S.M. Davis, K.J. Jacobs, T. Barnes & H.E. Fling. 2005. The use of conceptual ecological models to guide ecosystem restoration in South Florida. *Wetlands* 25:795–809.

[1.34] Davis, S., E.E. Gaiser, W. Loftus, & A. Huffman. 2005. Southern marl prairies conceptual ecological model. *Wetlands* 25:821-831.

[1.35] Gaiser, E.E. 2009. Periphyton as an indicator of restoration in the Everglades. *Ecol. Indic.* 9:S37-S45.

[1.36] Sargeant, B., E.E. Gaiser, & J. Trexler. 2011. Indirect and direct controls of macroinvertebrates and small fish by abiotic factors and trophic interactions in the Florida Everglades. *Freshw. Biol.* 56:2334-2346.

[1.37] Lee, S., E.E. Gaiser, & J. Trexler. 2013. Diatom-based models for inferring hydrology and periphyton abundance in a subtropical karstic wetland: Implications for ecosystem-scale bioassessment. *Wetlands* 33:157-173.

[1.38] Gregory, R. & R.L. Keeney. 1994. Creating policy alternatives using stakeholder values. *Manag. Sci.* 40:947-1068.

[1.39] Ladouceur, E., N. Shackelford, K. Bouazza, L. Brudvig, A. Bucharova, T. Conradi, T.E. Erickson, M. Barbowski, K. Garvy, W.S. Harpole, H.P. Jones, T. Knight, M.M. Nsikani, G. Paterno, K. Suding, V.M. Temperton. P. Török, D. Winkler, & J. Chase. 2022. Knowledge sharing for shared success in the decade of ecosystem restoration. *Ecol. Solut. Evid.* 3:e12117.

[1.40] Lowe, R.L. & Y. Pan. 1996. Benthic algal communities as biological monitors. Pp 705-733 In Stevenson, R.J., M.L. Bothwell, & R.L. Lowe. *Algal Ecology: Freshwater Benthic Ecosystems.* Academic Press. New York.

[1.41] Battarbee, R.W. 1999. The importance of palaeolimnology to lake restoration. In: Harper, D.M., Brierley, B., Ferguson, A.J.D., Phillips, G. (eds) *The Ecological Bases for Lake and Reservoir Management. Developments in Hydrobiology*, vol 136. Springer, Dordrecht.

[1.42] Battarbee, R. W., V. J. Jones, R. J. Flower, N. G. Cameron, H. Bennion, L. Carvalho, & S. Juggins. 2001. Diatoms. Pages 155–202 *in* J. P. Smol, H. J. B. Birks, and W. M. Last (editors). *Tracking environmental change using lake sediments. Volume 3: Terrestrial, algal and siliceous indicators.* Kluwer Academic Publishers, Dordrecht, The Netherlands.

[1.43] Smol, J.P. 2008. *Pollution of Lakes and Rivers: A Paleoenvironmental Perspective*, 2nd Edition. Blackwell Publ. 383 pp. Hoboken, NJ.

[1.44] Brenner, M., T. J. Whitmore, M.S. Flannery & M.W. Binford. 1993. Paleolimnological methods for defining target conditions in lake restoration: Florida case studies. *Lake Reserv. Manag.* 7:209-217.

[1.45] Emery Boeck, M., E. E. Gaiser, H. M. Swain, M. Brenner, J. H. Curtis, & W. F. Kenney. 2023. Cyclical browning in a subtropical lake inferred from diatom records. *Front. Ecol. Evol.* 11.

[1.46] Battarbee, R.W., G.L. Simpson, H. Bennion, V. Curtis. 2011. A reference typology of low alkalinity lakes in the UK based on pre-acidification diatom assemblages from lake sediment cores. *J. Paleolimnol.* 45:489–505.

[1.47] Bennion, H. & G.L. Simpson. 2011. The use of diatom records to establish reference conditions for UK lakes subject to eutrophication. *J. Paleolimnol.* 45:469–488.

[1.48] Bennion, H., G.L., Simpson, N.J. Anderson, G. Clarke, X. Dong, A. Hobæk, P. Guilizzoni, A. Marchetto, C. D. Sayer, H. Thies & M. Tolotti. 2011. Defining ecological and chemical reference conditions and restoration targets for nine European lakes. *J. Paleolimnol.* 45:415–431.

[1.49] Cunningham, L., K. Bishop, E. Mettävainio, & P. Rosén. 2011. Paleoecological evidence of major declines in total organic carbon concentrations since the nineteenth century in four nemoboreal lakes. *J. Paleolimnol.* 45:407–418.

[1.50] Costa-Böddeker, S., H. Bennion, T.A. de Jesus, A.L.S. Albuquerque, R. C. L. Figueira, & D. de C. Bicudo. 2012. Paleolimnologically inferred eutrophication of a shallow, tropical, urban reservoir in southeast Brazil. *J. Paleolimnol.* 48: 751–766.

[1.51] McGlue, M.M., A. Silva, M.L. Assine, J.C. Stevaux, & F.d.N. Pupim. (2015). Paleolimnology in the Pantanal: Using Lake Sediments to Track Quaternary Environmental Change in the World's Largest Tropical Wetland. In: Bergier, I., Assine, M. (eds) *Dynamics of the Pantanal Wetland in South America. The Handbook of Environmental Chemistry*, vol 37. Springer, NY.

[1.52] Bindler R., J. Rydberg, & I. Renberg. 2011. Establishing natural sediment reference conditions for metals and the legacy of long-range and local pollution on lakes in Europe. *J. Paleolimnol.* 45:519-531.

[1.53] Sanchez, C., E.E. Gaiser, C. Saunders, A. Wachnicka, & N. Oehm. 2013. Exploring siliceous subfossils as a tool for inferring past water level and hydroperiod in Everglades marshes. *J. Paleolimnol.* 49:45-66.

[1.54] Juggins, S., N.J. Anderson, J.M.R. Hobbs, & A.J. Heathcote. 2013. Reconstructing epilimnetic total phosphorus using diatoms: statistical and ecological constraints. *J. Paleolimnol.* 49: 373–390.

[1.55] Soininen, J. 2007. Environmental and spatial control of freshwater diatoms – a review. *Diatom Res.* 22:473-490.

[1.56] Battarbee, R.W., D.T. Monteith, S. Juggins, C.D. Evans, A. Jenkins, G.L. Simpson. 2005. Reconstructing pre-acidification pH for an acidified Scottish loch: a comparison of palaeolimnological and modelling approaches. *Environ. Pollut.* 137:135–149.

[1.57] Jeppesen, E., B. Moss, H. Bennion, L. Carvalho, L. DeMeester, H. Feuchtmayr, N. Friberg, M. O. Gessner, M. Hefting, T.L. Lauridsen, L.

Liboriussen, H.J. Malmquist, L. May, M. Meerhoff, J.S. Olafsson, M.B. Soons, & J.T.A. Verhoeven. 2010. Interaction of climate change and eutrophication. In Kernan, M., R. Battarbee, & B. Moss. Climate Change Impacts on Freshwater Ecosystems. Blackwell Publishing Ltd, Oxford, UK.

[1.58] Bennion, H., R.W. Battarbee, C.D. Sayer, G.L. Simpson, & T.A. Davidson. 2010. Defining reference conditions and restoration targets for lake ecosystems using palaeolimnology: a synthesis. *J. Paleolimnol.* 45:533-544.

[1.59] Canfield, D.E., Jr., R.W. Bachmann, & M.V. Hoyer. 2021. Restoration of Lake Okeechobee, Florida: Mission impossible? *Lake Res. Manag.* 37:95-111.

[1.60] Sayer, C.D., & N. Roberts. 2001. Establishing realistic restoration targets for nutrient-enriched shallow lakes: linking diatom ecology and palaeoecology at the Attenborough Ponds, U.K. *Hydrobiologia* 448:117–142.

[1.61] Simpson G.L. & N.J. Anderson. 2009. Deciphering the effect of climate change and separating the influence of confounding factors in sediment core records using additive models. *Limnol. Oceanogr.* 54:2529–2541.

[1.62] Vos, P.C. & H. de Wolf. 1993. Diatoms as a tool for reconstructing sedimentary environments in coastal wetlands; methodological aspects. *Hydrobiologia* 269:285–296.

[1.63] Conner, M.M., W.C. Saunders, N. Bouwes, & C. Jordan. 2016. Evaluating impacts using a BACI design, ratios, and a Bayesian approach with a focus on restoration. *Environ. Monit. Assess.* 188(555).

[1.64] Stevenson, R.J. & S.L. Rollins. 2017. Ecological assessment with benthic algae. Pp. 277-292 in Hauer, F. & G. Lamberti (eds) *Methods in Stream Ecology*. Elsevier, Amsterdam, The Netherlands.

[1.65] McCormick, P.V. & J. Cairns. 1994. Algae as indicators of environmental change. *J. Appl. Phycol.* 6:509-526.

[1.66] Pandey, L.K., E.A. Bergey, J. Lyu, J. Park, S. Choi, H. Lee, S. Depuydt, Y-T. Oh, S-M. Lee, & T. Han. 2017. The use of diatoms in ecotoxicology and bioassessment: Insights, advances and challenges. *Water Res.* 118: 39-58.

[1.67] Barbour, M. T., W. F. Swietlik, S.K. Jackson, & D.L. Courtemanch. 2000. Measuring the attainment of biological integrity in the USA: a critical element of ecological integrity. *Hydrobiologia* 422/423:453–64.

[1.68] Birk, S. & D. Hering. 2009. A new procedure for comparing class boundaries of biological assessment methods: a case study from the Danube basin. *Ecol. Indic.* 9:528-539.

[1.69] Bennion, H., M.G. Kelly, S. Juggins, M.L. Yallop, A. Burgess, J. Jamieson, & J. Krokowski. 2014. Assessment of ecological status in UK lakes using benthic diatoms. *Freshw. Sci.* 33.

[1.70] Steedman, R.J. 1994 Ecosystem health as a management goal. Freshwater Science 13:605-610.

[1.71] Hausmann, S., D. Charles, J. Gerritsen, & T.J. Belton. 2016. A diatom-based biological condition gradient (BCG) approach for assessing impairment and developing nutrient criteria for streams. *Sci. Total Environ.* 562:914-927.

[1.72] Davies, S.P. & S.K. Jackson. 2006. The biological condition gradient: a descriptive model for interpreting change in aquatic ecosystems. *Ecol. Appl.* 16:1251-1266.

[1.73] Gaiser, E.E., J. Richards, J. Trexler, R. Jones, & D. Childers. 2006. Periphyton responses to eutrophication in the Florida Everglades: Cross-system patterns of structural and compositional change. *Limnol. Oceanogr.* 51:617-630.

[1.74] Stevens, D.L. & A.R. Olsen. 2004. Spatially balanced sampling of natural resources. *J. Am. Stat. Assoc.* 99:262-278.

[1.75] Gaiser, E.E., P. Sullivan, F.A.C. Tobias, A.J. Bramburger, & J.C. Trexler. 2014. Boundary effects on benthic microbial phosphorus concentrations and diatom beta diversity in a hydrologically-modified, nutrient-limited wetland. *Wetlands* 34:55-64.

[1.76] Brandt, L.A., S.A. Balaguera-Reina, V. Briggs-Gonzalez, J.A. Browder, M. Cherkiss, N. Dorn, T. East, M. Ernest, A. Fine, P. Frederick, E. Gaiser, L. Garner, S. Geiger, S. Godfrey, J. Goldston, A. Huebner, N. Jennings, C. Kelble, J. Kline, N. LaSpina, J. Lorenz, C.J. Madden, F.J. Mazzotti, E. Montes, M. Parker, L. Rodgers, R. Sobczak, J. Spencer, J. Trexler, Z. Welch, & I. Zink. 2022. *System-wide Indicators for Everglades Restoration.* 2022 Report. Unpublished Technical Report. Science Coordination Group. South Florida Ecosystem Restoration Task Force. Pp 107.

[1.77] van Dam, H., A. Mertens, & J. Sinkeldam. 1994. A coded checklist and ecological indicator values of freshwater diatoms from The Netherlands. *Neth. J. Aquat. Ecol.* 28 117-133.

[1.78] Stevenson, R.J., B.H. Hill, A.T. Herlihy, L.L. Yuan, & S.B. Norton. 2008. Algae-P relationships, thresholds, and frequency distributions guide nutrient criterion development. *Freshw. Sci.* 27:783-799.

[1.79] Karr, J.R. 1981. Assessment of biotic integrity using fish communities. *Fisheries* 6:21-27.

[1.80] Daily, J.P., N.P. Hitt, D.R. Smith, & C.D. Snyder. 2012. Experimental and environmental factors affect spurious detections of ecological thresholds. *Ecology* 93: 17-23.

[1.81] Baker, M.E. & R.S. King. 2013. Of TITAN and straw men: an appeal for greater understanding of community data. *Freshw. Sci.* 32:489-506.

[1.82] González-Paz, L., C. Delgado, & I. Pardo. 2020. Understanding divergences between ecological status classification systems based on diatoms. *Sci. Total Environ.* 734:139418.

[1.83] Lavoie, I., S. Campeau, N. Zugic-Drakulic, J.G. Winter, & C. Fortin. 2014. Using diatoms to monitor stream biological integrity in Eastern Canada:

an overview of 10 years of index development and ongoing challenges. *Sci. Total Environ.* 475:187-200.

[1.84] João Feio, M.J., S.R.Q. Serra, A. Mortágua, A. Boucez, F. Rimeet, V. Vasselon, & S.F.P. Almeida. 2020. A taxonomy-free approach based on machine learning to assess the quality of rivers with diatoms. *Sci. Total Environ.* 722.

[1.85] Naghmeh, J., M.S. Ahmad, R. Taher, D.K. Afshin, & R. Maryam. 2020. Evaluation of river bio-resilience with artificial intelligence models (case study: Aliabad river). *Iranian Water Res. J.* 15:61-70.

[1.86] Stevenson, R.J., Y. Pan & P. Vaithiyanathan. 2002. Ecological assessment and indicator development in wetlands: the case of algae in the Everglades, USA. *Verh. - Int. Ver. Theor. Angew. Limnol.* 28:3.

[1.87] Stevenson, R.J., Y. Pan, K.M. Manoylov, C.A. Parker, D.P. Larsen, & A.T. Herlihy. 2008. Development of diatom indicators of ecological conditions for streams of the western US. *Freshw. Sci.* 27.

[1.88] Childers, D.L., R.D. Jones, J. Trexler, C. Buzzelli, J. Boyer, A. Edwards, E. Gaiser, K. Jayachandaran, D. Lee, J. Meeder, J. Pechmann, J. Richards, & L. Scinto. 2001. Quantifying the effects tween phosphorus inputs and wetland phosphorus concentrations of low level phosphorus enrichment on unimpacted Everglades wetlands with in situ flumes and phosphorus dosing. p. 127–152. In J.W. Porter and K.G. Porter (ed.) *The Everglades, Florida Bay, and coral reefs of the Florida Keys: An ecosystem sourcebook.* CRC Press, Boca Raton, FL.

[1.89] Gaiser, E.E., P. McCormick, & S. Hagerthey. 2011. Landscape patterns of periphyton in the Florida Everglades. *Crit. Rev. Environ. Sci. Technol.* 41(S1):92-120.

[1.90] Gaiser, E.E., A. Gottlieb, S. Lee, & J. Trexler. 2015. The importance of species-based microbial assessment of water quality in freshwater Everglades wetlands. In Entry, J., K. Jayachandrahan, A. Gottlieb, & A. Ogram (Eds.) *Microbiology of the Everglades Ecosystem.* Science Publishers. Pp. 115-130.

[1.91] Hagerthey, S., B. Bellinger, K. Wheeler, M. Gantar, & E.E. Gaiser. 2011. Everglades periphyton: A biogeochemical perspective. *Crit. Rev. Environ. Sci. Technol.* 41(S1):309-343.

[1.92] Dufrene, M. & P. Legendre. 1997. Species assemblages and indicator species: The need for a flexible asymmetrical approach. *Ecol. Monogr.*67:345-366.

[1.93] Stow, C.A., K. Glassner-Shwayder, D. Lee, Z. Whang, G. Arhonditsis, J.V. DePinto, & M.R. Twiss. 2020. Lake Erie phosphorus targets: an imperative for active adaptive management. *J. Great Lakes Res.* 46:672-676.

[1.94] Edwards, P.M., Y. Pan, L. Mork, & C. Thorne. 2020. Using diatoms to assess river restoration: A pilot study in Whychus Creek, Oregon, USA. *River Res. Appl.* 36:2089-2095.

[1.95] Zheng, L., R. J. Stevenson, & C. Craft. 2004. Changes in benthic algal attributes during salt marsh restoration. *Wetlands* 24:309-323.

[1.96] Bortolotti, L.E., R.D. Vinebrooke, & V.L. St. Louis. 2016. Prairie wetland communities recover at different rates following hydrological restoration. *Freshw. Biol.* 61:1874-1890.

[1.97] Diamond, S.E., R. Harvey, A. Heathcote, A. Lini & A.M. Morales-Williams. 2022. Decoupling of chemical and biological recovery from acidification in a montaine lake, Vermont, USA. *J. Paleolimnol.* 68:427-442.

[1.98] Janousek, C.N., C.A. Currin, & L.A. Levin. 2007. Succession of microphytobenthos in a restored coastal wetland. *Estuaries Coasts.* 30:265-276.

[1.99] Mayer, P. M. & S. M. Galatowitsch. 1999. Diatom communities as ecological indicators of recovery in restored prairie wetlands. *Wetlands* 19:765–774.

[1.100] Rimet, F., L. Ector, H -M. Cauchie & L. Hoffmann. 2009. Changes in diatom-dominated biofilms during simulated improvements in water quality: implications for diatom-based monitoring in rivers, *Eur. J. Phycol.* 44: 567-577.

[1.101] Arini, A., A. Feurtet-Mazel, R. Maury-Brachet, M. Coste, & F. Delmas. 2012. Field translocation of diatom biofilms impacted by Cd and Zn to assess decontamination and community restructuring capacities. *Ecol. Indic.* 18:520-531.

[1.102] Nicolosi Gelis, M.M., J. Cochero, J. Donadelli, & N. Gómez. 2020. Exploring the use of nuclear alterations, motility and ecological guilds in epipelic diatoms as biomonitoring tools for water quality improvement in urban impacted lowland streams. *Ecol. Indic.* 110: 105951.

[1.103] Naja, G.M., D.L. Childers, & E.E. Gaiser. 2017. Water quality implications of hydrologic restoration alternatives in the Florida Everglades, USA. *Restor. Ecol.* 25:S48-S58.

[1.104] Larras, F., R. Coulaud, E.Gautreau, E. Billoir, J. Rosebery, & P. Usseglio-Polatera. 2017. Assessing anthropogenic pressures on streams: A random forest approach based on benthic diatom communities. *Sci. Total Environ.* 586:1101-1112.

[1.105] Mazzei, V., & E.E. Gaiser. 2018. Diatoms as tools for inferring ecotone boundaries in a coastal freshwater wetland threatened by saltwater intrusion. *Ecol. Indic.* 88:190-204.

[1.106] Kotun, K. & A. Renshaw. 2014. Taylor Slough hydrology: fifty years of water management 1961-2021. *Wetlands* 34:S9-S22.

[1.107] Niemi, G.J. & M.E. McDonald. 2004. Application of ecological indicators. *Annu. Rev. Ecol. Evol. Syst.* 35:99-111.

[1.108] Schiller, A., C.T. Hunsaker, M.A. Kane, A.K. Wolfe, V.H. Dale, G.W. Suter, C.S. Russell, G. Pion, M.H. Jensen, & V.C. Konar. 2001. Communicating ecological indicators to decision makers and the public. *Conserv. Ecol.* 5:19.

[1.109] Rapport, D.J. & M. Hildén. 2013. An evolving role for ecological indicators: from documenting ecological conditions to monitoring drivers and policy responses. *Ecol. Indic.* 28:10-15.
[1.110] Haeckel, E., O. Breidbach, R. Hartmann, & I. Eibl-Eibesfeldt. 2008. *Art Forms in Nature: The Prints of Ernst Haeckel.* Prestel Publishing, Munich, DE.
[1.111] van Beynen, K. & T. Burress. 2018. Debris, diatoms, and dolphins: tracking child engagement at a public science festival. I*nt. J. Sci. Educ.* 8:355-365.
[1.112] Bahls, L. 2014. New diatoms from the American West—A tribute to citizen science. Proc. Acad. Nat. Sci. Phil. 163: 61-84.
[1.113] Hoyer, M.V., D.L. Bigham, R.W. Bachmann, & D.E. Canfield, Jr. 2014. Florida LAKEWATCH: Citizen Scientists protecting Florida's aquatic systems.i 77: 184-197.
[1.114] Cortada, X, "*Diatom Court*," hand-carved and hand-painted ceramic tile on coral rock, 2018. Pinecrest Gardens, Pinecrest, FL.
[1.115] Goldsmith, P. "*Amphora Proteus*," watercolor painting, 2016. Hartsfield-Jackson Atlantia International Airport, Altanta, GA.
[1.116] Cohen, R. "*Proschkinia browderiana*," glass sculpture, 2020. Dante Fascell Visitor Center Gallery, Homestead, FL.
[1.117] Gauthier, J., D. Walsh, D.T. Selbie, I. Domaizon, I. Gregory-Eaves. 2022. Sedimentary DNA of a human-impacted lake in Western Canada (Cultus Lake) reveals changes in micro-eukaryotic diversity over the past ~200 years. *Environ. DNA* 4:1106-1119.
[1.118] Kelly, M. 2019. Adapting the (fast moving) world of molecular ecology to the (slow-moving) world of environment regulation: lessons from the UK diatom metabarcoding exercise. *Metabarcoding Metagenom.* 3:127-135.
[1.119] Bailet, B., L. Apothéloz-Perret-Gentil, A. Baričević, T. Chonova, A. Franc, J. Frigerio, M. Kelly, D. Mora, M. Pfannkuchen, S. Proft, M. Ramon, V. Vasselon, J. Zimmermann, & M. Kahlert. 2020. Diatom DNA metabarcoding for ecological assessment: Comparison among bioinformatics pipelines used in six European countries reveals the need for standardization. *Sci. Total Environ.* 745:140948.
[1.120] Alers-Garcia, J., S.S. Lee, & S.A. Spaulding. 2021. Resources and practices to improve diatom data quality. *Limnol. Oceanogr. Bull.* 30:48-53.
[1.121] Spaulding, S.A., M.G. Potopova, I.W. Bishop, S.S. Lee, T.S. Gasperak, E. Jovanoska, P.C. Furey, & M.B. Edlund. 2021. Diatoms.org: supporting taxonomists, connecting communities. *Diatom Res.* 36:291-304.
[1.122] Gelis, M.M.N., M.B. Sathicq, J. Jupke, & J. Cochero. 2022. DiaThor: R package for computing diatom metrics and biotic indices. *Ecol. Model.* 465:109859.
[1.123] Gann, D., J. Richards, S. Lee, & E.E. Gaiser. 2015. Detecting and monitoring calcareous periphyton mats in the greater Everglades using passive remote sensing methods. In Entry, J., K. Jayachandrahan, A. Gottlieb,

& A. Ogram (Eds.) *Microbiology of the Everglades Ecosystem. Science Publishers*. pp. 350-372. New York, U.S.A.
[1.124] Awasthi, A., K. Singh, A. O'Grady, R. Courtney, A. Kalra, R. Singh, A. Cerdà, Y. Steinberger, & D. Patra. 2016. Designer ecosystems: a solution for the conservation-exploitation dilemma. *Ecol. Eng.* 93:73-75.
[1.125] Cabezas, A., M. Pallasch, I. Schönfelder, J. Gelbrecht, & D. Zak. Carbon, nitrogen, and phosphorus accumulation in novel ecosystems: shallow lakes in degraded fen areas *Ecol. Eng.* 66:63-71.
[1.126] Pandolfi, J.M., T.L. Staples, & W. Kiessling. 2020. Increased extinction in the emergence of novel ecological communities. *Science* 370:220-222.
[1.127] Sharma, N., D. Simon, A. Diaz-Garza, E. Fantino, A. Messaabi, F. Meddeb-Mouelhi, H. Germain, & I. Desgagné-Penix. 2021. Diatoms biotechnology: various industrial applications for a greener tomorrow. *Front. Mar. Sci.* 8.
[1.128] Castro-Roa, D. & G. Pinilla-Agudelo. 2014. Periphytic diatom index for assessing the ecological quality of the Colombian Andean urban wetlands of Bogatá. *Limnetica* 33:297-312.
[1.129] Jeppesen, E., M. Meerhoff, B.A. Jacobsen, R.S. Hansen, M. Søndergaard, J. P. Jensen, T. L. Lauridsen, N. Mazzeo & C. W. C. Branco. 2007. Restoration of shallow lakes by nutrient control and biomanipulation—the successful strategy varies with lake size and climate. *Hydrobiologia* 581: 269–285.
[1.130] Mann, D.G. & S.J.M. Droop. 1996. Biodiversity, biogeography and conservation of diatoms. *Hydrobiologia* 336:19-32.

2

Abundance-Occupancy Relationships in Freshwaters: The Case of Pond Diatoms

Saúl Blanco[1,2*], María Borrego-Ramos[1,2] and Óscar Fernández-Ramos[3]

[1]Diatom Lab, La Serna, León, Spain
[2]Ecology Unit, Department of Biodiversity and Environmental Management, Faculty of Biological and Environmental Sciences, University of León, León, Spain
[3]IHCantabria, Environmental Hydraulics Institute of Cantabria, Isabel Torres, Cantabria, Spain

Abstract
Interspecific abundance-occupancy relationships (AOR) have been addressed as one of the most important questions regarding the spatial ecology of diatoms. Whereas positive interspecific AOR is a common feature of ecological communities, little is known about its causes and consequences. Studies to date confirm that AOR depends mostly on the ecological features exhibited by the species pool analyzed, particularly niche characteristics and habitat guilds. The present study aims to test the persistence of such patterns using pond diatoms at a mesoscale level. Epiphytic and planktonic diatoms were collected in 29 shallow ponds throughout the Northern Iberian Plateau. Analysis of covariance (ANCOVA) tests revealed contrasting patterns when both assemblages are compared; for a given abundance level, epiphytic species were present in twice as many locations as planktonic ones, with noticeable effects also attributable to cell biovolume and motile guild. Contrary to our expectations, niche characteristics, measured in terms of species sensitivity and indicator value with respect to abiotic variables, had a negligible effect on AOR models. The consequences of such results are briefly discussed at a theoretical level.

Keywords: Bacillariophyta, ecological guilds, niche features, biovolume, Duero River basin

**Corresponding author*: saul.lanza@unileon.es

Nora I. Maidana, Magdalena Licursi and Eduardo Morales (eds.) Diatom Ecology: From Molecules to Metacommunities, (41–58) © 2024 Scrivener Publishing LLC

2.1 Introduction

The positive interspecific abundance-occupancy relationship (AOR) is one of the most well-known features of ecological communities, although it has only received the attention of ecologists in the last few decades [2.16]. The AOR refers to the patterns of species abundance and occupancy (geographic distribution) observed in ecological communities. Literature reviews (e.g., [2.16] [2.24]) suggest that these relationships are influenced by multiple mechanisms, such as species traits and habitat preferences, environmental filtering, species interactions and metapopulation dynamics. However, there are notable gaps in understanding the temporal dynamics and scale dependencies of this relationship. In any case, AOR is a widespread, but largely variable, ecological pattern found across groups of organisms of contrasting life-history characteristics, and understanding this variability is a key challenge for community ecologists [2.4] [2.56]. In aquatic ecosystems, AOR has been demonstrated at multiple spatial extents; that is, across ecoregions, within ecoregions and within drainage systems [2.44], although AOR models should only be applied across small, homogeneous areas [2.26]. The simplest explanation is the "rate of life" model [2.12], which suggests that habitat suitability varies spatially; thus, an increase in average habitat quality results in a simultaneous increment in the average density within the occupied area as well as the total habitable area. Therefore, in populations with very high colonization rates, new habitats are rapidly colonized and there will be a simple asymptotic relationship between abundance and occupancy [2.11]. This indicates that the abundance-occupancy relationship is most effectively understood as an outcome arising from the spatial dispersion of individual organisms [2.24] and not of the species themselves.

Previous studies have focused on analyzing the influence of species' ecological characteristics, as well as environmental factors, on AOR patterns. Particularly, geographic range, niche position and breadth, and resource availability, have been proposed, among others, as main drivers of this relationship [2.52]. Within freshwater ecosystems, multiple processes are most likely acting, depending on the spatial scale of the community [2.16]. For instance, according to Vilmi *et al.* [2.50], AOR in streams appears to be more structurally complex than in lakes, indicating that both types of aquatic habitats differ in their biodiversity patterning. This difference may be related to the different hydrological connections between lakes and rivers. To date, the only available study dealing specifically about AOR in lake diatoms is that of Lengyel *et al.* [2.31], who found interesting connections between this and niche breadth and position. However, no attempt has been made so far to confirm these findings with alternative datasets.

In this chapter we explore abundance-occupancy relationships using pond diatoms in a mesoscale-level study. By segregating diatom taxa by habitat type, ecological guilds and niche characteristics, we assess the relative importance of these predictors in explaining AOR parameters. Our contribution addresses one of the 15 important questions in the spatial ecology of diatoms [2.45], namely question 9, "Interspecific abundance-occupancy relationships in diatoms." These authors made a plea for new studies that consider how to describe the multiple dimensions of diatom niche breadths in order to properly reflect the influence of all factors of potential importance to diatoms.

2.2 Materials and Methods

2.2.1 Study Area

The study area (Figure 2.1, Table 2.1) is located in the Spanish part of the Duero River basin (NW Iberian Peninsula). The basin is crossed by the Duero River (895 km) and covers an extensive plateau of about 750 m a.s.l. (meters above sea level), formed mainly by Tertiary and Quaternary (alluvial or colluvial) materials, divided into two subsystems: a) the western plain, essentially composed of Paleozoic granites; and b) the central and eastern sedimentary catchment, composed of calcareous and muddy

Figure 2.1 Study area showing the location of the 22 studied ponds in the Duero River basin (NW Spain).

Table 2.1 Main limnological variables measured in the study sites.

	Method	Median	Minimum	Maximum
Area (ha)	GPS surveying of flooded perimeter	1.04	0.03	29.35
Elevation (m a.s.l.)	GPS data	776.00	755.00	946.00
Maximum depth (cm)	Transects with a graduated pole	70.00	20.00	290.00
Perimeter (m)	GPS surveying of flooded perimeter	374.00	66.00	3038.00
Secchi (cm)	Secchi disk	33.50	5.00	80.00
Bromides (ppb)	Ion chromatography	0.07	0.00	1.00
Chlorides (ppb)	Ion chromatography	17.40	2.06	490.00
Chlorophyll (ppb)	Spectrophotometry	11.21	1.73	269.98
Conductivity ($\mu S \cdot cm^{-1}$)	Field conductivimeter	372.00	90.50	2880.00
Fluorides (ppm)	Ion chromatography	0.05	0.04	0.55
Ammonium (ppm)	Ion chromatography	0.06	0.01	0.32
Nitrates (ppm)	Ion chromatography	0.03	0.00	0.34
Oxygen (%)	Field oxymeter	44.10	3.30	111.00
Oxygen (ppm)	Field oxymeter	3.89	0.31	8.85
pH	Field pH-meter	7.22	4.94	8.67
Ortophosphates (ppb)	Ion chromatography	8.22	3.56	94.50
Sulphates (ppb)	Ion chromatography	2.86	0.53	137.00
T (°C)	Field thermometer	18.60	14.00	27.30
Total N (ppm)	Ion chromatography	1.73	0.88	2.82
Total P (ppb)	Ion chromatography	72.21	9.25	573.15
Total suspended solids (ppm)	APHA 2540D	10.80	1.14	130.00

terrain with some intercalated evaporites. The surrounding mountains consist of heterogeneous siliceous material, except for the limestone in the central part of the Cantabrian Mountains [2.2]. The study site has suffered historically intense anthropogenic impacts.

2.2.2 Sample Collection and Processing

Samples were obtained during July 2018 from 29 shallow ponds, both temporary and permanent, at an average elevation of ~ 800 m a.s.l. on the western side of the basin. Limnological parameters were determined following standard protocols [2.1]. Epiphytic diatoms were collected from submerged stems of *Schoenoplectus lacustris* (L.) Palla randomly distributed across the surface of each system. At least 10–12 stems were randomly selected over the pond surface and cut at 10 cm below the water level in each pond. For each pond, the stems were placed together in a 1

Table 2.2 GLM ANCOVA results testing the contribution of different predictors to AOR variation.

Predictor	States/Range	Wald	Log-likelihood	χ^2	p
Abundance	[0.01–31.56%]	59.45	-470.82	139.38	< 0.001
Adnate	Yes/No	1.50	-470.32	1.00	0.31
Attached	No/Stalk/Pad	0.30	-470.66	0.32	0.85
Colony	No/Mucous Tube/ Ribon/Rosette/ Filament/ Arbuscular/Stellate	3.52	-469.32	3.01	0.81
Mobile	Yes/No	1.61	-470.18	1.27	0.26
Motile guild	Yes/No	3.83	-468.92	3.79	0.05
Pioneer	Yes/No	0.98	-470.82	139.38	0.46
Profile	No/Low/High	3.05	-469.62	2.41	0.30
Habitat	Planktonic/Epiphytic	62.73	-456.83	27.98	< 0.001
Biovolume	[41–38641 μm³]	0.10	-540.98	4.48	0.03
Indicator value	[1–3]	1.41	-543.22	0.61	0.43
Sensitivity	[0–1]	0.67	-273.60	2.00	0.16

L plastic bottle filled to 0.5 L with distilled water. Attached biofilm was removed by shaking for two minutes [2.55] in distilled water and preserved in 4% v/v formaldehyde. Details are available in Borrego-Ramos *et al.* [2.3]. Samples were processed and diatom species were identified and counted according to standardized methods [2.6] [2.9] by using an Olympus BX60 microscope (1000×) and the usual bibliography ([2.23] and references therein); 417±48 valves were counted for each sample. In order to collect planktonic diatoms, integrated phytoplankton samples were collected from the whole water column using a core sampler, and 125 mL subsamples were fixed with Lugol's solution. Diatoms were identified and counted by using Utermöhl's [2.48] method using an Olympus IX70 inverted microscope (1000×). Diatom taxa were classified according to their ecological traits (Table 2.2) following the criteria proposed by Rimet and Bouchez [2.41]. Individual biovolumes were obtained from OMNIDIA software v. 6 [2.30] database.

2.2.3 Statistical Analysis

Niche position has been commonly assessed as the multidimensional distance between the average habitat conditions occupied by a certain taxon and the average habitat conditions of the entire study region (e.g., [2.5] [2.8]). However, it is not advisable to use the same dataset to calculate AOR and niche features since taxa with large occupancies would also have low niche positions [2.33]. This is why it has been suggested (e.g., [2.21]) using an independent dataset to calculate niche characteristics. Therefore, we used the sensitivity parameter in the SPI (Specific Pollution sensitivity Index) metric as a proxy to measure niche position (marginality) [2.7]. This parameter derives from multivariate analyses on diatom and chemical data obtained from *ca.* 1000 sampling stations where the first correspondence analysis axis was divided into five segments and the taxa belonging to each segment were then attributed a value between 1 and 5 [2.29]. In our study, this score was rescaled to the (0–1) range so that species with sensitivity values of 3 would have a marginality of 0 and those with extreme values (1 or 5) would have a marginality of 1. Similarly, we used species indicator values used in this metric [2.7] as a surrogate of niche breadth, these scores range from 1 to 3 with eurytopic species having a value of 1 and the very specific species having a value of 3 [2.29].

To test the independent effects of predictors on diatom occupancy, we performed a generalized linear model (one-way Generalized Linear Model Analysis of Covariance with identity link and inverse normal distributions). The GLM ANCOVA allows analysis of the effects of categorically independent variables on an inverse, normally distributed dependent variable while controlling for the influence of continuous covariates, here diatom abundance (%). The identity link implies that the explanatory variables predict the expectation of the untransformed response variable. Independent variables were selected based on a "best subsets" routine [2.35], a variable selection method that systematically explores all possible combinations of predictor variables to find the subset that provides the best-fitting regression model according to a chosen criterion. Best-fit AOR functions were obtained for each resulting model using a Levenberg-Marquardt optimization. This algorithm is a numerical optimization technique commonly used for nonlinear least squares curve fitting and parameter estimation in mathematical models, especially effective when fitting models to data that may have complex nonlinear relationships [2.32]. To compare the performance of the different models generated, we used Akaike's Information Criterion with small-sample correction (AICc [2.27]). The following models were tested: linear, quadratic, power, exponential, von Bertalanffy, Michaelis-Menten, logistic, Gompertz, Gaussian and Hill's equation. Details are available in [2.20]. Statistical analyses were performed by using STATISTICA v. 10 for PC (StatSoft Inc., Minneapolis, USA) and PAST v. 4.11 [2.20].

2.3 Results

A total of 159 diatom species were identified in the analyzed systems, the epiphytic assemblage (β=6.44) being more diverse than the planktonic (β=4.86). The first dataset was dominated by the cosmopolitan *Achnanthidium minutissimum* (Kützing) Czarnecki, present in 27 out of 29 ponds, whereas *Fragilaria* sp. was the most abundant taxon in the plankton samples, occurring in 45% of the study sites. Guild breakdown within epiphytic diatoms is as follows: 85% motile, 98% non-pioneers, 91% non-adnate, 30% high-profile, 23% stalk-forming, 11% ribbon-forming species, whereas 81% motile, 92% non-pioneers, 91% non-adnate, 34% high-profile, 11% stalk-forming and 15% ribbon-forming species dominated the planktonic assemblage.

With respect to model fitting, the best function describing AOR in epiphytic diatoms is a Von Bertalanffy growth equation $O=99(1..93\cdot10^{..13A})$ (R^2=0.75), with an asymptotic occupancy level of ~99%, whereas planktonic forms fitted better to a logistic equation $O=46.7(1+7\cdot10^{..4A})^{-1}$ (R^2=0.87, Figure 2.2) saturating at ~47% of occupancy. Abundance-occupancy curve for taxa not assigned to the nonmotile guild also followed a Von Bertalanffy model $O=58(1..90\cdot10^{..25A})$ ($R^2=0.65$) peaking at ~58% of occupancy, whereas motile diatoms followed a Hill's sigmoidal AOR curve $O=3347.3/(1+A/19153)^{..54})$ ($R^2=0.74$) with a maximum occupancy of 100% (Figure 2.3).

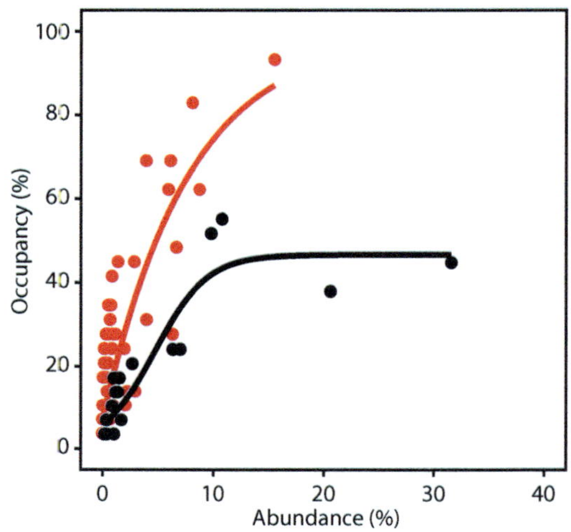

Figure 2.2 Best-fit curves describing AOR patterns for planktonic (black) and epiphytic (red) diatoms in the studied systems.

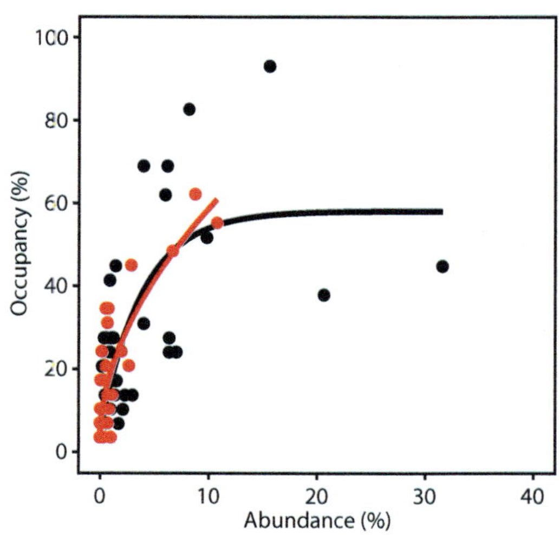

Figure 2.3 Best-fit curves describing AOR patterns for nonmotile (black) and motile (red) diatoms in the studied systems.

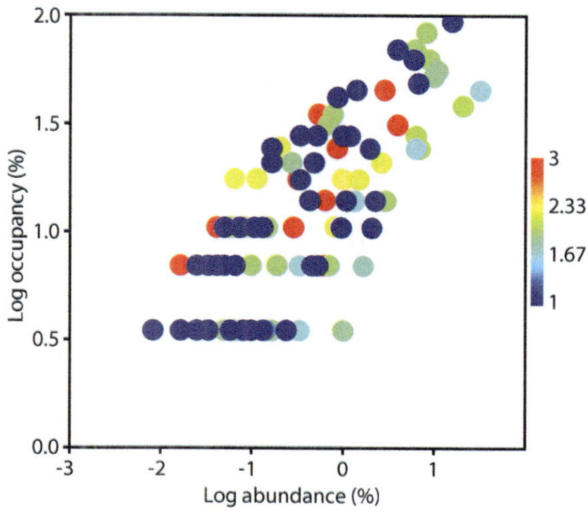

Figure 2.4 Abundance-occupancy relationship in pond diatoms. Dot colors represent species indicator value (see the scale on the right).

Diatom species occupancy has a significantly high correlation with abundance in the null model (with only the covariate) and all the subsequent ANCOVA models. Regarding quantitative predictors, relative abundance (%) and biovolume (Figure 2.4) were selected in all models (lowest AICc score of 949.6, l.ratio χ^2=143.4, p<0.001), but niche position (measured as sensitivity) or niche breadth (measured as indicator value) were selected in none of them (Table 2.2), with a negligible contribution to AOR models. Among qualitative independent features, only habitat (planktonic/epiphytic) was a significant explanatory variable (p<0.001), with also a marginally significant contribution of motile guild (Table 2.2).

2.4 Discussion

The study of abundance-occupancy relationships in pond diatoms elucidates the intricate interplay between species distribution patterns and their local abundances within aquatic ecosystems. These relationships offer valuable insights into the ecological dynamics governing diatom communities and their responses to environmental gradients.

Concerning the best-fit functions calculated for AOR models, despite its importance at the theoretical level [2.38], no comparative analysis with the literature data can be performed since in general AOR equations are not provided in similar studies, at most addressing an "asymptotic" increase in frequency of occurrence with population density. The few exceptions [2.10] [2.36] [2.53] use a Michaelis-Menten equation but are not tested against other models. This function was discarded in our case owing to its bad performance in terms of AICc and R^2 values. However, the ecological interpretation of these curves is not straightforward. It can be seen that, for a given abundance level, epiphytic species were present in twice as many locations as planktonic ones. Their respective model functions did not only differ in slope since occupancy for planktonic forms does not increase along with abundance for dominant species as in the case of epiphytic taxa. The opposite pattern has been observed in a study [2.13] where planktonic dispersing organisms showed an abundance-occupancy relationship with greater slope than that for non-planktonic organisms. Previous studies [2.49] suggest that taxa with wide niche breadths—generally found in such strong dispersers, better adapted to unpredictable habitats—should be more widespread but less abundant than habitat specialists. However, individual dispersal variables may fail by themselves to explain significant variance in AOR models [2.17]. Foggo *et al.* [2.13] confirm that no differences between the slopes of the abundance-occupancy relationship were apparent for different dispersal modes in marine macroinvertebrates.

Positive abundance-occupancy relationships are not merely reflecting the well-known pattern that widespread species tend to be abundant and narrowly distributed [2.16] but they can also mirror other structural features in diatom assemblages. In the case of the pond diatom populations studied in this paper, a positive, significant AOR was observed when all taxa are pooled. As previously observed [2.12], variance in occupancy is highest at low densities (Figure 2.4). However, contrary to our expectations, diatom specific traits had a negligible influence on diatom AOR patterns, with the exception of habitat and marginal significance, pertinent to the motile guild. Our results also show that niche features did not contribute significantly to AOR models in the pond diatom taxa studied. A similar study dealing with riverine fish communities found AOR curves differing relative to size guilds, but not relative to environmental tolerance [2.34]. Although niche-based variables have been observed as the main predictors of abundance-occupancy relationships in aquatic organisms [2.39] [2.42] [2.51], this effect is mostly related to niche position rather than niche breadth [2.19]. In this regard, a strong positive occupancy-abundance relationship has been observed in lake littoral macroinvertebrates [2.16], mostly explained by among-species variation in niche position, followed by niche breadth. Among a specific population size, marginal species exhibit more restricted ranges, as exemplified in subarctic pond diatoms [2.47]. In our case, marginality, measured in terms of species sensitivity, did not account either for the relationship observed between occupancy or abundance. This negative result may indicate the unsuitability of this metric to mirror species' marginality, although sensitivity to environmental variables had been already used to measure niche position (e.g., [2.46]). Alternatively, our findings may rather be confirming those reported by several authors [2.28] [2.43] who found no relationship between niche position, local abundance and/or regional occupancy. This can be attributed to the presence of neglected mechanisms proposed as possible determinants of AOR such as statistical artifacts, range position, and population dynamic explanations [2.15]. As in the case of Siqueira *et al.* [2.43] and for analogue reasons (see their discussion), none of these mechanisms seems to be applicable here.

On the contrary, cell size has been regarded as an important driver of AOR patterns in freshwaters [2.22] [2.47], and our results confirm a significant contribution of individual biovolume in explaining abundance-occupancy relationships, with larger organisms having both low populations and frequencies of occurrence among sampled ponds (Figure 2.5), as already predicted by Wilkinson [2.54]. This is explained by the fact that smaller-bodied taxa usually have larger populations—thus providing more propagules for colonization—and are more efficiently dispersed via passive ways, leading to a broader regional distribution. Heino and Soininen [2.22] did not find a direct correlation between diatom size and occupancy in Finnish streams,

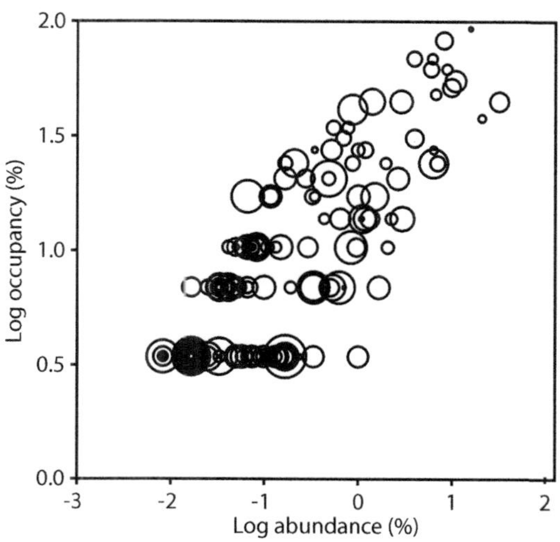

Figure 2.5 Abundance-occupancy relationship in pond diatoms. Circle diameter is proportional to log (cell biovolume)

and partially attributed this result to the limited size range present in these microalgae. Size range found in the present study covered ~3 orders of magnitude, similar to the range observed in that paper, although measured in terms of cell surface and not as biovolume, as in our case. In any case, this result highlights the importance of individual species size and dispersion abilities as drivers of diatom metacommunity structure in the ponds studied.

Our analysis also attributed marginally significant effects to the motile guild; this includes fast-moving raphid pennate taxa [2.37]. According to Rimet and Bouchez [2.41], the classification of diatom species as members or not of this guild is actually masking their assignation to high taxonomical levels, since 96% of the Mastogloiales, Naviculales, Bacillariales, Rhopalodiales, and Surirellales are members of this guild. Therefore, the different AOR patterns found may be merely reflecting taxonomic differences. This would contrast, however, with previous findings [2.13] [2.14] documenting a significant positive relationship between abundance and occupancy, but with the slope of this relationship not differing significantly between taxonomic groups. Given the lack of explanatory power of taxa ecological guilds here and in other studies [2.25], the positive variation regarding interspecific abundance-occupancy relationships does not seem to be explained in a simple way. The hypothesis that AOR is more dependent on abiotic factors may also be ruled out in light of several papers [2.18] [2.44] evidencing this relationship irrespective of disturbance level. Not explored here, invasiveness may account for some variability in AOR responses, with invasive species occurring at lower densities and at fewer sites [2.18] but showing significantly steeper slopes in abundance-occupancy space [2.4] than native ones (but see [2.40]). Our results thus also confirm that niche-based processes may not be a major cause for variation in the AOR dynamics of pond diatoms when compared to other features (e.g., cell size) with a larger predictive power.

In conclusion, our results may be summarized as follows: a) best-fit models describing AOR functions were saturation curves peaking at different occupancy levels, depending on the categorical descriptors used, b) contrary to previous findings dealing with lake diatoms [2.31], neither niche features nor ecological guilds contributed significantly to AOR variation, c) this may be caused by our choice of species' global sensitivity and indicator values as proxies of niche characteristics, and not calculated from the same dataset as usual, and d) on the contrary, species habitat (planktonic/epiphytic) and cell biovolume were the subset of variables that best described AOR models in the studied pond diatoms. Other mechanisms, not revealed in the scope of our analysis, should be explored in future studies, particularly those describing spatial processes such as dispersal ability, habitat connectiveness or mass effects.

Acknowledgments

We thank all the staff at the LIMNO group who helped to collect the samples. Financial support was provided by the project METAPONDS, funded by the Spanish Ministry of Economy and Industry, project CGL2017-84176R, and Fundación Biodiversidad, Ministerio para la Transición Ecológica y el Reto Demográfico (project BT-2019). This study was supported by Grant TED2021-131271B-I00 (DIATOMOL) funded by MCIN/AEI/https://doi.org/10.13039/501100011033 and by "European Union NextGenerationEU/PRTR." The authors are grateful to the reviewers who helped to improve this chapter.

References

[2.1] Andrew, E., Lenore S, C., y Arnold E, G. (2020) *Standard Methods for the Examination of Water and Wastewater.*

[2.2] Blanco, S., Romo, S., Cejudo-Figueiras, C., Gomà, J., Fernandez-Alaez, C., y Fernandez-Alaez, M. (2016) Modelling Primary Producers in Some Shallow Spanish Lakes. *Wetlands.*

[2.3] Borrego-Ramos, M., Bécares, E., García, P., Nistal, A., y Blanco, S. (2021) Epiphytic Diatom-Based Biomonitoring in Mediterranean Ponds: Traditional Microscopy versus Metabarcoding Approaches. *Water 2021 Vol 13 Page 1351*, **13** (10), 1351.

[2.4] Buckley, H.L., y Freckleton, R.P. (2010) Understanding the role of species dynamics in abundance–occupancy relationships. *J. Ecol.*, **98** (3), 645-658.

[2.5] Calenge, C., Dufour, A.B., y Maillard, D. (2005) K-select analysis: a new method to analyse habitat selection in radio-tracking studies. *Ecol. Model.*, **186** (2), 143-153.

[2.6] CEN (2003) Water quality-Guidance standard for the routine sampling and pretreatment of benthic diatoms from rivers.

[2.7] Coste, M. (1982) Étude des méthodes biologiques d'appréciation quantitative de la qualité des eaux. *Rapp. Cemagref QE Lyon-AF Bassin Rhône Méditerranée Corse.*

[2.8] Dolédec, S., Chessel, D., y Gimaret-Carpentier, C. (2000) Niche Separation in Community Analysis: A New Method. *Ecology*, **81** (10), 2914-2927.

[2.9] EN14407, C. (2004) Water quality: guidance standard for the identification, enumeration and interpretation of benthic diatom samples from running waters. *Int. Organ. Stand.*

[2.10] Faulks, L., Svanbäck, R., Ragnarsson-Stabo, H., Eklöv, P., y Östman, Ö. (2015) Intraspecific Niche Variation Drives Abundance-Occupancy Relationships in Freshwater Fish Communities. *https://doi.org/10.1086/682004*, **186** (2), 272-283.

[2.11] Freckleton, R.P., Gill, J.A., Noble, D., y Watkinson, A.R. (2005) Large-Scale Population Dynamics, Abundance-Occupancy Relationships and the Scaling from Local to Regional Population Size. *J. Anim. Ecol.*, **74** (2), 353-364.

[2.12] Freckleton, R.P., Noble, D., y Webb, T.J. (2015) Distributions of Habitat Suitability and the Abundance-Occupancy Relationship. *https://doi.org/10.1086/498655*, **167** (2), 260-275.

[2.13] Foggo, A., Bilton, D.T., y Rundle, S.D. (2007) Do developmental mode and dispersal shape abundance–occupancy relationships in marine macroinvertebrates? *J. Anim. Ecol.*, **76** (4), 695-702.

[2.14] Foggo, A., Frost, M.T., y Attrill, M.J. (2003) Abundance-occupancy patterns in British estuarine macroinvertebrates. *Mar. Ecol. Prog. Ser.*, **265**, 297-302.

[2.15] Gaston, K.J., Blackburn, T.M., y Lawton, J.H. (1997) Interspecific Abundance-Range Size Relationships: An Appraisal of Mechanisms. *J. Anim. Ecol.*, **66** (4), 579-601.

[2.16] Gaston, K.J., Blackburn, T.M., Greenwood, J.J.D., Gregory, R.D., Quinn, R.M., y Lawton, J.H. (2000) Abundance–occupancy relationships. *J. Appl. Ecol.*, **37** (SUPPL. 1), 39-59.

[2.17] Gaston, K.J., y Blackburn, T.M. (2003) Dispersal and the interspecific abundance-occupancy relationship in British birds. *Glob. Ecol. Biogeogr.*, **12** (5), 373-379.

[2.18] Gaston, K.J., Borges, P. a. V., FangLiang, H., y Gaspar, C. (2006) Abundance, spatial variance and occupancy: arthropod species distribution in the Azores. *J. Anim. Ecol.*, **75** (3), 646-656.

[2.19] Gregory, R.D., y Gaston, K.J. (2000) Explanations of commonness and rarity in British breeding birds: separating resource use and resource availability. *Oikos*, **88** (3), 515-526.

[2.20] Hammer, Ø., Harper, D.A.T., y Ryan, P.D. (2001) Past: Paleontological statistics software package for education and data analysis. *Palaeontol. Electron.*

[2.21] Heino, J. (2005) Positive relationship between regional distribution and local abundance in stream insects: a consequence of niche breadth or niche position? *Ecography*, **28** (3), 345-354.

[2.22] Heino, J., y Soininen, J. (2006) Regional occupancy in unicellular eukaryotes: a reflection of niche breadth, habitat availability or size-related dispersal capacity? *Freshw. Biol.*, **51** (4), 672-685.

[2.23] Hofmann, G., Werum, M., y Lange-Bertalot, H. (2011) *Diatomeen im Süßwasser-Benthos von Mitteleuropa: Bestimmungsflora Kieselalgen für die ökologische Praxis; über 700 der häufigsten Arten und ihrer Ökologie*, Gantner.

[2.24] Holt, A.R., Gaston, K.J., y He, F. (2002) Occupancy-abundance relationships and spatial distribution: A review. *Basic Appl. Ecol.*, **3** (1), 1-13.

[2.25] Holt, A.R., y Gaston, K.J. (2003) Interspecific abundance–occupancy relationships of British mammals and birds: is it possible to explain the residual variation? *Glob. Ecol. Biogeogr.*, **12** (1), 37-46.

[2.26] Hui, C., McGeoch, M.A., Reyers, B., Roux, P.C., Greve, M., y Chown, S.L. (2009) Extrapolating population size from the occupancy–abundance relationship and the scaling pattern of occupancy. *Ecol. Appl.*, **19** (8), 2038-2048.

[2.27] Hurvich, C.M., y Tsai, C.-L. (1993) A Corrected Akaike Information Criterion for Vector Autoregressive Model Selection. *J. Time Ser. Anal.*, **14** (3), 271-279.

[2.28] Kristiansen, T., Svenning, J.-C., Grández, C., Salo, J., y Balslev, H. (2009) Commonness of Amazonian palm (Arecaceae) species: Cross-scale links and potential determinants. *Acta Oecologica*, **35** (4), 554-562.

[2.29] Lavoie, I., Hamilton, P.B., Wang, Y.-K., Dillon, P.J., y Campeau, S. (2009) A comparison of stream bioassessment in Québec (Canada) using six European and North American diatom-based indices. *Nova Hedwig.*, **135**, 37-56.

[2.30] Lecointe, M.M.L., Coste, M., y Ector, L. (2016) OMNIDIA software, version 6: complete redesign.

[2.31] Lengyel, E., Szabó, B., y Stenger-Kovácse.g., C. (2020) Realized ecological niche-based occupancy–abundance patterns of benthic diatom traits. *Hydrobiologia*, **847** (14), 3115-3127.

[2.32] Levenberg, K. (1944) Method for the solution of certain problems in least squares siam. *J Numer Anal*, **16**, 588-A604.

[2.33] McCreaCie, J.W., y Adler, P.H. (2014) Abundance–occupancy relationships of larval black flies (Diptera: Simuliidae) in temperate Nearctic streams. *Insect Conserv. Divers.*, **7** (6), 523-532.

[2.34] Miranda, L.E., y Killgore, K.J. (2019) Abundance–occupancy patterns in a riverine fish assemblage. *Freshw. Biol.*, **64** (12), 2221-2233.

[2.35] Neter, J., Wasserman, W., y Kutner, M. (1990) Applied linear statistical models: regression, analysis of variance, and experimental designs.

[2.36] Östman, Ö., Östman, y Örjan (2011) Abundance-occupancy relationships in metapopulations: examples of rock pool Daphnia. *Oecol*, **165** (3), 687-697.

[2.37] Passy, S.I. (2007) Diatom ecological guilds display distinct and predictable behavior along nutrient and disturbance gradients in running waters. *Aquat. Bot.*, **2** (86), 171-178.

[2.38] Picard, N., y Favier, C. (2011) A Point-Process Model for Variance-Occupancy-Abundance Relationships. *Am. Nat.*, **178** (3), 383-396.

[2.39] Renner, S., Dalzochio, M.S., Périco, E., Sahlén, G., y Suhonen, J. (2020) Odonate species occupancy frequency distribution and abundance–occupancy relationship patterns in temporal and permanent water bodies in a subtropical area. *Ecol. Evol.*, **10** (14), 7525-7536.

[2.40] Rigal, F., Whittaker, R.J., Triantis, K.A., y Borges, P.A.V. (2013) Integration of non-indigenous species within the interspecific abundance–occupancy relationship. *Acta Oecologica*, **48**, 69-75.

[2.41] Rimet, F., y Bouchez, A. (2012) Life-forms, cell-sizes and ecological guilds of diatoms in European rivers. *Knowl. Manag. Aquat. Ecosyst.*, (406), 01.

[2.42] Rocha, M.P., Bini, L.M., Siqueira, T., Hjort, J., Grönroos, M., Lindholm, M., Karjalainen, S.-M., y Heino, J. (2017) Predicting occupancy and abundance by niche position, niche breadth and body size in stream organisms. *Oecologia 2017 1861*, **186** (1), 205-216.

[2.43] Siqueira, T., Bini, L.M., Cianciaruso, M.V., Roque, F.O., y Trivinho-Strixino, S. (2009) The role of niche measures in explaining the abundance-distribution relationship in tropical lotic chironomids. *Hydrobiologia*, **636**, 163-172.

[2.44] Soininen, J., y Heino, J. (2005) Relationships between local population persistence, local abundance and regional occupancy of species: distribution patterns of diatoms in boreal streams. *J. Biogeogr.*, **32** (11), 1971-1978.

[2.45] Soininen, J., y Teittinen, A. (2019) Fifteen important questions in the spatial ecology of diatoms. *Freshw. Biol.*, **64** (11), 2071-2083.

[2.46] Soto, C., y Palomares, F. (2015) Coexistence of sympatric carnivores in relatively homogeneous Mediterranean landscapes: functional importance of habitat segregation at the fine-scale level. *Oecologia*, **1** (179), 223-235.

[2.47] Teittinen, A., Weckström, J., y Soininen, J. (2018) Cell size and acid tolerance constrain pond diatom distributions in the subarctic. *Freshw. Biol.*, **63** (12), 1569-1578.

[2.48] Utermöhl, H. (1958) Methods of collecting plankton for various purposes are discussed. *SIL Commun. 1953-1996*, **9** (1), 1-38.

[2.49] Verberk, W.C.E.P., Velde, G.V.D., y Esselink, H. (2010) Explaining abundance–occupancy relationships in specialists and generalists: a case study on aquatic macroinvertebrates in standing waters. *J. Anim. Ecol.*, **79** (3), 589-601.

[2.50] Vilmi, A., Karjalainen, S.M., Wang, J., y Heino, J. (2019) Using traits to explain interspecific variation in diatom occupancy and abundance across lakes and streams. *J. Biogeogr.*, **46** (7), 1419-1428.

[2.51] Vilmi, A., Tolonen, K.T., Karjalainen, S.M., y Heino, J. (2019) Niche position drives interspecific variation in occupancy and abundance in a highly-connected lake system. *Ecol. Indic.*, **99**, 159-166.

[2.52] Warren, P.H., y Gaston, K.J. (1997) Interspecific Abundance-Occupancy Relationships: A Test of Mechanisms Using Microcosms. *J. Anim. Ecol.*, **66** (5), 730.

[2.53] Webb, T.J., Noble, D., y Freckleton, R.P. (2007) Abundance-Occupancy Dynamics in a Human Dominated Environment: Linking Interspecific and Intraspecific Trends in British Farmland and Woodland Birds. *J. Anim. Ecol.*, **76** (1), 123-134.

[2.54] Wilkinson, D.M. (2001) What is the upper size limit for cosmopolitan distribution in free-living microorganisms? *J. Biogeogr.*, **28** (3), 285-291.

[2.55] Zimba, P.V., y Hopson, M.S. (1997) Quantification of epiphyte removal efficiency from submersed aquatic plants. *Aquat. Bot.*, **58** (2), 173-179.

[2.56] Zuckerberg, B., Porter, W.F., y Corwin, K. (2009) The consistency and stability of abundance–occupancy relationships in large-scale population dynamics. *J. Anim. Ecol.*, **78** (1), 172-181.

3

The Ecology of Diatom Reproduction

David G. Mann[1,2*] and Mark B. Edlund[3]

[1]Royal Botanic Garden Edinburgh, Edinburgh, Scotland, UK
[2]Marine and Continental Waters Programme, Institute for Food and Agricultural Research and Technology (IRTA), Catalunya, Spain
[3]St. Croix Watershed Research Station, Science Museum of Minnesota, Marine on St. Croix, Minnesota, USA

Abstract

During the last 170 years, data have gradually accumulated on the diatom life cycle, documenting its unique pattern, in which slow size reduction during vegetative cell divisions in the diploid phase alternates with rapid size restitution via a special cell—the auxospore. However, while the cytological events of auxosporulation are quite well understood and recent studies have begun to reveal some of the underlying genetic, physiological and biochemical controls, progress in documenting the phenology of reproduction has lagged behind. Currently, the principal advances in understanding the ecology of diatom reproduction are coming from studies of selected phytoplankton species. They demonstrate tight coupling between the diatom life cycle, population densities, and seasonal changes in the characteristics of the water column. For benthic diatoms—comprising by far the majority of diatom species—there is far less information, much of it anecdotal but nevertheless valuable. Here we make a comprehensive review of the occurrence of auxosporulation in natural populations of diatoms, revealing an unexpectedly strong seasonal pattern, with N Hemisphere maxima in March–April and late summer, which deserves further study. Meanwhile, transcriptomics is beginning to show how reproduction is triggered and controlled and how it may affect population growth and structure.

**Corresponding author*: dmann@rbge.org.uk
David Mann: Google Scholar [https://scholar.google.com/citations?user=RbBz8RcAAAAJ&hl=en]; ORCID [https://orcid.org/0000-0003-0522-6802]
Mark Edlund: Google Scholar [https://scholar.google.com/citations?user=vTdNWHoAAAAJ&hl=en]; ORCID [http://orcid.org/0000-0002-0227-1117]

Nora I. Maidana, Magdalena Licursi and Eduardo Morales (eds.) Diatom Ecology: From Molecules to Metacommunities, (59–84) © 2024 Scrivener Publishing LLC

Keywords: Auxosporulation, Bacillariophyta, diatom reproduction, life cycle, phenology, seasonality, sexual reproduction, size spectra

3.1 Introduction

The first reports of sexual reproduction in diatoms are generally attributed to George Thwaites [3.93], who, in a breath-taking series of articles and addenda published in the *Annals and Magazine of Natural History*, documented "conjugation" first in *Epithemia turgida* (May 1847), subsequently in *Cymbella* and *Gomphonema* (June 1847), and finally in *Eunotia* and *Rhoicosphenia* (October 1847). The following year [3.94], he reported conjugation in a *Navicula*, in material observed in June 1847, and also "sporangia" (i.e., auxospores) in several centric diatoms now assigned to *Cyclotella*, *Melosira*, *Aulacoseira* and *Biddulphia* (N.B. in several cases, the names Thwaites gave to the species differ from their present ones); in the centrics, however, no information was given about when reproduction occurred. During the next few years, further observations of the process, occurrence and timing of auxosporulation were made by a number of British and continental European phycologists (e.g., [3.25]), especially William Smith, who summarized his own and Thwaites' data in a review of "Reproduction in the Diatomaceae" ([3.92], pp viii–xvii). This included a table of which species had been observed in conjugation or sporulation and when these stages were found (Table 3.1). So, right from the beginning of our understanding of the sexual phase in diatoms, there have been records of the phenology of diatom sex—one aspect of what we here call the "ecology of diatom reproduction." However, since 1856 there has been only one

Table 3.1 The 32 records of auxosporulation in natural material given by Smith [3.92], based on his own and Thwaites' observations: monthly totals, all species. Extra records up to 1856 but not tabulated by Smith are those of (a) Griffith [3.39] and (b) Carter [3.10]. However, Carter's observations were made on material kept for days or months in crude culture; hence, they may give a false impression of reproductive phenology in nature.

Jan	Feb	Mar	Apr	May	Jun	Jul	Aug	Sep	Oct	Nov	Dec
1	1	9	5	1	0	2	2	2	6	3	0
–	–	–	–	(a)	–	(b)	(b)	(b)	–	–	–

further attempt, as far as we know, to bring together everything that is known about the reproductive phenology of diatoms. This is a 150-year-old review by Pfitzer ([3.78], pp. 163, 164), which lists 72 dated reports known to him of auxosporulation in diatoms (see below).

The 150-year absence of an update suggests a general lack of interest in diatom phenology, which seems strange given the abundance and ecological importance of these organisms (e.g., [3.67]) and the fact that the majority of diatoms that have been studied in detail are obligately sexual organisms, with a life cycle that obliges cells to auxosporulate once they reach a critical minimal size (e.g., [3.11] [3.22] [3.85]) or die. Furthermore, this gap in our understanding is, if anything, getting bigger. During the 19th and 20th centuries many observations of auxosporulation were made from natural samples—and hence data on the physical and cytological processes were tied to particular samples made from particular places at particular times. In contrast, almost all recent studies have been laboratory-based: cells are isolated into culture and the sexual phase is induced by mixing compatible strains in heterothallic species or changing conditions in some way that is plausibly imagined to have environmental relevance. Quite often too, sexualization occurs by happenstance in cultures when the cells get small enough.

Laboratory-based work on auxosporulation is not new—the early, pivotal studies by Lüders and Karsten were largely based on temporary cultures of mixed material [3.51] [3.52] [3.53] [3.54] [3.55] [3.63] [3.78]—but in the last 20 years there has been an almost total shift in this direction. The result has been remarkable progress in our understanding of diatom mating systems and the genetic and biochemical mechanisms that underlie sexual reproduction (e.g., [3.1] [3.2] [3.4] [3.5] [3.11] [3.91]) but an almost complete loss of new insights into diatom phenology in nature, except in a few cases, which we will outline later. It is a curious feature of research into diatom sexuality and reproduction that we now have detailed answers to questions that were not prompted, or only scarcely, by natural phenomena, but instead principally by other laboratory observations and by pure intellectual curiosity. Examples include those just listed, but also more basic physiological studies such as the work on factors affecting the auxosporulation of *Rhabdonema adriaticum* by Rozumek [3.86] (still one of the most comprehensive investigations of the effects of environmental factors on auxosporulation in a single species), which seems not to have been prompted by any phenological observations of the species in nature (none have been reported), nor of other *Rhabdonema* species where phenological data are available (e.g., from [3.7] [3.8] [3.92]). This is in huge contrast to macroalgae, where there are well-studied patterns of reproduction and recruitment in natural populations (e.g., [3.6] [3.50]) and terrestrial plants, where the eons-old observation that flowering occurs in particular seasons and is followed sometime later by the production of fruits and seeds has led to investigations of the effects of daylength, temperature, and other factors on plant development. Hence, while there is a considerable and growing body of experimental data (reviewed by [3.9] [3.11] [3.80]) on the effects of various physical, chemical and biotic factors on sexual induction and auxosporulation—such as the factors just mentioned, together with light quality and intensity [3.2] and salinity [3.38], and associated microbiota [3.13] [3.71]—it is often not at all clear how these factors affect and control any particular diatom life cycle in nature.

In relation to diatom ecology, it is worth bearing in mind Thwaites' early comment [3.94] that "*physiologically* [his emphasis] we ought to consider the numerous frustules which have originated from the primordial frustule (the sporangium or product of conjugation) not as so many individuals of a species, but rather as parts of one individual, which, instead of, as takes place in the higher plants, cohering to form one structure, ... maintain an independent existence...[and] retain all the functions necessary for this independent life and for propagation of the species....The life of an individual Diatomaceous plant we may therefore consider to extend from the production of the sporangium to the period of the conjugation of the numerous frustules which have originated from it." Auxosporulation is thus an integral part of a diatom's life—an obligate stage in its development—in most but not all species. Hence, a full understanding of each species' occurrence, ecology and variation cannot be had without attention to this elusive stage. We echo the comment of Davis *et al.* [3.18] that laboratory experiments on sexualization should be coupled with "extensive field studies looking for the different sexual stages and the conditions they occur under." Another of their recommendations, that "the taxonomic description of every diatom species should include a description of the life cycle stages" is an unattainable ideal—many diatomists go for months or years, even their whole careers, without seeing a living auxospore or recently enlarged cell—but it is still something to work towards. The aim of this paper is therefore to bring together the patchy information that is available about when diatoms auxosporulate in nature (Figures 3.1 and 3.2) and to briefly explore how auxosporulation affects the size structure, ecology, and other characteristics of diatom populations.

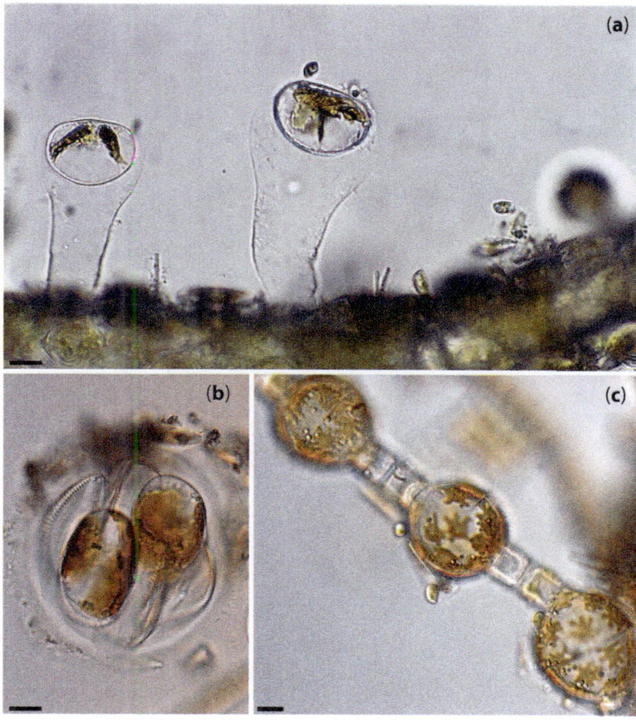

Figure 3.1 Many types of auxosporulation can be found in natural samples. (a) *Cocconeis pediculus* growing on the green alga *Cladophora* in West Lake Okoboji, Iowa, USA, produces a single auxospore using Geitler's Type II reproduction (Geitler's scheme is explained in [3.29]). (b) A rare example of paired auxospores produced via Type I sexuality in *Encyonema reimeri*, West Lake Okoboji, Iowa, June 10, 2015. (c) *Melosira varians* filament with three auxospores produced via oogamy in a small pond in Minnesota, USA, May 21, 2015. Several sperm cells continue to surround the middle auxospore. All scale bars = 10 μm.

3.2 Reports of Auxosporulation in Natural Populations

One of the difficulties in assembling a summary of published records of when auxosporulation occurs in nature is that authors do not always clearly state whether their observations were made on material freshly collected from nature or sometime later, perhaps after a few days in rough culture or incubation for a more extended period. For example, Pfitzer's list [3.78] gives the author of each report of auxosporulation, the species, and the month of observation, but in several cases careful reading of the original

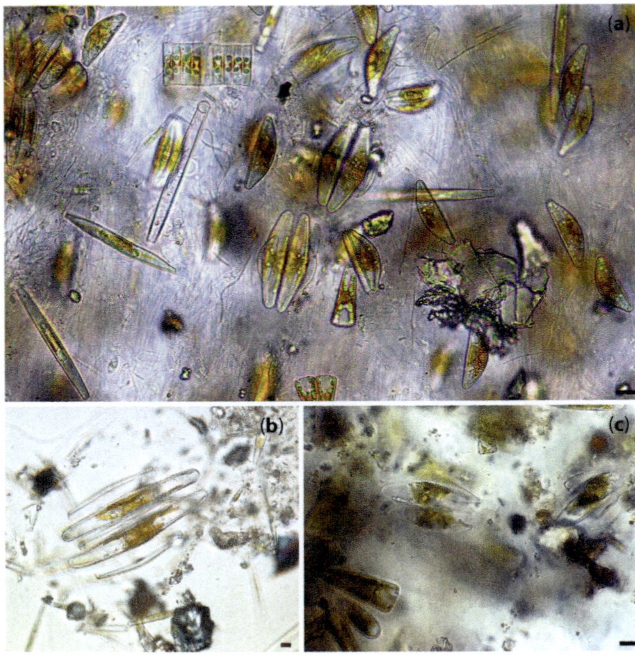

Figure 3.2 (a) *Cymbella affinis* undergoing synchronous Type I sexual reproduction within the periphyton mat on a navigation buoy, West Lake Okoboji, Iowa, USA, May 17, 2017. (b) Developing paired auxospores expand parallel to the empty gametangial valves of *Cymbella aspera*. The entire Type I sexual process takes place within a copulation envelope in the epilithon of Sugarloaf Cove, Lake Superior, Minnesota, USA, June 21, 2022. (c) *Gomphonella olivacea* undergoing synchronous Type I sexual reproduction in the epilithon of Fleming Creek, Michigan, USA, March 20, 1992. All scale bars = 10 μm.

publication suggests that auxosporulation was initiated after sample collection. Pfitzer's own observations fall into this category. We could find little explanation of how he treated samples and made his observations (16 auxosporulation records of 12 species, some species being observed twice), but in several places he refers to the use of cultures, which were presumably rough cultures of mixed material, maintained with periodic additions of fresh medium. For example, in relation to *Diploneis* ("*Navicula*") *elliptica*, he says that after he had made some scattered observations of developing auxospores, the culture failed ("Die Cultur gelang nicht": [3.78], p. 65), while under *Pinnularia* he indicates the use of coverslips as substrata ("Eine weitere Cultur auf dem Objectträger wurde vergeblich versucht"). He also mentions cultures in passing in his Introduction ([3.78], p. 3), though without specifying details of his own approach, and indicates elsewhere that he found it useful to aspirate samples (p. 168, footnote 5).

Given these hints about his practice, we discount all 16 of his records, and also the two by Focke [3.25]. All they indicate is that the samples used probably already contained cells that were within the size range in which sexual reproduction can be induced when they were removed from the natural environment. Within a short period, these cells realized their sexual potential but in nature the same might not have occurred. The distorting effect of shorter or longer periods of incubation in the laboratory is illustrated in Figure 3.3, which shows which epipelic species were observed in stages of auxosporulation on coverslips three or more days after the slips had been placed on the surface of mud from Blackford Pond, Scotland. None of them were auxosporulating when the coverslips were first observed a day after sample collection.

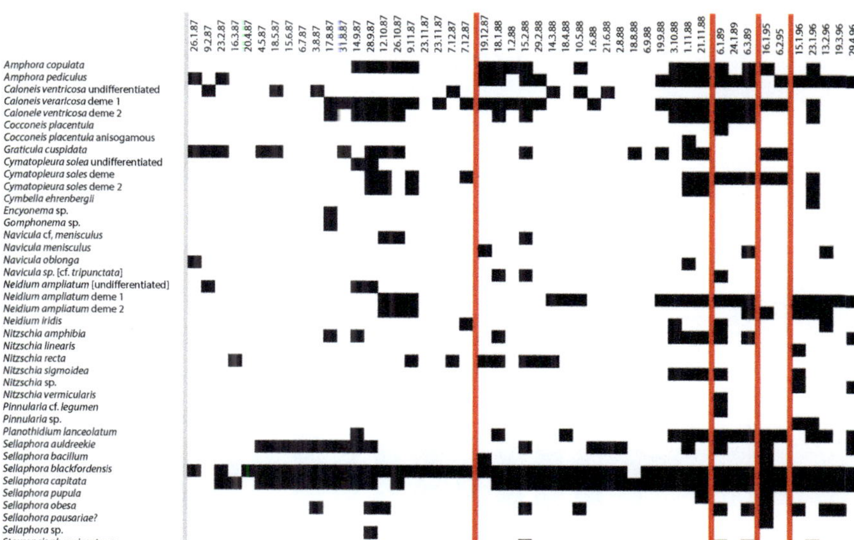

Figure 3.3 Auxosporulation recorded in epipelon from Blackford Pond, Edinburgh, two or more days after transfer to the laboratory, during two complete years (1987 and 1988) and at intervals thereafter. Black cells indicate observation of one or more stages in auxosporulation, from pairing through to initial cells. In three cases (*Caloneis ventricosa*, *Cymatopleura solea* and *Neidium ampliatum*), taxonomic interpretations were refined during the period of observation: each nominal species was treated initially as a single entity but later separated into two demes, according to cell dimensions and other characteristics and reproductive isolation [3.65]. Very little pattern can be detected (except perhaps a greater incidence in the autumn). Such observations indicate that potentially sexual cells are often present in epipelic species (throughout the year in the case of *Sellaphora capitata* and *S. blackfordensis*), which is important information because, unless cells are within the critical size thresholds, auxosporulation is impossible; however, it says very little otherwise about the natural phenology of species.

A search of the literature available to us, discounting records like Pfitzer's that are probably of cultured material, yielded over 490 dated records of auxosporulation in natural populations of diatoms, contained in over 150 publications together with some unpublished records of MBE and DGM, and these are graphed in Figure 3.4. Where a record stated that auxosporulation took place over a period of more than a month, we counted each month as a separate record. For example, Pérez-Martínez *et al.* [3.76] said that auxosporulation took place from June to September in *Pantocsekiella* (*Cyclotella*) *ocellata*; we counted this as four monthly records, one for each of June, July, August and September. There were a few reports for which vaguer information was given ("late summer," "beginning of spring," etc.) and these we omitted from Figure 3.4, because there was no easy way to

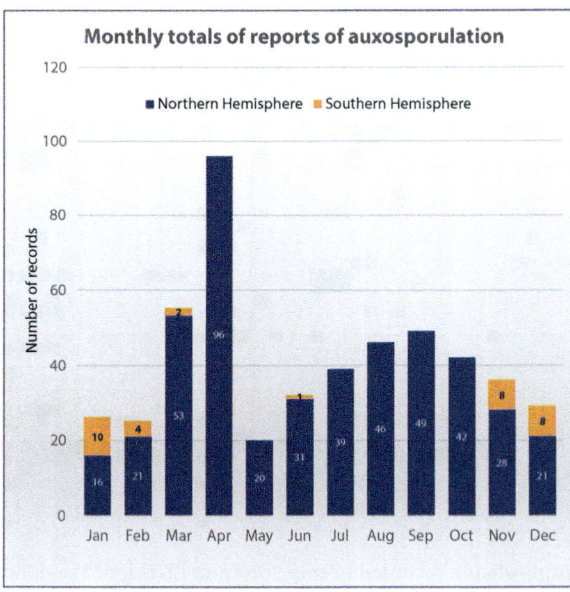

Figure 3.4 Frequencies of records of auxosporulation in nature (495 in total), by month, from a survey of all literature and personal observations known to us. Records were accepted only if it was clear from the source publication that auxosporulation was taking place in the sample at the time of collection. Any indication that the material was incubated in the laboratory for a few days, or that material was being kept in rough culture when auxosporulation occurred, disqualified the record. If authors recorded that auxosporulation took place over a period exceeding one month (e.g., September to November), we scored separate records for each month. Imprecise records (summer, late autumn) were not accepted. The raw data on which this figure is based are available in a supplementary spreadsheet available at https://websites.rbge.org.uk/algae/research/auxosporulation_in_natural_populations.html.

determine how this information should be interpreted. We did not accept estimates of when auxosporulation occurred that were based on size spectra in populations, because shifts in the frequency of a predominant size class (replacement of a class of small cells by a class of larger cells) do not necessarily mean that auxosporulation was occurring at that time (cf. [3.23] [3.64] [3.66]), though in several cases, such as for auxospores in *Rhizosolenia styliformis* reported in natural samples by Wimpenny [3.96], auxosporulation was accompanied by abrupt shifts in size classes. We did accept records of "microspores" (i.e., stages in the formation of sperm) as indicating that auxosporulation was occurring. Space does not permit us to tabulate all the records we found in the literature here. Instead, we will select a few illustrative examples for comment. The full database is available as a spreadsheet at https://websites.rbge.org.uk/algae/research/auxosporulation_in_natural_populations.html.

Observations in the spreadsheet noted as "personal observation" were samples examined immediately upon collection or in samples preserved at the time of collection.

Our survey is undoubtedly incomplete because it is virtually impossible to survey every publication on diatom occurrence and ecology for occasional records of auxosporulation. This is especially true for the period (cf. 1930 to 2000) in which a lot of literature is not available in digital form and so cannot be searched efficiently. For the period before 1930 we made extensive use of the *Biodiversity Heritage Library* (using the search term *auxosp**) and also of the ZOBODAT and *Gallica* databases (https://www.zobodat.at/ and https://gallica.bnf.fr). We are confident that our records for pennate diatoms are particularly good, partly because of the work already done by Geitler [3.29] [3.36] to collate records of pennate auxosporulation. For centric diatoms, however, we will almost certainly have missed records, either of auxospores or "microspores" (stages in spermatogenesis), that are probably contained in plankton surveys, in both lakes and oceans. We found a number of these but there must be more.

Somewhat surprisingly to us, given the wide nature of our survey (all diatoms: both centric and pennate, planktonic and benthic, lentic and lotic, Northern and Southern Hemispheres), Figure 3.4 shows a pattern in the monthly totals of auxosporulation records. Numbers increase dramatically to a sharp peak in the spring (April), followed by a collapse back to very low numbers in May and June. Then there is recovery to a broader, lower peak in the late summer and autumn (August to November), with a second maximum in September. Once the few Southern Hemisphere records (principally by Karsten [3.56] [3.57] [3.58]) are omitted, the pattern is even more striking. It broadly corresponds to frequently observed seasonal patterns of diatom abundance (though not necessarily cell division rates) in the temperate regions of the Northern Hemisphere, in lakes and oceans (e.g., [3.81] [3.84]).

Turning to individual species, the picture is far less clear. In most cases there are only one or two records for a particular species, and even where there are several records, these do not usually represent multiple records from the same place. Where there are several records for a particular species, they are often from different times of year, not from one month or a few consecutive months as one might expect from terrestrial plants. One of the best documented species is *Gomphonella olivacea* (synonym *Gomphonema olivaceum*; Figure 3.2c), studied by various authors ([3.12] [3.23] [3.29] [3.77] [3.87], MBE, personal observations) with records of auxosporulation in February (3), March (5), April (2), May (1), June (1), July (2) and September (1), thus reflecting more or less the general pattern of Figure 3.4. *Cymbella lanceolata* has records in January, March, April, May, June, October and November (one record each) and "summer" ([3.28] [3.31] [3.92] [3.93], DGM personal observation). *Pantocsekiella ocellata* records also span a long period in the year, from May to November, with the most occurrences (four out of 12 month-records) in September [3.40] [3.76] [3.83] [3.89]. These wide ranges (noted already by Pfitzer [3.78]) make it unlikely that these species are responding, at least in a simple way, to daylength, light intensity or ambient temperature.

In a few cases, auxosporulation is more restricted, with most or all records from at a particular time of year. Examples are *Meridion circulare*, with records apparently restricted to April (four records: [3.27] [3.30] [3.32] [3.33], plus an undated record probably from April [3.34]) and the *Gomphonema angustatum* species complex, also in April (ten records: [3.32] [3.33] [3.35], Edlund personal observations, with just one extra observation on a USA population from July). In both of these, auxosporulation has been suggested to be associated with snow-melt in the catchments of the streams where they were found (e.g., [3.32] [3.33]) and this could guide experiments to identify the particular factors responsible for triggering the sexual phase. Of course, both species also occur in streams that do not receive meltwater.

More complete data, set in the context of the whole annual cycle of growth and with accompanying environmental data, have been provided by Jewson and colleagues for three *Aulacoseira* species, *A. subarctica* [3.46], *A. skvortzowii* [3.48] and *A. baicalensis* [3.47], and also *Stephanodiscus neoastraea* [3.45]. These represent the most complete accounts to date of the phenology of reproduction in diatoms and the factors that likely control it in nature. The timing of auxosporulation and the controlling factors differ in each case. In *Stephanodiscus neoastraea*, a 4-year study of populations in Lough Neagh, Northern Ireland (1988–1991) showed auxosporulation is restricted to a 3–4-week period each year in autumn (late September–early December). The timing suggested that the trigger is release from a check to growth due to increased N availability. In *A. subarctica*, on the other hand, auxosporulation is spread over a much longer period. Jewson's data ([3.46], his Table 1) indicate that auxospores can be found in all months of the year except June and July, but since auxospores (as initial cells) can persist in populations for some time after their formation, it is possible that auxosporulation itself is more restricted in its occurrence. Certainly, it can occur over the whole winter, from November to February, and also in August. Jewson identified that "sexual differentiation of cells occurs after checks in growth during low irradiance," the effect being that the cost of auxosporulation in lost vegetative growth will be minimized (cf. [3.61] [3.62]). In *A. skvortzowii*, auxosporulation seems to be associated with low phosphate concentrations following a period of active growth [3.48], while auxospores of the third *Aulacoseira* species, *A. baicalensis*, occur towards the end of a period of growth under ice cover in Lake Baikal while light levels are high (snow cover on the ice shades the water column and inhibits auxosporulation); subsequently, the population forms resting cells in response to lower light "income" as the depth of the mixed water column increases [3.47].

In interpreting available data, and in planning future work, it is also important to recognize that many records of auxosporulation are not independent of the personal habits of the people making the records. In the case of the three *Aulacoseira* species and *Stephanodiscus astraea* mentioned above, this is not so: the studies of auxosporulation were undertaken as an integral part of planned seasonal sampling of the plankton, with back-up from laboratory experimentation. In many other cases, records have been accidental finds, or incidental records made as a byproduct of other activities, such as an interest in the process of auxosporulation (the number of gametes produced, configuration of gametangia and auxospores, etc.), rather than its timing. In these cases, records may partly reflect very human factors—greater freedom for university staff to travel and sample outside the teaching terms, timing of classes at field stations [3.23] [3.90], avoidance of mid-winter because of the weather, etc. For example, it is clear from the biography by Friedmann [3.26] that Geitler's year was a somewhat regimented one, with annual visits to particular places, so that the many records Geitler made cannot all be assumed to be unbiased samples. Furthermore, in some cases, patterns could conceivably be created because of the context of the studies being undertaken. The abundance of records of auxosporulation in *Gomphonema angustatum* partly reflects Geitler's interest in the species as an example of a complex comprising several "Sippen" with different reproductive and size characteristics, so that populations in which different Sippen were present simultaneously were particularly desirable. However, the consistency with which he found and recorded auxosporulation in April in this species probably reflects a real aspect of its phenology, since it is clear from his many publications that Geitler spent a lot of time observing samples at different times of the year from a variety of different sites, particularly in and around Vienna and Lunz in Austria, and the association with snow-melt ([3.35], p. 199) is worth further exploration.

To summarize, the overall pattern revealed by our survey seems too pronounced to be wholly an artifact. It seems that (Northern Hemisphere) diatoms avoid the "merry month of May" (*The Passionate Pilgrim*, W. Shakespeare) for sex and that, if we wish to find auxospore stages and are not fussy about the species, March, April and September are the best times. The agreement between the maximum overall occurrence of auxospores (Figure 3.3) and the generalized seasonality of diatoms—with maxima in spring and late summer-autumn—seems at first sight to reflect the adage that there is "no significant antagonism between factors promoting vegetative growth and those eliciting gametogenesis" [3.19]. Karsten ([3.54], p. 19) wrote "that auxospore formation always occurs at the time of the main development [of the species], sometimes at the beginning of it, sometimes at the very end" ("dass die Auxosporenbildung stets in die Zeit der Hauptentwickelung fällt, bald mehr im Anfang derselben liegt, bald ganz ans Ende gerückt erscheint"). It is worth remembering that dense populations may facilitate sexual reproduction, maximizing the possibility of pairing between potential gametangia in pennate diatoms and sperm–egg encounters in centric lineages. On the other hand, the cost to a population of switching to sexual reproduction from vegetative growth is maximal when the cells are in rapid exponential growth, so that one might expect auxosporulation to be triggered when cell densities are high but growth becomes limited (e.g., by phosphate in *A. skvortzowii* [3.48]). A complication revealed by recent laboratory studies is that sexually induced cells can arrest the growth of non-participating cells [3.59] [3.70], even when environmental conditions—specifically nutrient levels—are favorable for vegetative growth [3.1]. It is therefore critical to establish which of Karsten's possibilities is realized — does auxosporulation occur at the beginning or at the end of the period of abundance, or somewhere in the middle?

— and to establish exactly when abundant auxosporulation occurs in relation not only to population density *but also* the rate of cell division and environmental triggers. This information is nonexistent in most cases and our knowledge of individual species is abysmal. It is only in cases like the three *Aulacoseira* species and *Stephanodiscus neoastraea*, described above, that we can be confident that there is a repeated pattern and that we have enough information to be able to interpret it.

In passing, it is worth noting that in some other protists, such as volvocalean green algae and dinoflagellates, a clear association has been demonstrated between bloom termination, the induction of sexual reproduction, and the formation of dormant stages for perennation; examples are the freshwater phytoplankters *Eudorina* and *Ceratium* [3.82] [3.42] (and see [3.60]). However, the auxospore is not a dormant stage. Because of this and the unique size reduction—restitution cycle, the ecological dynamics of the sexual phase in diatoms cannot be expected to be like those of other unicells.

3.3 Ecological Significance of Auxosporulation

Edlund and Stoermer [3.22] extended Drebes' [3.19] and Karsten's [3.54] assertions further to address the ecological significance of auxosporulation. Two factors were identified in categorizing auxosporulation: whether populations underwent sexuality in a synchronous or asynchronous manner, and whether the trigger for sexuality was favorable or unfavorable for diatom growth. These four life history strategies help us better understand the handful of well-documented studies on sexuality as well as the overall patterns we find in our compiled data on auxosporulation in nature. The most common life history strategy is *synchronous sexuality under good growth conditions.* Jewson's [3.45] report on *Stephanodiscus neoastraea* undergoing mass auxosporulation when nitrate levels rise in Lough Neagh, as well as probably many of the instances of abundant auxospore production captured in our literature survey exemplify this strategy. In contrast, the reports of sexuality in *Aulacoseira skvortzowii* [3.48] and *A. herzogii* [3.49] represent cases of the second strategy—*synchronous sexuality under poor growth conditions.* In these two taxa, mass auxosporulation is triggered by environmental conditions that check the growth of the populations as phosphate levels are exhausted or as light become limiting for these two taxa, respectively, supporting Karsten's [3.54] earlier observation of mass auxosporulation "...sometimes at the very end" of the growth period.

With information on sexuality of diatoms known for only a few hundreds of species from among the tens of thousands of named species, researchers must ponder whether those rare singleton observations of pairing or auxospores represent one of the two types of asynchronous patterns of auxosporulation and whether these last two life history strategies are much more prevalent in nature than given credit. Classic examples of *asynchronous sexuality under favorable growth conditions* again look toward Jewson's studies on *Aulacoseira subarctica* [3.46] and *A. baicalensis* [3.47]. These two taxa have their major growth periods during the winter — under the ice for *A. baicalensis*! — with small portions of the populations undergoing asynchronous auxosporulation over the course of 2–3 months. Only through continual monitoring and careful observation of hundreds to thousands of cells could the significance of this ecological strategy be appreciated. The final and most rarely encountered life history strategy—*asynchronous sexuality under poor growth conditions*—has been invoked at times using negative observations from nature. Mizuno and Okuda [3.69] carefully studied the growth requirements for *Cocconeis scutellum* var. *ornata* in culture, including the contrasting culture conditions that triggered sexuality. However, when they took their lab observations into the field and tried to catch sexuality in nature, they were unsuccessful in documenting paired cells or auxosporulation, instead only being able to document that sexuality occurred with the reappearance of larger cell lines in their environmental samples. For the thousands of species for which we have no information on their sexuality, the asynchronous life history strategies may be where answers lie. Several studies report careful observation of multiple classes of cell sizes across many years or seasons where sexuality can only be inferred from the reappearance of large cell lines in taxa such as *Didymosphenia geminata* [3.3], *Cyclotella meneghiniana*, *Rhopalodia gibba* [3.9], *Diatoma vulgaris* [3.23], *Nitzschia sigmoidea* [3.64], and *Pseudo-nitzschia multistriata* [3.15]. For researchers, opportunity beckons and anyone encountering sexuality in their collections should note it, observe it, measure it, and document it (Figures 3.1 and 3.2). With much effort to understand sexuality taking place in the lab, the ecological importance of diatom sexuality remains an open field in our understanding of the global significance of diatoms.

Further incentives for studying when and where diatoms reproduce come from a consideration of what effects auxosporulation is likely to have on the growth and ecology of diatom populations. The first is that auxosporulation interrupts vegetative growth [3.61], potentially releasing niche space for other species, or non-auxosporulating cells of the same species. This is true even if auxosporulating cells suppress the growth of non-auxosporulating cells, as shown by Annunziata *et al.* [3.1], since many other losses are more or less inherent to sexual reproduction (aborted and superfluous gametes, lethal or maladapted gene combinations in the zygotes, etc. (e.g., [3.43]). Second, recently auxosporulated cells are much larger than their 'mother cells' (usually gametangia), with all that this might mean for surface area to volume ratios, and hence perhaps for nutrient uptake and intrinsic growth rate. One might therefore expect a tie-up between the timing of life-cycle events and the seasonal growth cycle, as found in *Diatoma* by Potapova and Snoeijs [3.79]. Curiously, however, in some species the growth rate remains remarkably similar during the life cycle, despite changes in cell size [3.74], while in other cases auxosporulation has been found to result in a greatly enhanced growth rate [3.14]. This "rejuvenation" is opposite to what one might expect from simple considerations of surface area to volume relationships. Third, in planktonic species, there will most likely be effects on the sedimentation rate, as a result of changes in the overall density of the cell (due to differences in the proportion of the cell comprised by the frustule) and form resistance (e.g., [3.16]). Other aspects that are likely to influence the development and fate of populations include susceptibility to parasites and grazing. Mann [3.64] noted that a shift in the size spectrum of a *Nitzschia sigmoidea* population occurred during infection by an oomycete and suggested that this might be because larger cells are a larger target for the oomycete zoospores. Size, shape and silicification will clearly be major determinants of whether cells can be ingested by grazers (see [3.45]), especially when the grazer is approximately the same size as the prey diatom, as is the case with some ciliates and amoebae. And lastly, of course, auxosporulation may create opportunities for populations to respond to their environment through reassortment of the genetic variation already present.

A final caution needs to be added: it is often assumed that auxosporulation is sexual, even when the process has not been directly observed, but there are cases (probably a minority, but perhaps more than is currently thought: [3.80], pp. 247, 248) where size is restored without allogamous sexual reproduction via automixis, parthenogenesis or vegetative enlargement (e.g., [3.36] [3.37]). Researchers can also find examples of autogamous auxosporulation within natural populations (e.g., *Muelleria peraustralis* [3.21]); some species are also noted to mix allogamous and autogamous or apomictic strategies for size restitution in natural populations (e.g., *Gomphonema*: MBE, personal observations; *Cocconeis czarneckii* [as *C. placentula*]: [3.20]).

3.4 Future Prospects for Phenological Studies

Research into the phenology of auxosporulation in diatoms has undoubtedly been affected historically by methodological and human factors and it is unrealistic to think that this can changed easily. Even in groups of organisms that can be studied more easily than diatoms (e.g., angiosperms), the long-term monitoring that phenology needs is often unavailable and when it is done, it has often been by particular individuals as an adjunct to their normal work (sometimes unauthorized) or as a hobby [3.95]. Long-term monitoring is not easily maintained in a scientific world dominated by three- or four-year project funding or PhD programs. In diatoms too, phenological observations have often been opportunistic (e.g., during the Iowa Lakeside Lab annual Diatoms class; Figures 3.1 and 3.2, [3.23] [3.90]) and the only systems where long-term data can be acquired relatively easily are the marine and freshwater phytoplankton of lakes and oceans, where long-term monitoring is sometimes maintained by research institutes. Here sampling is more or less straightforward and observations can be made alongside cell counts (e.g., [3.88]), as was also done in the past (e.g., [3.41] [3.73], etc.), even if the study was not autecological (contrast Jewson and colleagues' work).

The next most straightforward community is the epiphyton, which is relatively easily sampled through detachment of the substratum and where auxospores are often very obvious, being carried aloft on the stalks of vegetative cells in *Gomphonema*, *Gomphonella* or *Cymbella*, or on special pedicels produced by the auxospore in *Cocconeis* (e.g., [3.29], our Figures 3.1 and 3.2). Epilithon can also be studied relatively easily, since although brushing or scraping the rock will destroy the 3-dimensional structure of the community, auxospores are often surrounded by mucilage throughout their development and are likely to be kept together with the gametangia (Figures 3.1 and 3.2).

The most difficult communities to study are sediment-associated diatoms—epipelon and epipsammon—since here separation of diatoms from the substratum is difficult and in some cases the methods used to study them (e.g., harvesting epipelon on coverslips or via lens tissue) select very effectively against detection of auxospores. This is because the method depends on the motility of the cells, which is abandoned during the earliest stage of auxosporulation (i.e., pairing). Consequently, auxospores will simply be left behind and hidden in the substratum during sampling. Material harvested on coverslips in the laboratory and kept for a few days in natural water or culture medium often develops auxospores in a variety of species

(Figure 3.4), but all this demonstrates is that potentially sexual cells (i.e., cells in the right size range, within the critical thresholds for sexual reproduction [3.11]) were present in the community when it was sampled. To get around this, one of us tried for a year (autumn 1987 to autumn 1988) to study the auxosporulation of epipelic diatoms *in situ* in Blackford Pond, Edinburgh, by placing racks of microscope slides on the sediment surface at a water depth of c. 50 cm. The slides were placed vertically and sank into the very soft sediment by about half a centimeter, so that part of the slide could potentially sample the epipelic community inhabiting the top millimeters of sediment. The slides were left *in situ* for one or more weeks and then removed for immediate examination (and sometimes also for incubation in culture medium). Some auxosporulation was observed (Figure 3.5) but cases were few and it was noticeable that many were of attached diatoms (*Cocconeis*, *Gomphonema*, *Planothidium*) that had colonized the exposed upper parts of the slides. This illustrates, however, that use of artificial substrata could be an effective way forward to study phenology in some benthic diatoms (e.g., *Cocconeis czarneckii* [3.20]). Diatometer slides [3.75] have in some cases already given dated records of auxosporulation [3.20] [3.44].

In the end, however, the best hope of advances, whichever type of community is being studied, is probably the development of automated monitoring of communities via metatranscriptomics, if genes can be identified that are (1) associated unambiguously with auxosporulation and

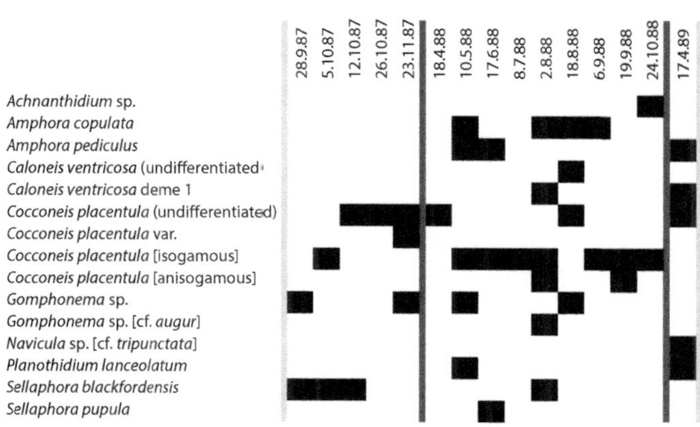

Figure 3.5 Auxosporulation on microscope slides placed vertically, spanning the mud–water interface in Blackford Pond. Colonization took place over two or more weeks of incubation *in situ*; records of auxosporulation were made on the day when the slides were retrieved from the pond. As in Figure 3.2, black cells indicate observation of one or more stages in auxosporulation, from pairing through to initial cells.

(2) exhibit sufficient variation to discriminate among coexisting species. Apparently suitable markers have been identified in a number of "model" organisms, e.g., genes associated with meiosis in *Skeletonema marinoi* [3.24] and with flagella (hence with male gametogenesis) in *Leptocylindrus danicus* [3.72], but the number of genomes known in sufficient detail is tiny, preventing wider application of this approach. Meanwhile, illustrative cases can be investigated without molecular monitoring, either as has been done in *Pseudo-nitzschia*, *Aulacoseira* and *Stephanodiscus* (see above: work by Sarno *et al.*, Jewson and colleagues) or via studies of size spectra, if the cardinal points of the species are known (e.g., [3.15] [3.17] [3.22] [3.65] [3.66]).

Finally, since it seems that no biological paper can be complete nowadays without reference to climate change, we would comment that, if we want to understand diatom species well enough to be able to predict in some fashion their responses to environmental perturbations (cf. the kinds of work currently being done with macroalgae [3.68]), we do need to have a better idea of when and why they auxosporulate in their natural habitats.

Acknowledgments

DGM thanks Christophe Legrand for the preliminary literature survey he undertook, back in 2009, at the suggestion of DGM and Dr Marina Montresor, to find literature records of auxosporulation in natural populations. Without this, it would have been a much more difficult task to complete the data summarized in Figure 3.3 and the spreadsheet supporting it. He also thanks the Royal Botanic Garden Edinburgh for support, especially the library staff for their help in building literature resources to support research into algae. MBE thanks Iowa Lakeside Laboratory for hosting 60 summers of Ecology and Systematics of Diatoms with a special thanks to the Lakeside students, assistants, and researchers who keep their eyes peeled every summer for diatom auxosporulation. MBE acknowledges funding from the Minnesota Department of Natural Resources Coastal Program (MLSCP 306-10), the Minnesota Environment and Natural Resources Trust Fund as recommended by the Legislative-Citizen Commission on Minnesota Resources (LCCMR), and subcontract PO 210254 from Global Algal Innovations, Inc., under Contract #DE-EE0008904, U.S. Department of Energy CDFA Number 81.049. Two anonymous reviewers made helpful comments on the manuscript and we thank them and the editors for their work and time; we appreciate it very much.

References

[3.1] Annunziata R., Mele, B.H., Marotta, P., Volpe, M., Entrambasaguas, L., Mager, S., Stec, K., d'Alcala, M.R., Sanges, R., Finazzi, G., Iudicone, D., Montresor, M., Ferrante, M.I. 2022. Trade-off between sex and growth in diatoms: Molecular mechanisms and demographic implications. *Sci. Adv.* 8, eabj9466.

[3.2] Bilcke, G., Van den Berge, K., De Decker, S., Bonneure, E., Poulsen, N., Bulankova, P., Osuna-Cruz, C.M., Dickenson, J., Sabbe, K., Pohnert, G., Vandepoele, K., Mangelinckx, S., Clement, L., De Veylder, L., Vyverman, W. 2021. Mating type specific transcriptomic response to sex inducing pheromone in the pennate diatom *Seminavis robusta. ISME Journal* 15, 562–576.

[3.3] Bishop, I.W., Spaulding, S.A. 2017. Life cycle size dynamics in *Didymosphenia geminata* (Bacillariophyceae). *J. Phycol.* 53, 652–663.

[3.4] Bondoc, K.G., Lembke, C., Vyverman, W., Pohnert G. 2016. Searching for a mate: pheromone-directed movement of the benthic diatom *Seminavis robusta. Microb. Ecol.* 72, 287–294.

[3.5] Bondoc, K.G.V., Lembke, C., Lang, S.N., Germerodt, S., Schuster, S., Vyverman, W., Pohnert, G. 2019. Decision-making of the benthic diatom *Seminavis robusta* searching for inorganic nutrients and pheromones. *ISME Journal* 13, 537–546.

[3.6] Brawley, S.H., Johnson, L.E. 1992. Gametogenesis, gametes and zygotes: an ecological perspective on sexual reproduction in the algae. *British Phycological Journal* 27, 233–252.

[3.7] Buffham, T.H. 1885. Newly observed phenomena in the conjugation of the diatom *Rhabdonema arcuatum. Journal of the Quekett Microscopical Club*, ser. 2, 2 131–137.

[3.8] Buffham, T.H. 1892. Conjugation of diatom: *Orthoneis binotata* Grunow. *Journal of the Quekett Microscopical Club*, ser. 2, 5, 27–30.

[3.9] Card, V.M., Carra, M. 2013. Investigation of evolutionary effects on the relative frequency of sexual reproduction in freshwater diatoms. *Phytotaxa* 127, 183–189.

[3.10] Carter, H.J. 1865. Conjugations of *Navicula serians, N. rhomboides*, and *Pinnularia gibba. Ann. Mag. Nat. Hist.* 15, 161-175.

[3.11] Chepurnov, V.A., Mann, D.G., Sabbe, K., Vyverman, W. 2004. Experimental studies on sexual reproduction in diatoms. *Int. Rev. Cytol.* 237, 91–154.

[3.12] Cholnoky, B. von. 1929. Beiträge zur Kenntnis der Auxosporenbildung. *Archiv für Protistenkunde* 68, 471–502.

[3.13] Cirri, E., De Decker, S., Bilcke, G., Markus, W., Osuna-Cruz, C.M., De Veylder, L., Vandepoele, K., Werz, O., Vyverman, W., Pohnert, G. 2019. Associated bacteria affect sexual reproduction by altering gene expression and metabolic processes in a biofilm inhabiting diatom. *Front. Microbiol.* 10, 1790.

[3.14] Costello, J.C., Chisholm, S.C. 1981. The influence of cell size on the growth rate of *Thalassiosira weissflogii*. *J. Plankton R.* 3, 415–419.

[3.15] D'Alelio, D., Ribera d'Alcala, M., Dubroca, L., Sarno, D., Zingone, A., Montresor, M. 2010. The time for sex: a biennial life cycle in a marine planktonic diatom. *Limnol. Oceanogr.* 55, 106–114.

[3.16] Davey, M.C. The relationship between size, density and sinking velocity through the life cycle of *Melosira granulata* (Bacillariophyta). *Diatom Res.* 1, 1–18, 1986.

[3.17] Davidovich, N. A. 2001. Species specific sizes and size range of sexual reproduction in diatoms. In: *Proceedings of the 16th International Diatom Symposium* (A. Economou-Amilli, ed.), 191-196. Amvrosiou Press, Athens.

[3.18] Davis, C.O., Harrison, P.J., Dugdale, R.C. 1973. Continuous culture of marine diatoms under silicate limitation. I. Synchronized life cycle of *Skeletonema costatum*. *J. Phycol.* 9, 175–180.

[3.19] Drebes, G. 1977. Sexuality. In *The Biology of Diatoms* (D. Werner, Ed.), Botanical Monographs, vol.13. Blackwell Scientific Publications, Oxford, pp. 250–283.

[3.20] Edlund, M.B., Bixby, R.J. 2001. Intra- and inter-specific differences in gametangial and initial cell size in diatoms. In *Proceedings of the 16th International Diatom Symposium* (A. Economou-Amilli, ed.). Amvrosiou Press, Athens, pp. 169–190.

[3.21] Edlund, M.B., Spaulding, S.A. 2006. Initial observations on uniparental auxosporulation in *Muelleria* (Frenguelli) Frenguelli and *Scoliopleura* Grunow (Bacillariophyceae). In: *Fossil and Recent Phycological Studies. Festschrift in honour of Dr. Dobrina Temniskova-Topalova* (N. Ognjanova-Rumenova & K. Manoylov, eds). Pensoft, Sofia & Moscow, pp. 211–223.

[3.22] Edlund, M.B., Stoermer, E.F. 1997. Ecological, evolutionary, and systematic significance of diatom life histories. *J. Phycol.* 33, 897–918.

[3.23] Edlund, M.B, Burge, D.R.L., Massa, E. Wu, S., Lee, S. 2018. Sex and size in diatoms. *25th International Diatom Symposium, Program with Abstracts.* Berlin, Germany.

[3.24] Ferrante, M.I., Entrambasaguas, L., Johansson, M., Töpel, M., Kremp, A., Montresor, M., Godhe, A. 2019. Exploring molecular signs of sex in the marine diatom *Skeletonema marinoi*. *Genes*, 10, 494.

[3.25] Focke, G.W. 1854. *Physiologische Studien*, vol. 2. C. Schünemann's Verlag, Bremen.

[3.26] Friedmann, I. 1996. Lothar Geitler (1899–1990). In *Prominent phycologists of the 20th century* (D.J. Garbary & M.J. Wynne, M.J., Eds. Lancelot Press, pp. 10–20.

[3.27] Geitler, L. 1927a. Über die Auxosporen von *Meridion circulare* und verwandten Diatomeen-Gattungen. *Mikrokosmos* 21, 79–82.

[3.28] Geitler, L. 1927b. Die Reduktionsteilung und Copulation von *Cymbella lanceolata*. *Arch. Protistenkunde* 58, 465–507.
[3.29] Geitler, L. 1932. Der Formwechsel der pennaten Diatomeen. *Arch. Protistenkunde* 78, 1–226.
[3.30] Geitler, L. 1940. Die Auxosporenbildung von *Meridion circulare*. *Arch. Protistenkunde* 94, 288–294.
[3.31] Geitler, L. 1949. Die Auxosporenbildung von *Nitzschia sigmoidea* und die Geschlechtsbestimmung bei den Diatomeen. *Port. Acta Biol. Ser. A*, R.B. Goldschmidt volume, 80–86.
[3.32] Geitler, L. 1960. Obligate pädogame Automixis bei einer Rasse von *Gomphonema angustatum*. *Österreichische Botanische Zeitschrift* 107, 275–280.
[3.33] Geitler, L. 1963. Rassenbildung bei *Gomphonema angustata*. *Österreichische Botanische Zeitschrift* 110, 481–483.
[3.34] Geitler, L. 1966. Anomalien der Auxosporen und Korrektionen während der Weiterentwicklung bei *Meridion circulare*. 113, 273–282.
[3.35] Geitler, L. 1970. Kleinsippen von *Gomphonema angustatum*. *Österreichische Botanische Zeitschrift* 118, 17–29.
[3.36] Geitler, L. 1973. Auxosporenbildung und Systematik bei pennaten Diatomeen und die Cytologie von *Cocconeis*-Sippen. *Österreichische Botanische Zeitschrift* 122, 299–321.
[3.37] Geitler, L. 1985. Automixis bei pennaten Diatomeen. *Plant Sys. Evol.* 150, 303–306.
[3.38] Godhe, A., Kremp, A., Montresor, M. 2014. Genetic and microscopic evidence for sexual reproduction in the centric diatom *Skeletonema marinoi*. *Protist* 165, 401–416.
[3.39] Griffith, J.W. 1855. On the conjugation of the Diatomaceae. *Annals and Magazine of Natural History*, ser. 2, 16, 92–94.
[3.40] Hegewald, E., Hindáková, A. 1997. Variabilität von einer natürlichen Population und von Klonen des *Cyclotella ocellata*-Komplexes (Bacillariophyceae) aus dem Gallbergweiher, Nordwestdeutschland. *Arch. Hydrobiol.*, Suppl. 86, 17–37.
[3.41] Hensen, V. 1887. Ueber die Bestimmung des Plankton's oder des im Meere treibenden Materials an Pflanzen und Thieren. *Bericht der Kommission zur Wissenschaftliche Untersuchung der deutschen Meere in Kiel, für die Jahre 1882-1886* 5, 1–108.
[3.42] Hickel, B. 1988. Sexual reproduction and life cycle of *Ceratium furcoides* (Dinophyceae) *in situ* in the lake Plußsee. *Hydrobiologia* 161, 41–48.
[3.43] Hiltz, M., Bates, S.S., Kaczmarska, I. 2000. Effect of light:dark cycles and cell apical length on the sexual reproduction of the pennate diatom *Pseudo-nitzschia multiseries* (Bacillariophyceae) in culture. *Phycologia* 39, 59–66.
[3.44] Hohn, M.H. 1959. Variability in three species of *Gomphonema* (Bacillariophyceae) undergoing auxospore formation. *Notulae Naturae of the Academy of Natural Science of Philadelphia* 316, 1–7.

[3.45] Jewson, D.H. 1992a. Life cycle of a *Stephanodiscus* species. (Bacillariophyta). *J. Phycol.* 28, 856-866.
[3.46] Jewson, D.H. 1992b. Size reduction, reproductive strategy and the life cycle of a centric diatom. *Philos. Trans. R. Soc. Lond. B* 336, 191–213.
[3.47] Jewson, D.H., Granin, N.G. 2015. Cyclical size change and population dynamics of a planktonic diatom, *Aulacoseira baicalensis*, in Lake Baikal. *Eur. J. Phycol.* 50, 1–19.
[3.48] Jewson, D.H., Granin, N.G., Zhdanov, A.A., Gorbunova, L.A., Bondarenko, N.A., Gnatovsky, R. Yu. 2008. Resting stages and ecology of the planktonic diatom *Aulacoseira skvortzowii* in Lake Baikal. *Limnol. Oceanogr.* 53, 1125–1136.
[3.49] Jewson, D.H., Khondker, M., Rahman, M.H., Lowry, S. 1993. Auxosporulation of the freshwater diatom *Aulacoseira herzogii* in Lake Banani, Bangladesh. *Diatom Res.* 8, 403–418.
[3.50] Kain, J.M. 1989. The seasons in the subtidal. *British Phycological Journal* 24, 203–215.
[3.51] Karsten, G. 1896. Untersuchungen über Diatomeen I. *Flora* 82, 286–296.
[3.52] Karsten, G. 1897a. Untersuchungen über Diatomeen II. *Flora* 83, 33–53.
[3.53] Karsten, G. 1897b. Untersuchungen über Diatomeen III. *Flora* 83, 203–222.
[3.54] Karsten, G. 1899. Die Diatomeen der Kieler Bucht. *Wissenschaftliche Meereruntersuchungen*, Kiel, N.F. 4, 17–205.
[3.55] Karsten, G. 1900. Die Auxosporenbildung der Gattungen *Cocconeis, Surirella* und *Cymatopleura*. *Flora* 87, 255–283.
[3.56] Karsten, G. 1905a. Das Phytoplankton des Antarktischen Meeres nach dem Material der Deutschen Tiefsee-Expedition. In *Wissenschaftliche Ergebnisse der Deutschen Tiefsee-Expedition auf dem Dampfer „Valdivia" 1898–1899*, C. Chun (Ed.). G. Fischer, Jena: 1–136.
[3.57] Karsten, G. 1905b. Das Phytoplankton des Atlantischen Oceans nach dem Material der Deutschen Tiefsee-Expedition 1898-1899. In *Wissenschaftliche Ergebnisse der Deutschen Tiefsee-Expedition auf dem Dampfer „Valdivia" 1898–1899*, C. Chun (Ed.). G. Fischer, Jena: 137–220.
[3.58] Karsten, G. 1907. Das Indische Phytoplankton. Nach dem Material der Deutschen Tiefsee-Expedition 1898–1899. In *Wissenschaftliche Ergebnisse der Deutschen Tiefsee-Expedition auf dem Dampfer „Valdivia" 1898–1899*, C. Chun (Ed.). G. Fischer, Jena: 221–548.
[3.59] Klapper, F., Audoor, S., Vyverman, W., Pohnert, G. 2021. Pheromone mediated sexual reproduction of pennate diatom *Cylindrotheca closterium*. *Journal of Chemical Ecology* 47, 504–512.
[3.60] Kremp, A., Heiskanen, A.S. 1999. Sexuality and cyst formation of the spring-bloom dinoflagellate *Scrippsiella hangoei* in the coastal northern Baltic Sea. *Mar. Biol.* 134, 771–777.
[3.61] Lewis, W.M. 1983. Interruption of synthesis as a cost of sex in small organisms. *Am. Nat.* 121, 825–833.

[3.62] Lewis, W.M. 1984. The diatom sex clock and its evolutionary significance. *Am. Nat.* 123, 73–80.

[3.63] Lüders, J.E. 1862. Beobachtungen über die Organisation, Theilung und Copulation der Diatomeen. *Botanische Zeitung* 20, 41–43, 49–52, 57–61, 65–69.

[3.64] Mann, D.G. 1988. Why didn't Lund see sex in *Asterionella*? A discussion of the diatom life cycle in nature. In *Algae and the Aquatic Environment*, F.E. Round (ed.). Biopress, Bristol, pp. 383–412.

[3.65] Mann, D.G. 1989. The species concept in diatoms: evidence for morphologically distinct, sympatric gamodemes in four epipelic species. *Plant Sys. Evol.* 164, 215–237.

[3.66] Mann, D.G. Size and sex. 2011. In *The Diatom World*, J. Seckbach & J.P. Kociolek (eds). Springer, Dordrecht, pp. 147–166.

[3.67] Mann, D.G., Crawford, R.M., Round, F.E. 2017. Bacillariophyta. In: *Handbook of the Protists*, J.M. Archibald, A.G.B. Simpson & C.H. Slamovits (eds), 62 pp. Springer, Cham.

[3.68] Martins, N., Barreto, Bartsch, I., Bernard, J., Serrão, E.A., Pearson, G.A. 2022. Daylength influences reproductive success and sporophyte growth in the Arctic kelp species *Alaria esculenta*. *Mar. Ecol. Prog. Ser.* 683, 37–52.

[3.69] Mizuno, M., Okuda, K. 1985. Seasonal change in the distribution of cell size of *Cocconeis scutellum* var. *ornata* (Bacillariophyceae) in relation to growth and sexual reproduction. *J. Phycol.* 21, 547–553.

[3.70] Moeys, S., Frenkel, J., Lembke, C., Gillard, J.T.F., Devos, V., Van den Berge, K., Bouillon, B., Huysman, M.J.J., De Decker, S., Scharf, J., Bones, A., Brembu, T., Winge, P., Sabbe, K., Vuylsteke, M., Clement, L., De Veylder, L., Pohnert, G., Vyverman, W. 2016. A sex-inducing pheromone triggers cell cycle arrest and mate attraction in the diatom *Seminavis robusta*. *Scientific Reports* 6, 19252.

[3.71] Nagai, S., Imai, I., Yamauchi, K., Manabe, T. 1999. Induction of sexuality in the diatom *Coscinodiscus wailesii* Gran by a marine bacterium *Alcaligenes* sp. in culture. In *Proceedings of the 14th International Diatom Symposium*, S. Mayama, M. Idei and I. Koizumi (eds.). Koeltz, Königstein, Germany, pp. 197–212.

[3.72] Nanjappa, D., Sanges, R., Ferrante, M.I., Zingone, A. 2017. Diatom flagellar genes and their expression during sexual reproduction in *Leptocylindrus danicus*. *BMC Genomics* 18, 813.

[3.73] Ostenfeld, C.H. 1903. Phytoplankton from the sea around the Faeröes. In *Botany of the Faeröes, based upon Danish investigations*, part II. Det Nordiske Forlag, Copenhagen, pp. 558–612.

[3.74] Paasche, E. 1973. The influence of cell size on growth rate, silica content, and some other properties of four marine diatom species. *Norwegian Journal of Botany* 20, 197–204.

[3.75] Patrick, R., Hohn, M.H., Wallace, J.H. 1954. A new method for determining the pattern of the diatom flora. *Notulae Naturae of the Academy of Natural Science of Philadelphia* 259, 1–12.
[3.76] Pérez-Martínez, C., Cruz-Pizarro, L., Sánchez-Castillo, P. 1992. Auxosporulation in *Cyclotella ocellata* (Bacillariophyceae) under natural and experimental conditions. *J. Phycol.* 28, 608–615.
[3.77] Peyer, H. 1972. *Beiträge zur Kenntnis der Entwicklungsgeschichte verschiedener Süßwasseralgen*. PhD Dissertation, University of Vienna.
[3.78] Pfitzer, E. 1871. Untersuchungen über Bau und Entwicklung der Bacillariaceen (Diatomaceen). In: *Botanische Abhandlungen*, J. Hanstein (ed.), vol. 2, 1-189. Bonn.
[3.79] Potapova, M., Snoeijs, P. 1997. The natural life cycle in wild populations of *Diatoma moniliformis* (Bacillariophyceae) and its disruption in an aberrant environment. *J. Phycol.* 33, 924–937.
[3.80] Poulíčková, A., Mann, D.G. 2019. Diatom sexual reproduction and life cycles. In *Diatoms: Fundamentals and Applications* (J. Seckbach & R. Gordon, eds). Scrivener, Beverley, Massachusetts, pp. 245–272.
[3.81] Reynolds, C.S. 1973. The phytoplankton of Crose Mere, Shropshire. *British Phycological Journal* 8, 153-162.
[3.82] Reynolds, C.S., Wiseman, S.M. 1982. Sinking losses of phytoplankton in closed limnetic systems. *J. Plankton Res.* 4, 489–522.
[3.83] Rojo, C., Kiss, K.T., Àlvarez-Cobelas, M., Rodrigo, M.A. 1999. Population dynamics of *Cyclotella ocellata* (Bacillariophyceae): endogenous and exogenous factors. *Arch. Hydrobiol.* 145, 479–495.
[3.84] Round, F.E. 1981. *The ecology of algae*. Cambridge University Press, Cambridge.
[3.85] Round, F.E., Crawford, R.M., Mann, D.G. 1990. *The diatoms. Biology and morphology of the genera.* Cambridge University Press, Cambridge.
[3.86] Rozumek, K.-E. 1968. Der Einfluß der Umweltfaktoren Licht und Temperatur auf die Ausbildung der Sexualstadien bei der pennaten Diatomee *Rhabdonema adriaticum* Kütz. *Beiträge zur Biologie der Pflanzen* 44, 365–388.
[3.87] Rudzki, B. 1965. Untersuchungen über periphere Nukleolen bei pflanzlichen Protisten, das interphasische Kernwachstum bei Dinophyceen und das Kopulationsverhalten von Gomphonemen. *Österreichische Botanische Zeitschrift* 112, 1–43.
[3.88] Sarno, D., Zingone, A., Montresor, M. 2010. A massive and simultaneous sex event of two *Pseudo-nitzschia* species. *Deep Sea Res. II* 57, 248–255.
[3.89] Schlegel, I., Scheffler, W. 1999. Seasonal development and morphological variability of *Cyclotella ocellata* (Bacillariophyceae) in the eutrophic Lake Dagow (Germany). *Int. Rev. Hydrobiol.* 84, 469–478.
[3.90] Schulte, N., Jordan, E., Woodell, J., Flack, M., Tunno, I., Kamae, S., Mendoza, S., Gauthier, J., Behnaz, B., Salas, E., Burge, D., Lee, S., Edlund, M. 2015. A closer look at *Encyonema reimeri*. *23rd North American*

Diatom Symposium, Program with Abstracts. Central Michigan University Biological Station, Beaver Island, Michigan.
[3.91] Shirokawa, Y., Shimada M. 2020. Synchronized emergence under diatom sperm competition. *Proceedings of the Royal Society of London* B 287, 20201074. http://dx.doi.org/10.1098/rspb.2020.1074.
[3.92] Smith, W. 1856. *A synopsis of British Diatomaceae*, vol. 2. J. van Voorst, London.
[3.93] Thwaites, G.H.K. 1847. On conjugation in the Diatomaceae. *Annals and Magazine of Natural History*, ser. 1, 20, 9–11, 343–344.
[3.94] Thwaites, G.H.K. 1848. Further observations on the Diatomaceae; with descriptions of new genera and species. *Annals and Magazine of Natural History*, ser. 2, 1, 161–172.
[3.95] Tooke, F., Battey, N.H. 2010. Temperate flowering phenology. *J. Exp. Bot.* 61, 2853–2862.
[3.96] Wimpenny, R. 1946. The size of diatoms II. Further observations on *Rhizosolenia styliformis* (Brightwell). *J. Mar. Biol. Assoc. U K* 26, 271–284.

4

Deterministic and Stochastic Effects on Freshwater Diatom Biodiversity and Community Composition

Xavier Benito[1*], Sophia I. Passy[2], Annika Vilmi[3], Aurélien Jamoneau[4], Juliette Tison-Rosebery[4], Maria Kahlert[5], Chad A. Larson[6], Joseph L. Mruzek[7], Janne Soininen[8] and Andrew Bramburger[9]

[1]*Marine and Continental Waters Programme, Institute for Agrifood and Technology Research (IRTA), La Ràpita, Spain*
[2]*Department of Biology, University of Texas at Arlington, Arlington, Texas, USA*
[3]*Freshwater Ecology, Nature Solutions, Finnish Environment Institute, Finland*
[4]*INRAE, UR EABX, Cestas, France*
[5]*Swedish University of Agricultural Sciences, Department of Aquatic Sciences and Assessment, Uppsala, Sweden*
[6]*Environmental Assessment Program, Washington State Department of Ecology, Olympia, Washington, USA*
[7]*Forestry and Environmental Conservation Department, Clemson University, South Carolina, USA*
[8]*Department of Geosciences and Geography, University of Helsinki, Finland*
[9]*Watershed Hydrology and Ecology Research Division of Environment and Climate Change Canada, Ontario, Canada*

Abstract

Recent research on diatom metacommunities has focused on disentangling the assembly mechanisms driving species and functional composition and biodiversity across space and time, including deterministic (environmental filtering and biotic interactions) and stochastic processes (dispersal and ecological drift). In this chapter, we provide an overview of this research and outline future directions. Environmental filtering and dispersal have received the most attention, while biotic interactions and ecological drift have remained comparatively understudied and require more investigations. We discuss diatom species and functional responses

**Corresponding author*: xavier.benito@irta.cat

Nora I. Maidana, Magdalena Licursi and Eduardo Morales (eds.) Diatom Ecology: From Molecules to Metacommunities, (85–122) © 2024 Scrivener Publishing LLC

to major environmental factors, operating at local scales, including inorganic and organic acidity, conductivity, and limiting nutrients, and at regional scales, namely land use and climate. Research has shown that both high rates of dispersal (mass effects) and low rates of dispersal (limited dispersal) are responsible for species and guild composition. We recommend further observational but also experimental investigations on the relative importance of assembly mechanisms across spatial and temporal scales and along latitudinal, longitudinal, and elevational gradients. Global change with respect to climate, land use, and dissolved organic matter has been recognized as an important driver of diatom compositional and biodiversity shifts. However, further modeling work and building harmonized global diatom databases that encompass spatial and temporal observations as well as morphological and molecular data may be necessary to forecast possible diatom community and functional responses in the decades ahead.

***Keywords*:** Biotic interactions, dispersal limitation, ecological drift, environmental filtering, guilds, mass effects, niche, priority effects

4.1 Introduction

The overarching goal of this chapter is to provide a review of recent investigations on the assembly mechanisms driving freshwater diatom species and functional composition and biodiversity (see Table 4.1 for definitions and examples), identify knowledge gaps, and suggest future directions. These mechanisms include deterministic (environmental filtering and biotic interactions) and stochastic processes (dispersal and ecological drift) as well as priority effects (e.g., when dispersal coupled with biotic interactions over time lead to spatial variability in community composition) [4.1] [4.2]. Deterministic abiotic and biotic processes constrain species niches and generate predictable species distributions, consistent with the classical work of Hutchinson [4.3]. Species distributions are also subject to stochastic processes, related to random dispersal and random variability in birth and death rates (i.e., ecological drift), as suggested by the influential theories of island biogeography [4.4] and the unified neutral theory of biodiversity [4.5]. Dispersal can have opposing effects on species communities. High rates of dispersal (mass effects) can offset environmental filtering via source-sink effects, i.e., when propagules from source locations with favorable environmental conditions are dispersed to sink locations with unfavorable environmental conditions [4.6]. Low rates of dispersal, on the other hand, may prevent species from occurring in suitable habitats.

Table 4.1 A glossary of the main concepts discussed in this chapter with supporting evidence.

Concept	Definition	Supporting evidence
Metacommunity	A set of local communities that are connected by dispersal. Metacommunities are governed by assembly mechanisms, encompassing deterministic and stochastic processes (defined below).	Recent diatom metacommunity research shows that connectivity among sites is a key factor underlying biodiversity patterns. For example, in both lotic and lentic systems, limited dispersal due to isolation is a major force structuring diatom metacommunities even when deterministic processes are expected due to environmental variability [4.164].
Deterministic (niche) processes	Processes that generate predictable community responses, including environmental filtering and biotic interactions. They are also referred to as niche processes because the biotic response depends on species' ecological niches, i.e., the set of environmental conditions necessary for positive growth.	
Environmental filtering	Abiotic factors dictate the occurrence of species in a particular location.	The available empirical evidence suggests that lotic and lentic diatoms can be filtered out from the regional pool by physical (e.g., light, mixing), hydrological (e.g., discharge, water depth), and chemical factors (e.g., pH, nutrients, conductivity) [4.51] [4.165]

(*Continued*)

Table 4.1 A glossary of the main concepts discussed in this chapter with supporting evidence. (*Continued*)

Concept	Definition	Supporting evidence
Biotic interactions	Positive (e.g., facilitation and commensalism) or negative interactions (e.g., competition and predation) constrain species assemblages.	Grazing pressure (e.g., by macroinvertebrates) may have comparatively little effect on diatom metacommunity patterns at larger scales [4.24]. Negative correlation between the abundances of two common stream diatoms, *Achnanthidium minutissimum* and *Cocconeis placentula*, was attributed to competitive exclusion at the local scale [4.108].
		Passy [4.106] developed a framework viewing guild density in benthic diatom metacommunities as a function of environmental gradients, space, and species interactions, both competitive and facilitative (measured as species richness of the competitor or facilitator guild, respectively). Testing of this framework revealed that guild density was controlled mostly by biotic interactions and/or the environment. However, at shorter environmental gradients in an experimental setting, biotic interactions did not appear to affect guild density [4.166]. Niche modeling approaches also concluded that biotic interactions did not structure species co-occurrences, which appeared to be "passive" [4.125].

(*Continued*)

Table 4.1 A glossary of the main concepts discussed in this chapter with supporting evidence. (*Continued*)

Concept	Definition	Supporting evidence
Stochastic (neutral) processes	These processes have a random nature and include mass effects, dispersal limitation, and ecological drift.	
Mass effects	Species may be present in an unfavorable (sink) habitat due to dispersal from a favorable (source) habitat. Mass effects can thus offset the effects of niche processes.	Jamoneau *et al.* [4.16] reported mass effects shaping diatom metacommunities from mid- and downstream habitats. Vilmi *et al.* [4.167] reported mass effects structuring the diatom metacommunity of a large lake system where niche conditions (e.g., trophic status) varied among 81 studied sites.
Dispersal limitation	Species may be prevented from reaching suitable habitats due to distance or other dispersal barriers. Dispersal limitation can mask the effects of niche processes.	In headwater lotic systems, which are relatively isolated within the river network, dispersal limitation can override the effect of environmental conditions [4.127]. In lentic systems, the lack of connectivity among lakes may limit the ability of planktonic diatoms to reach suitable pelagic conditions [4.25].
Ecological drift	Random birth and death events determine species composition. Small habitats with few individuals are more prone to ecological drift.	We know of a single study testing for ecological drift and reporting that periphytic diatom community composition in shallow arctic lakes was not structured by drift [4.123].

(*Continued*)

Table 4.1 A glossary of the main concepts discussed in this chapter with supporting evidence. (*Continued*)

Concept	Definition	Supporting evidence
Priority effects	Early colonizers either inhibit or facilitate later recruits. Compositional differences among early colonizers due to random dispersal can result in different successional trajectories under similar environmental conditions.	The scant available evidence shows contrasting priority effects through new colonization and *in-situ* speciation driving later successional stages in Antarctica during the Plio-Pleistocene (last 2.58 Ma) [4.143]. In contrast, priority effects were unimportant in a short-term experimental study of stream benthic diatoms [4.168].
Guild system	Classification of diatoms into ecological guilds, based on resource utilization and disturbance resistance [4.23] [4.169].	Guild responses to deterministic and stochastic effects are generally context dependent.
Low-profile	Species growing close to the substate (e.g., prostrate, adnate, short erect cells, solitary cells, slow moving cells) with low nutrient demands and high resistance to physical disturbance.	In small subarctic streams, low-profile diatoms were mostly structured by local environmental factors [4.104]. Oligotrophy and high discharge led to dominance of the low-profile guild in high-mountain streams [4.111].
High-profile	Species with tall stature (e.g., filamentous, branched, tube forming, and colonial) adapted to high nutrient concentrations and with low resistance to physical disturbance.	In the absence of grazing, the high-profile guild had greater relative abundance than in the presence of grazing and increased with the number of added nutrients [4.62].

(*Continued*)

Table 4.1 A glossary of the main concepts discussed in this chapter with supporting evidence. (*Continued*)

Concept	Definition	Supporting evidence
Motile	Fast-moving biraphid species with high nutrient demands and intermediate resistance to physical disturbance.	Motile diatoms had higher abundances in dammed rivers characterized by high total phosphorus concentrations and hydrological alterations (water flow, streambed lithology) [4.170].
Planktonic	Diatom species adapted to the water column but also occurring in the benthos due to sedimentation.	In Pyrenean lakes, planktonic diatoms were jointly affected by increased water pH in the hypolimnetic sediments and stochastic dispersal events during summer stratification [4.119].
Alpha diversity	A measure of species richness at a locality.	Environmental filtering prevailed over dispersal processes in controlling alpha diversity of stream algae; the environmental effect increased, while the dispersal effect decreased with latitude [4.150].
Beta diversity	A (dis)similarity measure of species composition (presence/absence or abundance) among samples (spatial and/or temporal).	Beta diversity of benthic diatoms from unimpacted streams in France and the US was driven by environmental filtering, while beta diversity in streams heavily enriched with nutrients was controlled by spatial processes across scales [4.129].

(*Continued*)

Table 4.1 A glossary of the main concepts discussed in this chapter with supporting evidence. (*Continued*)

Concept	**Definition**	**Supporting evidence**
Niche breadth	The range of environmental conditions where a species can maintain positive growth. According to the resource use hypothesis, the niche breadth of a species determines its local abundance and regional distribution.	At a subcontinental scale, stream diatoms with wider niche breadths had greater local abundance and regional distribution [4.88]. At inter-continental scale, diatom niche breadth was not conserved among different regions of the world [4.171].
Niche position	Measures if a species occurs in common or marginal environmental conditions within the study area. According to the resource abundance hypothesis, the niche position of a species determines its local abundance and regional distribution.	Support for the niche position hypothesis, whereby rare species had preference for marginal environmental conditions, was generally found at smaller scales [4.171].

Much freshwater research over the past decades has pursued determination of the relative roles of deterministic and stochastic processes at spatial scales ranging from individual stream reaches to the globe and at ecological to evolutionary temporal scales [4.7] [4.8] [4.9] [4.10] [4.11] [4.12] [4.13] [4.14]. The classical microbiological idea that "everything is everywhere but the environment selects" [4.15] suggests that microorganisms are shaped exclusively by environmental filtering because their large populations preclude dispersal limitation. However, mounting evidence reveals that diatoms experience both mass effects and dispersal limitation [4.16] [4.17], manifesting in species and genera endemism in individual lakes (e.g., Lakes Titicaca and Baikal [4.18] [4.19]), regionally constrained floras

[4.20] [4.21], and distinct biogeographic regions [4.22]. Following the creation of the diatom guild system for benthic river diatoms [4.23], new efforts were directed towards elucidating the diatom functional responses to environmental and dispersal controls [4.24] [4.25] [4.26].

Diatoms are excellent model organisms for investigating metacommunity assembly processes. First, diatoms are species-rich, representing the most speciose eukaryotic algal group in marine and freshwater environments with more than 200,000 estimated extant species [4.27]. Second, diatoms are easy to sample and are sensitive to environmental stimuli (e.g., water chemistry, climate), which makes them invaluable bioindicators [4.28] [4.29]. Third, diatoms have different modes of dispersal, including instream, aerial (e.g., by wind), and overland (e.g., zoochory), which contribute to broad distributions and enable reaching of suitable/preferred habitats [4.30]. Fourth, the preservation of diatom frustules in the sediment allows the assessment of historical processes of metacommunity assembly [4.31] and may provide new insights into patterns over time. Fifth, the high rates of speciation of ecologically important diatom taxa (e.g., *Nitzschia*) provide opportunities to study evolutionary processes, such as symbiotic relationships among algal groups [4.32]. The availability of new resources and tools, including species inventories of understudied regions (e.g., tropics) and ecosystems (e.g., islands) [4.33] [4.34], and ecological modeling [4.35] will broaden the research on the importance of different assembly mechanisms.

However, the presence of cryptic species or species complexes [4.36], represents a taxonomical challenge for measuring biodiversity and assessing the effects of assembly mechanisms in freshwater diatoms. At the same time, the continued development and refinement of genetic metabarcoding approaches hold great potential for advancing our understanding in this context [4.37] [4.38]. Recent studies have found conspecific variability in environmental preferences, which can further hamper the detection of an environmental signal in community composition. For instance, Pérez-Burillo *et al.* [4.39] demonstrated that genetic variants within some key freshwater diatom species (e.g., *Achanthidium minutissimum* and *Fistulifera saprophila*) differ in their ecological preferences. New spatially explicit databases of occurrence of molecularly distinct taxonomic units need to be built [4.40], which would allow distinguishing finer patterns in community response to environmental conditions [4.41]. Combining both established knowledge on morphotaxa and new knowledge derived from genetic data will be a way forward to resolving community assembly mechanisms within freshwater diatom species complexes [4.42].

There are also data and statistical challenges in assessing the roles of assembly mechanisms. First, the increasing amount of collected information by spatial ecology studies in diatoms (reviewed in Soininen and Teittinen [4.43]) often outpaces the capacity of researchers to progress towards metacommunity generalizations. Second, dispersal *per se* is difficult to measure because diatoms can disperse through water flow, air, human activities, and animal vectors, including introduced species [4.44] [4.45], and thus spatial proxies or classical statistical methods to account for dispersal effects (e.g., variance partitioning) may not be accurate. Third, context dependencies may cause assembly mechanisms to depend on geographic location. All these issues require synthesis if we are to use the available diatom information efficiently. This chapter is organized around five themes, which review deterministic and stochastic factors, their relative effects, interactions, and variability through time and space. Finally, a set of research questions and ways forward are suggested to guide future diatom metacommunity research.

4.2 Deterministic Effects on Diatom Metacommunities

Diatoms respond to environmental factors that operate at local and regional scales. Among the most influential local factors in streams and lakes are conductivity, pH, nutrient concentrations, and depth [4.46]. Conductivity has major effects on diatom distribution in both freshwater and brackish waters via ion composition and secondary salinization effects [4.47]. Both high and low pH can constrain diatom diversity. Acidification by inorganic acid deposition has adversely impacted diatom biodiversity in streams and lakes located in the northeastern United States and northern Europe [4.48] [4.49]. Conversely, Teittinen *et al.* [4.50] discovered a reduction in diatom richness with increasing pH between 4.4 and 8.5 in over 140 subarctic ponds, aligning with a growing body of evidence for diatoms adapted to acidic water conditions [4.51] [4.52]. Notably, diatoms are sensitive not only to the amount of acidity but also to the type of acidity. Thus, streams naturally acidified by dissolved organic carbon (DOC) possess greater diatom richness than inorganically acidified streams due to reduced aluminum toxicity [4.53] [4.54]. Dissolved organic carbon is on the rise in eastern North America and northern and central Europe, causing brownification of streams and lakes [4.55] [4.56]. As dissolved organic carbon can mitigate aluminum toxicity in acid streams and increase bioavailability of macro- and micronutrients, brownification may have a positive effect on diatom biodiversity [4.53]. However, brownification may also negatively impact algal biodiversity by promoting mixotrophic algae, competitively displacing diatoms and other phototrophic algae [4.57]. Research on the impacts of brownification on diatoms remains limited, necessitating further investigations in this area to achieve a comprehensive understanding of the contingencies in the diatom responses.

While many investigations have examined the responses of diatom community and trait richness and composition to macronutrient enrichment (reviewed in Poulíčková and Manoylov [4.58]), there is a notable paucity of research on the effects of the number of nutrients at high supply, including both macronutrients and micronutrients. In terrestrial systems, global studies have demonstrated that biodiversity declines as more nutrients become non-limiting because of fewer opportunities for tradeoff [4.59] [4.60]. In contrast, there are only a few aquatic investigations showing the opposite response, whereby benthic diatom and algal richness increased when macro- and micronutrients were available at high concentrations [4.61] [4.62] [4.63]. One of the potential reasons for this disparity is that in the spatially complex algal benthos, nutrient enrichment promotes the establishment of overstory species with high nutrient demand without negatively affecting the understory species tolerant of nutrient limitation. These results indicated that coexistence in terrestrial and benthic systems is driven by different processes, namely competition for nutrients and light in the former but complementarity and resource partitioning in the latter [4.64]. Yet, additional experimental research on this topic is needed to ascertain the generality of this conclusion.

Diatom species richness and composition are also controlled by regional environmental characteristics either directly or indirectly through their influence on local conditions [4.65] [4.66]. For instance, the proportion of wetlands in river catchments has a positive effect on diatom richness. This is because wetlands export DOC, which as discussed, reduces aluminum toxicity in streams acidified by inorganic acid deposition [4.53] and increases bioavailability of essential micronutrients such as iron [4.67] [4.68]. Land use is another factor with important, yet complex effects on diatom communities. On the one hand, it may promote diatom species richness by increasing nutrient availability due to agriculture [4.69] [4.70] and light due to deforestation [4.71]. On the other hand, higher urbanization intensity may result in a lower diatom richness driven by an increase of pollution-tolerant taxa [4.72].

Climate and its anthropogenically driven changes play a pivotal role as regional constraints on diatom communities. For instance, global warming has resulted in increased water temperature, reduced ice-cover, and enhanced stratification and wind strength in lakes [4.73]. A common response is the compositional shift in planktonic diatoms toward smaller-celled taxa, including many Cyclotelloid species in the last decades [4.74] [4.75] [4.76]. Changes in turbidity of the water column, often associated with alterations in stratification, can lead to changes in light penetration to the benthos, impacting photosynthetic processes [4.77]. Rühland *et al.* [4.78] reviewed evidence for increased diatom biomass following recent atmospheric temperature increase and anthropogenic nitrogen deposition in remote lakes. Indeed, diatom communities in these ecosystems have undergone direct climate-driven changes, primarily through the extension of ice-free seasons [4.79], though contrasting evidence has been found in tropical aquatic ecosystems [4.80].

Investigating the co-variation of environmental variables is critical for a comprehensive understanding of the deterministic processes shaping diatom biodiversity. For instance, nutrients co-vary with conductivity in many freshwater and brackish ecosystems, making it difficult to explain diatom biodiversity responses to individual factors [4.81] [4.82]. Experimental studies controlling for different environmental impacts (e.g., salinity and productivity) can be very informative in this respect [4.83]. Notably, climate and land use-driven eutrophication have strong interactive effects on aquatic communities [4.84]. Consequently, broad biogeographical gradients of diatom richness and abundance distribution cannot be explained with climate variability alone but require information on both climate and nutrients [4.85], and future studies should pursue this research direction.

Diatom response to environmental filtering depends on niche breadth and position. With respect to niche breadth, species are classified as specialists with narrow niches vs. generalists with broad niches. According to the resource use hypothesis [4.86], generalists are locally abundant and regionally prevalent because they can tolerate a broad range of conditions. In contrast, the resource abundance hypothesis suggests that species with high local abundance and regional prevalence are adapted to resources that are regionally common [4.87]. In this case, the species' niche position, i.e., the set of preferred environmental conditions, determines its abundance and distribution. Diatom research has been important in this debate by providing support for both hypotheses, although the resource use hypothesis tends to be more frequently corroborated at larger scales. Thus, at a subcontinental scale, stream diatoms with wider niche breadths had greater local abundance and regional distribution, consistent with the resource use hypothesis, while adaptation to common vs. rare conditions was inconsequential [4.88]. At comparable spatial scales, evidence also suggests that the broader the niche the wider the distribution of stream and lake diatoms [4.89] [4.90]. In contrast, studies at smaller scales have shown support for the resource availability hypothesis by demonstrating that species with more marginal niche position are rarer [4.91] [4.92] [4.93].

Species' environmental preferences determine the magnitude and direction of environmental filtering, and it is informative to explore the responses of individual ecological groups in conjunction with the total community. For example, acid stress severity and heterogeneity impacted diatom turnover (beta diversity) in Adirondack streams, but this response was driven mostly by acid-sensitive as opposed to acid-tolerant diatoms [4.48]. The distributional changes forced by global warming were projected to be greater in warm-water diatoms (positive) than in cold-water diatoms (negative), leading to higher overall local species richness (alpha diversity) but lower beta diversity [4.94], the latter phenomenon known as biotic homogenization [4.95]. Anthropogenic warming can also be the main driver of homogenization in the Arctic continent due to the loss of circumpolar diatom species [4.96]. In tropical reservoirs, diatom biotic homogenization was driven by eutrophication, whereby tolerant species inhabiting oligo-mesotrophic systems were able to expand to eutrophic systems [4.97]. As global change impacts intensify, diatoms can be used as an early warning system to understand impending environmental deterioration. For example, acidification can cause subtractive heterogenization when beta diversity increases due to the elimination of acid-sensitive species from the most impacted sites and this compositional shift can serve as an early warning of environmental degradation [4.48]. Diatom ecologists should take advantage of the available climatic (e.g., HydroATLAS [4.98], WorldClim [4.99], Global Forest Change [4.100]), land use, and human density projections [4.101] for the upcoming decades and investigate how richness and composition of diatom communities and ecological groups may respond to continued anthropogenic impacts and whether these responses vary geographically.

Environmental filtering triggers functional changes in diatom communities, which have been primarily evaluated using Passy's [4.23] guild classification [4.102] [4.103] [4.104]. An intercontinental study found that compared to diatom taxonomic composition, guild composition responded more strongly to environmental filtering, which suggests that guilds are particularly suitable for assessing global change effects [4.12]. Specifically, total phosphorus and species richness were positively correlated in motile taxa but negatively in high-profile and low-profile taxa. In addition to their distinct resource preference, ecological guilds also differ in niche breadth and thermal preferences. For instance, Lindholm *et al.* [4.105] found that planktonic diatoms are more specialized than the high-profile guild in boreal streams. A subcontinental stream study reported that low-profile and high-profile diatoms had preference for cold water, while motile diatoms occurred mostly in warmer waters, which has important implications for diatom distribution and biomass production under climate change [4.94].

The influence of biotic interactions on diatom composition has been poorly investigated (Table 4.1). However, a recent study suggested that stream diatom guilds and functional groups conform to three interaction modes: shared preference, distinct preference, and facilitative [4.106]. The shared preference mode encompasses species that have either broad tolerance to environmental conditions or competitive advantages in benign conditions, where they become dominant. Exemplifying this mode, the abundance of the tolerant low-profile guild was suppressed by higher richness of the dominant motile guild. In the distinct preference mode, species have different optima but can engage in negative interactions in areas where their environmental preferences overlap. Examples of this interaction mode are acidophilus and acidobiontic species, whose abundance was lower at high richness of the competitor. Finally, the facilitative mode includes species providing (donors) or receiving benefits (recipients) under stressful conditions. In favorable conditions, the recipients may competitively suppress the donors. Nitrogen fixing diatoms are donors of bioavailable nitrogen, and consistent with facilitation, higher richness of these diatoms was associated with increased abundance of the recipient motile guild, which has high nutrient demands. In contrast, greater richness of the motile guild decreased donor's abundance, consistent with competition. Other research examined the biotic effects of grazer abundance on diatom composition and documented only small effects compared to local environment and space [4.24] [4.107]. Álvarez-Cobelas *et al.* [4.108] found evidence of competitive exclusion within benthic diatom populations of the same genus and recommended consideration of biotic interactions at local scales before scaling up metacommunity processes to the catchment scale.

4.3 Stochastic Effects on Diatom Metacommunities

High dispersal rates can dampen the effect of local environmental conditions due to mass effects [4.1]. Mass effects maintain a flux of poorly adapted species and are challenging to measure in passive dispersers. Leboucher *et al.* [4.109] developed a new method for identification of diatoms influenced by mass effects. They suggested that benthic diatoms most prone to mass effects exhibit the following characteristics and patterns: strong spatial responses, narrower niche breadths (i.e., specialists), and weak negative co-occurrence. The sensitivity of species to mass effects was proposed as a new ecological trait that should be considered in both fundamental research, such as investigating the interplay between environmental factors and dispersal-in controlling species distribution, and applied research, such as water quality assessment [4.17] [4.109].

The lack of connectivity among sites or presence of spatial barriers may limit the ability of species to reach suitable habitats, leading to dispersal limitation. Although dispersal is a random process, in lotic systems, it is strongly influenced by flow directionality [4.110] [4.111]. Additionally, different dispersal processes may dominate along the watercourse with dispersal limitation being more prevalent upstream and mass effects, at mid- to downstream stretches [4.16].

Diatoms can exhibit variations in dispersal owing to differences in body size, attachment to the substrate, and their guild affiliation. Very few studies have assessed the importance of diatom body size for dispersal. In an investigation of diatom occurrences in over 200 locations along three 100-m reaches of a New York stream, Passy [4.113] demonstrated that the number of occurrences declined with body size, suggesting that smaller species, with a higher number of occurrences, were better dispersers. Attached diatoms appear to have lower dispersal ability than phytoplankton species [4.114]. Differences were also observed among diatom guilds. Thus, in US streams, low-profile taxa occurred in significantly more localities than high-profile and motile taxa, which was indicative of better dispersal of the low-profile guild [4.115]. In contrast, stronger spatial influence on the low-profile and high-profile guilds compared to the motile guild was attributed to stronger dispersal limitation [4.24]. Furthermore, motile diatoms have been shown to be strongly affected by mass effect in some systems [4.16], but less so in others [4.111]. In steep mountain streams, low-profile diatoms were influenced by mass effect, while limited dispersal was shown for high-profile diatoms [4.25]. Other studies have found non-significant differences in species distribution when comparing patterns across guilds in upstream refugia [4.116]. These conflicting findings imply that the guild responses to dispersal appear to be context dependent. It is also possible that different authors use different guild designations for the same species, e.g., by considering only the three original guilds (low-profile, high-profile, and motile) vs. distinguishing a fourth one (planktonic), by lumping species belonging to the same genus into a single guild, or due to subjectivity, given the lack of an international database on diatom guild affiliations. Although the use of diatom guilds for understanding community composition in relation to spatial factors has been criticized as too coarse [4.30], a growing number of studies have detected spatial differences among diatom guilds, suggesting untapped potential in evaluating guilds to identify spatial patterns. Future research should focus on unraveling why guild responses to dispersal vary idiosyncratically or depending on the specific context, as this understanding can lead to more accurate interpretations of spatial dynamics within diatom communities.

As dispersal effects are measured after accounting for environmental influences, omission of environmental variables with an impact on diatom distribution may lead to biased dispersal assessment. In rivers, for example, solar radiation, discharge, riparian shade, and groundwater inputs that can change over short distances [4.117] are rarely investigated. Likewise, in lakes, water level is often used as a surrogate for unquantified but ecologically important variables with potentially independent effects on diatoms, e.g., light penetration, vegetation type, and oxygen levels [4.118]. Microhabitat heterogeneity and characteristics are also difficult to evaluate [4.120] and can be strong determinants of species assemblages, e.g., within reach variability in flow velocity [4.14]. Within lakes, diatoms can also disperse through the movement of water, hence forming metacommunities inhabiting the open water vs. the littoral zone [4.119].

Ecological drift, stemming from random birth and death rates, has been among the most overlooked drivers of diatom community structure [4.43], despite its early recognition as an important community assembly process for diatoms [4.121]. In a set of 48 alpine lakes, Keck *et al.* [4.122] analyzed sedimentary micro-eukaryotic taxa, including diatoms, and reported a joint effect of long-term ecological drift and anthropogenic nutrient increase on the temporal change of species composition. Conversely, Remmer *et al.* [4.123] did not find evidence of ecological drift in the periphyton of Arctic lakes.

4.4 Relative Importance of Deterministic and Stochastic Effects on Diatom Metacommunities

An ongoing research effort is determining the relative contributions of environmental filtering and dispersal on diatom metacommunities. Some studies have found local factors to be more influential [4.24] [4.124], others have discovered dispersal to prevail [4.125], yet others have reported that the combination of these factors is important [4.16] [4.30] [4.116]. Lindholm *et al.* [4.104] examined benthic diatom communities from 55 small subarctic streams and found that the most important explanatory factors varied among diatom guilds and between specialists vs. generalists. For the low-profile and motile guilds, local environment was the most important driver, while the high-profile guild was equally structured by local environmental and spatial variables. Spatial variables dominated the assembly of the planktonic guild and specialists, while generalists were more strongly driven by coarser, catchment-level environmental conditions. The contributions of metacommunity processes at larger spatial and temporal scales have been less studied in diatoms than in macroorganisms [4.126]. However, substantial advances have been made over the past years, revealing that at subcontinental to intercontinental scales, diatom and guild richness and composition and diatom abundance distribution were driven by environmental filtering and dispersal constraints [4.12] [4.85].

The dominance of a particular metacommunity process depends on environmental settings such as isolation, trophic status, and scale. For instance, headwater streams are relatively isolated in the river network and dispersal limitation can either override the effect of environmental filtering [4.127] or operate in concert with it [4.24]. Dispersal limitation can also be a major driver of benthic diatom communities in isolated lakes [4.106] [4.128]. Using an innovative landscape approach and an intercontinental dataset (France and the US), Leboucher *et al.* [4.129] showed that across scales, beta diversity was driven by environmental filtering in unimpacted streams, but by spatial processes in heavily nutrient-enriched streams. This was attributed to a shift in species composition along the gradient of nutrient impact from specialists, with close association with the environment, to generalists, more strongly driven by dispersal. In 22 Mediterranean shallow ponds, Nistal-García *et al.* [4.130] found that epiphytic diatom communities generally varied in response to local environmental conditions and attributed this finding to the comparatively smaller scale. However, at larger spatial scales, diatom species composition may vary more closely with geographic distances among the study water bodies than with local environment, as shown in disparate aquatic habitats such as Antarctic lakes [4.22] and Atlantic salt marshes [4.131]. Future research should aim to identify environmental settings and scales where environmental filtering and dispersal have comparable vs. different importance and thus unravel how communities are shaped by the interplay of the environment and space.

Disturbance regimes can also exert substantial influence upon diatom community structure and determine the relative importance of deterministic and stochastic assembly processes. Hutchinson [4.3] proposed that the temporally and spatially variable nature of aquatic environments facilitate the coexistence of multiple taxa by limiting the intensity and duration of competitive interactions. Bramburger *et al.* [4.64] used a novel method of quantifying ecological specialization in groups of taxa (assemblages or genera) in analogous tropical and temperate systems. They demonstrated that tropical assemblages were characterized by higher species richness, greater ecological specialization, and resource partitioning associated with spatially diverse, temporally stable habitats. These findings agree with the time-stability hypothesis [4.132] [4.133]. This mechanism has been implicated in the development of species flocks among diatoms [4.134] and other groups. Further experimental perturbations of tropical benthic diatom communities induced responses that were inconsistent with competition but rather with ongoing niche partitioning [4.135].

Given the significance of diatoms as bioindicators, it is imperative to investigate how stochasticity may shape diatom metacommunities and interfere with biological and environmental influences. Traditionally, the relative importance of dispersal and niche processes in community assembly has been determined by variance partitioning, but Vilmi *et al.* [4.112] developed a new approach, which calculates the strength of these processes along a continuum, i.e., Dispersal-Niche Continuum Index (DNCI). Comparing stream diatom, macroinvertebrate, and fish communities from US Mid-Atlantic hydrological subregions, they found that smaller organisms were more strongly controlled by dispersal. This highlights the need to better account for spatial processes in biological monitoring, particularly in situations where dispersal is facilitated, e.g., dam removal [4.136], or non-point diffuse pollution from land use changes [4.137].

4.5 Assembly Processes Over Time

Historical contingencies, such as priority effects, defined as the competitive or facilitative effects of early colonizers on later recruits, have been rarely studied in a diatom metacommunity context. In an experimental study, Härnström *et al.* [4.138] resurrected *Skeletonema marinoi* resting cells and found intraspecific priority effects in the geographic distribution of open water and fjord populations. Priority effects through cell resurrection are still not well understood in freshwater diatoms. However, *Aulacoseira granulata* has been identified as a candidate species for experimental studies on priority effect because it may prevent establishment of other locally adapted species [4.139]. Using recent (ca. 150 years ago) paleolimnological observations, Benito *et al.* [4.140] did not find historical influences of past assemblage composition on current diatom assembly processes in a set of 14 lakes from the tropical Andes with varied levels of connectivity. Clearly, more research is necessary to determine the role of priority effects relative to other assembly processes as well as which species are responsible for the observed patterns within the metacommunity.

Unlike microorganisms such as bacteria, which mostly lack fossil records, diatoms provide a rich data source for studying historical events, including tectonic or paleoclimatic changes. Diatoms' short generation times enable tracing past events of environmental filtering and dispersal at shorter time scales, ranging from hundreds to thousands of years. This offers valuable insights into the historical dynamics of diatom communities and their responses to environmental changes over extended periods [4.141]. At centennial time scales, Winegardner *et al.* [4.142] observed an increase in diatom richness in modern sediments compared to pre-industrial times, likely associated with the arrival of new species via dispersal. At longer, geological scales, Pinseel *et al.* [4.143] reported that environmental filtering as a result of Miocene cooling and Plio-Pleistocene glaciations, drove multiple extinction events in Antarctic diatom flora. Community assembly of the diatom plankton in the 1.9-million-year-old Lake Ohrid was governed by environmental filtering, and to a lesser extent and in more recent times, by competition [4.13]. Diatom paleolimnological observations can also help elucidate whether past environmental filtering processes are consistent with the timing of diversification among and within lineages, which are typically inferred using genomic analyses of extant communities [4.144].

When spatially replicated diatom time series are available, which is comparatively rare, it is possible to determine the dominant community assembly processes over space and time. Bramburger *et al.* [4.145] demonstrated that dispersal was important in assembling local benthic diatom communities at annual time scales in individual Indonesian lakes (6-500 km^2), but as colonization progressed, niche-based processes began to dominate over dispersal through biotic interactions. At paleolimnological scales, Benito *et al.* [4.128] examined species richness in four high-elevation lakes in the tropical Andes and discovered consistent patterns of regional long-term dry/warm climatic variability significantly influencing local diatom diversity. These results suggest a lack of dispersal barriers over the last

2000 years and that regional "seed" sources are maintained over periods of millennia, enabling local recovery of diatoms in response to fluctuation between less favorable and favorable conditions. Similarly, 1-million-year-old cores from Lake Towuti (Sulawesi, Indonesia) showed repeated shifts between benthic and planktonic, mostly endemic assemblages persisting over tens of thousands of years [4.146]. These shifts were associated with long-term oscillations in mixing regimes and nutrient availability driven by global glacial/interglacial cycles [4.147]. The recurrent appearance of highly similar assemblages over such long periods further emphasizes the role of seed banks in providing long-term community resilience. It also underscores the powerful influence of deterministic mechanisms as drivers of community structure over geologic time scales.

Distinguishing the impacts of past dispersal events, both within and across geographically distant systems, from the influence of current stochastic and deterministic processes is a major challenge. This differentiation is essential to gain a comprehensive understanding of the underlying mechanisms shaping community dynamics across different temporal scales. Although not yet implemented for diatom metacommunity time series, recent studies applied null models to measure the relative strength of stochastic and deterministic processes by randomization of beta-diversity dissimilarity matrices [4.148] [4.149]. To our knowledge, however, no studies have adopted a null modeling approach to separate the effects of past from present dispersal effects, and thus, more research is needed.

4.6 Assembly Processes Over Space

An intriguing but not well researched question is how diatom metacommunity processes vary at broad spatial scales, e.g., across latitudes. It is generally predicted that the importance of environmental filtering would increase with latitude as climatic conditions become harsher, while the role of dispersal would be higher at low latitudes, where species are more specialized and with smaller ranges. An algal study (diatoms and soft algae) tested these predictions by examining the relative impacts of environmental filtering vs. spatial variables on richness and beta diversity along a latitudinal gradient [4.150]. Although niche processes dominated across latitudes, as predicted they became stronger at higher latitudes, while dispersal processes became stronger at lower latitudes but only for species richness, not beta diversity. However, this study encompassed only 19 latitudinal degrees and did not include tropical regions. An investigation of lake diatoms from lower latitudes (8°N–30°S), focusing on species composition at a guild level, found that consistent with expectations, at lower latitudes spatial effects explained more variability in the low- and high-profile guilds than the environment. Yet, motile and planktonic guilds were not structured by spatial factors near the equator [4.25]. Therefore, further research is needed at broader latitudinal ranges to determine how deterministic and stochastic processes control diatom composition at species and guild levels.

Elevational gradients also modulate the assembly processes driving diatom composition, but as with latitudinal gradients, research is limited. In Chinese rivers, He *et al.* [4.151] reported that dispersal effects were more prominent than the effects of the local environment across elevation. Conversely, Kurthen *et al.* [4.152] compared two groups of streams in a high mountain region and demonstrated that the group with lower network connectivity experienced stronger environmental filtering. An investigation of stream diatom composition at three spatial scales, within riffle, within basin, and across basins, revealed that environmental filtering was the strongest at the smallest scale, while the importance of dispersal processes increased with scale [4.149]. Given the results from these recent stream and lake studies, showing latitudinal, elevational, and scale dependence of diatom and algal community assembly, we advocate for further research on community assembly processes along broader latitudinal and scale gradients as well as along longitudinal and elevational gradients and in a wider range of aquatic ecosystems.

4.7 Concluding Remarks and Future Directions

In order to test hypotheses on deterministic and stochastic large-scale diatom community assembly process, it is necessary to build extensive and taxonomically and functionally harmonized databases. Freshwater diatom databases have been compiled in recent years at large spatial extents [4.33] [4.153], including those generated by national monitoring programs, such as the US National Water-Quality Assessment Program and the National Rivers and Streams Assessment, and the Canadian Circumpolar Diatom Database [4.154]. In Europe, some countries and regions have generated diatom databases to fulfill the European Water Framework Directive [4.155]. While some European databases are easily accessible through websites (e.g., https://environment.data.gov.uk/ecology/explorer/, https://datossuperficiales.chebro.es/WCASF/ or https://www.freshwaterecology.info)

or supplementary materials in published papers [4.156] [4.157], others have proven difficult to obtain (e.g., [4.158]). Diatom paleolimnological datasets from Pleistocene and Holocene have been complied into the Neotoma paleoecological database but are biased geographically toward northern latitudes [4.153]. Overall, most of the aforementioned databases were generated with a standardized sampling protocol but may require taxonomic harmonization. A key challenge is to generate platforms that would allow diatom researchers to add new data, harmonize them, and obtain existing data at scales from populations to biogeography [4.159]. The use of genetic metabarcoding approaches provides emerging opportunities to improve database harmonization through exploring taxonomic-free approaches [4.160]. However, community assembly research based on molecular-based datasets (e.g., Diat.barcode [4.161]) is still in its infancy [4.162]. A further challenge for metacommunity research is the lack of ecological information on diatom species particularly those identified by molecular approaches, as well as the absence of harmonization of functional attributes, such as guild affiliation. Combining both established knowledge on morphotaxa and new knowledge derived from genetic data might be one way forward [4.163].

In this review, we outlined current knowledge of the assembly mechanisms structuring diatom species and guild composition and biodiversity along spatial and temporal gradients. We showed that both deterministic and stochastic processes control diatom metacommunities, but their relative roles may vary spatially and temporally. Most of the research to date has focused on environmental filtering and dispersal. Further observational and experimental investigations are thus needed to determine how biotic interactions and ecological drift contribute to diatom assembly. We also recommend syntheses of the already substantial data on assembly mechanisms to elucidate their spatial (latitudinal, longitudinal, and elevational) and scale dependence across aquatic ecosystems. The advancement of diatom functional biogeography will be useful in discerning whether assembly mechanisms vary across levels of biological organization. Expansion of existing diatom databases by combining morphological, ecological, and molecular information as well the implementation of in silico experiments testing the roles of deterministic and stochastic processes, will further improve our understanding of the drivers of diatom metacommunities across ecosystems for the purposes of fundamental and applied science. Assessing whether the relative roles of assembly processes may shift with global change and if such shift is spatially dependent are important avenues of future research.

References

[4.1] Leibold, M. A.; Holyoak, M.; Mouquet, N.; Amarasekare, P.; Chase, J. M.; Hoopes, M. F.; Holt, R. D.; Shurin, J. B.; Law, R.; Tilman, D.; Loreau, M.; Gonzalez, A. The Metacommunity Concept: A Framework for Multi-Scale Community Ecology. *Ecol. Lett.* 2004, *7* (7), 601–613. https://doi.org/10.1111/j.1461-0248.2004.00608.x.

[4.2] Kraft, N. J. B.; Adler, P. B.; Godoy, O.; James, E. C.; Fuller, S.; Levine, J. M. Community Assembly, Coexistence and the Environmental Filtering Metaphor. *Funct. Ecol.* 2015, *29* (5), 592–599. https://doi.org/10.1111/1365-2435.12345.

[4.3] Hutchinson, G. E. The Paradox of the Plankton. *Am. Nat.* 1961, *95* (882), 137–145. https://doi.org/10.1086/282171.

[4.4] MacArthur, R.; Wilson, E. O. *The Theory of Island Biogeography*; Princeton University Press: Princeton, New Jersey, 1967.

[4.5] Hubbell, S. P. The Unified Neutral Theory of Biodiversity and Biogeography. *Monogr. Popul. Biol.* 2011, *32*, 1–448.

[4.6] Shmida, A.; Wilson, M. V. Biological Determinants of Species Diversity. *J. Biogeogr.* 1985, *12* (1), 1–20. https://doi.org/10.2307/2845026.

[4.7] Vyverman, W.; Verleyen, E.; Sabbe, K.; Vanhoutte, K.; Sterken, M.; Hodgson, D. A.; Mann, D. G.; Juggins, S.; Van de Vijver, B.; Jones, V.; Flower, R.; Roberts, D.; Chepurnov, V. A.; Kilroy, C.; Vanormelingen, P.; Wever, A. D. Historical Processes Constrain Patterns in Global Diatom Diversity. *Ecology* 2007, *88* (8), 1924–1931. https://doi.org/10.1890/06-1564.1.

[4.8] Leira, M.; Chen, G.; Dalton, C.; Irvine, K.; Taylor, D. Patterns in Freshwater Diatom Taxonomic Distinctness along an Eutrophication Gradient. *Freshw. Biol.* 2009, *54* (1), 1–14. https://doi.org/10.1111/j.1365-2427.2008.02085.x.

[4.9] Smucker, N. J.; Vis, M. L. Spatial Factors Contribute to Benthic Diatom Structure in Streams across Spatial Scales: Considerations for Biomonitoring. *Ecol. Indic.* 2011, *11* (5), 1191–1203. https://doi.org/10.1016/j.ecolind.2010.12.022.

[4.10] Stevenson, J. Ecological Assessments with Algae: A Review and Synthesis. *J. Phycol.* 2014, *50* (3), 437–461. https://doi.org/10.1111/jpy.12189.

[4.11] Blanco, S.; Olenici, A.; Ortega, F.; Jiménez-Gómez, F.; Guerrero, F. Identifying Environmental Drivers of Benthic Diatom Diversity: The Case of Mediterranean Mountain Ponds. *PeerJ* 2020, *8*, e8825. https://doi.org/10.7717/peerj.8825.

[4.12] Soininen, J.; Jamoneau, A.; Rosebery, J.; Passy, S. I. Global Patterns and Drivers of Species and Trait Composition in Diatoms. *Glob. Ecol. Biogeogr.* 2016, *25*, 940–950. https://doi.org/10.1111/geb.12452.

[4.13] Jovanovska, E.; Hauffe, T.; Stelbrink, B.; Cvetkoska, A.; Levkov, Z.; Wagner, B.; Lacey, J. H.; Ognjanova-Rumenova, N.; Hamilton, P. B.; Brandenburg,

K. M.; Albrecht, C.; Wilke, T. Environmental Filtering Drives Assembly of Diatom Communities over Evolutionary Time-Scales. *Glob. Ecol. Biogeogr.* 2022, *31* (5), 954–967. https://doi.org/10.1111/geb.13471.

[4.14] Passy, S. I. Spatial Paradigms of Lotic Diatom Distribution: A Landscape Ecology Perspective. *J. Phycol.* 2001, *37* (3), 370–378. https://doi.org/10.1046/j.1529-8817.2001.037003370.x.

[4.15] Baas-Becking, L. *Geobiologie of Inleiding Tot de Milieukunde*; Van Stockum and Zoon: The Hague, The Netherlands, 1934.

[4.16] Jamoneau, A.; Passy, S. I.; Soininen, J.; Leboucher, T.; Tison-Rosebery, J. Beta Diversity of Diatom Species and Ecological Guilds: Response to Environmental and Spatial Mechanisms along the Stream Watercourse. *Freshw. Biol.* 2018, *63*, 62–73. https://doi.org/10.1111/fwb.12980.

[4.17] Leboucher, T.; Mignien, L.; Wach, M.; Boutry, S.; Jamoneau, A.; Passy, S. I.; Tison-Rosebery, J. Consideration of Mass Effect Processes in Bioindication Allows More Accurate Bioassessment of Water Quality. *Ecol. Indic.* 2021, *127*, 107791. https://doi.org/10.1016/j.ecolind.2021.107791.

[4.18] Tapia, P. M.; Theriot, E. C.; Fritz, S. C.; Cruces, F.; Rivera, P. Distribution and Morphometry Analysis of Cyclostephanos Andinus Comb. Nov., a Planktonic Diatom from the Central Andes. *Diatom Res.* 2004, *19* (2), 311–327. https://doi.org/10.1080/0269249X.2004.9705877.

[4.19] Popovskaya, G. I.; Likhoshway, Y. V.; Genkal, S. I.; Firsova, A. D. The Role of Endemic Diatom Algae in the Phytoplankton of Lake Baikal. *Hydrobiologia* 2006, *568*, 87–94. https://doi.org/10.1007/s10750-006-0328-4.

[4.20] Bramburger, A. J.; Hamilton, P. B.; Hehanussa, P. E.; Haffner, G. D. Processes Regulating the Community Composition and Relative Abundance of Taxa in the Diatom Communities of the Malili Lakes, Sulawesi Island, Indonesia. *Hydrobiologia* 2008, *615*, 215–224. https://doi.org/10.1007/s10750-008-9562-2.

[4.21] Benito, X.; Fritz, S. C.; Steinitz-Kannan, M.; Tapia, P.; Kelly, M. A.; Lowell, T. V. Geo-Climatic Factors Drive Diatom Community Distribution in Tropical South American Freshwaters. *J. Ecol.* 2018, *106*, 1660–1672. https://doi.org/10.1111/1365-2745.12934.

[4.22] Verleyen, E.; Vijver, B.; Tytgat, B.; Pinseel, E.; Hodgson, D. A.; Kopalová, K.; Chown, S. L.; Ranst, E. V.; Imura, S.; Kudoh, S.; Nieuwenhuyze, W. V.; Sabbe, K.; Vyverman, W. Diatoms Define a Novel Freshwater Biogeography of the Antarctic. *Ecography* 2021, *44* (4), 548–560. https://doi.org/10.1111/ecog.05374.

[4.23] Passy, S. I. Diatom Ecological Guilds Display Distinct and Predictable Behavior along Nutrient and Disturbance Gradients in Running Waters. *Aquat. Bot.* 2007, *86* (2), 171–178. https://doi.org/10.1016/j.aquabot.2006.09.018.

[4.24] Göthe, E.; Angeler, D. G.; Gottschalk, S.; Löfgren, S.; Sandin, L. The Influence of Environmental, Biotic and Spatial Factors on Diatom

Metacommunity Structure in Swedish Headwater Streams. *PLoS One* 2013, *8* (8), e72237. https://doi.org/10.1371/journal.pone.0072237.

[4.25] Benito, X.; Fritz, S. C.; Steinitz-Kannan, M.; Vélez, M. I.; McGlue, M. M. Lake Regionalization and Diatom Metacommunity Structuring in Tropical South America. *Ecol. Evol.* 2018, *8* (16), 7865–7878. https://doi.org/10.1002/ece3.4305.

[4.26] Stenger-Kovács, C.; Lengyel, E.; Crossetti, L. O.; Uveges, V.; Padisák, J. Diatom Ecological Guilds as Indicators of Temporally Changing Stressors and Disturbances in the Small Torna-Stream, Hungary. *Ecol. Indic.* 2013, *24*, 138–147. https://doi.org/10.1016/j.ecolind.2012.06.003.

[4.27] Smol, J. P.; Stoermer, E. F. *The Diatoms: Applications for the Environmental and Earth Sciences*; Cambridge University Press, 2010.

[4.28] Pajunen, V.; Luoto, M.; Soininen, J. Stream Diatom Assemblages as Predictors of Climate. *Freshw. Biol.* 2016, *61* (6), 876–886. https://doi.org/10.1111/fwb.12750.

[4.29] Lobo, E. A.; Heinrich, C. G.; Schuch, M.; Wetzel, C. E.; Ector, L. Diatoms as Bioindicators in Rivers. *River Algae* 2016, 245–271.

[4.30] Keck, F.; Franc, A.; Kahlert, M. Disentangling the Processes Driving the Biogeography of Freshwater Diatoms: A Multiscale Approach. *J. Biogeogr.* 2018, *45* (7), 1582–1592. https://doi.org/10.1111/jbi.13239.

[4.31] Burge, D. R. L.; Edlund, M. B.; Frisch, D. Paleolimnology and Resurrection Ecology: The Future of Reconstructing the Past. *Evol. Appl.* 2018, *11* (1), 42–59. https://doi.org/10.1111/eva.12556.

[4.32] Mann, D. G.; Yamada, N.; Bolton, J. J.; Witkowski, A.; Trobajo, R. *Nitzschia captiva* sp.nov. (Bacillariophyta), the essential prey diatom of the kleptoplastic dinoflagellate *Durinskia capensis*, compared with *N. agnita, N. kuetzingioides* and other species. *Phycologia* 2023, *62* (2), 136–151. https://doi.org/10.1080/00318884.2023.2169024.

[4.33] Benito, X.; Feitl, M.; Carrevedo, M. L.; Vélez, M. I.; Escobar, J.; Tapia, P. M.; Steinitz-Kannan, M.; Fritz, S. C. Tropical South America Diatom Database: A Tool for Studying the Macroecology of Microorganisms. *Diatom Res.* 2022, in press, 1–13. https://doi.org/10.1080/0269249X.2022.2078429.

[4.34] Jamoneau, A.; Soininen, J.; Tison-Rosebery, J.; Boutry, S.; Budnick, W. R.; He, S.; Marquié, J.; Jyrkänkallio-Mikkola, J.; Pajunen, V.; Teittinen, A.; Tupola, V.; Wang, B.; Wang, J.; Blanco, S.; Borrini, A.; Cantonati, M.; Valente, A. C.; Delgado, C.; Dörflinger, G.; Gonçalves, V.; Kennedy, B.; Marques, H.; Papatheodoulou, A.; Raposeiro, P. M.; Ritter, C.; Serafim, A.; Van de Vijver, B.; Passy, S. I. Stream Diatom Biodiversity in Islands and Continents—A Global Perspective on Effects of Area, Isolation and Environment. *J. Biogeogr.* 2022, *49* (12), 2156–2168. https://doi.org/10.1111/jbi.14482.

[4.35] Sokol, E. R.; Barrett, J. E.; Kohler, T. J.; McKnight, D. M.; Salvatore, M. R.; Stanish, L. F. Evaluating Alternative Metacommunity Hypotheses for

Diatoms in the McMurdo Dry Valleys Using Simulations and Remote Sensing Data. *Front. Ecol. Evol.* 2020, *8*:295 https://doi.org/10.3389/fevo.2020.521668.

[4.36] Poulíčková, A.; Špačková, J.; Kelly, M. G.; Duchoslav, M.; Mann, D. G. Ecological Variation within *Sellaphora* Species Complexes (Bacillariophyceae): Specialists or Generalists? *Hydrobiologia* 2008, *614* (1), 373–386. https://doi.org/10.1007/s10750-008-9521-y.

[4.37] Lundholm, N.; Bates, S. S.; Baugh, K. A.; Bill, B. D.; Connell, L. B.; Léger, C.; Trainer, V. L. Cryptic and Pseudo-Cryptic Diversity in Diatoms—with Descriptions of *Pseudo-Nitzschia hasleana* sp. nov. and *P. fryxelliana* sp. nov. *J. Phycol.* 2012, *48* (2), 436–454. https://doi.org/10.1111/j.1529-8817.2012.01132.x.

[4.38] Sorhannus, U.; Ortiz, J. D.; Wolf, M.; Fox, M. G. Microevolution and Speciation in *Thalassiosira weissflogii* (Bacillariophyta). *Protist* 2010, *161* (2), 237–249. https://doi.org/10.1016/j.protis.2009.10.003.

[4.39] Pérez-Burillo, J.; Trobajo, R.; Leira, M.; Keck, F.; Rimet, F.; Sigró, J.; Mann, D. G. DNA Metabarcoding Reveals Differences in Distribution Patterns and Ecological Preferences among Genetic Variants within Some Key Freshwater Diatom Species. *Sci. Total Environ.* 2021, *798*, 149029. https://doi.org/10.1016/j.scitotenv.2021.149029.

[4.40] Cordier, T.; Alonso-Sáez, L.; Apothéloz-Perret-Gentil, L.; Aylagas, E.; Bohan, D. A.; Bouchez, A.; Chariton, A.; Creer, S.; Frühe, L.; Keck, F.; Keeley, N.; Laroche, O.; Leese, F.; Pochon, X.; Stoeck, T.; Pawlowski, J.; Lanzén, A. Ecosystems Monitoring Powered by Environmental Genomics: A Review of Current Strategies with an Implementation Roadmap. *Mol. Ecol.* 2021, *30* (13), 2937–2958. https://doi.org/10.1111/mec.15472.

[4.41] Apothéloz-Perret-Gentil, L.; Cordonier, A.; Straub, F.; Iseli, J.; Esling, P.; Pawlowski, J. Taxonomy-Free Molecular Diatom Index for High-Throughput eDNA Biomonitoring. *Mol. Ecol. Resour.* 2017, *17* (6), 1231–1242. https://doi.org/10.1111/1755-0998.12668.

[4.42] Keck, F.; Vasselon, V.; Rimet, F.; Bouchez, A.; Kahlert, M. Boosting DNA Metabarcoding for Biomonitoring with Phylogenetic Estimation of Operational Taxonomic Units' Ecological Profiles. *Mol. Ecol. Resour.* 2018, *18* (6), 1299–1309. https://doi.org/10.1111/1755-0998.12919.

[4.43] Soininen, J.; Teittinen, A. Fifteen Important Questions in the Spatial Ecology of Diatoms. *Freshw. Biol.* 2019, *64* (11), 2071–2083. https://doi.org/10.1111/fwb.13384.

[4.44] Schulte, N. O.; Khan, A. L.; Smith, E. W.; Zoumplis, A.; Kaul, D.; Allen, A. E.; Adams, B. J.; McKnight, D. M. Blowin' in the Wind: Dispersal, Structure, and Metacommunity Dynamics of Aeolian Diatoms in the McMurdo Sound Region, Antarctica. *J. Phycol.* 2022, *58* (1), 36–54. https://doi.org/10.1111/jpy.13223.

[4.45] Görgényi, J.; Boros, G.; Vitál, Z.; Mozsár, A.; Várbíró, G.; Vasas, G.; Borics, G. The Role of Filter-Feeding Asian Carps in Algal Dispersion. *Hydrobiologia* 2016, *764* (1), 115–126. https://doi.org/10.1007/s10750-015-2285-2.

[4.46] Soininen, J. Macroecology of Unicellular Organisms - Patterns and Processes. *Environ. Microbiol. Rep.* 2012, *4* (1), 10–22. https://doi.org/10.1111/j.1758-2229.2011.00308.x.

[4.47] Stenger-Kovács, C.; Béres, V. B.; Buczkó, K.; Tapolczai, K.; Padisák, J.; Selmeczy, G. B.; Lengyel, E. Diatom Community Response to Inland Water Salinization: A Review. *Hydrobiologia* 2023. https://doi.org/10.1007/s10750-023-05167-w.

[4.48] Pound, K. L.; Lawrence, G. B.; Passy, S. I. Beta Diversity Response to Stress Severity and Heterogeneity in Sensitive versus Tolerant Stream Diatoms. *Divers. Distrib.* 2019, *25* (3), 374–384. https://doi.org/10.1111/ddi.12865.

[4.49] Battarbee, R. W.; Simpson, G. L.; Shilland, E. M.; Flower, R. J.; Kreiser, A.; Yang, H.; Clarke, G. Recovery of UK Lakes from Acidification: An Assessment Using Combined Palaeoecological and Contemporary Diatom Assemblage Data. *Ecol. Indic.* 2014, *37*, 365–380. https://doi.org/10.1016/j.ecolind.2012.10.024.

[4.50] Teittinen, A.; Wang, J.; Strömgård, S.; Soininen, J. Local and Geographical Factors Jointly Drive Elevational Patterns in Three Microbial Groups across Subarctic Ponds. *Glob. Ecol. Biogeogr.* 2017, *26* (8), 973–982. https://doi.org/10.1111/geb.12607.

[4.51] Muñoz-López, C. L.; Rivera-Rondón, C. A. Diatom Response to Environmental Gradients in the High Mountain Lakes of the Colombia's Eastern Range. *Aquat. Sci.* 2021, *84* (1), 15. https://doi.org/10.1007/s00027-021-00838-z.

[4.52] Morales, E.; Vis, M. Epilithic Diatoms (Bacillariophyceae) from Cloud Forest and Alpine Streams in Bolivia, South America. *Proc. Acad. Nat. Sci. Philadelphia* 2007, *156*, 123–155. https://doi.org/10.1635/0097-3157(2007)156[123:EDBFCF]2.0.CO;2.

[4.53] Pound, K. L.; Lawrence, G. B.; Passy, S. I. Wetlands Serve as Natural Sources for Improvement of Stream Ecosystem Health in Regions Affected by Acid Deposition. *Glob. Change Biol.* 2013, *19* (9), 2720–2728. https://doi.org/10.1111/gcb.12265.

[4.54] Passy, S. I.; Legendre, P. Power Law Relationships among Hierarchical Taxonomic Categories in Algae Reveal a New Paradox of the Plankton. *Glob. Ecol. Biogeogr.* 2006, *15* (5), 528–535. https://doi.org/10.1111/j.1566-822x.2006.00246.x.

[4.55] Monteith, D. T.; Stoddard, J. L.; Evans, C. D.; De Wit, H. A.; Forsius, M.; Høgåsen, T.; Wilander, A.; Skjelkvåle, B. L.; Jeffries, D. S.; Vuorenmaa, J. Dissolved Organic Carbon Trends Resulting from Changes in

Atmospheric Deposition Chemistry. *Nature* 2007, *450* (7169), 537–540. https://doi.org/10.1038/nature06316.
[4.56] Jyväsjärvi, J.; Rajakallio, M.; Brüsecke, J.; Huttunen, K.-L.; Huusko, A.; Muotka, T.; Taipale, S. J. Dark Matters: Contrasting Responses of Stream Biofilm to Browning and Loss of Riparian Shading. *Glob. Change Biol.* 2022, *28* (17), 5159–5171. https://doi.org/10.1111/gcb.16279.
[4.57] Urrutia-Cordero, P.; Ekvall, M. K.; Ratcovich, J.; Soares, M.; Wilken, S.; Zhang, H.; Hansson, L.-A. Phytoplankton Diversity Loss along a Gradient of Future Warming and Brownification in Freshwater Mesocosms. *Freshw. Biol.* 2017, *62* (11), 1869–1878. https://doi.org/10.1111/fwb.13027.
[4.58] Poulíčková, A.; Manoylov, K. Ecology of Freshwater Diatoms – Current Trends and Applications. In *Diatoms: Fundamentals and Applications*; John Wiley & Sons, Ltd, 2019; pp 289–309. https://doi.org/10.1002/9781119370741.ch13.
[4.59] Harpole, W. S.; Sullivan, L. L.; Lind, E. M.; Firn, J.; Adler, P. B.; Borer, E. T.; Chase, J.; Fay, P. A.; Hautier, Y.; Hillebrand, H. Addition of Multiple Limiting Resources Reduces Grassland Diversity. *Nature* 2016, *537* (7618), 93–96. https://doi.org/10.1038/nature19324.
[4.60] Borer, E. T.; Grace, J. B.; Harpole, W. S.; MacDougall, A. S.; Seabloom, E. W. A Decade of Insights into Grassland Ecosystem Responses to Global Environmental Change. *Nat. Ecol. Evol.* 2017, *1* (5), 0118. https://doi.org/10.1038/s41559-017-0118.
[4.61] Passy, S. I. Continental Diatom Biodiversity in Stream Benthos Declines as More Nutrients Become Limiting. *Proc. Natl. Acad. Sci. USA* 2008, *105* (28), 9663–9667. https://doi.org/10.1073/pnas.0802542105.
[4.62] Passy, S. I.; Larson, C. A. Niche Dimensionality and Herbivory Control Stream Algal Biomass via Shifts in Guild Composition, Richness, and Evenness. *Ecology* 2019, *100* (11), e02831. https://doi.org/10.1002/ecy.2831.
[4.63] Zamorano, D., F.; Labra, A.; Matthaei, C. D.; Romero, Ú. Biogeographical Patterns of Species Richness in Stream Diatoms from Southwestern South America. Ecol. and Evol. 2024, e11156. https://doi.org/10.1002/ece3.11156
[4.64] Bramburger, A. J.; Hamilton, P. B.; Haffner, G. D.; Hehanussa, P. E. Variable Niche Breadth in Benthic Diatoms: Implications of Ecological Specialization and Generalization for Community Structure. *J. Great Lakes Res.* 2020, *46* (5), 1131–1139. https://doi.org/10.1016/j.jglr.2020.06.024.
[4.65] Passy, S. I. The Relationship between Local and Regional Diatom Richness Is Mediated by the Local and Regional Environment. *Glob. Ecol. Biogeogr.* 2009, *18* (3), 383–391. https://doi.org/10.1111/j.1466-8238.2009.00453.x.
[4.66] Gonzàlez-Trujillo, J. D.; Pedraza-Garzon, E.; Donato-Rondon, J. C.; Sabater, S. Ecoregional Characteristics Drive the Distribution Patterns of Neotropical Stream Diatoms. *J. Phycol.* 2020, *56* (4), 1053–1065. https://doi.org/10.1111/jpy.13005.

[4.67] Passy, S. I. A Distinct Latitudinal Gradient of Diatom Diversity Is Linked to Resource Supply. *Ecology* 2010, *91* (1), 36–41. https://doi.org/10.1890/09-0545.1.

[4.68] Dengg, M.; Stirling, C. H.; Safi, K.; Lehto, N. J.; Wood, S. A.; Seyitmuhhamedov, K.; Reid, M. R.; Verburg, P. Bioavailable Iron Concentrations Regulate Phytoplankton Growth and Bloom Formation in Low-Nutrient Lakes. *Sci. Total Environ.* 2023, 166399. https://doi.org/10.1016/j.scitotenv.2023.166399.

[4.69] Jüttner, I.; Chimonides, P. D. J.; Ormerod, S. J.; Cox, E. J. Ecology and Biogeography of Himalayan Diatoms: Distribution along Gradients of Altitude, Stream Habitat and Water Chemistry. *Fundam. Appl. Limnol. Für Hydrobiol.* 2010, *177* (4), 293–311. https://doi.org/10.1127/1863-9135/2010/0177-0293.

[4.70] Truchy, A.; Sponseller, R. A.; Ecke, F.; Angeler, D. G.; Kahlert, M.; Bundschuh, M.; Johnson, R. K.; McKie, B. G. Responses of Multiple Structural and Functional Indicators along Three Contrasting Disturbance Gradients. *Ecol. Indic.* 2022, *135*, 108514. https://doi.org/10.1016/j.ecolind.2021.108514.

[4.71] Lange, K.; Liess, A.; Piggott, J. J.; Townsend, C. R.; Matthaei, C. D. Light, Nutrients and Grazing Interact to Determine Stream Diatom Community Composition and Functional Group Structure. *Freshw. Biol.* 2011, *56* (2), 264–278. https://doi.org/10.1111/j.1365-2427.2010.02492.x.

[4.72] Teittinen, A.; Taka, M.; Ruth, O.; Soininen, J. Variation in Stream Diatom Communities in Relation to Water Quality and Catchment Variables in a Boreal, Urbanized Region. *Sci. Total Environ.* 2015, *530–531*, 279–289. https://doi.org/10.1016/j.scitotenv.2015.05.101.

[4.73] Woolway, R. I.; Kraemer, B. M.; Lenters, J. D.; Merchant, C. J.; O'Reilly, C. M.; Sharma, S. Global Lake Responses to Climate Change. *Nat. Rev. Earth Environ.* 2020, *1* (8), 388–403. https://doi.org/10.1038/s43017-020-0067-5.

[4.74] Rühland, K.; Paterson, A. M.; Smol, J. P. Hemispheric-Scale Patterns of Climate-Related Shifts in Planktonic Diatoms from North American and European Lakes. *Glob. Change Biol.* 2008, *14* (11), 2740–2754. https://doi.org/10.1111/j.1365-2486.2008.01670.x.

[4.75] Reavie, E. D.; Sgro, G. V.; Estepp, L. R.; Bramburger, A. J.; Shaw Chraïbi, V. L.; Pillsbury, R. W.; Cai, M.; Stow, C. A.; Dove, A. Climate Warming and Changes in Cyclotella Sensu Lato in the Laurentian Great Lakes. *Limnol. Oceanogr.* 2017, *62* (2), 768–783. https://doi.org/10.1002/lno.10459.

[4.76] Bramburger, A. J.; Reavie, E. D.; Sgro, G. V.; Estepp, L. R.; Chraïbi, V. L. S.; Pillsbury, R. W. Decreases in Diatom Cell Size during the 20th Century in the Laurentian Great Lakes: A Response to Warming Waters? *J. Plankton Res.* 2017, *39* (2), 199–210. https://doi.org/10.1093/plankt/fbx009.

[4.77] Dodds, W. K.; Bruckerhoff, L.; Batzer, D.; Schechner, A.; Pennock, C.; Renner, E.; Tromboni, F.; Bigham, K.; Grieger, S. The Freshwater Biome

Gradient Framework: Predicting Macroscale Properties Based on Latitude, Altitude, and Precipitation. *Ecosphere* 2019, *10* (7). https://doi.org/10.1002/ecs2.2786.

[4.78] Rühland, K. M.; Paterson, A. M.; Smol, J. P. Lake Diatom Responses to Warming: Reviewing the Evidence. *J. Paleolimnol.* 2015, *54* (1), 1–35. https://doi.org/10.1007/s10933-015-9837-3.

[4.79] Hobbs, W. O.; Telford, R. J.; Birks, H. J. B.; Saros, J. E.; Hazewinkel, R. R.; Perren, B. B.; Saulnier-Talbot, É.; Wolfe, A. P. Quantifying Recent Ecological Changes in Remote Lakes of North America and Greenland Using Sediment Diatom Assemblages. *PLoS One* 2010, *5* (4), e10026. https://doi.org/10.1371/journal.pone.0010026.

[4.80] Michelutti, N.; Wolfe, A.; Cooke, C.; Hobbs, W.; Vuille, M. Climate Change Forces New Ecological States in Tropical Andean Lakes. *PLoS One* 2015, *10* (2), e0115338. https://doi.org/10.1371/journal.pone.0115338.

[4.81] Trobajo R.; Sullivan, M.J. Applied diatom studies in estuaries and shallow coastal environments. In: Smol JP, Stoermer EF, eds. *The Diatoms: Applications for the Environmental and Earth Sciences.* Cambridge University Press; 2010:309-323.

[4.82] Ryves, D. B.; Clarke, A. L.; Appleby, P. G.; Amsinck, S. L.; Jeppesen, E.; Landkildehus, F.; Anderson, N. J. Reconstructing the Salinity and Environment of the Limfjord and Vejlerne Nature Reserve, Denmark, Using a Diatom Model for Brackish Lakes and Fjords. *Can. J. Fish. Aquat. Sci.* 2004, *61* (10). https://doi.org/10.1139/f04-127.

[4.83] Teittinen, A.; Soininen, J.; Virta, L. Studying Biodiversity–Ecosystem Function Relationships in Experimental Microcosms among Islands. *Ecology* 2022, *103* (5), e3664. https://doi.org/10.1002/ecy.3664.

[4.84] Jeppesen, E.; Moss, B.; Bennion, H.; Carvalho, L.; DeMeester, L.; Feuchtmayr, H.; Friberg, N.; Gessner, M. O.; Hefting, M.; Lauridsen, T. L.; Liboriussen, L.; Malmquist, H. J.; May, L.; Meerhoff, M.; Olafsson, J. S.; Soons, M. B.; Verhoeven, J. T. A. Interaction of Climate Change and Eutrophication. In *Climate Change Impacts on Freshwater Ecosystems*; Kernan, M., Battarbee, R. W., Moss, B., Eds.; Wiley-Blackwell: Oxford, UK, 2010; pp 119–151. https://doi.org/10.1002/9781444327397.ch6.

[4.85] Passy, S. I.; Larson, C. A.; Jamoneau, A.; Budnick, W.; Heino, J.; Leboucher, T.; Tison-Rosebery, J.; Soininen, J. Biogeographical Patterns of Species Richness and Abundance Distribution in Stream Diatoms Are Driven by Climate and Water Chemistry. *Am. Nat.* 2018, *192* (5), 605–617. https://doi.org/10.1086/699830.

[4.86] Brown, J. H. On the Relationship between Abundance and Distribution of Species. *Am. Nat.* 1984, *124* (2), 255–279. http://www.jstor.org/stable/2461494.

[4.87] Hanski, I.; Kouki, J.; Halkka, A. Three Explanations of the Positive Relationship between Distribution and Abundance of Species. In *Species diversity in ecological communities: historical and geographical perspectives* (eds Ricklefs & Schluter); The University of Chicago Press: Chicago, USA, 1993; pp 108–116.

[4.88] Passy, S. I. A Hierarchical Theory of Macroecology. *Ecol. Lett.* 2012, *15* (9), 923–934. https://doi.org/10.1111/j.1461-0248.2012.01809.x.

[4.89] Rocha, M. P.; Bini, L. M.; Siqueira, T.; Hjort, J.; Grönroos, M.; Lindholm, M.; Karjalainen, S.-M.; Heino, J. Predicting Occupancy and Abundance by Niche Position, Niche Breadth and Body Size in Stream Organisms. *Oecologia* 2018, *186* (1), 205–216. https://doi.org/10.1007/s00442-017-3988-z.

[4.90] Lengyel, E.; Szabó, B.; Stenger-Kovács, C. Realized Ecological Niche-Based Occupancy–Abundance Patterns of Benthic Diatom Traits. *Hydrobiologia* 2020, *847* (14), 3115–3127. https://doi.org/10.1007/s10750-020-04324-9.

[4.91] Heino, J.; Soininen, J. Regional Occupancy in Unicellular Eukaryotes: A Reflection of Niche Breadth, Habitat Availability or Size-Related Dispersal Capacity? *Freshw. Biol.* 2006, *51* (4), 672–685. https://doi.org/10.1111/j.1365-2427.2006.01520.x.

[4.92] Soininen, J.; Heino, J. Variation in Niche Parameters along the Diversity Gradient of Unicellular Eukaryote Assemblages. *Protist* 2007, *158*, 181–191. https://doi.org/10.1016/j.protis.2006.11.002.

[4.93] Vilmi, A.; Karjalainen, S. M.; Karjalainen, S. M.; Wang, J.; Heino, J. Using Traits to Explain Interspecific Variation in Diatom Occupancy and Abundance across Lakes and Streams. *J. Biogeogr.* 2019, *46* (7), 1419–1428. https://doi.org/10.1111/jbi.13584.

[4.94] Pound, K. L.; Larson, C. A.; Passy, S. I. Current Distributions and Future Climate-Driven Changes in Diatoms, Insects and Fish in US Streams. *Glob. Ecol. Biogeogr.* 2021, *30* (1), 63–78. https://doi.org/10.1111/geb.13193.

[4.95] Olden, J. D.; Rooney, T. P. On Defining and Quantifying Biotic Homogenization. *Glob. Ecol. Biogeogr.* 2006, *15* (2), 113–120. https://doi.org/10.1111/j 1466-822X.2006.00214.x.

[4.96] Pla-Rabés, S.; Hamilton, P. B.; Ballesteros, E.; Gavrilo, M.; Friedlander, A. M.; Sala, E. The Structure and Diversity of Freshwater Diatom Assemblages from Franz Josef Land Archipelago: A Northern Outpost for Freshwater Diatoms. *PeerJ* 2016, *4*, e1705. https://doi.org/10.7717/peerj.1705.

[4.97] Zorzal-Almeida, S.; Bartozek, E. C. R.; Bicudo, D. C. Homogenization of Diatom Assemblages Is Driven by Eutrophication in Tropical Reservoirs. *Environ. Pollut.* 2021, *288*, 117778. https://doi.org/10.1016/j.envpol.2021.117778.

[4.98] Linke, S.; Lehner, B.; Ouellet Dallaire, C.; Ariwi, J.; Grill, G.; Anand, M.; Beames, P.; Burchard-Levine, V.; Maxwell, S.; Moidu, H.; Tan, F.; Thieme, M. Global Hydro-Environmental Sub-Basin and River Reach

Characteristics at High Spatial Resolution. *Sci. Data* 2019, *6* (1), 283. https://doi.org/10.1038/s41597-019-0300-6.
[4.99] Fick, S. E.; Hijmans, R. J. WorldClim 2: New 1-Km Spatial Resolution Climate Surfaces for Global Land Areas. *Int. J. Climatol.* 2017, *37* (12), 4302–4315. https://doi.org/10.1002/joc.5086.
[4.100] Hansen, M. C.; Potapov, P. V.; Moore, R.; Hancher, M.; Turubanova, S. A.; Tyukavina, A.; Thau, D.; Stehman, S. V.; Goetz, S. J.; Loveland, T. R.; Kommareddy, A.; Egorov, A.; Chini, L.; Justice, C. O.; Townshend, J. R. G. High-Resolution Global Maps of 21st-Century Forest Cover Change. *Science* 2013, *342* (6160), 850–853. https://doi.org/10.1126/science.1244693.
[4.101] Goldewijk, K. K.; Beusen, A.; Drecht, G. van; Vos, M. de. The HYDE 3.1 Spatially Explicit Database of Human-Induced Global Land-Use Change over the Past 12,000 Years. *Glob. Ecol. Biogeogr.* 2011, *20* (1), 73–86. https://doi.org/10.1111/j.1466-8238.2010.00587.x.
[4.102] Berthon, V.; Bouchez, A.; Rimet, F. Using Diatom Life-Forms and Ecological Guilds to Assess Organic Pollution and Trophic Level in Rivers: A Case Study of Rivers in South-Eastern France. *Hydrobiologia* 2011, *673* (1), 259–271. https://doi.org/10.1007/s10750-011-0786-1.
[4.103] Abonyi, A.; Horváth, Z.; Ptacnik, R. Functional Richness Outperforms Taxonomic Richness in Predicting Ecosystem Functioning in Natural Phytoplankton Communities. *Freshw. Biol.* 2018, *63* (2), 178–186. https://doi.org/10.1111/fwb.13051.
[4.104] Mouillot, D.; Graham, N. A. J.; Villéger, S.; Mason, N. W. H.; Bellwood, D. R. A Functional Approach Reveals Community Responses to Disturbances. *Trends Ecol. Evol.* 2013, *28* (3), 167–177. https://doi.org/10.1016/j.tree.2012.10.004.
[4.105] Lindholm, M.; Grönroos, M.; Hjort, J.; Karjalainen, S. M.; Tokola, L.; Heino, J. Different Species Trait Groups of Stream Diatoms Show Divergent Responses to Spatial and Environmental Factors in a Subarctic Drainage Basin. *Hydrobiologia* 2018, *816*, 213–230. https://doi.org/10.1007/s10750-018-3585-0.
[4.106] Passy, S. I. Framework for Community Functioning: Synthesis of Stress Gradient and Resource Partitioning Concepts. *PeerJ* 2017, *5*, e3885. https://doi.org/10.7717/peerj.3885.
[4.107] Vilmi, A.; Karjalainen, S. M.; Heino, J. Ecological Uniqueness of Stream and Lake Diatom Communities Shows Different Macroecological Patterns. *Divers. Distrib.* 2017, *23* (9), 1042–1053. https://doi.org/10.1111/ddi.12594.
[4.108] Álvarez-Cobelas, M.; Rojo, C. Linking the Community and Metacommunity Perspectives: Biotic Relationships Are Key in Benthic Diatom Ecology. *Water* 2022, *14* (23), 3805. https://doi.org/10.3390/w14233805.
[4.109] Leboucher, T.; Tison-Rosebery, J.; Budnick, W. R.; Jamoneau, A.; Vyverman, W.; Soininen, J.; Boutry, S.; Passy, S. I. A Metacommunity

Approach for Detecting Species Influenced by Mass Effect. *J. Appl. Ecol.* 2020, *57* (10), 2031–2040. https://doi.org/10.1111/1365-2664.13701.

[4.110] Liu, J.; Soininen, J.; Han, B. P.; Declerck, S. A. J. Effects of Connectivity, Dispersal Directionality and Functional Traits on the Metacommunity Structure of River Benthic Diatoms. *J. Biogeogr.* 2013, *40* (12), 2238–2248. https://doi.org/10.1111/jbi.12160.

[4.111] Dong, X.; Li, B.; He, F.; Gu, Y.; Sun, M.; Zhang, H.; Tan, L. Flow Directionality, Mountain Barriers and Functional Traits Determine Diatom Metacommunity Structuring of High Mountain Streams. *Sci Rep* 2016, *6*, 24711. https://doi.org/10.1038/srep24711.

[4.112] Vilmi, A.; Gibert, C.; Escarguel, G.; Happonen, K.; Heino, J.; Jamoneau, A.; Passy, S. I.; Picazo, F.; Soininen, J.; Tison-Rosebery, J.; Wang, J. Dispersal–Niche Continuum Index: A New Quantitative Metric for Assessing the Relative Importance of Dispersal versus Niche Processes in Community Assembly. *Ecography* 2020, *44* (3), 370–379. https://doi.org/10.1111/ecog.05356.

[4.113] Passy, S. I. Differential Cell Size Optimization Strategies Produce Distinct Diatom Richness–Body Size Relationships in Stream Benthos and Plankton. *J. Ecol.* 2007, *95* (4), 745–754. https://doi.org/10.1111/j.1365-2745.2007.01248.x.

[4.114] Wetzel, C. E.; de Bicudo, D. C.; Ector, L.; Lobo, E. A.; Soininen, J.; Landeiro, V. L.; Bini, L. M. Distance Decay of Similarity in Neotropical Diatom Communities. *PLoS One* 2012, *7* (9), 10–11. https://doi.org/10.1371/journal.pone.0045071.

[4.115] Passy, S. I. Abundance Inequality in Freshwater Communities Has an Ecological Origin. *Am. Nat.* 2016, *187* (4), 502–516. https://doi.org/10.1086/685424.

[4.116] González-Trujillo, J. D.; Petsch, D. K.; Córdoba-Ariza, G.; Rincón-Palau, K.; Donato-Rondon, J. C.; Castro-Rebolledo, M. I.; Sabater, S. Upstream Refugia and Dispersal Ability May Override Benthic-Community Responses to High-Andean Streams Deforestation. *Biodivers. Conserv.* 2019, *28* (6), 1513–1531. https://doi.org/10.1007/s10531-019-01739-2.

[4.117] Johnson, R. K.; Angeler, D. G. Effects of Agricultural Land Use on Stream Assemblages: Taxon-Specific Responses of Alpha and Beta Diversity. *Ecol. Indic.* 2014, *45*, 386–393. https://doi.org/10.1016/j.ecolind.2014.04.028.

[4.118] Leira, M.; Cantonati, M. Effects of water-level fluctuations on lakes: an annotated bibliography. *Hydrobiologia* 2008, 613, 171–184. https://doi.org/10.1007/s10750-008-9465-2

[4.119] Pla-Rabés, S.; Catalan, J. Diatom Species Variation between Lake Habitats: Implications for Interpretation of Paleolimnological Records. *J. Paleolimnol.* 2018, *60*, 169–187. https://doi.org/10.1007/s10933-018-0017-0.

[4.120] Taxböck, L.; Karger, D. N.; Kessler, M.; Spitale, D.; Cantonati, M. Diatom Species Richness in Swiss Springs Increases with Habitat Complexity and Elevation. *Water* 2020, *12* (2), 449. https://doi.org/10.3390/w12020449.

[4.121] Patrick, R. Benthic Stream Communities: A Discussion of the Factors That Affect Their Structure and How They Function. *Am. Sci.* 1970, *58* (5), 546–549. https://www.jstor.org/stable/27829244.

[4.122] Keck, F.; Millet, L.; Debroas, D.; Etienne, D.; Galop, D.; Rius, D.; Domaizon, I. Assessing the Response of Micro-Eukaryotic Diversity to the Great Acceleration Using Lake Sedimentary DNA. *Nat. Commun.* 2020, *11* (1), 3831. https://doi.org/10.1038/s41467-020-17682-8.

[4.123] Remmer, C. R.; Robichaud, C. D.; Polowyk, H.; Rooney, R. The Role of Ecological Drift in Structuring Periphytic Diatom Communities. *J. Freshw. Ecol.* 2019, *34* (1), 363–377. https://doi.org/10.1080/02705060.2019.1614104.

[4.124] Verleyen, E.; Vyverman, W.; Sterken, M.; Hodgson, D. A.; De Wever, A.; Juggins, S.; Van De Vijver, B.; Jones, V. J.; Vanormelingen, P.; Roberts, D.; Flower, R.; Kilroy, C.; Souffreau, C.; Sabbe, K. The Importance of Dispersal Related and Local Factors in Shaping the Taxonomic Structure of Diatom Metacommunities. *Oikos* 2009, *118* (8), 1239–1249. https://doi.org/10.1111/j.1600-0706.2009.17575.x.

[4.125] Bottin, M.; Soininen, J.; Alard, D.; Rosebery, J. Diatom Cooccurrence Shows Less Segregation than Predicted from Niche Modeling. *PLoS One* 2016, *11* (4), e0154581. https://doi.org/10.1371/journal.pone.0154581.

[4.126] Barberán, A.; Casamayor, E. O.; Fierer, N. The Microbial Contribution to Macroecology. *Front. Microbiol.* 2014, *5*, 203.

[4.127] Oliveira, P. H. F. de; Machado, K. B.; Teresa, F. B.; Heino, J.; Nabout, J. C. Spatial Processes Determine Planktonic Diatom Metacommunity Structure of Headwater Streams. *Limnologica* 2020, *84*, 125813. https://doi.org/10.1016/j.limno.2020.125813.

[4.128] Benito, X.; Vilmi, A.; Luethje, M.; Carrevedo, M. L.; Lindholm, M.; Fritz, S. C. Spatial and Temporal Ecological Uniqueness of Andean Diatom Communities Are Correlated With Climate, Geodiversity and Long-Term Limnological Change. *Front. Ecol. Evol.* 2020, *8*, 260. https://doi.org/10.3389/fevo.2020.00260.

[4.129] Leboucher, T.; Budnick, W. R.; Passy, S. I.; Boutry, S.; Jamoneau, A.; Soininen, J.; Vyverman, W.; Tison-Rosebery, J. Diatom β-Diversity in Streams Increases with Spatial Scale and Decreases with Nutrient Enrichment across Regional to Sub-Continental Scales. *J. Biogeogr.* 2019, *46* (4), 734–744. https://doi.org/10.1111/jbi.13517.

[4.130] Nistal-García, A.; García-García, P.; García-Girón, J.; Borrego-Ramos, M.; Blanco, S.; Bécares, E. DNA Metabarcoding and Morphological Methods Show Complementary Patterns in the Metacommunity Organization of Lentic Epiphytic Diatoms. *Sci. Total Environ.* 2021, *786*, 147410. https://doi.org/10.1016/j.scitotenv.2021.147410.

[4.131] Plante, C. J.; Hill-Spanik, K.; Cook, M.; Graham, C. Environmental and Spatial Influences on Biogeography and Community Structure of Saltmarsh Benthic Diatoms. *Estuaries Coasts* 2021, *44* (1), 147–161. https://doi.org/10.1007/s12237-020-00779-0.

[4.132] Wallace, A. R. *Tropical Nature, and Other Essays*; Macmillan and Company, 1878.

[4.133] Sanders, H. L. Marine Benthic Diversity: A Comparative Study. *Am. Nat.* 1968, *102* (925), 243–282. https://doi.org/10.1086/282541.

[4.134] Kociolek, J. P.; Hamsher, S. E.; Kulikovskiy, M.; Bramburger, A. J. Are There Species Flocks in Freshwater Diatoms? A Review of Past Reports and a Look to the Future. *Hydrobiologia* 2017, *792* (1), 17–35. https://doi.org/10.1007/s10750-016-3075-1.

[4.135] Bramburger, A. J.; Hamilton, P. B.; Haffner, G. D. Effects of a Simulated Upwelling Event on the Littoral Epilithic Diatom Community of an Ancient Tropical Lake (Lake Matano, Sulawesi Island, Indonesia). *Hydrobiologia* 2014, *739* (1), 133–143. https://doi.org/10.1007/s10750-013-1598-2.

[4.136] Patrick, C. J.; Anderson, K. E.; Brown, B. L.; Hawkins, C. P.; Metcalfe, A.; Saffarinia, P.; Siqueira, T.; Swan, C. M.; Tonkin, J. D.; Yuan, L. L. The Application of Metacommunity Theory to the Management of Riverine Ecosystems. *WIREs Water* 2021, *8* (6), e1557. https://doi.org/10.1002/wat2.1557.

[4.137] Turunen, J.; Muotka, T.; Vuori, K.-M.; Karjalainen, S. M.; Rääpysjärvi, J.; Sutela, T.; Aroviita, J. Disentangling the Responses of Boreal Stream Assemblages to Low Stressor Levels of Diffuse Pollution and Altered Channel Morphology. *Sci. Total Environ.* 2016, *544*, 954–962. https://doi.org/10.1016/j.scitotenv.2015.12.031.

[4.138] Härnström, K.; Ellegaard, M.; Andersen, T. J.; Godhe, A. Hundred Years of Genetic Structure in a Sediment Revived Diatom Population. *Proc. Natl. Acad. Sci. USA* 2011, *108* (10), 4252–4257. https://doi.org/10.1073/pnas.1013528108.

[4.139] Poister, D.; Schaefer, A.; Baert, A.; Tracey, J.; Richards, K. Stimulated Rejuvenation of Dormant *Aulacoseria granulata* (Bacillariophyta) by *Gloeocystis planctonica* (Chlorophyta) in a Eutrophic River. *J. Phycol.* 2015, *51* (2), 332–342. https://doi.org/10.1111/jpy.12277.

[4.140] Benito, X.; Feitl, M.; Fritz, S. C.; Mosquera, P. V.; Schneider, T.; Hampel, H.; Quevedo, L.; Steinitz-Kannan, M. Identifying Temporal and Spatial Patterns of Diatom Community Change in the Tropical Andes over the Last ~150 Years. *J. Biogeogr.* 2019, *46* (8), 1889–1900. https://doi.org/10.1111/jbi.13561.

[4.141] Korhonen, J. J.; Soininen, J.; Hillebrand, H. A Quantitative Analysis of Temporal Turnover in Aquatic Species Assemblages across Ecosystems. *Ecology* 2010, *91* (2), 508–517. https://doi.org/10.1890/09-0392.1.

[4.142] Winegardner, A. K.; Legendre, P.; Beisner, B. E.; Gregory-Eaves, I. Diatom Diversity Patterns over the Past c. 150 Years across the Conterminous United States of America: Identifying Mechanisms behind Beta Diversity. *Glob. Ecol. Biogeogr.* 2017, *26* (11), 1303–1315. https://doi.org/10.1111/geb.12640.

[4.143] Pinseel, E.; Van de Vijver, B.; Wolfe, A. P.; Harper, M.; Antoniades, D.; Ashworth, A. C.; Ector, L.; Lewis, A. R.; Perren, B.; Hodgson, D. A.; Sabbe, K.; Verleyen, E.; Vyverman, W. Extinction of Austral Diatoms in Response to Large-Scale Climate Dynamics in Antarctica. *Sci. Adv.* 2021, *7* (38), eabh3233. https://doi.org/10.1126/sciadv.abh3233.

[4.144] Cuenca-Cambronero, M.; Courtney-Mustaphi, C. J.; Greenway, R.; Heiri, O.; Hudson, C. M.; King, L.; Lemmen, K. D.; Moosmann, M.; Muschick, M.; Ngoepe, N.; Seehausen, O.; Matthews, B. An Integrative Paleolimnological Approach for Studying Evolutionary Processes. *Trends Ecol. Evol.* 2022, *37* (6), 488–496. https://doi.org/10.1016/j.tree.2022.01.007.

[4.145] Bramburger, A. J.; Hamilton, P. B.; Hehanussa, P. E.; Haffner, G. D. Processes Regulating the Community Composition and Relative Abundance of Taxa in the Diatom Communities of the Malili Lakes, Sulawesi Island, Indonesia. *Hydrobiologia* 2008, *615* (1), 215–224. https://doi.org/10.1007/s10750-008-9562-2.

[4.146] Ageli, M. K.; Hamilton, P. B.; Bramburger, A. J.; Weidman, R. P.; Song, Z.; Russell, J.; Vogel, H.; Bijaksana, S.; Haffner, G. D. Benthic-Pelagic State Changes in the Primary Trophic Level of an Ancient Tropical Lake. *Palaeogeogr. Palaeoclimatol. Palaeoecol.* 2022, *594*, 110937. https://doi.org/10.1016/j.palaeo.2022.110937.

[4.147] Russell, J. M.; Vogel, H.; Bijaksana, S.; Melles, M.; Deino, A.; Hafidz, A.; Haffner, D.; Hasberg, A. K.; Morlock, M.; von Rintelen, T. The Late Quaternary Tectonic, Biogeochemical, and Environmental Evolution of Ferruginous Lake Towuti, Indonesia. *Palaeogeogr. Palaeoclimatol. Palaeoecol.* 2020, *556*, 109905. https://doi.org/10.1016/j.palaeo.2020.109905.

[4.148] Ossyssek, S.; Hofmann, A. M.; Geist, J.; Raeder, U. Diatom Red List Species Reveal High Conservation Value and Vulnerability of Mountain Lakes. *Diversity* 2022, *14* (5), 389. https://doi.org/10.3390/d14050389.

[4.149] Soininen, J. Are Diatom Community Assembly Processes Scale Invariant in Streams? *Freshw. Biol.* 2023, *68* (3), 502–508. https://doi.org/10.1111/fwb.14042.

[4.150] Mruzek, J. L.; Budnick, W. R.; Larson, C. A.; Luc, D. K.; Passy, S. I. Stronger Niche than Dispersal Effects on α- and β-Diversity of Stream Algae, Insects, and Fish across Latitudes in the United States. *Glob. Ecol. Biogeogr.* 2022, *31* (12), 2453–2462. https://doi.org/10.1111/geb.13595.

[4.151] He, F.; Wu, N.; Dong, X.; Tang, T.; Domisch, S.; Cai, Q.; Jähnig, S. C. Elevation, Aspect, and Local Environment Jointly Determine Diatom and Macroinvertebrate Diversity in the Cangshan Mountain, Southwest China. *Ecol. Indic.* 2020, *108*, 105618. https://doi.org/10.1016/j.ecolind.2019.105618.

[4.152] Kurthen, A. L.; He, F.; Dong, X.; Maasri, A.; Wu, N.; Cai, Q.; Jähnig, S. C. Metacommunity Structures of Macroinvertebrates and Diatoms in High Mountain Streams, Yunnan, China. *Front. Ecol. Evol.* 2020, *8*, 330.

[4.153] Williams, J. W.; Grimm, E. C.; Blois, J. L.; Charles, D. F.; Davis, E. B.; Goring, S. J.; Graham, R. W.; Smith, A. J.; Anderson, M.; Arroyo-Cabrales, J.; Ashworth, A. C.; Betancourt, J. L.; Bills, B. W.; Booth, R. K.; Buckland, P. I.; Curry, B. B.; Giesecke, T.; Jackson, S. T.; Latorre, C.; Nichols, J.; Purdum, T.; Roth, R. E.; Stryker, M.; Takahara, H. The Neotoma Paleoecology Database, a Multiproxy, International, Community-Curated Data Resource. *Quat. Res.* 2018, *89* (1), 156–177. https://doi.org/10.1017/qua.2017.105.

[4.154] Pienitz, R.; Cournoyer, L. Circumpolar Diatom Database (CDD): A New Database for Use in Paleolimnology and Limnology. *J. Paleolimnol.* 2017, *57* (2), 213–219. https://doi.org/10.1007/s10933-016-9932-0.

[4.155] Tornés, E.; Alández-Rodríguez, J.; Corrochano, A.; Nolla-Querol, P.; Trapote, M. C.; Sabater, S. Impacts of Climate Change on Stream Benthic Diatoms—a Nation-Wide Perspective of Reference Conditions. *Hydrobiologia* 2022, *849*, 1821–1837. https://doi.org/10.1007/s10750-022-04829-5.

[4.156] Juggins, S.; Kelly, M.; Allott, T.; Kelly-Quinn, M.; Monteith, D. A Water Framework Directive-Compatible Metric for Assessing Acidification in UK and Irish Rivers Using Diatoms. *Sci. Total Environ.* 2016, *568*, 671–678. https://doi.org/10.1016/j.scitotenv.2016.02.163.

[4.157] Álvarez-Blanco, I.; Blanco, S.; Cejudo-Figueiras, C.; Bécares, E. The Duero Diatom Index (DDI) for River Water Quality Assessment in NW Spain: Design and Validation. *Environ. Monit. Assess.* 2013, *185* (1), 969–981. https://doi.org/10.1007/s10661-012-2607-z.

[4.158] Syrykh, L.; Subetto, D.; Nazarova, L. Paleolimnological Studies on the East European Plain and Nearby Regions: The PaleoLake Database. *J. Paleolimnol.* 2021, *65* (3), 369–375. https://doi.org/10.1007/s10933-020-00172-8.

[4.159] Michener, W. K.; Jones, M. B. Ecoinformatics: Supporting Ecology as a Data-Intensive Science. *Trends Ecol. Evol.* 2012, *27* (2), 85–93. https://doi.org/10.1016/j.tree.2011.11.016.

[4.160] Feio, M. J.; Serra, S. R. Q.; Mortágua, A.; Bouchez, A.; Rimet, F.; Vasselon, V.; Almeida, S. F. P. A Taxonomy-Free Approach Based on Machine Learning to Assess the Quality of Rivers with Diatoms. *Sci. Total Environ.* 2020, *722*, 137900. https://doi.org/10.1016/j.scitotenv.2020.137900.

[4.161] Rimet, F.; Gusev, E.; Kahlert, M.; Kelly, M. G.; Kulikovskiy, M.; Maltsev, Y.; Mann, D. G.; Pfannkuchen, M.; Trobajo, R.; Vasselon, V.; Zimmermann, J.; Bouchez, A. Diat.Barcode, an Open-Access Curated Barcode Library for Diatoms. *Sci. Rep.* 2019, *9* (1), 15116. https://doi.org/10.1038/s41598-019-51500-6.

[4.162] Rimet, F.; Canino, A.; Chonova, T.; Guéguen, J.; Bouchez, A. Environmental Filtering and Mass Effect Are Two Important Processes Driving Lake Benthic Diatoms: Results of a DNA Metabarcoding Study in a Large Lake. *Mol. Ecol.* 2023, *32* (1), 124–137. https://doi.org/10.1111/mec.16737.

[4.163] Kahlert, M.; Kelly, M. G.; Mann, D. G.; Rimet, F.; Sato, S.; Bouchez, A.; Keck, F. Connecting the Morphological and Molecular Species Concepts to Facilitate Species Identification within the Genus *Fragilaria* (Bacillariophyta). *J. Phycol.* 2019, *55* (4), 948–970. https://doi.org/10.1111/jpy.12886.

[4.164] Jyrkänkallio-Mikkola, J.; Siljander, M.; Heikinheimo, V.; Pellikka, P.; Soininen, J. Tropical Stream Diatom Communities – The Importance of Headwater Streams for Regional Diversity. *Ecol. Indic.* 2018, *95*, 183–193. https://doi.org/10.1016/j.ecolind.2018.07.030.

[4.165] Teittinen, A.; Kallajoki, L.; Meier, S.; Stigzelius, T.; Soininen, J. The Roles of Elevation and Local Environmental Factors as Drivers of Diatom Diversity in Subarctic Streams. *Freshw. Biol.* 2016, *61*, 1509–1521. https://doi.org/10.1111/fwb.12791.

[4.166] Passy, S. I.; Larson, C. A. Succession in Stream Biofilms Is an Environmentally Driven Gradient of Stress Tolerance. *Microb. Ecol.* 2011, *62* (2), 414–424. https://doi.org/10.1007/s00248-011-9879-7.

[4.167] Vilmi, A.; Karjalainen, S. M.; Hellsten, S.; Heino, J. Bioassessment in a Metacommunity Context: Are Diatom Communities Structured Solely by Species Sorting? *Ecol. Indic.* 2016, *62*, 86–94. https://doi.org/10.1016/j.ecolind.2015.11.043.

[4.168] Dunck, B.; Rodrigues, L.; Lima-Fernandes, E.; Cássio, F.; Pascoal, C.; Cottenie, K. Priority Effects of Stream Eutrophication and Assembly History on Beta Diversity across Aquatic Consumers, Decomposers and Producers. *Sci. Total Environ.* 2021, *797*, 149106. https://doi.org/10.1016/j.scitotenv.2021.149106.

[4.169] Rimet, F.; Bouchez, A. Life-Forms, Cell-Sizes and Ecological Guilds of Diatoms in European Rivers. *Knowl. Manag. Aquat. Ecosyst.* 2012, 406, https://doi.org/10.1051/kmae/2012018.

[4.170] Falasco, E.; Bona, F.; Risso, A. M.; Piano, E. Hydrological Intermittency Drives Diversity Decline and Functional Homogenization in Benthic Diatom Communities. *Sci. Total Environ.* 2021, *762*, 143090. https://doi.org/10.1016/j.scitotenv.2020.143090.

[4.171] Soininen, J.; Jamoneau, A.; Rosebery, J.; Leboucher, T.; Wang, J.; Kokociński, M.; Passy, S. I. Stream Diatoms Exhibit Weak Niche Conservation along Global Environmental and Climatic Gradients. *Ecography* 2019, *42* (2), 346–353. https://doi.org/10.1111/ecog.03828.

5

Recent Insights into Diatom Distributions and the Contributions of Molecular Approaches

Jana Kulichová[1]* and Rosa Trobajo[2]

[1]*Department of Botany, Charles University, Faculty of Science, Prague, Czech Republic*
[2]*Marine and Continental Waters Program, Institute of Agrifood Research and Technology (IRTA), La Ràpita, Catalonia, Spain*

Abstract

Biogeography is a foundational discipline of modern biology with roots in the 18th and 19th centuries. However, both then and for most of the 20th century, biogeography focused on metazoans and vascular plants. Microorganisms, including diatoms, were relatively neglected, largely because of methodological challenges and disagreement about the basic units being mapped, since species concepts developed for larger organisms were difficult to apply. Consequently, views on diatom biogeography have shifted about whether species are predominantly cosmopolitan, exhibit disjunct patterns related to paleogeographic (e.g., tectonic) changes, or are endemic to a restricted region (lake, island, climate zone, sea). In parallel, there has been discussion of the roles of dispersal and ecological specialization. This chapter reviews factors affecting diatom distributions, both marine and freshwater, focusing on proximate factors rather than longer-term historical influences. We pay particular attention to the impact of the molecular revolution. We note that DNA metabarcoding can largely eliminate inconsistencies in identification, allow many more samples to be processed, and generate vast numbers of DNA reads. The way is opened to study the rare biosphere and cryptic species,

**Corresponding author*: vesela6@natur.cuni.cz
Jana Kulichová: https://orcid.org/0000-0001-8035-4770, https://botany.natur.cuni.cz/janakulichova.html
Rosa Trobajo: https://orcid.org/0000-0001-9498-3797, https://www.irta.cat/en/personal/rosa-trobajo/

Nora I. Maidana, Magdalena Licursi and Eduardo Morales (eds.) Diatom Ecology: From Molecules to Metacommunities, (123–162) © 2024 Scrivener Publishing LLC

and to use large matched metabarcoding and environmental datasets to explore biodiversity in the context of evolutionary and ecological processes.

***Keywords*:** Biogeography, ecology, metabarcoding, metacommunity

5.1 Introduction

In the last 300 years it has increasingly been realized that humans share this planet with a vast number of different types of organisms. The Swedish biologist Carl Linnaeus in the 1700s provided a new nomenclatural scheme for organizing information about this diversity and many expeditions took place, reporting new species and the communities in which they occurred (from the expeditions of James Cook, Carl Peter Thunberg, Alexander von Humboldt, and many others: e.g., [5.22] [5.30] [5.56]). Obvious differences and discontinuities in the distributions of species and in their abundances prompted "why" and "how" questions such as the following (quoted in [5.73]) from *The Origin of Species* of 1859, in which Darwin asked: "Who can explain why one species ranges widely and is very numerous, and why another allied species has a narrow range and is rare? Yet these relations are of the highest importance, for they determine the present welfare and, as I believe, the future success and modification of every inhabitant of this world.". In the present context (of global and climate change), this importance has only increased.

To attempt to understand patterns of occurrence and rarity (i.e., the large- or small-scale biogeography of organisms) is at the same time to try to understand community assembly, which has been extensively studied in macroorganisms (vascular plants and metazoans), but more recently and increasingly in microorganisms too, especially bacteria (e.g., [5.57] [5.58] [5.86]).Turning to diatoms, biogeographical patterns in the distributions of some species were noted when the study of diatoms was still in its infancy. For example, in the 1850s, before publication of the *Origin*, Ehrenberg noted (from examination of expedition and other samples sent to him) that a number of diatom species now placed in the genus *Tetracyclus* seemed to be restricted to a circum-Pacific distribution (see [5.130] [5.132]). Since then, many works touching on or reviewing diatom biogeography have been published (e.g., [5.48] [5.61] [5.63]), often with arguments based on collections made in remote or understudied regions, such as the Antarctic continent and subantarctic islands [5.126]; and there have been shifting views about the species concept and the relative importance of endemism and cosmopolitanism (see [5.122]), often presenting these as incompatible opposites. For example, [5.26] [5.27], following Baas Becking (who wrote "Everything is everywhere, but, the environment selects" [5.20]), it was argued that diatoms, like other protists, are unlikely to be subject to dispersal limitation. The argument—sometimes referred to as the ubiquity hypothesis—is that, due to the huge numbers of cells in their populations, which drive ubiquitous dispersal, microbial species can occur everywhere provided their niche requirements are met.

Before the development of molecular survey techniques (metabarcoding: Section 5.3), studies of diatom biogeography depended almost wholly on microscopical studies, supplemented in some cases by spot-checks of identity using Sanger sequencing. In diatoms, as in most protists, determining the geographical distribution of a diatom species based on microscopy techniques is challenging because, first, the task requires geographically extensive and locally intensive sampling conducted over a long enough period to ensure that seasonal taxa are recorded. Such exhaustive sampling of diatoms is practically impossible (see [5.104]), especially in view of a second challenge, which is that to define the distribution of something we need to be able to identify it with certainty. Reliable identification of species from frustule morphology requires a high level of skill, a thorough knowledge of the relevant literature, and is time-consuming and expensive. Furthermore, comprehensive diatom floras do not exist for some areas, e.g., many parts of Africa, central Asia, South America, Oceania. A third challenge is that in several cases the taxonomy is still in a state of flux, with no agreement about species boundaries in several "species complexes," which are almost inevitably thought to have broad ecological requirements and distributions (referred to as aggregates in Table 5.1).

One approach that has been advocated in other groups of protists where sampling, detection, and identification are challenging is to focus on "flagship" species [5.28], i.e., species that are so large or distinctive that "we cannot have missed it so it cannot be here." Convincing examples among diatoms are the genera *Eunophora* and *Veigaludwigia*, which show not only a contemporary restriction to the Southern Hemisphere but also a Miocene presence in Antarctica before later global cooling [5.80]. The flagship approach is even stronger if backed by appropriate environmental analyses to identify the biotopes in which a species might be expected to be found, based on its known occurrences. But its plausibility is weakened somewhat by reports like those by [5.9] and [5.36], where "flagship" ciliates were discovered on continents away from their original discovery.

Progress in understanding distributions requires, among other things, parallel information on the environmental factors that might limit the distributions of individual diatom species. The role of dispersal can be

Table 5.1 Summary of recent molecular-based studies on factors affecting distributions of diatom species and lineages/haplotypes of species complexes (abbreviated as aggregates - agg.).

Analyzed taxa	Geogr. region	Habitat type	Molecular marker	Data analysis	Important factor(s) and/or patterns of distribution	References
Achnanthidium minutissimum agg.	[1]moreR	[2]FWsb	[3]SSU, LSU, rbcL	[4]Sseq	several lineages originating from a single region, one widely distributed lineage (throughout the northern hemisphere)	[5.81]
A. minutissimum agg.	twoR	FWsb	rbcL	eDNA	three of the five putative species were identified in both France and Georgia and showed higher levels of haplotype diversity (almost sixty haplotypes) than the two species with restricted distributions (one haplotype confined to Georgia and three to France)	[5.90]
A. minutissimum, *Fistulifera saprophila*, *Nitzschia inconspicua*, and *N. soratensis*	oneR	FWrb	rbcL	eDNA	several ecological groupings of haplotypes within morphospecies // haplotypes with scattered distribution (Europe and i) North America, ii) South Korea, iii) volcanic Islands) or with restricted distribution (only in one catchment, only from tropics, a saline habitat, or volcanic islands)	[5.76]

(*Continued*)

Table 5.1 Summary of recent molecular-based studies on factors affecting distributions of diatom species and lineages/haplotypes of species complexes (abbreviated as aggregates - agg.). (*Continued*)

Analyzed taxa	Geogr. region	Habitat type	Molecular marker	Data analysis	Important factor(s) and/or patterns of distribution	References
Aneumastus spp.	oneR	FWsb	SSU, tufA, rbcL	Sseq	endemic lineages in ancient lake, two colonization events leading to intralacustrine speciation	[5.109]
Asterionella formosa	twoR	FWp	GENw	GENt	a complex population structure (evidence for both restricted and moderate to high gene flow), genetic differentiation within lakes much stronger than among lakes	[5.118]
A. formosa and *Aulacoseira* spp.	oneR	FWp	SSU, LSU, rbcL	Sseq	most taxa sampled in temperate southern Chile (*Asterionella formosa, Aulacoseira ambigua, A. granulata, A. pusilla*) found to be similar or identical to Northern Hemisphere strains	[5.117]

(*Continued*)

Table 5.1 Summary of recent molecular-based studies on factors affecting distributions of diatom species and lineages/haplotypes of species complexes (abbreviated as aggregates - agg.). (*Continued*)

Analyzed taxa	**Geogr. region**	**Habitat type**	**Molecular marker**	**Data analysis**	**Important factor(s) and/or patterns of distribution**	**References**
Chaetoceros curvisetus agg.	moreR	MAp	SSU	eDNA	warm temperate to tropical distribution, strictly tropical lineages or climate-independent distribution of lineages (from Arctic to tropics) // widely distributed or scattered distribution (e.g., Florida and Singapore)	[5.18]
C. socialis agg.	moreR	MAp	LSU, MTBOL	ECOPH	climatic but not seasonal differences between lineages (differences between southern and northern strains in synthesized metabolites and growth at low temperatures)	[5.21] [5.38]

(*Continued*)

Table 5.1 Summary of recent molecular-based studies on factors affecting distributions of diatom species and lineages/haplotypes of species complexes (abbreviated as aggregates - agg.). (*Continued*)

Analyzed taxa	Geogr. region	Habitat type	Molecular marker	Data analysis	Important factor(s) and/or patterns of distribution	References
Chaetoceros spp.	moreR	Map	mGEN	eDNA	each of the MAGs (metagenome-assembled genomes) in a relatively small number of samples with an uneven distribution (potential habitat specialists), one MAG distributed in samples across the Arctic Ocean (a potential pan-Arctic species) // the genetic differentiation between populations seemed to correlate with abiotic factors (only a minor correlation with geographic distance)	[5.71]
Cylindrotheca closterium agg.	moreR	MAp, MAb	LSU, ITS, COI, psbA, rbcL	Sseq, ECOPH	the difference in the optimum temperature and thermal niche width related to the climatic conditions at the sites from which the strains were isolated // no obvious geographic pattern (except Antarctic clade)	[5.110]

(*Continued*)

Table 5.1 Summary of recent molecular-based studies on factors affecting distributions of diatom species and lineages/haplotypes of species complexes (abbreviated as aggregates - agg.). (*Continued*)

Analyzed taxa	Geogr. region	Habitat type	Molecular marker	Data analysis	Important factor(s) and/or patterns of distribution	References
C. closterium, Navicula phyllepta, and *Seminavis robusta*	oneR	MAb	BACT	eDNA, ECOPH, COEX	decreased or increased growth of strains dependent on presence or absence of bacteria and/or diatoms	[5.49]
Didymosphenia geminata	oneR	FWrb	SSU, ITS, CHLreg	Sseq	recent dispersal of lineages to New Zealand // widely distributed lineages (New Zealand, Canada, USA) and genetically distinct lineage from Iran	[5.45] [5.47]
D. geminata	oneR	FWrb	rbcL	eDNA, qPCR	early stage of invasion of non-native species to mid-Atlantic streams of United States // scattered distribution of lineages (USA and i) New Zealand, Canada, ii) Italy)	[5.43]

(*Continued*)

Table 5.1 Summary of recent molecular-based studies on factors affecting distributions of diatom species and lineages/haplotypes of species complexes (abbreviated as aggregates - agg.). (*Continued*)

Analyzed taxa	Geogr. region	Habitat type	Molecular marker	Data analysis	Important factor(s) and/or patterns of distribution	References
Ditylum brightwellii	oneR	MAp	ITS, MSAT	Sseq, GENt	physiological variation between metapopulations (each bloom at a different time of year) // two of thc three populations have different light preferences // no genetic signature between two of the three populations occupying adjacent locations despite daily mixing	[5.95] [5.96]
Eunotia bilunaris 'robust' and *Sellaphora capitata*	oneR	FWsb	MSAT	GENt	isolation-by-distance pattern in population structure on relatively small spatial scales (little or no differentiation within ponds and interconnected ponds, highly differentiated isolated ponds)	[5.120]

(*Continued*)

Table 5.1 Summary of recent molecular-based studies on factors affecting distributions of diatom species and lineages/haplotypes of species complexes (abbreviated as aggregates - agg.). (*Continued*)

Analyzed taxa	Geogr. region	Habitat type	Molecular marker	Data analysis	Important factor(s) and/or patterns of distribution	References
Fragilaria spp. and *Ulnaria* spp.	oneR	FWsb, FWrb	rbcL	eDNA	difference in the realized niches of haplotypes (habitat, nutrients, alkalinity) // widely distributed haplotypes (over the region; Japan and Europe), some with restricted distribution (within the region)	[5.42]
Fragilariopsis kerguelensis	oneR	MAp	SSU, LSU, ITS, oGEN	Sseq, ddRAD	one genotype present across sites, two genotypes (north and south) separated by longitudinal water circulation // differences in ecophysiology of the three genotypes (south: better performance at 0°C, north: at 6°C, and reduced performance of the third genotype in all treatments)	[5.83]
Frustulia spp.	moreR	FWsb	SSU, LSU, COI, rbcL	Sseq	widely distributed clades (Europe, USA and New Zealand), lineages with unique phylogenetic positions and restricted distribution (from New Zealand, USA or Europe only)	[5.8] [5.116]

(*Continued*)

Table 5.1 Summary of recent molecular-based studies on factors affecting distributions of diatom species and lineages/haplotypes of species complexes (abbreviated as aggregates - agg.). (*Continued*)

Analyzed taxa	**Geogr. region**	**Habitat type**	**Molecular marker**	**Data analysis**	**Important factor(s) and/or patterns of distribution**	**References**
Frustulia spp.	oneR	FWsb	LSU, rbcL	Sseq	widely distributed closely related lineages with wide niche breadth and lineages with narrow realized niches (pH, habitat type)	[5.115] [5.128]
Gomphonema parvulum agg.	moreR	FWrb	SSU, LSU, ITS, COI, rbcL	Sseq, eDNA	geographical origin (clades from Korea, Faroe Islands or Europe only), climatic conditions (clade from Mexico and Mayotte Island), trophy (two clades from clean waters and two from anthropogenically polluted waters)	[5.1] [5.46]
Hantzschia amphioxys	twoR	TER, Moist	SSU, tufA, rbcL	Sseq	scattered distribution of lineages (from Antarctica and Europe, from Canada and Mongolia, from Europe and Mongolia)	[5.59] [5.106]
H. amphioxys agg. and *P. borealis* agg.	moreR	TER, FWsb	LSU, rbcL	Sseq, ECOPH	Antarctic lineages grow well at lower temperatures than most lineages from temperate regions (lower optimum and lethal temperatures)	[5.106]

(*Continued*)

Table 5.1 Summary of recent molecular-based studies on factors affecting distributions of diatom species and lineages/haplotypes of species complexes (abbreviated as aggregates - agg.). (*Continued*)

Analyzed taxa	Geogr. region	Habitat type	Molecular marker	Data analysis	Important factor(s) and/or patterns of distribution	References
Leptocylindraceae	oneR	MAp	SSU	eDNA	rare occurrence of all species (except one); seasonal variability in species detection // species already known from the Mediterranean (Gulf of Naples) are distributed across European seas; scattered distribution (Canada, Europe)	[5.70]
Navicula cryptocephala agg.	moreR	FWsb	LSU, ITS	Sseq	widely distributed lineages (Scotland, continental Europe, Australia)	[5.84]
N. phyllepta agg.	oneR	ESTb	SSU, ITS, rbcL	qPCR, ECOPH	the two genotypes differ in tolerance to low salinity, seasonal changes of abiotic factors affect the abundances of both genotypes	[5.15] [5.119]
Nitzschia anatoliensis	oneR	FWsb	SSU, LSU, rbcL, CHLreg	Sseq	discovered in a high alkalinity lake in Turkey (isolate closely related to strains from high pH Siberian lakes)	[5.105]

(*Continued*)

Table 5.1 Summary of recent molecular-based studies on factors affecting distributions of diatom species and lineages/haplotypes of species complexes (abbreviated as aggregates - agg.). (*Continued*)

Analyzed taxa	Geogr. region	Habitat type	Molecular marker	Data analysis	Important factor(s) and/or patterns of distribution	References
N. inconspicua agg. and *N. soratensis* agg.	oneR	FWrb, ESTb	LSU, rbcL	Sseq, ECOPH	differences in growth of genotypes on salinity gradient (tolerance to freshwater conditions); preferences of species for conductivity and/or alkalinity (catchment geology, i.e., stream type)	[5.44] [5.92]
N. palea agg.	moreR	FWsb, FWrb	LSU, COI, rbcL	Sseq	some widely distributed genotypes; positive correlation between genetic and geographical distances (relatively high local diversity, substantial increase in genetic distances from local to regional scale)	[5.91]
Pinnularia borealis agg.	moreR	TER, FWsb	SSU, LSU, COI, psbA, psbC, rbcL	Sseq, eDNA	few lineages inhabit a different habitat type (submerged in water) // relatively restricted distribution of most species (e.g., Tasmania, Sub-Antarctica) and bipolar distribution of two species; the importance of allopatric speciation	[5.78] [5.79]

(*Continued*)

Table 5.1 Summary of recent molecular-based studies on factors affecting distributions of diatom species and lineages/haplotypes of species complexes (abbreviated as aggregates - agg.). (*Continued*)

Analyzed taxa	Geogr. region	Habitat type	Molecular marker	Data analysis	Important factor(s) and/or patterns of distribution	References
P. gibba-subgibba agg.	moreR	FWsb	SSU, LSU, COI, psbA, rbcL	Sseq	closely related lineages found on different volcanic islands (locations are younger than estimated age of lineages) // the delimited species seem mostly restricted to a single location; the only three species were found in multiple regions (none of which was shared between hemispheres)	[5.50] [5.51]
Pseudo-nitzschia arctica	oneR	MAp	LSU, ITS, rbcL	Sseq	lineage restricted to arctic (Arctic Sea, Northeast Atlantic)	[5.74]
P. pungens agg.	moreR	MAp	ITS	Sseq	one clade from warm waters (Vietnam, China, Mexico) and two clades from temperate areas // one globally distributed clade in temperate coastal seas and one clade with restricted distribution (northeast Pacific)	[5.12]

(*Continued*)

Table 5.1 Summary of recent molecular-based studies on factors affecting distributions of diatom species and lineages/haplotypes of species complexes (abbreviated as aggregates - agg.). (*Continued*)

Analyzed taxa	Geogr. region	Habitat type	Molecular marker	Data analysis	Important factor(s) and/or patterns of distribution	References
P. pungens	oneR	MAp	MSAT	GENt	weak genetic differentiation of isolates over the spatial and temporal scales (North Sea) // widespread population (similar genetic variability of strains between the North Sea and the Northeast Atlantic)	[5.24]
Pseudo-nitzschia spp.	oneR	MAp	LSU	eDNA	variable temporal patterns of species occurrence (Mediterranean Sea); some previously identified species from a given location not found // detection of genotypes found in distant regions (South Africa, East Atlantic coast, Australia)	[5.65] [5.93]
Sellaphora capitata	twoR	FWsb	ITS, COI, MSAT	Sseq, GENt	clustering of populations by country origin (Scotland closer to Belgium but not to England) // some lineages composed of European and Australian isolates	[5.23]

(*Continued*)

Table 5.1 Summary of recent molecular-based studies on factors affecting distributions of diatom species and lineages/haplotypes of species complexes (abbreviated as aggregates - agg.). (*Continued*)

Analyzed taxa	Geogr. region	Habitat type	Molecular marker	Data analysis	Important factor(s) and/or patterns of distribution	References
Seminavis robusta	twoR	MAb	LSU, ITS, rbcL	Sseq	three clades co-occurring within the same biofilm community	[5.16]
Skeletonema marinoi	oneR	MAp	BACT	ECOPH, COEX	bacteria influence the growth of *Skeletonema* strains on a gradient of abiotic factors (iron, temperature, wavelength, dimethylsulfoniopropionate)	[5.40]
S. marinoi	twoR	MAp	ITS, MSAT	Sseq, GENt	local populations have remained genetically distinct for over a century (despite mixing of water masses between the open sea and the fjord)	[5.35]
S. marinoi	oneR	MAp	MSAT	COEX, GENt	intraspecific interactions affect growth on a salinity gradient	[5.103]
S. marinoi	oneR	MAp	MSAT	GENt, ECOPH, COEX	competitive advantage of local population due to adaptation to local condition (higher fitness in native than in foreign waters), priority effect (growth of the first inoculated strain was higher than the strain added later)	[5.98] [5.99] [5.102]

(*Continued*)

Table 5.1 Summary of recent molecular-based studies on factors affecting distributions of diatom species and lineages/haplotypes of species complexes (abbreviated as aggregates - agg.). (*Continued*)

Analyzed taxa	Geogr. region	Habitat type	Molecular marker	Data analysis	Important factor(s) and/or patterns of distribution	References
Skeletonema spp.	oneR	MAp	LSU	eDNA	community structure and species richness influenced by seasonal changes in water temperature (bay in the Northeast Atlantic)	[5.10]
Skeletonema spp.	moreR	MAp	LSU	Sseq	species preferences for water temperature (e.g., temperate waters at lower latitudes in winter, cold currents, upwelling sites) // mostly widespread species	[5.52]
Skeletonema spp.	oneR	MAp	SSU, LSU, COI	Sseq	endemic cold season subclade (Mediterranean Sea in winter) in a sister position to worldwide all-season subclade	[5.134]
Skeletonema spp. and *Thalassiosira rotula*	oneR	MAp	TRSCR	MTTR	seasonal differences in nitrogen and phosphorus metabolism enable coexistence of marine planktonic diatoms	[5.2]

(*Continued*)

Table 5.1 Summary of recent molecular-based studies on factors affecting distributions of diatom species and lineages/haplotypes of species complexes (abbreviated as aggregates - agg.). (*Continued*)

Analyzed taxa	Geogr. region	Habitat type	Molecular marker	Data analysis	Important factor(s) and/or patterns of distribution	References
T. pseudonana	oneR	MAp	a single strain	ECOPH	thermal adaptation of clones *in vitro* (change in growth rate and shape of thermal response)	[5.72]
Thalassiosira spp.	oneR	MAp	SSU	eDNA	community structure and species richness influenced by seasonal changes (bay in the Northeast Atlantic), differences in thermal niches of species (cold adapted, warm adapted, intermediate and eurythermal species)	[5.94]
aeolian diatoms	oneR	Air	SSU	eDNA	wind dispersal in arid polar desert (imprint of the locality, enhanced dispersal in winter)	[5.101]
marine planktonic diatoms	moreR	MAp	SSU	eDNA	cosmopolitan ribotypes, effect of semi-enclosed basins, water currents, upwelling; genera diversify locally to varying environmental conditions (ecotype differentiation)	[5.60]

(*Continued*)

Table 5.1 Summary of recent molecular-based studies on factors affecting distributions of diatom species and lineages/haplotypes of species complexes (abbreviated as aggregates - agg.). (*Continued*)

Analyzed taxa	Geogr. region	Habitat type	Molecular marker	Data analysis	Important factor(s) and/or patterns of distribution	References
marine planktonic diatoms	oneR	MAp	SSU	eDNA	regional environmental conditions (i.e., site-specific communities), large-scale geographic pattern not obvious	[5.82]
marine planktonic diatoms	moreR	MAp	SSU, BACT	eDNA	distinct patterns and relationships to environmental variables between sampling sites (Australian coastal waters from temperate to tropical latitudes) // co-occurrence between specific diatoms and bacteria (on a spatiotemporal scale)	[5.54]

[1]**Area under study**: (oneR) - one region (e.g., Europe, Indian Ocean, river basin, European coast), (twoR) - two regions, (moreR) - more than two regions. [2]**Habitat type**: (MAp) - marine plankton (including estuaries, coastal waters), (MAb) - marine benthos, (FWsb) - freshwater standing waters, benthic, (FWrb) - running waters, benthic, (TER) - soils, (ESTb) - benthos in estuaries, (FWp) - freshwater, plankton, (Air) - airborne material, (Moist) - moist biofilms. [3]**Sequenced molecular marker**: (SSU) - the small subunit rRNA gene, nucleus, (LSU) - the large subunit rRNA gene, nucleus, (ITS) - ribosomal internal transcribed spacer regions, nucleus, (COI) - cytochrome oxidase subunit 1 gene, mitochondria, (psbA) - gene encoding the D1 protein of photosystem II, chloroplast, (tufA) - protein synthesis elongation factor Tu gene, chloroplast, (psbC) - CP43 proteins of the photosystem II gene, chloroplast, (rbcL) - the large subunit of RuBisCO gene, chloroplast, (MSAT) - microsatellites, (TRSCR) - transcriptome, (BACT) - bacterial DNA, (MTBOL) - metabolomics, (CHLreg) – markers for more than three chloroplast regions, (oGEN) - organelle genome (chloroplast, mitochondria), (mGEN) - metagenome, (GENw) - genome wide. [4]**Data analysis**: (Sseq) - Sanger sequencing of isolates, (eDNA) - metabarcoding of environmental DNA (including HTS), (GENt) - genotyping, (qPCR) - real-time quantitative PCR of environmental DNA, (ECOPH) - ecophysiological experiment with cultures, (COEX) - coexistence of cultures, (MTTR) - metatranscriptome of environmental DNA, (ddRAD) - double digest restriction site-associated DNA.

evaluated only when the influence of environmental factors is taken into account. Of course, it is important to have in mind that factors other than the most commonly measured ones (e.g., pH, temperature, conductivity) could be the most influential, such as selective grazers and or parasitoids, and hydrological fluctuations. It is also important to remember that spatial variation in community composition may often correlate with distance simply because sites that are close together tend to have more similar environmental conditions than sites that are farther apart [5.104]. For a more in-depth discussion see Chapter 4, "Deterministic and Stochastic Effects on Freshwater Diatom Biodiversity and Community Composition."

In the next sections we outline the factors that shape diatom distributions and communities, and also how the advent of high-throughput DNA sequencing is changing the way in which diatom species (or lineages/haplotypes of species complexes) distributions can be investigated. We focus on the interplay between ecology and biogeography and pay less attention to historical factors, which are covered by other reviews (e.g., [5.131] [5.133]). Hence, wider issues in diatom biogeography, such as vicariance, narrow endemism, or the occurrence and interpretation of species flocks, are not addressed.

5.2 A Metacommunity Framework

To understand the mechanisms that have played an important role in the assembly of diatom communities around the world, it is useful to adopt a metacommunity concept [5.55] [5.127]. The premise of this concept is connectivity among communities through dispersal, which is at least theoretically fulfilled in the case of diatoms because some diatom lineages/haplotypes seem to be globally distributed in marine [5.12] [5.18] [5.60] [5.134], freshwater [5.91], and terrestrial ecosystems [5.79]. Despite the possibility of ubiquitous dispersal, however, lineages/haplotypes within species complexes show high variability in distribution patterns. Some lineages/haplotypes are widely or globally distributed, while others have a scattered or restricted distribution [5.12] [5.42] [5.81] [5.91]. Some lineages/haplotypes are confined to a relatively narrow range of environmental conditions (e.g., occurrence in a particular season and/or climatic zone) whereas others appear to be ecological generalists [5.1] [5.18] [5.52] [5.76] [5.83] [5.134]. Genetic differentiation can occur between local populations that are interconnected by movements of water masses [5.35] [5.95] [5.96], between spatially isolated populations of freshwater sites [5.23] [5.120], or different lineages of species complexes can live in sympatry [5.15] [5.16]

[5.79] [5.121] [5.128]. Such diverse patterns suggest that multiple mechanisms influence community assembly differently depending on spatial and temporal scales. The mechanisms include: *selection* – responses to local abiotic conditions (independent of competition) and biotic interactions; *dispersal* – passive movements from one location to another; *ecological drift* – random changes in species relative abundances due to stochastic processes such as birth and death; and *mutations* – evolutionary diversification (both via mutations underlying the adaptive evolution of traits, including speciation, and via genetic drift) [5.34] [5.73] [5.113] [5.125]. In Table 5.1 we summarize recent studies that evaluate the contributions of abiotic conditions, biotic interactions, and dispersal in determining the occurrence of selected lineages/haplotypes of species complexes, including populations of species. We use "haplotype" in the same sense as De Luca *et al.* [5.18], for a sequence variant in a metabarcoding marker gene; "lineages" refer to species or populations separated by sequencing of isolates. We have restricted this compilation to studies that use molecular characterizations of the diatoms present and focus on the role of proximate factors in determining their distribution patterns. Molecular characterization helps minimize the noise introduced by differing taxon concepts [5.3]. Proximate factors are the abiotic conditions, biotic interactions, and dispersal that directly influence diatom community composition over a short time scale (years), as opposed to the broader factors, such as vicariance, that operate over millennia.

5.2.1 Abiotic Factors

To determine the relative importance of environmental filters in the distribution of diatoms, we need to know the ecological niche requirements of individual diatom species and, ideally, those species would be genetically delimited units (lineages). The knowledge about the autecology of diatom morphospecies that has resulted from empirical evidence acquired for over a century, can be transferred to lineages, but only to a certain extent. Phylogenetic units within species complexes usually originate from habitats typical for a given morphospecies (e.g., epilithon of running waters, epipelon of standing waters, temperate coastal plankton, semi-terrestrial habitats, etc.). However, finer resolution of the ecological requirements of lineages needs to be assessed de novo at multiple spatiotemporal scales because morphological similarity is not always associated with genetic and phylogenetic similarity. Morphological features characteristic of a morphospecies can evolve multiple times, making the morphospecies polyphyletic or paraphyletic. For example, phylogenetically not closely related lineages resembling the same morphospecies [5.8] [5.25] [5.64] [5.68] [5.74], or morphologically distinct lineage(s) occurring within an otherwise morphologically homogeneous set of closely related lineages [5.13] [5.77] [5.84], disrupt monophyly. Such morphospecies can no longer be considered as evolutionarily meaningful units. It is not known how often convergence in morphology of lineages meets divergence of niches, but its occurrence may lead to misconception of wide niche breadth of diatom morphospecies with respect to temperature in marine plankton [5.21] [5.74], salinity in brackish habitats [5.15] [5.92], and trophy in freshwaters [5.1] [5.76].

Inconsistent success in recognizing ecological signal in phylogenetic data may be related to variability in diversification rates of ecological traits. At different levels of resolution within species complexes (i.e., at the level of species or populations), ecological specialization has been found in some lineages/haplotypes, while others may occur in sympatry with unknown ecological differentiation or be euryvalent. Specific trait differences have been detected in lineages and/or haplotypes with respect to temperature in different climatic zones [5.12] [5.21] [5.38] [5.107] [5.110], habitat type [5.4] [5.77], spatial heterogeneity of local conditions in freshwaters [5.15] [5.76] [5.92], and seasonal changes in marine plankton [5.10] [5.65] [5.93] [5.94] [5.134].

Estimation of ecological niches within species complexes is not always straightforward. Experimental cultivation of clones representing different lineages along environmental gradients may show overlapping responses, which in some cases may erroneously indicate functional neutrality because realized niches may be related to survival under extreme conditions of the gradient. For example, lineages of two brackish species complexes (*Navicula phyllepta* agg., *Nitzschia inconspicua* agg.) grew over a wide range of salinities, but only some of them survived freshwater conditions [5.92] [5.119]. And along a temperature gradient, the responses of lineages of the marine planktonic species complex (*Chaetoceros socialis* agg.) did not overlap in the extreme experimental treatments [5.21]. Our ability to estimate realized niches improves as the number of sequenced isolates within and between sampling sites increases, because detection of a lineage/haplotype at a sampling site does not necessarily imply that the conditions for its survival and growth are met [5.6]; it is instead the presence and abundances of a lineage/haplotype at multiple sampling sites that indicate its ecological requirements. Quantitative assessments to infer realized niches can be done by sequencing isolates from well-studied genera (e.g., *Skeletonema*, *Pseudo-nitzschia*, *Sellaphora*, Table 5.1), morphological identification of species supported by molecular data in field samples [5.14]

[5.85] [5.115], quantitative real-time PCR [5.15] [5.43], and sequencing of environmental DNA [5.18] [5.42] [5.76] (see also Section 5.3).

5.2.2 Biotic Interactions

Interspecific interactions as possible factors shaping diatom communities have mostly been considered theoretically, especially when abiotic factors fail to explain distribution patterns. Empirical evidence for biotic interactions that affect diatom distributions is scarce. What little is known has shown that long-term persistence of species and populations at locations may be enhanced by escape from competitive exclusion through differences in seasonal preferences (e.g., [5.2] [5.93]) and by competitive advantage through local adaptation of populations [5.33] [5.98] and/or site-specific biotic interactions [5.49] [5.98].

5.2.3 Dispersal

Given the changing perception of diatom diversity, from morphospecies to evolutionary units, there is a need to reconsider not only the role of selection, but also the role of dispersal in community assembly. Consistent with observations derived from morphology [5.122], individuals belonging to the same lineage/haplotype of a species complex may occur either at spatially distant locations or in a relatively small geographic area [5.12] [5.76] [5.83] [5.91]. Passive movements of diatom cells can occur relatively easily within the aquatic realm [5.24] [5.120] or via vectors that do not expose cells to desiccation and/or freezing [5.108]. However, transport of cells through conditions beyond their tolerance is generally not possible without the formation of resting stages [5.41] [5.106]. Diatoms adapted to terrestrial conditions are intrinsically more tolerant to temperature extremes and desiccation and are therefore more easily dispersed [5.108], but their colonization success may be limited by small population sizes [5.78]. In view of the fact that diatoms make up only a small fraction of living cells found in material transmitted by air or via waterfowl [5.31] [5.53] [5.101] [5.111], human-mediated introductions have received increased attention in explaining the wide geographic ranges of some freshwater diatoms [5.5] [5.47] [5.62].

Phylogenetic reconstructions have revealed that isolates from the same geographic area can form several lineages that are not very closely related. For example, isolates of terrestrial *Pinnularia borealis* agg. from (Sub-) Antarctica [5.78] and freshwater *Gomphonema parvulum* agg. from a tropical island [5.46] represented several lineages, some of which were most closely related to European lineages. This pattern, together with widely distributed haplotypes and lineages of species complexes, suggests that genetic distances are not spatially correlated. But this conclusion may be premature without the use of statistical analysis. A common freshwater species complex (*Nitzschia palea* agg.) had relatively high local genetic distances and zero minimum genetic distance at all spatial scales (from less than 1 km to more than 1000 km), but genetic variation increased significantly from local to large spatial scales, with the highest median and maximum genetic distance at scales greater than 100 or 1000 km [5.91].

Variability in the distribution patterns among haplotypes/lineages of species complexes could be the result of multiple colonization events (natural and/or human-mediated dispersal) in which the priority effect may have played an important role. In this concept, the early-arriving colonists adapt to local conditions and prevent subsequent arrivals from establishing in the site through monopolization of resources [5.19]. The priority effect has been supported by high adaptability of populations and clones [5.33] [5.72], experiments with coexisting cultures [5.98] [5.99] [5.102], and a field study analyzing historical biogeography [5.78]. Another aspect to consider when evaluating geographical distribution of diatoms is artifacts caused by undersampling (Section 5.4).

5.3 Metabarcoding

Metabarcoding combines DNA barcoding (taxon identification through a short DNA region) with high-throughput sequencing (HTS) of environmental samples to determine the taxonomic composition of samples and, to some extent, the abundance of the genetic units present. The outcome is tens of thousands to millions of short DNA sequences (normally <400 base pairs, for Illumina MiSeq technology). A bioinformatic pipeline is then used to process the sequence data, applying appropriate filtering to minimize artefacts and to be as certain as possible that the sequences we work with reflect true biological entities rather than being products of the method. The sequences are then assigned to named organisms through comparison with reference sequences, a step that is essential for linking the genetic information to the vast wealth of information and knowledge (such as species ecology, biology, life history, and fossil record) that has been built up in the last two centuries. Of paramount importance is the maintenance of reference sequence databases of properly identified sequences that are as comprehensive as possible. Such databases require evaluation and curation of the sequences deposited in public databases (GenBank, BOLD, etc.) and the names attached to them, as well as continuous updating to incorporate new sequences.

For freshwater diatoms (both living and subfossil), one of the most commonly used barcode markers is a region of the chloroplast-encoded gene *rbcL* (the gene coding for the large subunit of RuBisCO). A well-curated open-access reference database (Diat.barcode) for *rbcL* has been established [5.89] and is continuously revised and updated [5.88]. In addition, initiatives exist (e.g., SILVA [5.32] and PR2 [5.124]) for the construction and maintenance of reference sequence databases of another commonly used marker for metabarcoding diatoms, a short region of the nuclear 18S rDNA.

Since the introduction of metabarcoding (circa 10 years ago) the methodology has already improved considerably and advances will certainly continue to be made (e.g., better quality filters; use of longer sequences; phylogenetic placement of the sequences; one example is [5.39]). However, even in its present form, diatom metabarcoding is already a reliable tool for many types of research in which it is important to characterize community composition and abundance; these include ecological, biogeographical and biomonitoring studies (e.g., [5.17] [5.37] [5.54] [5.60] [5.76] [5.98] [5.123] [5.129]). The outcomes of these studies illustrate how some of the problems inherent in ecological and biogeographical studies of diatoms (Section 5.1) can be minimized or overcome. Firstly, in comparison with microscopical analysis, metabarcoding greatly reduces the time and the cost of effective analysis of large numbers of samples (e.g., [5.97]) at high taxonomic resolution (often at species level or below). Secondly, a sequence of base pairs is a sequence of base pairs: there is no species concept to communicate and no description to interpret. So, there are no ambiguities, which facilitates comparison between studies and regions. If the taxonomy changes, the sequences remain the same and can be reallocated to a different named diatom taxon, as appropriate. For example, De Luca *et al.* [5.18] were able to combine two complementary large global metabarcoding marine datasets, one focused on open ocean waters (Tara Oceans) and the other more on coastal waters (Ocean Sampling Day) in order to elucidate the genetic variability and biogeography of a very common planktonic diatom (*Chaetoceros curvisetus* agg.). This was possible because both datasets included metabarcoding sequences of the V9 18S rDNA and therefore were fully comparable. For rivers, there are now many datasets from biomonitoring and diversity studies in Eurasia and North America, all based on a partial *rbcL* marker and already comprising more than 3000 samples, which can potentially be combined (after due data wrangling) and interrogated to study the genetic diversity and distribution of benthic freshwater species [5.75].

The HTS genetic markers currently used are short and provide limited (phylo)genetic resolution. In some cases, one commonly used *rbcL* marker cannot discriminate between species in some Surirellaceae [5.75] while 18S rDNA cannot separate some groupings assessed to be species-level lineages in *Pinnularia* [5.78]. Nevertheless, conclusions relevant to understanding diatom biogeography do seem to be emerging, especially when it has been possible to make direct comparisons between DNA metabarcoding and light microscopy observations for the same samples. An example is the question of whether freshwater or marine diatoms tend to show high levels of intraspecific variation. High levels of variation had previously been found using Sanger sequencing of clonal isolates for some selected morphospecies (e.g., *Nitzschia palea* [5.114] and *Nitzschia inconspicua* [5.92], *Pinnularia borealis* [5.79], *Gomphonema parvulum* [5.46]). Metabarcoding has shown the phenomenon is very common.

A second conclusion based on HTS metabarcoding relates to the observation that some haplotypes (normally the most abundant ones) are widespread whereas others have restricted distributions. For example, Pérez Burillo [5.75] observed that the single *rbcL* haplotypes found of *Navicula tripunctata* and *Gomphonema truncatum*, and the two of *Navicula radiosa* were all widespread, occurring in freshwater samples spanning western and eastern North America and Eurasia. In the marine phytoplankton too, quite widespread haplotypes have been documented, such as *Chaetoceros curvisetus* sp. 2, which occurs from the poles to the tropics [5.18]. On the other hand, examples of restricted or heterogeneous distributions include *rbcL* haplotypes of some very common freshwater diatoms such as *Achnanthidium minutissimum* agg., *Fistulifera saprophila* [5.75] [5.76] and some V4 18S rDNA haplotypes of marine phytoplankton species that are restricted to Arctic waters [5.112]. Overall, diatoms seem to conform to the "moderate endemicity model" of microbial biogeography [5.7] [5.29], rather than the "ubiquity model" [5.26].

A third example is that analysis of the occurrence of haplotypes within a species complex in relation to environmental factors has shown that, in some cases, different haplotypes have different ecological preferences. This is what is found in *Achnanthidium minutissimum* agg. and *Fistulifera saprophila* [5.76].

Also, interestingly, recent work to simultaneously characterize the dynamics of diatoms and bacteria in Australian coastal waters using 16S and 18S rDNA [5.54] has shown that diatom communities and distributions are not only shaped by abiotic parameters (e.g., temperature, nutrients) but can also be strongly influenced by specific bacterial populations. Without metabarcoding this kind of investigation would be almost impossible.

5.4 The Problem of Undersampling

Conclusions regarding the biogeography of diatoms may be flawed due to undersampling of geographical areas and habitats. This is true whether studies are based on traditional microscope-based surveys, in-depth studies of Sanger-sequenced isolates, or metabarcoding: in all cases it is never possible to be sure that an area has been sampled sufficiently well to establish that a species (lineage/haplotype) is absent. Undersampling bias can be minimized by intensive sampling of (apparently) suitable habitats in different geographical areas [5.18] [5.52] [5.78] [5.91] and by seasonal sampling [5.10] [5.93] [5.134]. Overall, however, individual explorations of particular, previously unsampled regions, or intensive sampling campaigns in a local area, commonly result in both the loss of claimed endemic status of some lineages or haplotypes of the species complex and new claims of the likely endemicity of newly discovered genetic units [5.1] [5.46] [5.59] [5.87] [5.93] [5.116], which does not really improve our understanding of large-scale distribution patterns. For these, it is necessary to assemble truly global datasets, which is why the Tara Oceans project has been so successful and why freshwater diatom studies have not yet achieved anything comparable.

Biogeographic inferences about diatom distributions are also complicated by the fact that, even where habitats have been well sampled, some lineages/haplotypes of species complexes may escape detection during sampling because of their rarity. The number of rare taxa can highly exceed the diversity of common taxa [5.66], especially in the microeukaryotic world [5.73], increasing the likelihood of false negative results. Furthermore, the rarity of a taxon in a sample does not necessarily mean that its low abundance is spatially or temporally stable, because rare taxa can become common if conditions allow their growth, and common taxa can become rare under unfavorable conditions [5.11] [5.57] [5.100]. Empirical evidence based on diatom DNA sequences from marine plankton has shown that some species and lineages of species complexes can track seasonal variability in abiotic factors by remarkable changes in their abundances [5.10] [5.93], while other species and lineages may be consistently rare across samples [5.10] [5.52] [5.70] [5.93].

5.5 Conclusions

Consistent with previous reviews of diatom biogeography (e.g., [5.48] [5.61] [5.63] [5.122]), which were based principally on morphospecies, the current distributions of diatom species and lineages/haplotypes of species complexes cannot be explained by a single scenario (e.g., using the ubiquity hypothesis [5.26] [5.27]), owing to the dynamic nature of the mechanisms underlying community assembly. The biogeography of diatoms is shaped not only by differences in their predisposition to disperse [5.106] [5.108], but also by differences in their ecological tolerances [5.92] [5.119], local population sizes [5.52] [5.70] [5.78], competitive abilities [5.33] [5.49] [5.98], and diversification rates [5.67] [5.69]. The future lies in linking distribution patterns and assembly processes using large datasets—both organismal (morphological and metabarcoding, and increasingly metagenomic) and abiotic—to explain the ecology and evolution of diatom biodiversity (see [5.97]).

Acknowledgment

We are very grateful to two reviewers for helpful comments and David Mann for comments and editorial help. RT acknowledges support from the CERCA Programme/Generalitat de Catalunya and JK thanks the Charles University institutional grant (Cooperatio, Biology) for its support.

References

[5.1] Abarca, N., Jahn, R., Zimmermann, J. and Enke, N. 2014. Does the cosmopolitan diatom *Gomphonema parvulum* (Kutzing) Kutzing have a biogeography? *PLoS One* 9(1), #e86885.

[5.2] Alexander, H., Jenkins, B.D., Rynearson, T.A. and Dyhrman, S.T. 2015. Metatranscriptome analyses indicate resource partitioning between diatoms in the field. *PNAS* 112(17), 2182-2190.

[5.3] Alverson, A.J. 2008. Molecular systematics and the diatom species. *Protist* 159(3), 339-353.

[5.4] Alverson, A.J., Beszteri, B., Julius, M.L. and Theriot, E.C. 2011. The model marine diatom *Thalassiosira pseudonana* likely descended from a freshwater ancestor in the genus *Cyclotella*. *BMC Evol. Biol.* 11(1), 1-8.

[5.5] Alverson, A.J., Chafin, T.K., Jones, K.A., Manoylov, K.M., Johnson, H., Julius, M.L., Nakov, T., Ruck, E.C., Theriot, E.C. and Yeager, K.M. 2021.

Microbial biogeography through the lens of exotic species: the recent introduction and spread of the freshwater diatom *Discostella asterocostata* in the United States. *Biol. Invasions* 23(7), 2191-2204.

[5.6] Bass, D. and Boenigk, J. 2011. Everything is everywhere: a twenty-first century de-/reconstruction with respect to protists. In: *Biogeography of Microscopic Organisms: Is Everything Small Everywhere?* D. Fontaneto, (ed.) Cambridge University Press, Cambridge: 88-110.

[5.7] Bass, D., Richards, T.A., Matthai, L., Marsh, V. and Cavalier-Smith, T. 2007. DNA evidence for global dispersal and probable endemicity of protozoa. *BMC Evol. Biol.* 7, 162.

[5.8] Bouchard, A.J., Hamilton, P.B., Savoie, A.M. and Starr, J.R. 2019. Molecular and morphological data reveal hidden diversity in common North American *Frustulia* species (Amphipleuraceae). *Diatom Res.* 34(4), 205-223.

[5.9] Bourland, W. 2017. How far do ciliate flagships sail? A proposed Gondawanaland endemic species at anchor in Idaho soils. *Protist* 168(3), 352-361.

[5.10] Canesi, K.L. and Rynearson, T.A. 2016. Temporal variation of *Skeletonema* community composition from a long-term time series in Narragansett Bay identified using high-throughput DNA sequencing. *Mar. Ecol. Prog. Ser.* 556, 1-16.

[5.11] Caron, D.A. and Countway, P.D. 2009. Hypotheses on the role of the protistan rare biosphere in a changing world. *Aquat. Microb. Ecol.* 57(3), 227-238.

[5.12] Casteleyn, G., Chepurnov, V.A., Leliaert, F., Mann, D.G., Bates, S.S., Lundholm, N., Rhodes, L., Sabbe, K. and Vyverman, W. 2008. *Pseudo-nitzschia pungens* (Bacillariophyceae): a cosmopolitan diatom species? *Harmful Algae* 7(2), 241-257.

[5.13] Cerino, F., Orsini, L., Sarno, D., Dell'Aversano, C., Tartaglione, L. and Zingone, A. 2005. The alternation of different morphotypes in the seasonal cycle of the toxic diatom *Pseudo-nitzschia galaxiae. Harmful Algae* 4(1), 33-48.

[5.14] Cipolletta, F., RUSSO, A., D'Alelio, D., Margiotta, F., Sarno, D., Zingorne, A. and Montresor, M. 2022. Vertical distribution of *Pseudo-nitzschia* in the Gulf of Naples across the seasons. *Mediterr. Mar. Sci.* 23(3), 525–535.

[5.15] Créach, V., Ernst, A., Sabbe, K., Vanelslander, B., Vyverman, W. and Stal, L.J. 2006. Using quantitative PCR to determine the distribution of a semicryptic benthic diatom, *Navicula phyllepta* (Bacillariophyceae). *J. Phycol.* 42(5), 1142-1154.

[5.16] De Decker, S., Vanormelingen, P., Pinseel, E., Sefbom, J., Audoor, S., Sabbe, K. and Vyverman, W. 2018. Incomplete reproductive isolation between genetically distinct sympatric clades of the pennate model diatom *Seminavis robusta. Protist* 169(4), 569-583.

[5.17] De Luca, D., Kooistra, W.H., Sarno, D., Gaonkar, C.C. and Piredda, R. 2019. Global distribution and diversity of *Chaetoceros* (Bacillariophyta, Mediophyceae): integration of classical and novel strategies. *PeerJ* 7, #e7410.

[5.18] De Luca, D., Piredda, R., Sarno, D. and Kooistra, W.H. 2021. Resolving cryptic species complexes in marine protists: phylogenetic haplotype networks meet global DNA metabarcoding datasets. *ISME J.* 15(7), 1931-1942.

[5.19] De Meester, L., Gómez, A., Okamura, B. and Schwenk, K. 2002. The Monopolization Hypothesis and the dispersal - gene flow paradox in aquatic organisms. *Acta Oecol.* 23(3), 121-135.

[5.20] De Wit, R. and Bouvier, T. 2006. 'Everything is everywhere, but, the environment selects'; what did Baas Becking and Beijerinck really say? *Environ. Microbiol.* 8(4), 755-758.

[5.21] Degerlund, M., Huseby, S., Zingone, A., Sarno, D. and Landfald, B. 2012. Functional diversity in cryptic species of *Chaetoceros socialis* Lauder (Bacillariophyceae). *J. Plankton Res.* 34(5), 416-431.

[5.22] Ehrenberg, C.G. 1832. Die geographische Verbreitung der Infusionsthierchen in Nord-Africa und West-Asien: beobachtet auf Hemprich und Ehrenbergs Reisen. *Abh. Königl. Akad. Wiss., physikalische Klasse* 1829, 1-20.

[5.23] Evans, K.M., Chepurnov, V.A., Sluiman, H.J., Thomas, S.J., Spears, B.M. and Mann, D.G. 2009. Highly differentiated populations of the freshwater diatom *Sellaphora capitata* suggest limited dispersal and opportunities for allopatric speciation. *Protist* 160(3), 386-396.

[5.24] Evans, K.M., Kühn, S.F. and Hayes, P.K. 2005. High levels of genetic diversity and low levels of genetic differentiation in North Sea *Pseudo-nitzschia pungens* (Bacillariophyceae) populations. *J. Phycol.* 41(3), 506-514.

[5.25] Evans, K.M., Wortley, A.H., Simpson, G.E., Chepurnov, V.A. and Mann, D.G. 2008. A molecular systematic approach to explore diversity within the *Sellaphora pupula* species complex (Bacillariophyta). *J. Phycol.* 44(1), 215-231.

[5.26] Finlay, B.J. 2002. Global dispersal of free-living microbial eukaryote species. *Science* 296(5570), 1061-1063.

[5.27] Finlay, B.J., Monaghan, E.B. and Maberly, S.C. 2002. Hypothesis: the rate and scale of dispersal of freshwater diatom species is a function of their global abundance. *Protist* 153(3), 261-273.

[5.28] Foissner, W. 2006. Biogeography and dispersal of micro-organisms: a review emphasizing protists. *Acta Protozool.* 45(2), 111-136.

[5.29] Foissner, W. 2008. Protist diversity and distribution: some basic considerations. *Biodivers. Conserv.* 17, 235–242.

[5.30] Francisco-Ortega, J., Santos-Guerra, A., Romeiras, M.M., Carine, M.A., Sánchez-Pinto, L. and Duarte, M.C. 2015. The botany of the three voyages of Captain James Cook in Macronesia: an introduction. *Revista de la Academia Canaria de Ciencias* 27, 357-410.

[5.31] Genitsaris, S., Moustaka-Gouni, M. and Kormas, K.A. 2011. Airborne microeukaryote colonists in experimental water containers: diversity, succession, life histories and established food webs. *Aquat. Microb. Ecol.* 62(2), 139-152.

[5.32] Glöckner, F.C., Yilmaz, P., Quast, C., Gerken, J., Beccati, A., Ciuprina, A., Bruns, G., Yarza, P., Peplies, J. and Westram, R. 2017. 25 years of serving the community with ribosomal RNA gene reference databases and tools. *J. Biotechnol.* 261, 169-176.

[5.33] Godhe, A. and Rynearson, T. 2017. The role of intraspecific variation in the ecological and evolutionary success of diatoms in changing environments. *Philos. Trans. R Soc. Lond. B Biol. Sci.* 372, #20160399.

[5.34] Hanson, C.A., Fuhrman, J.A., Horner-Devine, M.C. and Martiny, J.B. 2012. Beyond biogeographic patterns: processes shaping the microbial landscape. *Nat. Rev. Microbiol.* 10(7), 497-506.

[5.35] Härnström, K., Ellegaard, M., Andersen, T.J. and Godhe, A. 2011. Hundred years of genetic structure in a sediment revived diatom population. *PNAS* 108(10), 4252-4257.

[5.36] Hines, H.N., McCarthy, P.J. and Esteban, G.F. 2020. First records of 'flagship'soil ciliates in North America. *Protist* 171(3), #125739.

[5.37] Huang, S., Herzschuh, U., Pestryakova, L., Zimmermann, H., Davydova, P., Biskaborn, B., Shevtsova, I. and Stoof-Leichsenring, K. 2020. Genetic and morphologic determination of diatom community composition in surface sediments from glacial and thermokarst lakes in the Siberian Arctic. *J. Paleolimnol.* 64(3), 225-242.

[5.38] Huseby, S., Degerlund, M., Zingone, A. and Hansen, E. 2012. Metabolic fingerprinting reveals differences between northern and southern strains of the cryptic diatom *Chaetoceros socialis*. *Eur. J. Phycol.* 47(4), 480-489.

[5.39] Jamy, M., Foster, R., Barbera, P., Czech, L., Kozlov, A., Stamatakis, A., Bending, G., Hilton, S., Bass, D. and Burki, F. 2020. Long-read metabarcoding of the eukaryotic rDNA operon to phylogenetically and taxonomically resolve environmental diversity. *Mol. Ecol. Resour.* 20(2), 429-443.

[5.40] Johansson, O.N., Pinder, M.I., Ohlsson, F., Egardt, J., Töpel, M. and Clarke, A.K. 2019. Friends with benefits: Exploring the phycosphere of the marine diatom *Skeletonema marinoi*. *Front. Microbiol.* 10, #1828.

[5.41] Kaczmarska, I., Poulíčková, A., Sato, S., Edlund, M.B., Idei, M., Watanabe, T. and Mann, D.G. 2013. Proposals for a terminology for diatom sexual reproduction, auxospores and resting stages. *Diatom Res.* 28(3), 263-294.

[5.42] Kahlert, M., Karjalainen, S.M., Keck, F., Kelly, M., Ramon, M., Rimet, F., Schneider, S., Tapolczai, K. and Zimmermann, J. 2022. Co-occurrence, ecological profiles and geographical distribution based on unique molecular identifiers of the common freshwater diatoms *Fragilaria* and *Ulnaria*. *Ecol. Indic.* 141, #109114.

[5.43] Keller, S.R., Hilderbrand, R.H., Shank, M.K. and Potapova, M. 2017. Environmental DNA genetic monitoring of the nuisance freshwater diatom, *Didymosphenia geminata*, in eastern North American streams. *Divers. Distrib.* 23(4), 381-393.

[5.44] Kelly, M.G., Trobajo, R., Rovira, L. and Mann, D.G. 2015. Characterizing the niches of two very similar *Nitzschia* species and implications for ecological assessment. *Diatom Res.* 30(1), 27-33.

[5.45] Kelly, S.R. 2009. *The origin, genetic diversity and taxonomy of the invasive diatom Didymosphenia geminata (Bacilliariophyceae) in New Zealand.* Master Thesis, The University of Waikato, New Zealand.

[5.46] Kermarrec, L., Bouchez, A., Rimet, F. and Humbert, J.-F. 2013. First evidence of the existence of semi-cryptic species and of a phylogeographic structure in the *Gomphonema parvulum* (Kützing) Kützing complex (Bacillariophyta). *Protist* 164(5), 686-705.

[5.47] Kilroy, C. and Novis, P. 2018. Is *Didymosphenia geminata* an introduced species in New Zealand? Evidence from trends in water chemistry, and chloroplast DNA. *Ecol. Evol.* 8(2), 904-919.

[5.48] Kociolek, J.P. and Spaulding, S.A. 2000. Freshwater diatom biogeography. *Nova Hedwigia* 71(1-2), 223-241.

[5.49] Koedooder, C., Stock, W., Willems, A., Mangelinckx, S., De Troch, M., Vyverman, W. and Sabbe, K. 2019. Diatom-bacteria interactions modulate the composition and productivity of benthic diatom biofilms. *Front. Microbiol.* 10, #1255.

[5.50] Kollár, J., Pinseel, E., Vanormelingen, P., Poulíčková, A., Souffreau, C., Dvořák, P. and Vyverman, W. 2019. A polyphasic approach to the delimitation of diatom species: a case study for the genus *Pinnularia* (Bacillariophyta). *J. Phycol.* 55(2), 365-379.

[5.51] Kollár, J., Pinseel, E., Vyverman, W. and Poulickova, A. 2021. A time-calibrated multi-gene phylogeny provides insights into the evolution, taxonomy and DNA barcoding of the *Pinnularia gibba* group (Bacillariophyta). *Fottea* 21(1), 62-72.

[5.52] Kooistra, W.H., Sarno, D., Balzano, S., Gu, H., Andersen, R.A. and Zingone, A. 2008. Global diversity and biogeography of *Skeletonema* species (Bacillariophyta). *Protist* 159(2), 177-193.

[5.53] Kristiansen, J. 1996. Dispersal of freshwater algae - a review. *Hydrobiologia* 336(1), 151-157.

[5.54] Le Reun, N., Bramucci, A., O'Brien, J., Ostrowski, M., Brown, M.V., Van de Kamp, J., Bodrossy, L., Raina, J.-B., Ajani, P. and Seymour, J. 2022. Diatom biogeography, temporal dynamics, and links to bacterioplankton across seven oceanographic time-series sites spanning the Australian continent. *Microorganisms* 10(2), #338.

[5.55] Leibold, M.A., Holyoak, M., Mouquet, N., Amarasekare, P., Chase, J.M., Hoopes, M.F., Holt, R.D., Shurin, J.B., Law, R., Tilman, D., Loreau, M.

and Gonzalez, A. 2004. The metacommunity concept: a framework for multi-scale community ecology. *Ecol. Lett.* 7(7), 601-613.

[5.56] Linnaeus, C.t.Y. 1782. *Supplementum plantarum.* Orphanotropheum, Brunswick.

[5.57] Logares, R., Audic, S., Bass, D., Bittner, L., Boutte, C., Christen, R., Claverie, J.-M., Decelle, J., Dolan, J.R. and Dunthorn, M. 2014. Patterns of rare and abundant marine microbial eukaryotes. *Curr. Biol.* 24(8), 813-821.

[5.58] Logares, R., Deutschmann, I.M., Junger, P.C., Giner, C.R., Krabberød, A.K., Schmidt, T.S., Rubinat-Ripoll, L., Mestre, M., Salazar, G. and Ruiz-González, C. 2020. Disentangling the mechanisms shaping the surface ocean microbiota. *Microbiome* 8(1), 1-17.

[5.59] Maltsev, Y., Maltseva, S., Kociolek, J.P., Jahn, R. and Kulikovskiy, M. 2021. Biogeography of the cosmopolitan terrestrial diatom *Hantzschia amphioxys* sensu lato based on molecular and morphological data. *Sci. Rep.* 11(1), 1-19.

[5.60] Malviya, S., Scalco, E., Audic, S., Vincent, F., Veluchamy, A., Poulain, J., Wincker, P., Iudicone, D., de Vargas, C. and Bittner, L. 2016. Insights into global diatom distribution and diversity in the world's ocean. *PNAS* 113(11), 1516-1525.

[5.61] Mann, D.G. 1999. The species concept in diatoms. *Phycologia* 38(6), 437-495.

[5.62] Mann, D.G. 2022. Determining geographical range and alien status in diatoms: three instructive case histories of species newly recorded in the British Isles, including a non-native marine species from the Pacific, *Diademoides luxuriosa. Diatom Res.*, DOI 10.1080/0269249X.2022.2078428.

[5.63] Mann, D.G. and Droop, S.J.M. 1996. Biodiversity, biogeography and conservation of diatoms. In: *Biogeography of Freshwater Algae.* Kluwer Academic Publishers: 19-32.

[5.64] Mann, D.G., Trobajo, R., Sato, S., Li, C., Witkowski, A., Rimet, F., Ashworth, M.P., Hollands, R.M. and Theriot, E.C. 2021. Ripe for reassessment: A synthesis of available molecular data for the speciose diatom family Bacillariaceae. *Mol. Phylogenet. Evol.* 158, #106985.

[5.65] McDonald, S.M., Sarno, D. and Zingone, A. 2007. Identifying *Pseudo-nitzschia* species in natural samples using genus-specific PCR primers and clone libraries. *Harmful Algae* 6(6), 849-860.

[5.66] McGill, B.J., Etienne, R.S., Gray, J.S., Alonso, D., Anderson, M.J., Benecha, H.K., Dornelas, M., Enquist, B.J., Green, J.L. and He, F. 2007. Species abundance distributions: moving beyond single prediction theories to integration within an ecological framework. *Ecol. Lett.* 10(10), 995-1015.

[5.67] Nakov, T., Beaulieu, J.M. and Alverson, A.J. 2018. Accelerated diversification is related to life history and locomotion in a hyperdiverse lineage of microbial eukaryotes (Diatoms, Bacillariophyta). *New Phytol.* 219(1), 462-473.

[5.68] Nakov, T., Beaulieu, J.M. and Alverson, A.J. 2018. Insights into global planktonic diatom diversity: The importance of comparisons between phylogenetically equivalent units that account for time. *ISME J.* 12(11), 2807-2810.

[5.69] Nakov, T., Beaulieu, J.M. and Alverson, A.J. 2019. Diatoms diversify and turn over faster in freshwater than marine environments. *Evolution* 73(12), 2497-2511.

[5.70] Nanjappa, D., Audic, S., Romac, S., Kooistra, W.H. and Zingone, A. 2014. Assessment of species diversity and distribution of an ancient diatom lineage using a DNA metabarcoding approach. *PLoS One* 9(8), #e103810.

[5.71] Nef, C., Madoui, M.-A., Pelletier, É. and Bowler, C. 2022. Whole-genome scanning reveals environmental selection mechanisms that shape diversity in populations of the epipelagic diatom *Chaetoceros*. *PLoS Biol.* 20(11), #e3001893.

[5.72] O'Donnell, D.R., Hamman, C.R., Johnson, E.C., Kremer, C.T., Klausmeier, C.A. and Litchman, E. 2018. Rapid thermal adaptation in a marine diatom reveals constraints and trade-offs. *Global Change Biol.* 24(10), 4554-4565.

[5.73] Pascoal, F., Costa, R. and Magalhães, C. 2021. The microbial rare biosphere: current concepts, methods and ecological principles. *FEMS Microbiol. Ecol.* 97(1), #fiaa227.

[5.74] Percopo, I., Ruggiero, M.V., Balzano, S., Gourvil, P., Lundholm, N., Siano, R., Tammilehto, A., Vaulot, D., Sarno, D. and Mock, T. 2016. *Pseudo-nitzschia arctica* sp. nov., a new cold-water cryptic *Pseudo-nitzschia* species within the *P. pseudodelicatissima* complex. *J. Phycol.* 52(2), 184-199.

[5.75] Pérez Burillo, J. 2022. *Benthic diatom metabarcoding: developing new approaches to research and biomonitoring in aquatic ecosystems.* Doctoral Thesis, Universitat Rovira i Virgili, Tarragona, Catalonia, Spain.

[5.76] Pérez-Burillo, J., Trobajo, R., Leira, M., Keck, F., Rimet, F., Sigró, J. and Mann, D.G. 2021. DNA metabarcoding reveals differences in distribution patterns and ecological preferences among genetic variants within some key freshwater diatom species. *Sci. Total Environ.* 798, #149029.

[5.77] Pinseel, E., Hejduková, E., Vanormelingen, P., Kopalová, K., Vyverman, W. and Van de Vijver, B. 2017. *Pinnularia catenaborealis* sp. nov. (Bacillariophyceae), a unique chain-forming diatom species from James Ross Island and Vega Island (Maritime Antarctica). *Phycologia* 56(1), 94-107.

[5.78] Pinseel, E., Janssens, S.B., Verleyen, E., Vanormelingen, P., Kohler, T.J., Biersma, E.M., Sabbe, K., Van de Vijver, B. and Vyverman, W. 2020. Global radiation in a rare biosphere soil diatom. *Nat. Commun.* 11(1), 1-12.

[5.79] Pinseel, E., Kulichová, J., Scharfen, V., Urbánková, P., Van de Vijver, B. and Vyverman, W. 2019. Extensive cryptic diversity in the terrestrial diatom *Pinnularia borealis* (Bacillariophyceae). *Protist* 170(2), 121-140.

[5.80] Pinseel, E., Van de Vijver, B., Wolfe, A.P., Harper, M., Antoniades, D., Ashworth, A.C., Ector, L., Lewis, A.R., Perren, B. and Hodgson, D.A. 2021. Extinction of austral diatoms in response to large-scale climate dynamics in Antarctica. *Sci. Adv.* 7(38), #eabh3233.

[5.81] Pinseel, E., Vanormelingen, P., Hamilton, P.B., Vyverman, W., Van de Vijver, B. and Kopalova, K. 2017. Molecular and morphological characterization of the *Achnanthidium minutissimum* complex (Bacillariophyta) in Petuniabukta (Spitsbergen, High Arctic) including the description of *A. digitatum* sp. nov. *Eur. J. Phycol.* 52(3), 264-280.

[5.82] Piredda, R., Claverie, J.-M., Decelle, J., de Vargas, C., Dunthorn, M., Edvardsen, B., Eikrem, W., Forster, D., Kooistra, W.H. and Logares, R. 2018. Diatom diversity through HTS-metabarcoding in coastal European seas. *Sci. Rep.* 8(1), 1-12.

[5.83] Postel, U., Glemser, B., Salazar Alekseyeva, K., Eggers, S.L., Groth, M., Glöckner, G. John, U., Mock, T., Klemm, K. and Valentin, K. 2020. Adaptive divergence across Southern Ocean gradients in the pelagic diatom *Fragilariopsis kerguelensis*. *Mol. Ecol.* 29(24), 4913-4924.

[5.84] Poulickova, A., Vesela, J., Neustupa, J. and Skaloud, P. 2010. Pseudocryptic diversity versus cosmopolitanism in diatoms: a case study on *Navicula cryptocephala* Kutz. (Bacillariophyceae) and morphologically similar taxa. *Protist* 151(3), 353-369.

[5.85] Poulíčková, A., Špačková, J., Kelly, M.G., Duchoslav, M. and Mann, D.G. 2008. Ecological variation within *Sellaphora* species complexes (Bacillariophyceae): specialists or generalists? *Hydrobiologia* 614, 373-386.

[5.86] Ramond, P., Siano, R., Sourisseau, M. and Logares, R. 2023. Assembly processes and functional diversity of marine protists and their rare biosphere. *Environmental Microbiome* 18(1), 1-14.

[5.87] Rimet, F., Abarca, N., Bouchez, A., Kusber, W.-H., Jahn, R., Kahlert, M., Keck, F., Kelly, M.G., Mann, D.G. and Piuz, A. 2018. The potential of high throughput sequencing (HTS) of natural samples as a source of primary taxonomic information for reference libraries of diatom barcodes. *Fottea* 18(1), 37-54.

[5.88] Rimet, F., Gusev, E., Kahlert, M., Kelly, M.G., Kulikovskiy, M., Maltsev, Y., Mann, D.G., Pfannkuchen, M., Trobajo, R. and Vasselon, V. 2019. Diat. barcode, an open-access curated barcode library for diatoms. *Sci. Rep.* 9(1), 1-12.

[5.89] Rimet, F., Chaumeil, P., Keck, F., Kermarrec, L., Vasselon, V., Kahlert, M., Franc, A. and Bouchez, A. 2016. R-Syst::diatom: an open-access and curated barcode database for diatoms and freshwater monitoring. *Database* 2016, #baw016.

[5.90] Rimet, F., Pinseel, E., Bouchez, A., Japoshvili, B. and Mumladze, L. 2023. Diatom endemism and taxonomic turnover: Assessment in high-altitude alpine lakes covering a large geographical range. *Sci. Total Environ.* 871, #161970.

[5.91] Rimet, F., Trobajo, R., Mann, D.G., Kermarrec, L., Franc, A., Domaizon, I. and Bouchez, A. 2014. When is sampling complete? The effects of geographical range and marker choice on perceived diversity in *Nitzschia palea* (Bacillariophyta). *Protist* 165(3), 245-259.

[5.92] Rovira, L., Trobajo, R., Sato, S., Ibáñez, C. and Mann, D.G. 2015. Genetic and physiological diversity in the diatom *Nitzschia inconspicua*. *J. Eukaryot. Microbiol.* 62(6), 815-832.

[5.93] Ruggiero, M.V., Sarno, D., Barra, L., Kooistra, W.H., Montresor, M. and Zingone, A. 2015. Diversity and temporal pattern of *Pseudo-nitzschia* species (Bacillariophyceae) through the molecular lens. *Harmful Algae* 42, 15-24.

[5.94] Rynearson, T.A., Flickinger, S.A. and Fontaine, D.N. 2020. Metabarcoding reveals temporal patterns of community composition and realized thermal niches of *Thalassiosira* spp. (Bacillariophyceae) from the Narragansett Bay long-term plankton time series. *Biology* 9(1), #19.

[5.95] Rynearson, T.A., Lin, E.O. and Armbrust, E.V. 2009. Metapopulation structure in the planktonic diatom *Ditylum brightwellii* (Bacillariophyceae). *Protist* 160(1), 111-121.

[5.96] Rynearson, T.A. and Virginia Armbrust, E. 2004. Genetic differentiation among populations of the planktonic marine diatom *Ditylum brightwellii* (Bacillariophyceae). *J. Phycol.* 40(1), 34-43.

[5.97] Santoferrara, L., Burki, F., Filker, S., Logares, R., Dunthorn, M. and McManus, G.B. 2020. Perspectives from ten years of protist studies by high-throughput metabarcoding. *J. Eukaryot. Microbiol.* 67(5), 612-622.

[5.98] Sefbom, J., Kremp, A., Hansen, P.J., Johannesson, K., Godhe, A. and Rengefors, K. 2023. Local adaptation through countergradient selection in northern populations of *Skeletonema marinoi*. *Evol. Appl.* 16(2), 311-320.

[5.99] Sefbom, J., Sassenhagen, I., Rengefors, K. and Godhe, A. 2015. Priority effects in a planktonic bloom-forming marine diatom. *Biol. Lett.* 11(5), #20150184.

[5.100] Ser-Giacomi, E., Zinger, L., Malviya, S., De Vargas, C., Karsenti, E., Bowler, C. and De Monte, S. 2018. Ubiquitous abundance distribution of non-dominant plankton across the global ocean. *Nat. Ecol. Evol.* 2(8), 1243-1249.

[5.101] Schulte, N.O., Khan, A.L., Smith, E.W., Zoumplis, A., Kaul, D., Allen, A.E., Adams, B.J. and McKnight, D.M. 2022. Blowin' in the wind: Dispersal, structure, and metacommunity dynamics of aeolian diatoms in the McMurdo Sound region, Antarctica. *J. Phycol.* 58(1), 36-54.

[5.102] Sildever, S., Sefbom, J., Lips, I. and Godhe, A. 2016. Competitive advantage and higher fitness in native populations of genetically structured planktonic diatoms. *Environ. Microbiol.* 18(12), 4403-4411.
[5.103] Sjöqvist, C.O. and Kremp, A. 2016. Genetic diversity affects ecological performance and stress response of marine diatom populations. *ISME J.* 10(11), 2755-2766.
[5.104] Soininen, J., Tupola, V., Voutilainen, I., Cantonati, M. and Teittinen, A. 2021. Diatom biogeography in freshwaters – new insights from between-region comparisons and the role of unmeasured environmental factors. *Diatom Res.*, DOI 10.1080/0269249X.2021.1999859.
[5.105] Solak, C.N., Castineau, R., Lemieux, C., Turmel, M., Gorecka, E., Trobajo, R., Rybak, M., Yılmaz, E. and Witkowski, A. 2021. *Nitzschia anatoliensis* sp. nov., a cryptic diatom species from the highly alkaline Van Lake (Turkey). *PeerJ* 9, #e12220.
[5.106] Souffreau, C., Vanormelingen, P., Sabbe, K. and Vyverman, W. 2013. Tolerance of resting cells of freshwater and terrestrial benthic diatoms to experimental desiccation and freezing is habitat-dependent. *Phycologia* 52(3), 246-255.
[5.107] Souffreau, C., Vanormelingen, P., Van de Vijver, B., Isheva, T., Verleyen, E., Sabbe, K. and Vyverman, W. 2013. Molecular evidence for distinct Antarctic lineages in the cosmopolitan terrestrial diatoms *Pinnularia borealis* and *Hantzschia amphioxys*. *Protist* 164(1), 101-115.
[5.108] Souffreau, C., Vanormelingen, P., Verleyen, E., Sabbe, K. and Vyverman, W. 2010. Tolerance of benthic diatoms from temperate aquatic and terrestrial habitats to experimental desiccation and temperature stress. *Phycologia* 49(4), 309-324.
[5.109] Stelbrink, B., Jovanovska, E., Levkov, Z., Ognjanova-Rumenova, N., Wilke, T. and Albrecht, C. 2018. Diatoms do radiate: Evidence for a freshwater species flock. *J. Evol. Biol.* 31(12), 1969-1975.
[5.110] Stock, W., Vanelslander, B., Rüdiger, F., Sabbe, K., Vyverman, W. and Karsten, U. 2019. Thermal niche differentiation in the benthic diatom *Cylindrotheca closterium* (Bacillariophyceae) complex. *Front. Microbiol.* 10, #1395.
[5.111] Szabó, B., Szabó, A., Vad, C.F., Boros, E., Lukić, D., Ptacnik, R., Márton, Z. and Horváth, Z. 2022. Microbial stowaways: Waterbirds as dispersal vectors of aquatic pro-and microeukaryotic communities. *J. Biogeogr.*, DOI 10.1111/jbi.14381.
[5.112] Šupraha, L., Klemm, K., Gran-Stadniczeňko, S., Hörstmann, C., Vaulot, D., Edvardsen, B. and John, U. 2022. Diversity and biogeography of planktonic diatoms in Svalbard fjords: The role of dispersal and Arctic endemism in phytoplankton community structuring. *Elem. Sci. Anth.* 10(1), #00117.
[5.113] Thompson, P.L., Guzman, L.M., De Meester, L., Horváth, Z., Ptacnik, R., Vanschoenwinkel, B., Viana, D.S. and Chase, J.M. 2020. A process-based

metacommunity framework linking local and regional scale community ecology. *Ecol. Lett.* 23(9), 1314-1329.

[5.114] Trobajo, R., Clavero, E., Chepurnov, V.A., Sabbe, K., Mann, D.G., Ishihara, S. and Cox, E.J. 2009. Morphological, genetic and mating diversity within the widespread bioindicator *Nitzschia palea* (Bacillariophyceae). *Phycologia* 48(6), 443-459.

[5.115] Urbánková, P., Scharfen, V. and Kulichová, J. 2016. Molecular and automated identification of the diatom genus *Frustulia* in northern Europe. *Diatom Res.* 31(3), 217-229.

[5.116] Urbánková, P. and Veselá, J. 2013. DNA-barcoding: A case study in the diatom genus *Frustulia* (Bacillariophyceae). *Nova Hedwigia, Beiheft* 142(3-4), 147-162.

[5.117] Van de Vyver, E., Pinseel, E., Verleyen, E., Vanormelingen, P., Van Wichelen, J., de Jong, R., Urrutia, R. and Vyverman, W. 2022. Planktonic diatom communities in temperate South-Central Chilean lakes with a focus on *Asterionella formosa* and the genus *Aulacoseira*. *J. Paleolimnol.* 68(3), 279-296.

[5.118] Van den Wyngaert, S., Möst, M., Freimann, R., Ibelings, B.W. and Spaak, P. 2015. Hidden diversity in the freshwater planktonic diatom *Asterionella formosa*. *Mol. Ecol.* 24(12), 2955-2972.

[5.119] Vanelslander, B., Créach, V., Vanormelingen, P., Ernst, A., Chepurnov, V.A., Sahan, E., Muyzer, G., Stal, L.J., Vyverman, W. and Sabbe, K. 2009. Ecological differentiation between sympatric pseudocryptic species in the estuarine benthic diatom *Navicula phyllepta* (Bacillariophyceae). *J. Phycol.* 45(6), 1278-1289.

[5.120] Vanormelingen, P., Evans, K.M., Mann, D.G., Lance, S., Debeer, A.E., D'hondt, S., Verstraete, T., De Meester, L. and Vyverman, W. 2015. Genotypic diversity and differentiation among populations of two benthic freshwater diatoms as revealed by microsatellites. *Mol. Ecol.* 24(17), 4433-4448.

[5.121] Vanormelingen, P., Chepurnov, V.A., Mann, D.G., Sabbe, K. and Vyverman, W. 2008. Genetic divergence and reproductive barriers among morphologically heterogeneous sympatric clones of *Eunotia bilunaris* sensu lato (Bacillariophyta). *Protist* 159(1), 73-90.

[5.122] Vanormelingen, P., Verleyen, E. and Vyverman, W. 2008. The diversity and distribution of diatoms: from cosmopolitanism to narrow endemism. *Biodivers. Conserv.* 17, 393–405.

[5.123] Vasselon, V., Rimet, F., Kálmán, T. and Bouchez, A. 2017. Assessing ecological status with diatoms DNA metabarcoding: Scaling-up on a WFD monitoring network (Mayotte island, France). *Ecol. Indic.* 82, 1-12.

[5.124] Vaulot, D., Sim, C.W.H., Ong, D., Teo, B., Biwer, C., Jamy, M. and Lopes dos Santos, A. 2022. metaPR2: a database of eukaryotic 18S rRNA metabarcodes with an emphasis on protists. *Mol. Ecol. Resour.* 22(8), 3188-3201.

[5.125] Vellend, M. 2010. Conceptual synthesis in community ecology. *Q. Rev. Biol.* 85(2), 183-206.

[5.126] Verleyen, E., Van de Vijver, B., Tytgat, B., Pinseel, E., Hodgson, D.A., Kopalová, K., Chown, S.L., Van Ranst, E., Imura, S. and Kudoh, S. 2021. Diatoms define a novel freshwater biogeography of the Antarctic. *Ecography* 44(4), 548-560.

[5.127] Verleyen, E., Vyverman, W., Sterken, M., Hodgson, D.A., De Wever, A., Juggins, S., Van de Vijver, B., Jones, V.J., Vanormelingen, P., Roberts, D., Flower, R., Kilroy, C., Souffreau, C. and Sabbe, K. 2009. The importance of dispersal related and local factors in shaping the taxonomic structure of diatom metacommunities. *Oikos* 118(8), 1239-1249.

[5.128] Veselá, J., Urbánková, P., Černá, K. and Neustupa, J. 2012. Ecological variation within traditional diatom morphospecies: diversity of *Frustulia rhomboides* sensu lato (Bacillariophyceae) in European freshwater habitats. *Phycologia* 51(5), 552-561.

[5.129] Wang, J., Liu, Q., Zhao, X., Borthwick, A.G., Liu, Y., Chen, Q. and Ni, J. 2019. Molecular biogeography of planktonic and benthic diatoms in the Yangtze River. *Microbiome* 7(1), 1-15.

[5.130] Williams, D.M. 1996. Fossil species of the diatom genus *Tetracyclus* (Bacillariophyta, 'ellipticus' species group): morphology, interrelationships and the relevance of ontogeny. *Philos. Trans. R. Soc. Lond. B Biol. Sci.* 351(1348), 1759-1782.

[5.131] Williams, D.M. 2011. Historical biogeography, microbial endemism and the role of classification: everything is endemic. In: *Biogeography of Microscopic Organisms. Is Everything Small Everywhere?* D. Fontaneto, (ed.) Cambridge University Press, Cambridge: 11-31.

[5.132] Williams, D.M. and Huxley, R. 1998. Christian Gottfried Ehrenberg (1795-1876): the man and his legacy. An introduction. In: *Christian Gottfried Ehrenburg (1795-1876): the Man and His Legacy*. D.M. Williams and R. Huxley, (eds.) Linnean Society of London, London: 1-13.

[5.133] Williams, D.M. and Kociolek, J.P. 2017. Historical biogeography of diatoms in Australasia: A preliminary assessment. In: *Handbook of Australasian Biogeography*. CRC Press, Boca Raton: 17-46.

[5.134] Yamada, M., Otsubo, M., Tsutsumi, Y., Mizota, C., Nakamura, Y., Takahashi, K. and Iwataki, M. 2017. Utility of mitochondrial-encoded cytochrome c oxidase I gene for phylogenetic analysis and species identification of the planktonic diatom genus *Skeletonema*. *Phycol. Res.* 65(3), 217-225.

6

Diatom Taxonomy: Disentangled from Ecology?

Eileen J. Cox

Life Sciences, The Natural History Museum, Cromwell Road, London, England

Abstract

Over the last 200 years innumerable species and varieties of diatoms have been described based on wall morphology and, given the persistence of their silica walls in sediments, diatoms have also been used to document changes in aquatic systems over time. But accurate predictions of environmental conditions are predicated on having reliable data on the ecological range and requirements of individual taxa tied to accurate identification of those taxa. Even with the development of molecular tools, sequences must be linked to described taxa to provide more than simply a list of OTUs. Unfortunately, there has often been a divide between diatom taxonomy and ecology; some workers focus on questions around the identity and relationships of diatoms, while others are interested in environmental questions, what occurs when and where, or what that tells us about the environment.

This chapter will review some of the issues around the recognition and description of diatom taxa, including questions of how to handle cases of morphological variability and phenotypic plasticity. It provides a brief overview of changing emphases in diatom ecology, the use of diatoms as bioindicators and assessing ecological "preferences" of individual taxa from field and experimental data. In bringing these areas together it important to recognise that diatoms are not simply static objects under the microscope, but organisms with the potential to adapt. The increasing use of molecular techniques offers potential complementarity but also requires a willingness to challenge existing paradigms and openness to new interpretations.

***Keywords*:** Diatoms, ecological preferences, experimental studies, integrating genetic data, morphology, taxonomic discrimination, variability

Email: E.Cox@nhm.ac.uk

Nora I. Maidana, Magdalena Licursi and Eduardo Morales (eds.) Diatom Ecology: From Molecules to Metacommunities, (163–202) © 2024 Scrivener Publishing LLC

6.1 Introduction

As scientific disciplines, taxonomy and ecology have different aims and objectives, the former focusing on naming, describing and classifying organisms, while the latter is concerned with the functional relationships of different organisms to each other and their environment. Yet ecology is dependent on taxonomy, because it is important to know what entities (organisms) are being dealt with, and taxonomy can be informed by ecology, insofar as studying organisms in their environment provides information on their life history strategies and responses to different conditions that may be expressed morphologically. Unfortunately, within research into diatoms, these disciplines have tended to develop with little reference to each other, detailed taxonomic works contain limited ecological information, while ecological studies rarely keep pace with taxonomic revisions. Both lack the underpinning of detailed life history studies or long-term investigations of taxa in particular ecosystems. And while heavy reliance is placed on features of their siliceous cell walls, there is relatively little knowledge of the factors affecting the formation of those walls.

Over 20 years ago, Kociolek and Stoermer [6.130] produced an opinion piece entitled "Taxonomy and Ecology: A Marriage of Necessity," which was followed by another contribution by Kociolek [6.129] arguing for the development of online resources to support the dissemination of taxonomic information to both taxonomists and ecologists. However, despite their arguments that taxonomy and ecology should be linked [6.130], efforts seem to have been focused more on making taxonomic information more accessible via online resources [6.129], useful as these are, rather than integrating ecological and life history information into taxonomic investigations. Seeking to bring taxonomy and ecology together, Stoermer [6.225] argued that appropriate use of computer technology could allow the development of integrated databases, linking specimens, images, locality, ecological information, and diatom analyses, improving consistency in identification alongside occurrence and abundance information. Sterrenburg [6.229] meanwhile emphasized the challenges in describing temporal and spatial conditions at the scale experienced by diatoms. Computer technology has undoubtedly facilitated the collation of data and made it more widely accessible, particularly via online resources, such as Diatoms of North America (https://diatoms.org) and the Diatom Flora of Britain and Ireland (https://naturalhistory.museumwales.ac.uk/Diatom). These provide critical taxonomic information and validated occurrence records, with references to relevant literature, but detailed accompanying ecological information remains relatively sparse.

Molecular studies are playing an increasing role in exploring taxonomic relationships, defining taxa and clarifying diatom phylogeny [6.241], and are also being increasingly used to characterize diatom assemblages in the field, thanks to barcoding methods [6.122]. However, whether they represent a fundamentally new approach to diatom taxonomy or ecology is less clear (see below).

This chapter will first review aspects of diatom taxonomy and some of the challenges that should be addressed, followed by a review of diatom ecology. It will then seek to outline where the challenges lie and how research into diatoms could be better integrated to reduce the separation between these apparently disparate areas.

6.2 Diatom Taxonomy

6.2.1 Overview of Species Concepts and Changing Emphases

Discussions of diatom species concepts and systematics have taken place at various diatom symposia [6.59] [6.61] [6.159] [6.197]. Mann [6.156] provided a long review on the topic, while Cox [6.42] [6.46] [6.48] [6.51] [6.52] [6.54] [6.56] [6.58] has addressed various issues around diatom systematics, as well as approaches to taxon recognition and designation, and Medlin [6.163] [6.164] and Alverson [6.9] have addressed the use of molecular techniques for delimiting diatom species.

Early discussion [6.197] focused around problems in delimiting taxa based on morphological differences, given the peculiarities of the diatom life history and the clonal nature of many gatherings, plus the relative paucity of characters compared to multicellular organisms. Mann and Kociolek [6.159] provided a list of species concepts that have been recognized: typological, nominalist, biological, phenetic, phylogenetic and evolutionary, followed by a discussion of deme terminology [6.88] adopted by Mann [6.153] [6.154] as an informal recognition system following his work on mating behavior in *Sellaphora* Mereschkowsky and other raphid genera [6.155].

Subsequent workshops [6.59] [6.61] focused on generic delimitation and character interpretation, and the type method, respectively. It was re-emphasized that, while the rules around the type method are to enable reference to a taxonomic group and indicate its taxonomic rank [6.114] [6.196], they do not provide any guidance on how to recognize that group. Type specimens do not represent all morphologies within the taxon and therefore should not be used as the only morphological information for that taxon [6.53].

Following his work [6.153] [6.154] on some raphid diatoms, and earlier work by Geitler, e.g., [6.83] [6.84], Mann promoted the biological species concept for diatoms, recognizing several new species within *Sellaphora pupula* (Kützing) Mereschkowsky sensu lato [6.160], some of which were treated as varieties or forms by earlier workers, e.g., Hustedt [6.109], Krammer and Lange-Bertalot [6.137]. Lange-Bertalot and Krammer [6.137] adopted a broad morphological species concept, seen in the range of specimens recognized as a single species. However, Mann [6.156] concluded that minor morphological differences, rather than a function of phenotypic plasticity, are genetically determined and faithfully transmitted through the life history. But, establishing this requires detailed observation of live material of many diatom populations and has relatively rarely been undertaken (but see [6.39] [6.66] [6.267] [6.268]). Nevertheless, there is now an increasing trend to discriminate taxa on very fine morphological differences, seen by the proliferation of new species and genera, often from single sites, without any supporting study of changes over their life history [6.22] [6.138] [6.139] [6.166] [6.167] [6.168] [6.169] [6.203] [6.223] [6.257] [6.259] [6.260] [6.261] [6.262] [6.263] [6.264] [6.265].

On the other hand, many ecologists observed variation in wall morphology of similar diatoms under different environmental conditions, particularly in relation to salinity, nutrients, and temperature. Variants were referred to as different "morphotypes" of a single species, e.g., of *Stephanodiscus hantzschii* Grunow in Cleve & Grunow by Håkansson and Bailey-Watts [6.93], or as "Sippen" of the *Fragilaria capucina* sensu Krammer & Lange-Bertalot by Hürlimann and Straub [6.108], who linked the different morphologies to the conditions (natural and experimental) under which they occurred. Changes with salinity were shown for *Thalassiosira* Cleve by Fryxell [6.81], for *Skeletonema* Greville by Balzano *et al.* [6.18] and for *Nitzschia* Hassall by Trobajo *et al.* [6.249] [6.251], with silica concentrations for *Stephanodiscus* Ehrenberg by Bailey-Watts [6.16], and with temperature for *Cyclotella comensis* Grunow by Hausmann and Lotter [6.98], who inferred different temperature optima for the different morphological forms. But such observations did not lead to taxonomic revisions at the species level. It is therefore arguable that, only with the wider dissemination (and influence) of Mann's work [6.154] [6.156] and his interpretation of the significance of small morphological differences, has such variation been considered indicative of speciation.

With the development of molecular tools, process streamlining and cost reduction per sample for use, more genetic sequences have been obtained for many more taxa, enabling comparisons between morphologically similar clones, as well as phylogenetic analyses, to be undertaken (see Section 6.2.3

below). The question then becomes, how much genetic difference indicates a discontinuity between species, and how does this relate to morphology?

6.2.2 Naming and the Type Method

Following the International Code of Nomenclature [6.252], "The application of names of taxonomic groups is determined by means of nomenclatural types." In other words, a name is linked to the first specimen (s) to which the name was applied, with the oldest name having priority. However, as pointed out [6.196], statements such as "the type of the species" is shorthand for "the type of the specific name" and should not be interpreted as meaning that the type specimen is the most representative (typical) example of the taxon [6.53] [6.56]. This applies not only to species, but also to taxa chosen or designated as the "type" of a genus.

This becomes particularly problematic for some of the earliest described genera, the understanding of which has shifted over time, particularly when, until the mid-20th century, there was little reference to type specimens. Thus, some generic names can no longer be applied to some "well-known" widely distributed species, as in the case of *Synedra ulna* (Nitzsch) Ehrenberg; now *Ulnaria ulna* (Nitzsch) Compère [6.271]. On the other hand, when describing new genera for groups of known species, it may be possible to designate a generitype which lies at the center of the variation within the new taxon; early generitypes may not be "typical" [6.56].

Although the early diatomists retained and distributed parts of their collections, with Ehrenberg in particular linking illustrations of his new taxa to specimens [6.73] [6.74], it seems that only from the 1970s onwards did comparisons with type material become common (e.g., [6.42] [6.97] [6.99] [6.100] [6.198]). It is not clear what precisely triggered this practice, but it marks an increased focus on diatom collections, e.g., BM, Bremerhaven, Philadelphia, with individuals such as Robert Ross, Reimer Simonsen and Ruth Patrick playing a key role. In preparing their Diatoms of the United States [6.180] [6.181], the authors "made several extended trips to many herbaria...to locate and examine type specimens and/or type material of many original authors" [6.193], explicitly only describing and illustrating taxa they have been able to verify [6.180]. Interestingly, they also stated that they sought to illustrate only specimens that were typical (i.e., representative) of the taxon. At the same time, use of scanning electron microscopy was becoming important and the contemporary focus of paleolimnological studies into the effects of acid rain (in both Europe and N. America), which depended on consistent application of names and establishing the "correct" name for taxa, was driving increased interest in diatom taxonomy.

Checking modern specimens against historical types aimed to ensure consistency in the application of names, but also revealed problems [6.92], showing how historical allocations to genera based on illustrations and brief descriptions in the 19th century can mislead and conflict with modern concepts of particular genera, then requiring nomenclatural revision. But it is also important to remember that any taxonomic concept is open to modification as new discoveries are made, and definitions can be adjusted in their light. As I pointed out [6.56], if the interpretation of a genus is too tightly tied to the generitype, related taxa may be unnecessarily separated off, with the creation of new genera. This can result in monotypic genera (e.g., *Pseudencyonema* Krammer [6.135], *Diprora* Main [6.152], *Gomphocymbellopsis* Krammer [6.136], *Talaroneis* Kooistra *et al.* [6.133], *Pseudostriatella* Sato *et al.* [6.209], *Skeletomastus* Harper *et al.* [6.956], *Celebesia* Kapustin *et al.* [6.120], *Karthickia* Gluschenko *et al.* [6.89], *Alveocymba* Kapustin *et al.* [6.121], *Chinia* Lin *et al.* [6.143]) being recognized, potentially perpetuating focus on the generitype to define the genus. Of course, it is always possible that other taxa will be discovered that belong in these new monotypic genera, but again, this will depend on how the genus has been defined.

6.2.3 Impact of Molecular Studies

In addition to their use in revealing phylogenetic relationships within the diatoms [6.158] [6.241], molecular studies have led to the description of new species, especially within what were previously considered widespread planktonic species, e.g., *Skeletonema* Greville and *Pseudo-nitzschia* H. Peragallo [6.150] [6.151] [6.165] [6.207] [6.208] [6.274]. In these cases, sequence discrepancies stimulated closer morphological examination particularly using SEM, revealing consistent subtle character differences, although some "new" species were subsequently abandoned [6.134]. Thus, following Medlin *et al.*'s discrimination [6.165] of *Skeletonema pseudocostatum* Medlin, further studies on strains of *Skeletonema* led to the recognition of five new species [6.207] [6.274], and two years later, another new species [6.208]. Valve and girdle band structure were used as morphological criteria, but variability in girdle band ultrastructure which had discriminated *S. dohrnii* Sarno & Kooistra and *S. marinoi* Sarno & Zingone showed that these taxa could not be separated morphologically, questioning their recognition as discrete species. It was also found that *S. dohrnii* band structure could be found in genetically defined specimens of *S. marinoi* [6.134]. Further studies on 184 *Skeletonema* strains revealed that, for all but one species, morphologically defined taxa contained multiple, albeit similar, genotypes, concluding that "no single set of criteria currently defines species in *Skeletonema*" [6.134]. Similarly, based on both morphology and DNA, new taxa (cryptic and pseudocryptic species) were described for *Pseudo-nitzschia* [6.150] [6.151]. Cryptic species are genetically different but morphologically the same, whereas pseudocryptic species show minor morphological differences only detected with detailed examination. Difficulties in identifying cryptic or pseudocryptic *Pseudo-nitzschia* species can be problematic for monitoring programs because of the potential toxicity of different strains [6.150], although it has also been a spur to the investigation of both the taxonomy and physiology of this genus.

Whereas morphological discrimination of genetically discrete taxa has been possible for *Skeletonema* and *Pseudo-nitzschia*, albeit with additional diversity being found within some species [6.134], Kaczmarska *et al.* [6.119] found that the genetic divergence among 24 *Asterionellopsis* F.E. Round in Round *et al.* clones was not matched by their morphological variation, although they did describe four new species. Similarly, Pinseel *et al.* [6.185] found that distinct molecular lineages of *Pinnularia borealis* could not be separated using LM or SEM. On the other hand, work on *Sellaphora* [6.23] [6.78] [6.153] [6.155] [6.160] might have suggested that genetically or reproductively discrete taxa are invariably morphologically distinct. However, it is worth noting that, traditionally, LM allocation to *S. pupula* (as *Navicula pupula* Kützing in Krammer & Lange-Bertalot [6.137]) rested primarily on the presence of a few distinctive features, e.g., polar bars, resulting in more broadly defined species than for some other naviculoid taxa. Thus, while Mann and colleagues' work [6.23] [6.78] [6.1523] [6.155] [6.160] is elegant in its combined use of morphology, breeding patterns and molecular studies, *Sellaphora* may have been a serendipitous choice and not necessarily representative of all raphid diatoms.

Medlin [6.163] [6.164] discussed the use of molecular techniques to define species limits, suggesting that molecular substitutions in slow evolving regions of morphologically well-defined species, e.g., small subunit rRNA, could suggest the existence of cryptic or sibling spp. Other genes have been used with different diatoms to explore species and population level relationships. Trobajo *et al.* [6.248] investigated relationships within geographically dispersed strains of *Nitzschia palea* (Kützing) W. Smith using LSU rDNA and later cox1, rbcL and LSU rDNA [6.250], concluding that this comprises several species, although these have not been formally described. Elsewhere, Evans *et al.* [6.77] had shown 2.9–3.7% divergences were indicative of species differences, but rates of gene evolution differ between genes and taxa, so simple extrapolation from the *Sellaphora* example is probably unwise. In addition, whereas genetic analyses are performed on diatom clones, i.e., single genotypes, species should be expected to contain multiple genotypes, which can confer ecological or physiological adaptability [6.90]. How much genetic diversity can be expected in a diatom species remains unanswered, and objective criteria allowing the definition of a precise molecular threshold to discriminate species remain to be found, if indeed this is possible.

6.2.4 Morphological Variation Within Clones

The regular inclusion of images covering the (presumed) entire size range of many taxa shows that diatomists are aware that changing size and valve proportions should be considered when defining the limits of species, but if specimens in a particular sample do not cover the entire morphological range, and other gatherings have not been observed, species boundaries may be drawn too narrowly. Thus, auxosporulation in *Dickieia ulvacea* Berkeley ex Kützing, as *Navicula ulvacea* (Berkeley ex Kützing) Cleve, was found to produce cells that fitted *Navicula subinflatoides* Hustedt [6.45], indicating that the latter should not be treated as a separate species. Elsewhere, life history studies of some clones of *Gomphonema parvulum* (Kützing) Kützing sensu lato revealed that cells at the maximum end of the size range would be assigned to other species [6.195], while Meyer *et al.* [6.171] concluded that *Cyclotella wulfiae* Meyer & Håkansson is the auxospore stage of *C. meneghiniana* Kützing. *Cyclotella wulfiae* had previously been described from a single plankton sample in which *C. meneghiniana* was the most abundant taxon [6.170], and although whether *C. wulfiae* represented the auxospore stage of *C. meneghiniana* was discussed [6.171], it was discounted based on the hemispherical shape of a *C. meneghiniana* initial cell [6.104, fig. 24] compared with the concentrically undulate central part of the valve of *C. wulfiae*. Later experimental studies [6.171] showed that those characters which earlier seemed sufficient to define *C. wulfiae* were found in the initial valves of *C. meneghiniana*, linking these as parts of the same taxon.

Whilst it is possible to link morphological variants within clones when working with cultured material, discontinuity in the size ranges of field specimens from different localities or collection dates may lead to their recognition as different taxa, whereas observations of changes with decreasing cell size or over time would show probable conspecificity [6.43] [6.46] [6.172]. One of the problems with size reduction in pennate diatoms is that change is allometric, but shape perception is scale independent [6.43] [6.49]. Thus, for cells of the same width, the eye interprets shorter cells as broader than longer ones. Valve outlines can also change in different ways, e.g., apices may become less pronounced in some species, but retain their distinctiveness in others.

The use of shape analysis was proposed to help determine whether populations showing shape variation from different sites [6.230] [6.232] or sampling occasions [6.224] were conspecific, or to enable discrimination of non-identity in taxa with few morphological characters, such as *Asterionella* Hassall [6.179] and *Fragilariforma* Williams & Round [6.127], in the absence of experimental studies. More recently, shape analyses are being used when discriminating new species [6.190] and to separate morphologically similar taxa with different environmental distributions [6.118] [6.189].

6.2.5 Phenotypic Plasticity – Implications

Heterovalvy and phenotypic variation have been recognized in a range of species [6.58] [6.72] [6.227], but their significance has probably been neglected because of the persistent, widely repeated dogma that diatom valves are replicated precisely at each cell division. Thus, Mann's [6.156] suggestion that "although more studies are certainly necessary and despite some clear demonstrations of polymorphism and phenotypic plasticity, enough is known to make it unlikely that future research will greatly undermine the reliance taxonomists have placed on shape, wall structure, striation density, and pattern" is widely accepted. Nevertheless, Mann [6.156] does note that "Schmid [6.211] takes a different view, claiming that, because ontogeny is 'highly sensitive to the environment,' there are in fact many fewer species than are currently recognized." However, studying ontogeny and development, or following entire life histories, usually requires a more experimental approach, obtaining and maintaining clones in culture [6.45] [6.55] [6.172] [6.195] [6.244], although serendipity may provide some information (e.g., [6.242] [6.243]).

Heterovalvy in chain-forming centric diatoms, e.g., *Aulacoseira* Thwaites, *Skeletonema*, and *Chaetoceros* Ehrenberg [6.2] [6.64] [6.65], with the development of linking or separation valves, allows chain length to be regulated and presumably thereby also the position in the water column. But this is rarely recognized as an expression of phenotypic plasticity, potentially mediated by the environment [6.26] [6.64] [6.186] or the presence of grazers [6.10] [6.27] [6.29]. For example, chain length in *Skeletonema* increased with increasing temperature and nutrients, thereby maintaining its position in the water column [6.236], whereas reduced chain length reduced grazing losses [6.29].

Heterovalvy in chain-forming diatoms and in Janus cells clearly indicates that a single genotype can generate different morphologies under different environmental conditions. However, the production of araphid valves in monoraphid taxa, and different spine types in some centric taxa, such as *Corethron* Castracane [6.62] [6.63], occurs regularly with each cell division. There must therefore be some internal control determining which forming valve infills its developing raphe fissure or develops particular spine types, i.e., how the appropriate hypovalve is formed in sibling cells.

Heterovalvy (other than within monoraphid diatoms) has been observed in wild populations of different diatoms (Janus cells) [6.13] [6.55] [6.72] [6.161] [6.227] [6.240], but the taxonomic implications have rarely been pursued. Thus, although it has been recognized that the individual valves of a single frustule would be assigned to different species [6.13] [6.83], the necessary taxonomic revisions have seldom followed. However, recognizing that specimens attributed to *Mastogloia grevillei* Wm. Smith and *M. danseyi* (Thwaites) Thwaites ex Smith are phenotypic expressions of the same entity, Edlund and Burge [6.72] made the necessary taxonomic revision, with the former taxon being recognized at forma level as *M. danseyi* f. *grevillei* (Wm. Smith) Edlund & Burge (*M. danseyi* has priority). Experimental manipulation on this taxon [6.107] showed that changing the conductivity of the medium triggered the switch between phenotypes.

Variability in *Stephanodiscus hantzschii* Grunow in Cleve & Grunow [6.93] [6.94] [6.128], with the nominate form having more heavily silicified valves, straighter fascicles and lower areola density compared to *S. hantzschii fo. tenuis* has been linked to silica availability. Stoermer *et al.* [6.229] suggested that the latter is an ecophenotype produced under silica limitation, also suggested [6.12] to have a higher phosphorus optimum than the nominate form. Again, heterovalvy indicates that morphological variation is linked to environmental conditions [6.12].

The link between morphology and temperature was shown for *Thalassiosira gravida* Cleve and *T. rotula* Meunier [6.235] but, despite showing that these were morphological forms of the same entity, no taxonomic revisions were made. Sar *et al.* [6.206] later recognized these morphs as part of the same species, with *T. gravida* having nomenclatural priority and the correct name for the taxon. However, no use of the epithet *rotula* to indicate the environment which generated that morphology (e.g., as *T. gravida* fo. *rotula*) was proposed. The effect of changing salinity on valve morphology may be great enough for specimens grown under contrasting conditions to correspond to different published taxon descriptions, e.g., *Cyclotella cryptica* Reimann, Lewin & Guillard v. *Cyclotella meneghiniana* Kützing [6.213], *Thalassiosira tumida* (Janisch) Hasle v. *Thalassiosira eccentrica* (Ehrenberg) Cleve-like areolation [6.81]. Other studies have also reported significant morphological variation in different salinities, but have not suggested correspondence with other taxa [6.162] [6.251].

6.2.6 Use of Infraspecific Categories for Diatom Taxa

The definition of subspecific categories (ICN) follows a classificatory hierarchy, without designating criteria on the degree of dissimilarity warranting separation [6.252]. Thus, infraspecific categories have been variously used for diatoms, their use often seeming largely a matter of personal taste [6.48]. No consistent criteria have been applied [6.201], and some individuals have argued that "varieties and forms have little scientific status" [6.202] or that their use should be avoided [6.157]. Others considered that recognition as a variety implied genetic differentiation [6.101] [6.111], although despite this, Hustedt [6.111] was not in favor of recognizing varieties or forms. On the other hand, labelling subtly different morphological variants allows differences to be registered and avoids loss of information, particularly in an ecological context [6.48].

Varietal status has been used for morphological variants of a (presumed single) species under different ecological conditions, stressing their indicator value (e.g., [6.95]), whereas the above use of form [6.72] identifies morphological variants that are clearly reversible phenotypic expressions of a single genotype under different conditions, which can provide information on the conditions under which they were formed. However, diatomists generally seem reluctant to designate varieties, preferring to refer to cryptic, pseudocryptic or sibling species when genetic differences are found within morphologically similar clones [6.10] [6.119] [6.150] [6.163] [6.164] [6.185] [6.266]. Thus, Vanelslander *et al.* [6.266] did not formally (nomenclaturally) discriminate two clades of *Navicula phyllepta* Kützing, despite the morphological, genetic and ecophysiological differences between them. On the other hand, Abarca *et al.* [6.1] described two varieties of *Gomphonema subclavatum* Grunow on the basis of morphological and genetic (3 base pairs in *rbcL*-marker) differences (albeit relatively small differences). This allows their new taxa to be discriminated while reflecting their close relationship and could be a useful approach for handling other pseudocryptic or sibling species.

6.3 Diatom Ecology

6.3.1 History of Diatom Ecology

Diatom ecology has included a range of studies and techniques, the focus of which have gradually changed over the years. Studies of the spatial and seasonal patterns of diatoms in different ecosystems, marine and freshwater, planktonic and benthic were a major focus in the mid-20th century. Links between environmental variables such as nutrients, temperature and light, and fluctuations in species occurrences in the plankton [6.17] [6.145] [6.146] [6.147] [6.194] [6.210] [6.218] [6.220] [6.231] and in benthic communities [6.5] [6.200] [6.226] were shown and species interactions and growth strategies explored [6.126] [6.194] [6.220] [6.245] [6.246]. Despite experimental studies to determine some of the physiological adaptations or constraints contributing to the field observations (see below), on the whole, ecological studies of diatoms have been based on field observations linked to measured environmental variables. More recently, as concern over the problems of the eutrophication and pollution of natural waters has grown, given their widespread distribution and ecological diversity, diatoms have become an important tool to evaluate water quality, as well as to trace past histories.

6.3.1.1 Phytoplankton

Lund and co-workers [6.145] [6.146] [6.147] [6.148] [6.149] showed how seasonal patterns of freshwater phytoplankton growth were linked to changes in dissolved nutrients in the epilimnion, suggesting that species had different requirements and competitive abilities, helping elucidate Hutchinson's "paradox of the plankton" [6.113]. Other long-term studies [6.85] [6.86] showed the contrasting responses of taxa to silica inputs. Field manipulations [6.149] and experimental studies using cultures [6.126] [6.245] [6.246] confirmed many of the predictions, demonstrating causal relationships between nutrients (silicon and phosphorus) and phytoplankton growth [6.194] [6.245]. Contrasting uptake kinetics of individual species revealed their ecophysiological differences and were used to assess their competitive ability [6.126] [6.245] [6.246], explaining some of the seasonal phenomena [6.145] [6.146] [6.147] [6.148] [6.247]. Similar experimental studies on marine phytoplankton [6.35] [6.176] [6.177] [6.178] [6.216] [6.217] [6.218] showed how salinity, silicon, light and temperature could control diatom phytoplankton succession.

6.3.1.2 Benthic Ecology

During the 1960s and 1970s many studies focused on patterns of diatom distribution on both marine and freshwater sediments [6.35] [6.71] [6.234], including spatial and seasonal aspects [6.11] [6.105] [6.106] [6.173] and their role in primary production [6.102]. Revealing the complexity of the environment and species responses, Admiraal's [6.5] comprehensive review of the ecology of sediment-dwelling estuarine diatoms summarizes much of the earlier work. While field observations revealed potential interactions between species and the fluctuating physical and chemical characteristics at the sediment-water interface, complementary experimental studies were used to test these correlations [6.3] [6.4] [6.6]. This reflected the contemporary interest in exploring how cell physiology could explain ecological phenomena (cf. work on phytoplankton by [6.126] [6.246] [6.247]) even in the more complex heterogeneous benthic environment.

The complexity of investigating growth and productivity in benthic environments is seen in the divergence in approach between freshwater and estuarine systems. Whereas the regular emersion of sediments in estuaries has allowed *in-situ* studies of primary productivity [6.30] [6.184] [6.255], these have been more challenging in consistently immersed freshwater habitats [6.8] [6.226]. Although some laboratory studies have been performed to understand how nutrients, light and temperature can control the growth of benthic diatoms, the use of artificial substrata [6.36] and field-based manipulations have been more popular. These have included transplantation experiments and the use of diffusing substrata to mimic different nutrient or pH regimes [6.79] [6.103], whereas laboratory culture experiments have been more restricted (but see [6.49] [6.50] [6.249] [6.251]). There has also always been a tension between autecological investigations versus whole assemblage studies under "natural" conditions. Whole assemblage studies integrate the effects of all environmental factors but cannot disentangle the effects of co-varying factors, e.g., light intensity and temperature [6.144], or demonstrate more than a correlation between species and environmental factors.

Whereas growth and primary productivity studies have related these to total assemblages (biomass), most ecological studies today are largely focused on the links between species occurrences and environment via spot sampling. There is less focus on investigating the spatial and seasonal patterns of assemblages (but see [6.222]) or the ecophysiology of the component taxa. It is therefore arguable that most contemporary studies are essentially directed towards monitoring rather than ecology.

6.3.1.3 *Paleolimnology*

The possession of resistant silica frustules and their preservation in sediments makes diatoms useful tools to investigate past environments, both fossil and sub-fossil. In particular, studies of Holocene lake deposits provided evidence of changing climate, salinity and trophic status. Diatoms were used to trace marine-freshwater transitions [6.32] and changing lake trophic history [6.99] [6.199], but their use as indicators of acidification, coinciding with growing awareness of the impact of acid rain, markedly increased their use (e.g., PIRLA and SWAP programs [6.19] [6.38]) and later to climate change [6.20]. Initially weighted averaging equations were employed, using allocations to pH categories, then transfer functions [6.28], based on correspondence analysis approaches to explain relationships between species and environmental variables were developed. Alongside studies of modern assemblages and corresponding physical and chemical analyses, these so-called training sets allowed paleolimnologists to infer environmental optima for individual species, and thereby historical conditions, in lake systems [6.21] [6.24] [6.34] [6.70] [6.117] [6.219].

One notable outcome was a push to harmonize taxonomy in the large multi-institutional programs [6.19] [6.38]. Workshops brought contributors together to compare specimens and nomenclature, and to reach agreement on their naming to ensure consistency. However, this did not necessarily result in the "taxonomically correct" name, i.e., based on the original material and descriptions, which would have required time-consuming visits to historical collections and detailed comparative work (cf. [6.272]). Ecological information was linked to "agreed names" (consensus taxonomy) within each program, but others might apply those names to quite different entities, with the information content of the name therefore becoming unreliable. That this is still an issue is demonstrated by a recent paper [6.253] that proposes a revised protocol for counting diatoms to reduce taxonomic bias.

6.3.2 Diatoms as Bioindicators

With growing concern over the eutrophication and pollution of natural waters, diatoms became a recognized group of organisms with which to evaluate contemporary water quality. They have been promoted as useful bioindicators of aquatic ecosystems, both past and present [6.233], thanks to their wide distributions and diverse assemblages, and to the resistance of their siliceous walls (on which many are identified) to damage and degradation. But, if diatoms (or any organism) are to provide information about environmental conditions at the sampling site, precise taxonomic identification linked to knowledge of their ecological distribution is essential.

Over 100 years ago, Hustedt [6.110] formalized the pH relationships of some diatoms as acidobiontic, acidophilic, indifferent, alkaliphilic and alkalibiontic, according to their occurrence in different pH regimes. These were then used to develop various indices [6.40] [6.116] [6.174] to estimate pH based on the diatom assemblage that could be used for both modern and historical/fossil assemblages. A comparable approach was used for salinity regimes [6.131] and trophic status developed from saprobic indices [6.132] [6.141] [6.142] [6.256].

Whereas some of the early work to infer environmental conditions used so-called indicator species, e.g., *Eunotia* spp. occur under acidic conditions [6.80] [6.112] [6.132], this was superseded by the development of indices based on the full assemblage at a site [6.33] [6.68], particularly to evaluate trophic status or pollution (e.g., DAIpo [6.270]). Again, weighted averaging methods [6.28] led to correspondence analyses [6.238] [6.239], which have been used extensively to explore taxon-environment relationships in modern material. However, these are based on correlations between occurrences (relative abundance) and measured environmental variables, unable to separate the effects of co-varying factors or to demonstrate causality [6.49] [6.50].

Incentives to monitor water quality, including the Water Framework Directive in Europe, stimulated work on diatoms in rivers and streams, leading to the development of various schemes, such as the Specific Pollution Sensitivity Index (IPS), Trophic Diatom Index (TDI), and Diatom Biological Index (IBD), in different regions for evaluating impact [6.37] [6.41] [6.123]. Although they may differ in the optimum and tolerance values determined *a priori* to individual taxa in the different European regions, their calculations of water quality are all based on a weighting average equation. On the other hand, in North America and Canada, indices of biotic integrity were developed, involving a range of metrics, inference models and ordination techniques [6.142].

Despite an enormous increase in the number of new diatom species described, many ecological studies still rely on a limited number of standard texts to identify taxa. Furthermore, because routine monitoring is not invariably performed by specialists, but relies on consistency, taxon identification may not be to the lowest possible taxonomic level but a compromise between precision and practicality [6.124].

6.3.3 Problems with Establishing Species 'Preferences'

Most of the analytical methods use species relative abundances in their calculations, which assume that the most numerous taxon is best suited to the particular environment, ignoring any effects of relative cell sizes (biovolume) and assemblage density [6.60], let alone successional events.

Percentage abundance is a relative measure that does not provide any information on density (cell numbers per unit area or volume) and does not allow between-site comparisons of a taxon's growth [6.69]. A taxon can have a high percentage abundance but low density if it is one of very few taxa in the sample [6.60], a particular problem for studies on benthic communities or sediment core samples. Whereas collecting known volumes of water to evaluate phytoplankton density is relatively easy, obtaining appropriate quantitative samples of benthic assemblages to relate biomass and taxon counts to surface area is more difficult [6.71] [6.102]. The use of artificial substrata overcomes some of these issues but relies on colonization being representative of the resident flora [6.36] [6.212]. And even if quantitative samples have been taken, variation in the assemblages may have been presented as percentage abundances [6.205] because this was how assemblage data were presented.

Behind the assumption that the most numerous taxon is best suited to the environment, is (implicitly) the correlate that that taxon grows "best," or competes most successfully, under those conditions, whereas in fact it may be that its apparent dominance simply reveals that it is one of the few taxa able to grow under those conditions. Thus, *Eunotia exigua* (Brébisson ex Kützing) Rabenhorst, despite being considered an acidobiontic taxon (optimum 4.0-4.5) [6.242], can grow at neutral pH values [6.48], and was found "quite abundant in waters of pH 7.0" [6.7]. Experimental studies permit more rigorous evaluation of tolerances and optima, also allowing the effects of co-varying factors to be separated [6.50] (see Section 6.3.4 below).

Whereas nutrients can become depleted in the epilimnion, and seasonal succession in phytoplankton taxa may be determined by limiting nutrients, this may not be the case within the benthos, where sediments or moving water replenish nutrients as they are used [6.31]. In addition, measuring the environment in which benthic taxa are living is more problematic as this is essentially an interface between the substratum and overlying water, or between sediment and its surface. It is arguable that epilithic taxa growing on essentially inert surfaces respond primarily to the overlying water, and therefore can indicate its chemistry, the situation for epipelic and epiphytic taxa is more complex. Even in the same water, sediment chemistry will vary with differences in rock type or sediment grain size, affecting the epipelic flora [6.31] [6.44] [6.47], and even the epiphyton may vary depending on host and water chemistry [6.75]. Thus, simple measurements of overlying water chemistry may not accurately record the environment in which the diatoms are growing, or therefore provide information on their ecological requirements.

Diatom preparation and counting techniques introduce additional potential errors. Because these techniques remove organic material before slide preparation, the resulting sample potentially includes autochthonous and allochthonous material, both live and dead specimens at the time of sampling. Even fossil material can be found in modern sediment samples [6.15]; extinct *Aulacoseira canadensis* (Hustedt) Simonsen has been found in recent benthic samples due to erosion of diatomites in catchment. Discriminating between live and dead cells in a sample may also be important [6.87] [6.192] [6.175] [6.272] particularly if the diatoms concerned occur under different conditions. Routine counting techniques can also significantly underestimate large specimens [6.188] [6.254] unless a two-stage protocol is adopted, counting large cells over the entire slide, followed by transect counts at higher magnification for smaller specimens.

6.3.4 Implications of Experimental Work

Experimental studies offer a means to investigate diatom responses to defined variables under controlled conditions. They cannot precisely replicate the natural world, but can separate the effects of co-varying factors (e.g., light and temperature [6.49]), and provide a test of optima derived from field-based observations [6.50] [6.214]. Thus, growing *Meridion circulare* (Greville) C.Agardh and *U. ulna* at different temperatures and light regime, showed that their occurrences in the field reflected their tolerance or otherwise of high light and high temperature, factors that often co-vary, effects that are difficult to separate in the field [6.144]. Experiments also revealed that *E. exigua*, usually considered acidobiontic, can grow at neutral pH [6.50]. It may not have a "preference" for acidic conditions but is one of the few taxa that can grow under such conditions.

Studies of salinity tolerance in several *Skeletonema* strains revealed broader salinity tolerances under experimental conditions than in the field, i.e., their potential salinity range is wider than their realized niche, presumably because additional environmental factors come into play [6.18]. Experimental studies can demonstrate the ecophysiological capacity of individual strains under defined conditions, but interacting factors in the natural environment will modify this. Most field observations also measure standing crop rather than productivity, i.e., a snapshot of performance rather than revealing how well taxa are growing (a dynamic process).

Using the results of culture experiments, Sime [6.214] determined the phosphorus optima of several planktonic diatoms and compared them to optima derived from diatom training set transfer functions [6.25]. Optima obtained experimentally were markedly lower than those from the training sets, although the rank ordering of taxa was similar. Experimental data revealed the physiological requirements for phosphorus, for which training sets could only provide correlative information using field measurements. In addition, because field taxonomic data are integrated over the whole year, any seasonal fluctuations, when interspecific physiological differences can be critical, are lost.

6.3.5 What Do Molecular Studies Tell Us?

In addition to their taxonomic value, molecular techniques have also been used to study seasonal and spatial variation in diatom populations, particularly for planktonic species [6.90] [6.91] [6.204] [6.215]. Because diatoms predominantly reproduce asexually, large numbers of genetically identical individuals can potentially be produced at a particular site, with the possibility that clonal and genetic diversity at a site is very low. However, genetic diversity between clonal populations can allow those populations to become dominant under different conditions at a single site [6.90]. Clones from different populations of a single species can show variable responses to temperature [6.91], light [6.204], and salinity [6.215], each with particular phenotypic characteristics that allow them to dominate under specific conditions. As conditions change, different clones will become dominant [6.90], allowing the species to survive over a wider range of conditions than an individual clone. Genetic diversity may also confer resistance to parasites [6.67]. Studies on clonal diversity in benthic taxa are much more limited [6.76] [6.269] and focused more on spatial patterns and population differentiation, but again reveal clonal diversity within species, as originally suspected from allozyme studies [6.82] [6.221].

Given the need for, but general lack of, taxonomic expertise required for monitoring methods using diatoms, along with the reduction in costs for DNA analysis, the use of eDNA and barcoding techniques to evaluate water quality are being investigated and developed [6.14] [6.122] [6.182] [6.183] [6.237]. While eDNA differences between sites can be readily observed, their interpretation requires sequences to be reliably and precisely linked to known taxa, and reliable DNA reference databases to be available. As with morphological methods, the link between taxon presence and environmental signal must also have been established. Alternatively, using the ecological profiles of OTUs (operational taxonomic units) rather than species identifications has been advocated to overcome the problem of incomplete DNA reference libraries [6.237].

The difficulties of linking eDNA results to traditional identification-based analyses have been shown by Khan-Bureau *et al.* [6.125], who compared morphological and DNA barcoding methods. Comparing taxa identified morphologically and from eDNA sequences, they [6.125] showed that only 15 of the 63 morphologically identified taxa could be linked to DNA sequences, and seven DNA sequences had no morphological diatom matches. Enlarging the DNA reference databases would reduce the problem but is an intensive endeavour [6.274] and relies on accurate identification of sequenced samples. As with traditional diatom monitoring methods, accurate taxon identification and knowledge of their ecological ranges remain a critical issue.

6.4 Where Are the Challenges?

Bringing diatom taxonomy and ecology together requires not only how to discriminate and identify taxa, but also knowledge of how that morphology may vary over the life cycle and with environmental conditions. Thus, diatoms should be understood as dynamic organisms, not simply static objects under a microscope. Even using molecular techniques requires questions around the degree of genetic variation within and between species to be clarified. Simple percentage differences, however tempting that approach might be, cannot be relied on to determine taxonomic boundaries. There is also the question of how much variation in gene expression plays a role in phenotypic responses to fluctuating environmental conditions.

Similarly, when considering their ecology, thinking about the precise environment in which species are living, as well as an awareness of spatial and temporal variation is critical. It is also imperative that identifications are based on the most well-supported evidence, and that, if not illustrated, the relevant texts are referenced to allow cross-checking. The importance of documenting morphology and/or maintaining voucher specimens is also critical in linking molecular sequence and barcode data to taxon names. From both an ecological and phylogenetic perspective, assigning the "wrong" name means that misleading conclusions can be drawn and therefore the potential information content of the name cannot be relied upon by subsequent workers.

With any branch of science there is a danger that published statements become established as paradigms and data that contradict those paradigms are quietly laid aside. Thus, the assumption that diatom morphology is replicated precisely with each cell division can lead to unnecessary taxon splitting, unless tempered with an understanding that some of the constraints on wall formation or contrasting environmental conditions may lead to change. The role of time, with respect to both life history changes and the semi-conservative nature of frustule formation, should not be forgotten. Clearly primarily ecological studies may not allow the opportunity for investigations of life histories or experimental tests of morphological variability to be undertaken, but it is important to look for evidence to support any conclusions, taking advantage of complementary approaches, e.g., shape analysis to compare and contrast specimens from different sites can aid taxon discrimination, while ordination methods can be used to explore taxon relationships with the environment [6.118]. This is particularly valuable for groups of morphologically similar, but ecologically contrasting taxa [6.118]. Similarly, Poulíčková *et al.* [6.191] demonstrated that different taxa within the *Sellaphora pupula*/*S. bacillum*/*S. laevissimum* complexes have different ecological distributions, and that their discrimination is ecologically informative.

6.5 Integrating Information

The ongoing requirement to monitor freshwater systems (e.g., Water Framework Directive) sits at the interface between diatom taxonomy and ecology and continues to focus attention on the identity and indicator value of many diatom taxa. It is therefore essential that these disciplines come together, rather than operating independently. With the development of molecular barcoding techniques and their uptake, arguably to reduce error from morphological identifications by different analysts or the cost of specialist knowledge, some interesting convergences and potential synergies are emerging. Morphological approaches are showing that some previously considered widespread taxa with broad ecological tolerances, such as *Achnanthidium minutissimum* (Kützing) Czarnecki, comprise discrete units with narrower ecological ranges, the recognition of which would increase the information content of environmental analyses [6.189] [6.190], whilst barcoding methods are revealing genetic diversity linked to morphological and ecological variation in this complex [6.182] [6.183]. Similarly contrasting ecological groupings have been found for genetic variants in other species [6.183], indicating that ecotypic differentiation is not an isolated phenomenon within diatoms, borne out by population studies on planktonic taxa [6.90].

It is impossible today for any individual to be proficient in all areas of diatom research, but greater awareness by the individual of the findings in different areas is essential. Diatoms are living organisms and therefore should be expected to show variability and adaptability. To what extent working with cleaned frustules has subconsciously created an illusion of morphological constancy is an unanswered question, but combined with an emphasis on quantitative, measurable morphological characters (size, diameter, length, breadth, stria density, etc.) has probably contributed to a tendency to split rather than link morphological variants.

6.6 Conclusions

Personal experience plays a large role in how we approach scientific questions and will also affect the conclusions we reach. Conclusions are based on observations and learnt experience, and observations depend on the material we have observed. As Potapova [6.187] commented in relation to various diatom halobion systems, "Each classification was based on its author's experience with diatoms inhabiting fresh, saline, and intermediate salinity waters of a particular type (estuaries, coastal areas, and inland waters) and of a particular geographic area." It is important, while learning from what we have observed ourselves, that we also have an open mind to the findings from other disciplines, and work towards a more synthetic approach to diatom taxonomy and ecology, seeing them as intrinsically linked and mutually informative rather than independent, unrelated disciplines.

Acknowledgments

I am grateful to all the organizations in which I have worked, the University of Bristol, the University of Oxford, the Biologische Anstalt Helgoland, the Max Planck Institute for Limnology, the University of Sheffield and the Natural History Museum, London, for providing different opportunities to study many aspects of diatoms and to discuss and collaborate with other individuals, not only diatomists. In particular, I would like to thank Iain Sime, Dawn Rose, Katie Bentley, Ingrid Jüttner and Rosa Trobajo who, through their various projects, have contributed to my thinking on diatoms.

References

[6.1] Abarca, N., Zimmermann, J., Kusber, W.-H., Mora, D., Van, D.T., Skibbe, O. & Jahn, R. 2020. Defining the core group of the genus *Gomphonema* Ehrenberg with molecular and morphological methods. *Botany Letters* 167: 1-46.

[6.2] Abe, K., Lam, D.W., Ashworth, M. & Jordan, R. Observations on *Praecorethron crawfordii* gen. et sp. nov. (Corethrales, Bacillariophyceae) from Upper Cretaceous marine sediments, southwest Pacific. *Nova Hedw.* 151:273-314. 2021.

[6.3] Admiraal, W. Influence of light and temperature on the growth rate of estuarine benthic diatoms in culture. *Mar. Biol.* 39, 1-9. 1977.

[6.4] Admiraal, W. Tolerance of estuarine benthic diatoms to high concentrations of ammonia, nitrite ion and orthophosphate. *Mar. Biol.* 43, 307-315. 1977.

[6.5] Admiraal, W. The ecology of estuarine sediment-inhabiting diatoms. *Prog. Phycol. Res.,* 3, 269-322. 1984.

[6.6] Admiraal, W. & Peletier, H. Sulphide tolerance of benthic diatoms in relation to their distribution in an estuary. *Br. phycol. J.* 14, 185-196. 1979.

[6.7] Alles, E., Nörpel-Schempp, M. & Lange-Bertalot, H. Taxonomy and ecology of characteristic *Eunotia* species in headwaters with low electric conductivity. *Nova Hedw.* 53, 171-213. 1991.

[6.8] Aloi, J.E. A critical review of recent freshwater periphyton field methods. *Can. J. Fish. Aquat. Sci.* 47, 656-670. 1990.

[6.9] Alverson, A. Molecular systematics and the diatom species. *Protist* 59: 339-353. 2008.

[6.10] Amato, A., Sabatini, V., Nylund, G.M., Bergkvist, J., Basu, S., Andersson, M.X., Sanges, R., Godhe, A., Kiørbo, T., Selander, E. & Ferrante, M.I. Grazer-induced transcriptomic and metabolomic response of the chain-forming diatom *Skeletonema marinoi*. *The ISME Journal* 12, 1594–1604, 2018.

[6.11] Amspoker, M.C. & McIntire, C.D. Distribution of intertidal diatoms associated with sediments in the Yaquina estuary, Oregon. *J. Phycol.* 14, 387-395. 1978.

[6.12] Anderson, N.J. The biostratigraphy and taxonomy of small *Stephanodiscus* and *Cyclostephanos* species (Bacillariophyceae) in a eutrophic lake, and their ecological implications. *Br. phycol. J.* 25, 217-235. 1990.

[6.13] Andrejić, J.Z., Spaulding, S.A., Manoylov, K.M. & Edlund, M.B. Phenotypic plasticity in diatoms: Janus cells in in four *Gomphonema* taxa. *Diatom Res.* 33, 453-470. 2019.

[6.14] Apothéloz-Perret-Gentil, L., Bouchez, A., Cordier, T., Cordonier, A., Guéguen, J., Rimet, F., Vasselon, V. & Pawlowski, J. Monitoring the ecological status of rivers with diatom eDNA metabarcoding: A comparison of taxonomic markers and analytical approaches for the inference of a molecular diatom index. *Mol. Ecol.* 30, 2959-2968. 2020.

[6.15] Bahls, L., Potapova, M., Fallu, M.-A. & Pienitz, R. *Aulacoseira canadensis* and *Aulacoseira crassipuntata* (Bacillariophyta) in North America. *Nova Hedw.* 135, 167-184. 2009.

[6.16] Bailey-Watts, A.E. Studies on the control of the early spring diatom maximum in Loch Leven 1981. In: Round F.E. (ed.), *Algae and the aquatic environment*, 53-87. 1988.

[6.17] Bailey-Watts, A.E. The abundance, size distribution and species composition of unicellular Centrales assemblages at late winter-early spring maxima in Loch Leven [6.Kinross, Scotland] 1968-1985. In: Round, F.E. (ed.) *Proc. 9th Int. Diat. Symp.*, 1-16. 1988.
[6.18] Balzano, S., Kooistra, W.H.C.F. & Sarno, D. Effects of salinity on the growth rate and morphology of ten *Skeletonema* strains. *J. Plankton Res.*, 33, 937–945. 2011.
[6.19] Battarbee, R.W. Diatoms, lake acidification and the Surface Water Acidification Programme (SWAP): a review. *Hydrobiologia* 274, 1-7. 1994.
[6.20] Battarbee, R.W. Palaeolimnological approaches to climate change, with special regard to the biological record. *Quat. Sci. Rev.* 19, 107-124. 2000.
[6.21] Battarbee, R.W., Monteith, D.T., Juggins, S., Evans, C.D., Jenkins, A. & Simpson, G.L. Reconstructing pre-acidification pH for an acidified Scottish loch: a comparison of palaeolimnological and modelling approaches. *Env. Poll.* 137, 135-49. 2005.
[6.22] Beauger, A., Wetzel, C.E., Voldaire, o., Allain, E., Breton, V., Miallier, D. & Ector, L. *Fontina* gen. nov. (Bacillariophyta): a new diatom genus from a thermos-mineral spring of the French Massif Central [6.France]. *Diatom Res.* 37, 51-61. 2022.
[6.23] Behnke, A.T., Friedl, T., Chepurnov, V.A. & Mann, D.G. Reproductive compatibility and rDNA sequence analyses in the *Sellaphora pupula* species complex. *J. Phycol.* 40, 193-208. 2004.
[6.24] Bennion, H. & Battarbee, R.W. The European Union Water Framework Directive: opportunities for palaeolimnology. *J. Paleolimnol.* 38, 285-95. 2007.
[6.25] Bennion, H., Juggins, S. & Anderson, N.J. Predicting epilimnetic phosphorus concentrations using an improved diatom-based transfer function and its application to lake eutrophication management. *Environ. Sci. Technol.*, 30, 2004-2. 1996.
[6.26] Bentley, K. Clack, C. & Cox, E.J. Diatom colony formation: a computational study predicts a single mechanism can produce both linkage and separation valves due to an environmental switch. *J. Phycol.*, 48: 716-728. 2012.
[6.27] Bergkvist, J., Thor, P., Jakobsen, H., Wängberg, S.-Å. & Selander, E. Grazer-induced chain length plasticity reduces grazing risk in a marine diatom. *Limnol. Oceanogr.*, 57, 318-324. 2012.
[6.28] Birks, H.J.B., Line, J.M., Juggins, S., Stevenson, A.C. and ter Braak, C.J.F. Diatoms and pH reconstruction. *Phil. Trans. R. Soc. Lond. Series* B 327, 263-78. 1990.
[6.29] Bjaerke, O., Jonsson, P.R., Alam, A. & Selander, E. Is chain length in phytoplankton regulated to evade predation? *J. Plankton Res.*, 37, 1110-1119. 2015.
[6.30] Brotas, V. & Catarino, F. Microphytobenthos primary production of Tagus estuary intertidal flats (Portugal). *Neth. J. Aquat. Ecol.* 29, 333-339. 1995.

[6.31] Burkholder, J.M. Interactions of benthic algae with their substrata. In: Stevenson, R.J., Bothwell, M.L. & Lowe, R.L. (eds) *Algal ecology. Freshwater benthic ecosystems*, 253-297. 1996.

[6.32] Buzer, J.S. 1981. Diatom analysis of sediments from Lough Ine, Co. Cork, Southwest Ireland. *New Phytol.*, 89, 511-533.

[6.33] Cairns, J. Indicator species vs. the concept of community structure as an index of pollution. *Water Resour. Bull.* 10, 338-347. 1974.

[6.34] Cameron, N.G. The representation of diatom communities by fossil assemblages in a small acid lake. *J. Paleolimnol.* 14, 185-223. 1995.

[6.35] Castenholz, R.W. An experimental study of the vertical distribution of littoral marine diatoms. *Limnol. Oceanogr.* 8, 450-462. 1963.

[6.36] Cattaneo, A., & Amireault, M.C. How artificial are artificial substrata for periphyton? *J. North Am. benthol. Soc.* 11, 244-256. 1992.

[6.37] CEMAGREF. Etude des méthodes biologiques d'appréciation quantitative de la qualité des eaux. In: Rapport Q.E. & Lyon A.F. (eds), *Bassin Rhône-Mediterrannée-Corse* (Lyon: CSIRO), 218pp. 1982.

[6.38] Charles, D.F. & Whitehead, D.R. The PIRLA project: Paleoecological investigations of recent lake acidification. *Hydrobiologia* 143/*Dev. Hydrobiol.* 37, 13-20. 1986.

[6.39] Chepurnov, V.A., Mann, D.G., Sabbe, K. & Vyverman, W. Experimental studies on sexual reproduction in diatoms. *Int. Rev. Cytol.*, 237, 91-154. 2004.

[6.40] Cholnoky, B.J. *Die Ökologie der Diatomeen in Binnengewässern*. Lehre: J Cramer. 1968.

[6.41] Coste, M., Boutry, S., Tison-Rosebery, J., & Delmas, F. Improvements of the Biological Diatom Index (BDI): description and efficiency of the new version [6.BDI-2006]. *Ecol. Indic.* 9, 621–650. 2009.

[6.42] Cox, E. J. A reappraisal of the diatom genus *Amphipleura* Kütz. using light and electron microscopy. *Br. phycol. J.*, 10, 1-12. 1975.

[6.43] Cox, E.J. Observations on the diatom genus *Donkinia* Ralfs in Pritchard. II. Frustular studies and intra-specific variation. *Bot. Mar.*, 36, 553-566. 1983.

[6.44] Cox, E.J. Observations on some benthic diatoms from North German lakes. The effect of substratum and light regime. *Verh. Int. Verein. Limnol.* 22, 924-928. 1984.

[6.45] Cox, E.J. Auxosporulation by a naviculoid diatom and the taxonomic implications. *Br. Phycol. J.*, 20, 169-179. 1985.

[6.46] Cox, E.J. Some taxonomic and ecological considerations of morphological variation within natural populations of benthic diatoms. In: Ricard, M. [6.ed.], *Proc. 8th Int. Diatom Symp.*, 163-172. O. Koeltz, Koeningstein. 1986.

[6.47] Cox, E.J. Studies on the algae of a small softwater stream. III. Interaction between discharge, sediment composition and diatom flora. *Arch. Hydrobiol. Suppl.* 83, 567-584. 1990.

[6.48] Cox, E.J. Diatom systematics – a review of past and present practice and a personal view for future development. *Nova Hedw. Beih.*, 106, 1-20. 1993.

[6.49] Cox, E.J. Freshwater diatom ecology: developing an experimental approach as an aid344 to ecological interpretation. *Hydrobiologia* 269/270, 447-452. 1993.

[6.50] Cox, E.J. Ecological tolerances and optima – real or imaginary? *Verh. Verein. Limnol.* 25, 2238-2241. 1994.

[6.51] Cox, E.J. Morphological variation in widely distributed diatom taxa: taxonomic and ecological considerations. In Marino, D. & Montresor, M. (eds) *Proc. 13th Int. Diat. Symp.*, 335-345. Bristol, Biopress. 1995.

[6.52] Cox, E.J. Assessing and designating taxa at or below the species level – a consideration of current status and some suggested guidelines for the future. *Nova Hedw.*, 65, 13-36. 1997.

[6.53] Cox E.J. Studies on the diatom genus *Navicula* Bory. VIII. Variation in valve morphology in relation to the generic diagnosis based on *Navicula tripunctata* (O.F. Müller) Bory. *Diatom Res.*, 14, 207-237. 1999.

[6.54] Cox, E.J. Generic concepts and intrageneric variation in diatoms with particular reference to *Navicula* Bory. In: John, J. (ed) *Proc. 15th Int. Diat. Symp.*, 289-304. Ruggell, A.R.G. Gantner. 2002.

[6.55] Cox, E.J. Raphe loss and spine formation in *Diadesmis gallica* (Bacillariophyta) – an intriguing example of phenotypic polymorphism in a diatom. *Nova Hedw., Beih.*, 130, 163-176, 2006.

[6.56] Cox, E.J. What's in a name? Diatom classification should reflect systematic relationships. *Acta Bot. Croat.* 68, 443-454. 2009.

[6.57] Cox, E.J. Morphogenetic information and the selection of taxonomic characters for raphid diatom systematics. *Plant Ecol. Evol.*, 143, 271-277, 2010.

[6.58] Cox, E.J. Diatom identification in the face of changing species concepts and evidence of phenotypic plasticity. *J. Micropal.*, 33, 111-120. 2014.

[6.59] Cox, E.J. & Gogorev, R. Report of a workshop on 'Systematics'. In: (A. Economou-Amilli, ed.), *Proc. 16th Int. Diatom Symp. 25 Aug – 1 Sept 2000, Athens and Aegean Islands*, 567-571. University of Athens, Greece. 2001.

[6.60] Cox, E.J., Marxsen, J. & Horvath, P. Primary producers. In: Wagner, R., Marxsen, J., Zwick, P. & Cox, E.J. [6.eds] *Central European stream ecosystems. The long term study of the Breiten*bach: 99-129. Weinheim, Wiley-VCH. 2011.

[6.61] Cox, E.J. & Witkowski, A. Workshop Report. Taxonomy – types, nomenclature, new combinations. In: M. Poulin, ed., *Proc. 17th Int. Diatom Symp. 2002*, 463-466. Biopress Limited, Bristol. 2004.

[6.62] Crawford, R.M. & Hinz, F. The spines of the centric diatom *Corethron criophilum*: light microscopy of vegetative cell division. *Eur. J. Phycol.*, 30, 95-105. 1995.

[6.63] Crawford, R.M., Hinz, F. & Honeywill, C. Three species of the diatom genus *Corethron castracane*: structure, distribution and taxonomy. *Diatom Res.,* 13, 1-28. 1998.

[6.64] Davey, M.C. Seasonal variation in the filament morphology of the freshwater diatom *Melosira granulata* (Ehrenb.) Ralfs. *Freshw. Biol.* 18, 5-16. 1987.

[6.65] Davey, M.C. & Crawford R.M. The relationship between filament formation and the separation valve index in *Melosira granulata* (Bacillariophyta). *J. Phycol.* 22, 144-150. 1986.

[6.66] Davidovich, N.A. & Bates, S.S. Sexual reproduction in the pennate diatoms *Pseudo-nitzschia multiseries* and *P. pseudodelicatissima* [6.Bacillariophyceae]. *J. Phycol.*, 34, 126–137. 1998.

[6.67] De Bruin, A., Ibelings, B.W., Rijkeboer, M., Brehm, M. & van Donk, E. Genetic variation in *Asterionella formosa* (Bacillariophyceae): is it linked to frequent epidemics of host-specific parasitic fungi?*J Phycol.* 40, 823–830. 2004.

[6.68] Descy, J.-P. A new approach to water quality estimation using diatoms. *Nova Hedw. Beih.* 64, 305-323. 1979.

[6.69] Descy, J.-P. Ecologie et distribution de diatomées benthiques dans le basin belge de la Meuse. *Inst. Roy. des sci. nat. de Belgiques* 18. 70pp. 1984.

[6.70] Douglas, M.S.V. & Smol, J.P. Freshwater diatoms as indicators of environmental change in the High Arctic. In: Stoermer, E.F. & Smol, J.P. (eds) *The diatoms. Applications for the environmental and earth sciences*, 227-244. Cambridge University Press. 1999.

[6.71] Eaton, J.W. & Moss, B. The estimation of numbers and pigment content in epipelic algal populations. *Limnol. Oceanogr.* 11, 584-595. 1966.

[6.72] Edlund, M.B. & Burge, D.R. Polymorphism in *Mastogloia* (Bacillariophyceae) revisited. *Plant Ecol. Evol.* 152, 351-357. 2019.

[6.73] Ehrenberg, C.G. *Die Infusionsthierchen als vollkommene Organismen.* Leipzig, Leopold Voss, pp. 548. 1838.

[6.74] Ehrenberg, C.G. *Mikrogeologie. Das Erden und Felsen schaffende Wirken des unsichtbar kleinen selbststandigen Lebens auf der Erde.* Leipzig: Leopold Voss, pp1854.

[6.75] Eminson D. & Moss, B. The composition and ecology of periphyton communities in freshwaters. *Br. Phycol. J.*, 15, 429-446. 1980.

[6.76] Evans, K.M., Chepurnov, V.A., Sluiman, H.J., Thomas, S.J., Spears, B.M. & Mann, D.G. Highly differentiated populations of the freshwater diatom *Sellaphora capitata* suggest limited dispersal and opportunities for allopatric speciation. *Protist* 160, 386-396. 2009.

[6.77] Evans, K.M., Wortley, A.H., & Mann, D.G. 2007. An assessment of potential diatom "barcode" genes (cox1, rbcL, 18S and ITS rDNA) and their effectiveness in determining relationships in *Sellaphora* (Bacillariophyta). *Protist* 158, 349-364.

[6.78] Evans, K.M., Wortley, A.H., Simpson, G.E., Chepurnov, V.A. & Mann, D.G. A molecular systematic approach to explore diversity within the *Sellaphora pupula* species complex (Bacillariophyta). *J. Phycol.* 44, 215-231. 2008.

[6.79] Fairchild, G.W., Lowe, R.L. & Richardson, W.B. Algal periphyton growth on nutrient-diffusing substrates; an in situ bioassay. *Ecology*, 66, 465-472. 1985.

[6.80] Fjerdingstad, E. Pollution of stream estimated by benthal phytomicro-organisms. I. A saprobic system based on communities of organisms and ecological factors. *Int. Rev. Gesamt. Hydrobiol.* 49, 63–131. 1964.

[6.81] Fryxell, G.A. Polymorphism in relation to environmental conditions as exemplified by clonal cultures of *Thalassiosira tumida* (Janisch) Hasle. In Round, F.E. (ed.), *Proc. 9th Int. Diatom Symp., Bristol 1986*. Biopress, Bristol, 61–73. 1988.

[6.82] Gallagher, J.C. Physiological variation and electrophoretic banding patterns of genetically different seasonal populations of *Skeletonema costatum* (Bacillariophyceae). *J. Phycol.* 18, 148-162. 1982.

[6.83] Geitler, L. Der Formwechsel der pennaten Diatomeen. *Arch. Protistenk.*, 1-226. 1932.

[6.84] Geitler, L. Formwechsel, sippenspezifischer Paarungsmodus und Systematik bei einigen pennaten Diatomeen. *Pl. Syst. Evol.*, 124, 7-30. 1975.

[6.85] Gibson, C.E. Silica budgets and the ecology of planktonic diatoms in an unstratified lake (Lough Neagh, N. Ireland) *Int. Rev. gesamt. Hydrobiol.*, 66, 641-644. 1981.

[6.86] Gibson, C.E., Wang, G. & Foy, R.H. Silica and diatom growth in Lough Neagh: the importance of internal cycling. *Freshw. Biol.*, 45, 285-293. 2000.

[6.87] Gillett, N., Pan, Y. & Parker, C. Should only live diatoms be used in the bioassessment of small mountain streams? *Hydrobiologia* 620, 135-147. 2009.

[6.88] Gilmour, J.S.L. & Gregor, J.W. Demes: A suggested new terminology. *Nature* 144, 333. 1939.

[6.89] Gluschenko, A., Kuznetsova, I., Kociolek, J.P. & Kulikovskiy, M. *Karthickia verestigmata* gen. et sp. nov. – an interesting diatom with frustular morphology similar to several different cymbelloid diatom genera. *Phycologia* 58, 605-613. 2019.

[6.90] Godhe A. & Rynearson T The role of intraspecific variation in the ecological and evolutionary success of diatoms in changing environments. *Phil. Trans. R. Soc. B* 372, 20160399. 2017.

[6.91] Gsell, A., de Senerpont Domis, L.N., Przytulska-Bartosiewicz, A., Mooij, W.M., van Donk, E. & Ibelings, B.W. Genotype-by-temperature interactions may help to maintain clonal diversity in *Asterionella formosa* (Bacillariophyceae). *J. Phycol.* 48, 1197-1208. 2012.

[6.92] Håkansson, H. Examination of diatom type material of C.A. Agardh. *Nova Hedw. Beih.*, 64, 163-168. 1979.

[6.93] Håkansson, H. & Bailey-Watts, A.E. A contribution to the taxonomy of *Stephanodiscus hantzschii* Grunow, a common freshwater planktonic diatom. *Diatom Res.* 8, 317-332. 1993.

[6.94] Håkansson, H. & Stoermer, E.F. Observations on the type material of *Stephanodiscus hantzschii* Grunow in Cleve and Grunow. *Nova Hedw.* 39, 477–495. 1984.

[6.95] Hamilton, P.B., Poulin, M. & Taylor, M.C. *Neidium alpinum* var. quadripunctatum (Hustedt) comb. nov., an important acidobiontic taxon from Northeastern North America. *Diat. Res.* 5, 289-299. 1990.

[6.96] Harper, M.E., Patterson, J.E. & Harper, J.F. New diatom taxa from the world's first marine Bioblitz held in New Zealand: *Skeletomastus* a new genus, *Skeletomastus coelatus* nov. comb. and *Pleurosigma inscriptura* a new species. *Acta Bot. Croat.* 68, 339-349. 2009.

[6.97] Hasle, G.R. & Fryxell, G. A. The genus *Thalassiosira*: species with a linear areola array. *Nova Hedw. Beih.* 54, 15-66. 1977.

[6.98] Hausmann, S. & Lotter, A.F. Numerical *Cyclotella comensis* taxonomy and its importance for quantitative temperature reconstruction. *Freshw. Biol.*, 46: 1323–1333. 2001.

[6.99] Haworth, E.Y. The recent diatom history of Loch Leven, Kinross. *Freshw. Biol.*, 2, 131-141. 1972.

[6.100] Haworth, E.Y. A scanning electron microscope study of some different frustule forms of the genus *Fragilaria* found in Scottish late-glacial sediments. *Br. Phycol. J.* 10, 73-80. 1975.

[6.101] Hendey N.I. *An Introductory Account of the Smaller Algae of British Coastal Waters.* Her Majesty's Stationery Office, London. 317 pp. 1964.

[6.102] Hickman, M. Standing crops and primary productivity of the epipelon of two small ponds in North Somerset, U.K. *Oecologia (Berl.)* 6, 238-253. 1971.

[6.103] Hirst, H., Chaud, F., Delabie, C., Jüttner, I. & Ormerod, S.J. Assessing the short-term response of stream diatoms to acidity using inter-basin transplantations and chemical diffusing substrates. *Freshw. Biol.*, 49, 1072-1088. 2004.

[6.104] Hoops, H.J. & Floyd, G.L. Ultrastructure of the centric diatom *Cyclotella meneghiniana*: vegetative cell and auxospore development. *Phycologia* 18, 424-435. 1979.

[6.105] Hopkins, J.T. A study of the diatoms of the Ouse Estuary, Sussex. II. The ecology of the mud-flat diatom flora. *J. mar. biol. Ass. U.K.* 44, 333-341. 1964.

[6.106] Hopkins, J.T. A study of the diatoms of the Ouse Estuary, Sussex. III. The seasonal variation in the littoral epiphytic flora and the shore plankton. *J. mar. biol. Ass. U.K.* 44, 613-644. 1964.

[6.107] Hurley, C. *Polymorphism in the diatom genus* Mastogloia. Senior thesis, Loras College, Dubuque, Iowa, USA. 1987.
[6.108] Hürlimann, J. & Straub, F. Morphologische und ökologische Charakterisierung von Sippen um den *Fragilaria capucina*-Komplex sensu Lange-Bertalot 1980. *Diatom Res.*, 6, 21–47. 1991.
[6.109] Hustedt, F. Bacillariophyta (Diatomeae). In: Pascher, A. (ed.) *Die Süsswasserflora Mitteleuropas, Heft 10.* 2nd edition. Stuttgart, Gustav Fischer Verlag. 1930.
[6.110] Hustedt, F. Systematische und ökologische Untersuchungen über den Diatomeen-Flora von Java, Bali, Sumatra. *Arch. Hydrobiol. (Suppl.)* 15 & 16. 685 pp. 1937-1939.
[6.111] Hustedt, F. Die Systematik der Diatomeen in ihren Beziehungen zur Geologie und Ökologie nebst einer Revision des Halobien-systems. *Sv. Bot. Tidskr.* 47, 274–344. 1953.
[6.112] Hustedt, F. Die Diatomeenflora des Fluß-systems der Weser im Gebiet der Hansestadt Bremen. *Abh. Naturwiss. Ver. Bremen* 34, 181-440. 1957.
[6.113] Hutchinson, G.E. The paradox of the plankton. *Amer. Nat.*, 95, 137-145. 1961.
[6.114] Jahn, R. & Kusber, W.-H. A key to diatom nomenclature. *Diatom Res.*, 24, 101-111. 2009.
[6.115] Jordan, R.W., Ligowski, R., Nöthig, E.-M., & Priddle, E.J. The diatom genus *Proboscia* in Antarctic waters. *Diatom Res.*,6, 79-107. 1991.
[6.116] Jørgensen, E.G., *Diatoms in some Danish lakes and ponds.* K. *dansk. Vidensk. Selsk. Skr.* 5, 1-140. 1948.
[6.117] Juggins, S., Flower, R.J. & Battarbee, R.W. Palaeolimnological evidence for recent chemical and biological changes in UK acid waters monitoring network sites. *Freshw. Biol.* 36, 203-19. 1996.
[6.118] Jüttner, I., Hamilton, P.B., Wetzel, C. E., Van de Vijver, B., King, L., Kelly, M.G., Williams, D.M. & Ector, L. A study of the morphology and distribution of four *Achnanthidium* Kütz. species (Bacillariophyta), implications for ecological status assessment, and description of two new European species. *Crypt. Algol.* 43, 147-176. 2022.
[6.119] Kaczmarska, I., Mather, L., Luddington, I.A., Muise, F. & Ehrman, J.M. Cryptic diversity in a cosmopolitan diatom known as *Asterionellopsis glacialis* (Fragilariaceae): Implications for ecology, biogeography, and taxonomy. *Am. J. Bot.*, 101, 267–286. 2014.
[6.120] Kapustin, D.A., Kociolek, J.P.; Glushchenko, A.M. & Kulikovskiy, M.S. A rediscovery of *Cymbella mirabilis* Hustedt, a rare endemic diatom, and description of *Alveocymba* gen. nov. *Diatom Res.* 35, 281-187. 2020.
[6.121] Kapustin, D. Kulikovskiy, M. & Kociolek, J.P. *Celebesia* gen. nov., a new cymbelloid diatom genus from the ancient Lake Matano (Sulawesi Island, Indonesia). *Nova Hedw.* 146, 147-155. 2017.
[6.122] Kelly, M.G., Juggins, S., Mann, D.G., Sato, S., Glover, R., Boonham, N., Sapp, M., Lewis, E., Hany, U., Kille, P., Jones, T. & Walsh, K. Development

of a novel metric for evaluating diatom assemblages in rivers using DNA metabarcoding. *Ecol. Indic.*, 118, 106725. 2020.

[6.123] Kelly, M.G. & Whitton, B.A. The trophic diatom index: a new index for monitoring eutrophication in rivers. *J. Appl. Phycol.* 7, 433–444. 1995.

[6.124] Kelly, M.G. & Yallop, M. *A streamlined taxonomy for the Trophic Diatom Index*. Science report SC070034/TR1, Environment Agency, Bristol. 24 pp. 2012.

[6.125] Khan-Bureau, D.A., Ector, L., Morales, E.A., Wade, E.J. & Lewis, L.A. Contrasting morphological and DNA barcoding methods for diatom (Bacillariophyta) identification from environmental samples in the Eightmile River in Connecticut, USA. *Nova Hedw. Beih.* 146, 279-302. 2017.

[6.126] Kilham, S.S. Kinetics of silicon-limited growth in the freshwater diatom *Asterionella formosa*. *J. Phycol.*, 11, 396-399. 1975.

[6.127] Kingston, J.C. & Pappas, J.L. Quantitative shape analysis as a diagnostic and prescriptive tool in determining *Fragilariaforma* (Bacillariophyta) taxon status. *Nova Hedw. Beih.*, 135, 103-119. 2009.

[6.128] Kobayasi, H., Inoue, H. & Kobayashi, H. Fine structure and taxonomy of the small and tiny *Stephanodiscus* (Bacillariophyceae) species in Japan. 2. *Stephanodiscus hantzschii* Grun. form. *tenuis* (Hust.) Håk. et Stoermer. *Jap. J. Phycol.* 33, 233-238. 1985.

[6.129] Kociolek, J.P. Taxonomy and ecology: further considerations. *Proc. Cal. Acad. Sci.* 56, 99-106. 2005.

[6.130] Kociolek, J.P. & Stoermer, E.F. Taxonomy and ecology; A marriage of necessity. *Diatom Res.*, 16, 433-442. 2001.

[6.131] Kolbe, R.W. *Zur Ökologie, Morphologie und Systematik der Brackwasser-diatomeen*. Jena: Fischer. 1927.

[6.132] Kolkwitz, R. & Marsson, M. Ökologie der pflanzlichen Saprobien. *Ber. deutsch. bot. Ges.* 26, 505-519. 1908.

[6.133] Kooistra, W., Forlani, G., Sterrenburg, F.A.S. & De Stefano, M. Molecular phylogeny and morphology of the marine diatom *Talaroneis posidoniae* gen. sp. nov. (Bacillariophyta) advocate the return of the Plagiogrammaceae to the pennate diatoms. *Phycologia*, 43, 58-67. 2004.

[6.134] Kooistra, W.H.C.F., Sarno, D., Balzano, S., Gu, H., Andersen, R.A. & Zingone, A. Global diversity and biogeography of *Skeletonema* species (Bacillariophyta). *Protist* 159, 177-193. 2008.

[6.135] Krammer, K. Die cymbelloiden Diatomeen - Eine Monographie der weltweit bekannten Taxa. Teil 1. Allgemeines und *Encyonema* Part. *Bibliotheca Diatomologica* 36. 382 pp. 1997.

[6.136] Krammer, K. *Cymbopleura, Delicata, Navicymbula, Gomphocymbellopsis, Afrocymbella*. *Diatoms of Europe* 4, 1-530. 2003.

[6.137] Krammer, K. & Lange-Bertalot, H. Bacillariophyceae 1. Teil: Naviculaceae. In: Ettl. H., Gerloff, J., Heynig, H. & Mollenhauer, D. (eds). *Süsswasserflora von Mitteleuropa, Vol. 2/1*. G. Fischer, Stuttgart & New York. 1986.

[6.138] Kulikovskiy, M., Maltsev, Y., Andreeva, S., Glushchenko, A., Gusev, E., Podunay, Y., Ludwig, T.V., Tusset, E. & Kociolek, J.P. Description of a new diatom genus *Dorofeyukea* gen. nov. with remarks on phylogeny of the family Stauroneidaceae. *J. Phycol.* 55, 173-185. 2019.
[6.139] Kulikovskiy, M., Maltsev, Y., Glushchensko, A., Kuznetsova, I., Kapustin, D., Gusev, E., Lange-Bertalot, H., Genkal, S. & Kociolek, J.P. *Gogorevia*, a new monoraphid diatom genus for *Achnanthes exigua* and allied taxa (Achnanthidiaceae) described on the basis of an iIntegrated molecular and morphological approach. *J. Phycol.* 56, 1601-1613. 2020.
[6.140] Lange-Bertalot, H. Diatomeen-Differentialarten anstelle von Leitformen: ein geeigneteres Kriterium der Gewässerbelastung. *Arch. Hydrobiol. Suppl.* 51, 393-427. 1978.
[6.141] Lange-Bertalot, H. Toleranzgrenzen und Populationsdynamik benthischer Diatomeen bei unterschiedlich starker Abwasserbelastung. *Arch. Hydrobiol. Suppl.* 56, 184-219. 1979.
[6.142] Lavoie, I., Campeau, S., Darchambeau, F., Cabana, G. & Dillon, P.J. Are diatoms good integrators of temporal variability in stream water quality? *Freshw. Biol.* 53, 827-841. 2008.
[6.143] Lin, H., Wu, W., Sun, L., Witkowski, A., Li, X., Patil, V., Liang, J., Li, X., Gao, Y. & Chen, C. *Chinia* gen. nov. - the second diatom genus simonsenioid raphe from mangroves in Fujian, *China. J. Oceanogr. Limnol.* 40. DOI: 10.1007/s00343-021-1067-0. 2021.
[6.144] Lowe, R.L., Golladay, S. & Webster, J.R. Periphyton response to nutrient manipulation in streams draining clearcut and forested basins. *J. North Am. Benthol. Soc.* 5, 221-229. 1986.
[6.145] Lund, J.W.G. Studies on *Asterionella*. I. The origin and nature of the cells producing seasonal maxima. *J. Ecol.*, 37, 389-419. 1949.
[6.146] Lund, J.W.G. Studies on *Asterionella* Hass. II. Nutrient depletion and the spring maximum. *J. Ecol.*, 38, 1-35. 1950.
[6.147] Lund, J.W.G. The seasonal cycle of the planktonic diatom *Melosira italica* (Ehr.) Kütz. subsp. *subarctica* O. Mull. *J. Ecol.*, 42, 151-179. 1954.
[6.148] Lund, J.W.G., Mackereth, F.J.H. & Mortimer, M.H. Changes in depth and time of certain chemical and physical conditions and of the standing crop of *Asterionella formosa* Hass. in the north basin of Windermere in 1947. *Phil. Trans. Roy. Soc., Lond. B*, 246, 255-290. 1963.
[6.149] Lund, J.W.G. & Reynolds, C.S. The development and operation of large limnetic enclosures in Blelham Tarn, English Lake District, and their contribution to phytoplankton ecology. *Prog. Phycol. Res.*, 1, 1-65. 1982.
[6.150] Lundholm, N., Bates, S.S., Baught, K.A., Bill, B.D., Connell, L.B., Léger, C. & Trainer, V.L. Cryptic and pseudo-cryptic diversity in diatoms – With descriptions of *Pseudo-nitzschia hasleana* sp. nov. and *P. fryxelliana* sp. nov. *J. Phycol.*, 48: 436–454. 2012.
[6.151] Lundholm, N., Moestrup, O., Kotaki, Y., Hoef-Emden, K., Scholin, C. & Miller, P. Inter- and intraspecific variation of the *Pseudo-nitzschia*

delicatissima complex (Bacillariophyceae) illustrated by rRNA probes, morphological data and phylogenetic analyses. *J. Phycol.* 42, 464-481. 2006.

[6.152] Main, S.P. *Diprora haenaensis* gen. et sp. nov., a filamentous, pseudoaerial, araphid diatom from Kaua'i (Hawaiian Islands). *Diatom Res.* 18, 259-272. 2003.

[6.153] Mann, D.G. Observations on copulation in *Navicula pupula* and *Amphora ovalis* in relation to the nature of diatom species. *Ann. Bot.* 54, 429-438. 1984.

[6.154] Mann, D.G. The nature of diatom species: analyses of sympatric populations. In: Round, F.E. (ed.) *Proc. 9th Int. Diat. Symp.*, 317-327. 1988.

[6.155] Mann, D.G. The species concept in diatoms: evidence for morphologically distinct, sympatric gamodemes in four epipelic species. *Pl. Syst. Evol.* 164, 215-237. 1989.

[6.156] Mann, D.G. The species concept in diatoms. *Phycologia*, 38, 437-495, 1999.

[6.157] 157 Mann, D.G. Shifting sands: the use of the lower taxonomic ranks in diatoms. *Diatom* 13, 13-17. 1999.

[6.158] Mann, D.G. & Evans, K.M. Molecular genetics and the neglected art of diatomics. In: Brodie, J. & Lewis, J. (eds), *Unraveling the algae: the past, present, and future of algal systematics.* The Systematics Association Special Volume Series 75. 231-265. 2007.

[6.159] Mann, D.G. & Kociolek, J.P. The species concept in diatoms: report on a workshop. In: Simola, H. (ed), *Proc. 10th Int. Diat. Symp.*, 585-592. 1990.

[6.160] Mann, D.G., McDonald, S.M., Bayer, M.M., Droop, S.J.M., Chepurnov, V.A., Loke, R.E., Ciobanu, A. & de Buf, J.M.F. 2004. Morphometric analysis, ultrastructure and mating data provide evidence for five new species of *Sellaphora* (Bacillariophyceae). *Phycologia* 43, 459-482.

[6.161] McBride, S.A. & Edgar, R.K. Janus cells unveiled: frustular morphometric variability in *Gomphonema angustatum. Diatom Res.*, 13, 293-310. 1998.

[6.162] McMillan, M. & Johansen, J.R. Changes in valve morphology of *Thalassiosira decipiens* (Bacillariophyceae) cultured in media of four different salinities. *Br. Phycol. J.* 23, 307-316. 1988.

[6.163] Medlin, L.K. Can molecular techniques help define species limits? *Diatom* 12: 19-23. 1997.

[6.164] Medlin, L.K. Mini review: Diatom species as seen through a molecular window. *Brazilian J. Bot.* 41, 457–469. 2018.

[6.165] Medlin, L.K., Elwood, H.J., Stickel, S. & Sogin, M.L. Morphological and genetic variation within the diatom *Skeletonema costatum* (Bacillariophyta): Evidence for a new species, *Skeletonema pseudocostatum. J. Phycol.*, 27: 514–524. 1991.

[6.166] Metzeltin, D. & Lange-Bertalot, H. Diatoms from the "Island Continent" Madagascar. Iconographia Diatomologica 11: 1–286. 2002.

[6.167] Metzeltin D. & Lange-Bertalot, H. Tropical Diatoms of South America II. *Iconographia Diatomologica* 18, 1-877. 2007.

[6.168] Metzeltin, D., Lange-Bertalot, H. & García-Rodríguez, F. Diatoms of Uruguay. Compared with other taxa from South America and elsewhere. *Iconographia Diatomologica* 15, 1-736. 2005.

[6.169] Metzeltin, D., Lange-Bertalot, H. & Soninkhishig, N. Diatoms in Mongolia. *Iconographia Diatomologica* 20, 1-691. 2009.

[6.170] Meyer, B. & Håkansson, H. *Cyclotella wulfiae*, a new diatom from a brackish lake in Germany. *Diatom Res.* 12, 279-285. 1997.

[6.171] Meyer, B., Wulff, M. & Håkansson, H. Phenotypic variation of life-cycle stages in clones of three similar *Cyclotella* species after induced auxospore production. *Diatom Res.* 16, 343-363. 2001.

[6.172] Mohamad, H., Mora, D., Skibbe, O., Abarca, N., Deutschmeyer, V., Enke, N., Kusber W.-H., Zimmermann, J. & Jahn, R. Morphological variability and genetic marker stability of 16 monoclonal pennate diatom strains cultivated within a two-year period. *Diatom Res.* 37, 307-328. 2022.

[6.173] Moore, J.W. & McIntire, C.D. Spatial and seasonal distribution of littoral diatoms in Yaquina Estuary, Oregon (U.S.A.). *Bot. Mar.* 20, 99-109. 1977.

[6.174] Nygaard, G. Ancient and recent flora of diatoms and Chrysophyceae in Lake Gribso. In Berg, K. & Peterson, I. (eds.), Studies on Humic, Acid Lake Gribso. *Folia Limnological Scandinavia* 8: 32-94, pl. 1-12. 1956.

[6.175] Owen, B.B., Afzal, M. & Cody, W.R. Distinguishing between live and dead diatoms in periphyton communities. In: Wetzel, R.L. (ed.), *Methods and measurements of periphytic communities: A review. ASTM, STP* 690, 70-76. 1979.

[6.176] Paasche, E. Silicon and the ecology of marine planktonic diatoms. I. *Thalassiosira pseudonana* (*Cyclotella pseudonana*) grown in a chemostat with silicate as a limiting nutrient. *Mar. Biol.*, 19, 117-126. 1973.

[6.177] Paasche, E. Silicon and the ecology of marine planktonic diatoms. II. Silicate-uptake kinetics in five diatom species. *Mar. Biol.*, 19, 262-269. 1973.

[6.178] Paasche, E. Silicon. In: Morris, I. (ed.) *The physiological ecology of phytoplankton*, pp. 259-284, Blackwell, Oxford. 1980.

[6.179] Pappas, J.L. & Stoermer, E.F. Fourier shape analysis and fuzzy measure shape group differentiation of Great Lakes *Asterionella* Hassall (Heterokontophyta, Bacillariophyceae). In: Economou-Amilli, A. (ed.), *Proc. 16th Int. Diat. Symp., 25 Aug – I Sept 2000, Athens and Aegean Islands*, 485-501. 2001.

[6.180] Patrick R. & Reimer C.W. The Diatoms of the United States, exclusive of Alaska and Hawaii, Volume 2, Part 1. *Monographs Acad. Nat. Sci. Philad.* 13, 212 pp. 1975.

[6.181] Patrick, R. & Reimer, C W. The diatoms of the United States exclusive of Alaska and Hawaii. Volume 1. *Monographs Acad. Nat. Sci. Philad.*, 13, 688 pp. 1966.

[6.182] Pérez-Burillo, J., Trobajo, R., Leira, M., Keck, F., Rimet, F., Sigró, J. & Mann, D.G. DNA metabarcoding reveals differences in distribution patterns and ecological preferences among genetic variants within some key freshwater diatom species. *Sci. Total. Environ.* 798, 149029. 2021.

[6.183] Pérez-Burillo, J., Trobajo, R., Vasselon, V., Rimet, F., Bouchez, A. & Mann, D.G. Evaluation and sensitivity analysis of diatom DNA metabarcoding for WFD bioassessment of Mediterranean rivers. *Sci. Total. Environ.* 727, 138445. 2020.

[6.184] Perkins, R.G., Underwood, G.J.C., Brotas, V., Snow, G.D., Jesus, B. & Ribeiro, L. Responses of microphytobenthos to light: primary production and carbohydrate allocation over an emersion period. *Mar. Ecol. Prog. Ser.* 223, 101-112. 2001.

[6.185] Pinseel, E., Kulichová, J., Scharfen, V., Urbánková, P., Van de Vijver, B. & Vyvermann, W. Extensive cryptic diversity in the terrestrial diatom *Pinnularia borealis* (Bacillariophyceae). *Protist* 170, 121-140. 2019.

[6.186] Poister, D., Kurth, A., Farrell, A. & Gray, S. Seasonality of *Aulacoseira ambigua* abundance and filament length: biogeochemical implications. *Plankton Benthos Res.*, 7, 55-63. 2012.

[6.187] Potapova, M. Patterns of diatom distribution in relation to salinity. In: Seckbach, J. & Kociolek, J.P. [6.eds] *The diatom world*, pp. 315-332. 2011.

[6.188] Potapova, M., Gallen, C., & Friedman, N. *Stauroneis terryi* D.B. Ward ex T.C. Palmer: a forgotten diatom from northeastern North America. *Diatom Res.* 37, 155-163. 2022.

[6.189] Potapova, M. & Hamilton, P.B. Morphological and ecological variation within the *Achnanthidium minutissimum* (Bacillariophyceae) species complex. *J. Phycol.* 43, 561-575. 2007.

[6.190] Poulíčková, A., Neustupa, J., Hasler, P., Tomanec, O. & Cox, E.J. A new species, *Navicula lothargeitleri* sp. nov., within the *Navicula cryptocephala* complex (Bacillariophyceae). *Phytotaxa* 273, 23-33. 2016.

[6.191] Poulíčková, A., Špačková, J., Kelly, M.G., Duchoslav, M. & Mann, D.G. Ecological variation within *Sellaphora* species complexes (Bacillariophyceae): specialists or generalists? *Hydrobiologia* 614, 373-386. 2008.

[6.192] Pryfogle, P.A. & Lowe, R.L. Sampling and interpretation of epilithic diatom lotic communities. In: Wetzel, R.L. (ed.), Methods and measurements of periphytic communities: A review. *ASTM, STP* 690, 77-89. 1979.

[6.193] Reimer, C.W. & Henderson, M.V. Bibliography, addenda, and corrigenda for The Diatoms of the United States, Volume 1 and Volume 2, Part 1. *Proc. Acad. Sci. Philadelphia* 151, 129-155. 2001.

[6.194] Reynolds, C.S. *The ecology of freshwater phytoplankton*. Cambridge University Press. 384 pp. 1984.

[6.195] Rose, D.T. & Cox, E.J. What constitutes *Gomphonema parvulum*? Long-term culture studies show that some varieties of *G. parvulum* belong with other *Gomphonema* species. *Plant Ecol. Evol.*, 147-366-373. 2014.

[6.196] Ross R. Nomenclature for diatomists. *Diatom Res.* 8, 429-438. 35 Krammer, K & Lange-Bertalot, H. 1986. Bacillariophyceae I. Teil: Naviculaceae. In: Ettl, H., Gerloff, J., Heynig, H. & Mollenhauer, D. (eds.). *Süsswasserflora von Mitteleuropa, Vol. 2/1.* G. Fischer, Stuttgart & New York. 1993.

[6.197] Ross, R. & Mann, D.G. What is a diatom species? In: M. Ricard (ed.), *Proc. 8th Int. Diat. Symp.*, 743-744. Koeltz Scientific Books, Koenigstein. 1986.

[6.198] Ross, R., Sims, P.A. & Hasle, G.R. Observations on some species of the Hemiauloideae. *Nova Hedw. Beih.* 54, 179-214. 1977.

[6.199] Round, F.E. The late-glacial and post-glacial diatom succession on the Kentmere Valley deposit. Part I. Introduction, methods and flora. *New Phytol.* 56, 98-126. 1957.

[6.200] Round, F.E. Benthic marine diatoms. *Oceanogr. Mar. Biol. Ann. Rev.*, 9, 83-139. 1971.

[6.201] Round, F.E. Morphology, ecology and taxonomy, two examples of their interaction in diatoms. In: Ricard, M. (ed.), *Ouvrage dédié à la mémoire du Professeur Henri Germain (1903-1989)*, 223-234. Koeltz Scientific Books, Koeningstein. 1990.

[6.202] Round, F.E. Genera, species and varieties – are problems real or imagined? *Diatom* 13, 25-29. 1997.

[6.203] Rumrich, U., Lange-Bertalot, H. & Rumrich, M. Diatoms of the Andes. from Venezuela to Patagonia/Tierra del Fuego and two additional contributions. In: Lange-Bertalot, H. (ed.), *Iconographia Diatomologica. Annotated Diatom Micrographs. Vol. 9.* Koeltz Scientific Books, Königstein, Germany. 673pp. 2000.

[6.204] Rynearson, T.A. & Armbrust, E.V. Genetic differentiation among populations of the planktonic diatom *Ditylum brightwellii* (Bacillariophyceae). *J. Phycol.* 40, 34-43. 2004.

[6.205] Sabbe, K. Short-term fluctuations in benthic diatom numbers on an intertidal sandflat in the Westerschelde estuary (Zeeland, The Netherlands). *Hydrobiologia* 269, 275-284. 1993.

[6.206] Sar, E.A., Sunesen, I, Lavigne, A.S. & Lofeudo, S. *Thalassiosira rotula*, a heterotypic synonym of *Thalassiosira gravida*: morphological evidence. *Diatom Res.* 25, 109-119. 2011.

[6.207] Sarno, D., Kooistra, W.H.C.F., Hargraves, P.E. & Zingone, A. Diversity in the genus *Skeletonema* (Bacillariophyceae). III. Phylogenetic position and morphology of *Skeletonema costatum* and *Skeletonema grevillei*, with the description of *Skeletonema* ardens sp. nov. *J. Phycol.*, 43, 156-170. 2007.

[6.208] Sarno, D., Kooistra, W.H.C.F., Medlin, L.K., Percopo, I. & Zingone, A. Diversity in the genus *Skeletonema* (Bacillariophyceae). II. An assessment of the taxonomy of *S. costatum*-like species with the descripton of four new species. *J. Phycol.*, 41, 151-176. 2005.

[6.209] Sato, S., Mann, D.G., Matsumoto, S. & Medlin, L.K. *Pseudostriatella* (Bacillariophyta): a description of a new araphid diatom based on observations of frustule and auxospore structure and 18SrDNA phylogeny. *Phycologia* 47, 371-391. 2008.

[6.210] Schelske, C.L. & Stoermer, E.F. Eutrophication, silica and predicted changes in algal quality in Lake Michigan. *Science* 173, 423–424& 1971.

[6.211] Schmid, A.-M.M. Aspects of morphogenesis and function of diatom cell walls with implications for taxonomy. *Protoplasma* 181, 43-60. 1994.

[6.212] Schneck, F., Schwartzbold, A. & Melo, A.S. Substrate roughness affects stream benthic algal diversity, assemblage composition, and nestedness. *J. N. Am. Benthol. Soc.* 30, 1049-1056. 2011.

[6.213] Schultz, M.E. Salinity-related polymorphism in the brackish-water diatom *Cyclotella cryptica*. *Can. J. Bot.*, 49: 1285–1289. 1971.

[6.214] Sime, I. *An eco-physiological investigation of nutrient limited growth and competition between five species of planktonic diatoms.* Thesis submitted for the degree of Doctor of Philosophy, Department of Geography, University College London. 278 pp. 1997.

[6.215] Sjöqvist, C. & Kremp, A. Genetic diversity affects ecological performance and stress response of marine diatom populations. *ISME J.* 10, 2755-2766. 2016.

[6.216] Smayda, T.J. Phytoplankton studies in Lower Narragansett Bay. *Limnol. Oceanogr.*, 2, 299-380. 1957.

[6.217] Smayda, T.J. Experimental studies on the influence of temperature, light and salinity on cell division of the marine diatom, *Detonula confervacea* (Cleve) Gran. *J. Phycol.*, 5, 150–157. 1969.

[6.218] Smayda, T.J. Phytoplankton species succession. In: Morris, I. (ed.) *The physiological ecology of phytoplankton*, pp. 493-570. Blackwell, Oxford. 1980.

[6.219] Smol, J., Wolfe, P.B., Birks, H.J.B., Douglas, M.S.V., Jones, V.J., Korhola, A., Pienitiz, R., Rühland, K., Sorvari, S., Antoniades, D., Brooks, S.J., Fallu, M.-A., Hughes, M., Keatley, B.E., Laing, T.E., Michelutti, N., Nazarova, L., Nyman, M., Paterson, A.M., Perren, B., Quinlan, R., Rautio, M., Saulnier-Talbot, E., Siltonen, S., Solovieva, N. & Weckström, J. Climate-driven regime shifts in the biological communities of arctic lakes. *PNAS* 102, 4397-4402. 2005.

[6.220] Sommer, U. 1990. Phytoplankton nutrient competition – from laboratory to lake. In: J.B. Grace & Tilman (eds.), *Perspectives on plant competition*, pp. 193-213. Academic Press, San Diego, CA.

[6.221] Soudek, D. & Robinson, G.G.C. Electrophoretic analysis of the species and population structure of the diatom *Asterionella formosa*. *Can. J. Bot.* 61, 418-433. 1983.

[6.222] Špačková, J., Hašler, P., Štěpánkova, J. & Poulíčkova, A. Seasonal succession of epipelic algae: a case study on a mesotrophic pond in a temperate climate. *Fottea* 9, 121-133. 2009.

[6.223] Spaulding, S.A., Kociolek, J.P. & Davis, D.R. A new diatom (Bacillariophyceae) genus with two new species from New Mexico, USA. *Eur. J. Phycol.* 37, 135-143. 2002.
[6.224] Steinman, A. & Ladewski, T. Quantitative shape analysis of *Eunotia pectinalis* (Bacillariophyceae) and its application to seasonal distribution patterns. *Phycologia* 26, 467-477. 1987.
[6.225] Sterrenburg, F.A.S. Taxonomy and ecology: an inseparable pair. *Proc. Cal. Acad. Sci.* 56, 156-161. 2005.
[6.226] Stevenson, R.J., Bothwell, M.J. & Lowe, R.L. *Algal ecology. Freshwater benthic ecosystems.* Academic Press, 753 pp. 1996.
[6.227] Stoermer, E.F. Polymorphism in *Mastogloia*. *J. Phycol.*, 3, 73-77. 1967.
[6.228] Stoermer, E.F. Marrying taxonomy and ecology: an attempt. *Proc. Cal. Acad. Sci.* 56, 146-155. 2005.
[6.229] Stoermer, E.F., Kingston, J.C. & Sicko-Goad, L. The morphology and taxonomic relationships of *Stephanodiscus binderanus* var. *oestrupii* (A. Cl.) A. Cl. *Nova Hedw. Beih.* 64, 65-78. 1979.
[6.230] Stoermer, E.F. & Ladewski, T.B. Quantitative analysis of shape variation in type and modern populations of *Gomphoneis herculeana*. *Nova Hedw.*, 73, 347-386. 1982.
[6.231] Stoermer, E.F., Ladewski, B.G., & Schelske, C.L. Population response of Lake Michigan phytoplankton to nitrogen and phosphorus enrichment. *Hydrobiologia* 57, 249-265. 1978.
[6.232] Stoermer, E.F., Qi, Y.-Z. & Ladewski, T.B. A quantitative investigation of shape variation in *Didymosphenia* (Lyngbye) M. Schmidt (Bacillariophyta). *Phycologia* 25, 494-502. 1986.
[6.233] Stoermer, E.F. & Smol, J.P. (eds) *The Diatoms: Applications for the environmental and earth Sciences*. Cambridge University Press. 469pp. 1999.
[6.234] Sullivan, M.J. Diatom community structure: taxonomic and statistical analyses of a Mississippi salt marsh. *J. Phycol.* 14, 468-475. 1978.
[6.235] Syvertsen, E. *Thalassiosira gravida* and *T. rotula*: ecology and morphology. *Nova Hedw. Beih.*, 54, 99–112. 1977.
[6.236] Takabayashi, M., Lew, K., Johnson, A., Marchi, A. & Dugdale, R. The effect of nutrient availability and temperature on chain length of the diatom *Skeletonema costatum*. *J. Plankton Res.*, 28, 831-840. 2006.
[6.237] Tapolczai, K., Vasselon, V., Bouchez, A., Stenger-Kovács, C., Padisák, J. & Rimet, F. The impact of OUT sequence similarity threshold on diatom-based bioassessment: A case study of the rivers of Mayotte (France, Indian Ocean). *Ecol. Evol.* 9, 166-179. 2018.
[6.238] ter Braak, C.J.F. Canonical correspondence analysis: a new eigenvector technique for multivariate direct gradient analysis. *Ecology* 67, 1167–1179. 1986.
[6.239] ter Braak, C.J.F. & Schaffers, A.P. Co-correspondence analysis: a new ordination method to relate two community compositions. *Ecology* 85, 834-846. 2004.

[6.240] Teubner, K. A light microscopical investigation and multivariate statistical analyses of heterovalvar cells of *Cyclotella*-species from lakes of the Berlin-Brandenberg region. *Diatom Res.*, 10: 191–205. 1995.

[6.241] Theriot, E.C., Ruck, E., Ashworth, M., Nakov, T. & Jansen, R.K. Status of the pursuit of the diatom phylogeny: are traditional views and new molecular paradigms really that different? In: Seckbach, J. & Kociolek, J.P. (eds), *The diatom world*, pp. 119-142. Springer. 2011.

[6.242] Tiffany, M.A. Valve development in *Aulacodiscus*. *Diatom Res.* 23, 185-212. 2008.

[6.243] Tiffany, M.A. Valve and girdle band morphogenesis in the pseudocellate diatom species *Biddulphia biddulphiana* JE Smith (Boyer) and *Isthmia nervosa* Kütz. *Nova Hedw. Beih.* 144, 61-95. 2015.

[6.244] Tiffany, M.A. & Hernández-Becceril, D.U. Valve development in the diatom family Asterolampraceae H. L Smith 1872. *Micropaleontology*, 51, 217-258. 2005.

[6.245] Tilman, D. Resource competition between planktonic algae: an experimental and theoretical approach. *Ecology* 58, 338-348. 1977.

[6.246] Tilman, D. & Kilham, S.S. Phosphate and silicate growth and uptake kinetics of the diatoms *Asterionella formosa* and *Cyclotella meneghiniana* in batch and semicontinuous culture. *J. Phycol.*, 12, 375-373. 1976.

[6.247] Tilman, D., Mattson, M., & Langer, S. Competition and nutrient kinetics along a temperature gradient: An experimental test of a mechanistic approach to niche theory. *Limnol. Oceanogr.*, 26, 1020-1033. 1981.

[6.248] Trobajo, R., Clavero, E., Chepurnov, V.A., Sabbe, K., Mann, D.G., Ishihara, S., & Cox, E.J. Morphological, genetic and mating diversity within the widespread bioindicator *Nitzschia palea* (Bacillariophyceae). *Phycologia* 48, 443-459. 2009.

[6.249] Trobajo, R., Cox, E.J. & Quintana, X.D. The effects of some environmental variables on the morphology of *Nitzschia frustulum* (Bacillariohyta) in relation to its use as a bioindicator. *Nova Hedw.* 79, 433-445. 2004.

[6.250] Trobajo, R., Mann, D.G., Clavero, E., Evans, K.M., Vanormelingen, P. & McGregor, R. The use of partial *cox*1, *rbc*L and LSU rDNA sequences for phylogenetics and species identification within the *Nitzschia palea* species complex (Bacillariophyceae). *Eur. J. Phycol.*, 45, 413-425. 2010.

[6.251] Trobajo, R., Rovira, L., Mann, D.G. & Cox, E.J. Effects of salinity on growth and on valve morphology of five estuarine diatoms. *Phycol. Res.*, 59, 83-90. 2011.

[6.252] Turland, N. J., Wiersema, J. H., Barrie, F. R., Greuter, W., Hawksworth, D. L., Herendeen, P. S., Knapp, S., Kusber, W.-H., Li, D.-Z., Marhold, K., May, T. W., McNeill, J., Monro, A. M., Prado, J., Price, M. J. & Smith, G. F. (eds.) *International Code of Nomenclature for algae, fungi, and plants (Shenzhen Code) adopted by the Nineteenth International Botanical Congress Shenzhen, China, July 2017*. Regnum Vegetabile 159. Glashütten: Koeltz Botanical Books. DOI https://doi.org/10.12705/Code.2018. 2018.

[6.253] Tyree, M.A., Bishop, I.W., Hawkins, C.P., Mitchell, R. & Spaulding, S.A. Reduction of taxonomic bias in diatom species data. *Limnol. Oceanogr.: Methods* 18, 271-279. 2020.

[6.254] Tyree, M.A., Carlisle, D.M. & Spaulding, S.A. Diatom enumeration method influences biological assessments of southeastern USA streams. *Freshw. Sci.* 39, 183-195. 2020.

[6.255] Underwood, G.J.C. & Kromkamp, J. Primary production by phytoplankton and microphytobenthos in estuaries. *Adv. Ecol. Res.* 29, 93-253. 1999.

[6.256] Van Dam, H., Mertens, A. & Sinkeldam, J. A coded checklist and ecological indicator values of freshwater diatoms from The Netherlands. *Neth. J. Aquat. Ecol.* 28, 117-133. 1994.

[6.257] Van de Vijver, B. & Cocuqyt, C. Four new diatom species from La Calera Hot Spring in the Peruvian Andes (Colca Canyon). *Diatom Res.* 24, 209-223. 2009.

[6.258] Van de Vijver, B. & Lange-Bertalot, H. *Cymbella amelieana* sp nov. a new large *Cymbella* species from Swedish rivers. *Diatom Res.* 23, 511-518. 2008.

[6.259] Van de Vijver, B., Beyens, L. & Lange-Bertalot, H. The genus *Stauroneis* in the Arctic and (sub-)Antarctic regions. *Bibl. Diatomol.* 51, 1-371. 2004.

[6.260] Van de Vijver, B., Frenot, Y. & Beyens, L. Freshwater diatoms from Île de la Possession (Crozet Archipelago, Subantarctica). *Bibl. Diatomol.* 46, 1-412. 2002.

[6.261] Van de Vijver, B., Jarlman, A. & Lange-Bertalot, H. Four new *Navicula* species (Bacillariophyta) from Swedish rivers. *Crypt. Algol.* 31, 355-367. 2010.

[6.262] Van de Vijver, B., Jarlman, A., Lange-Bertalot, H., Mertens, A., de Haan, M. & Ector, L. Four new European *Achnanthidium* species (Bacillariophyceae). *Algol. Stud.* 136/137, 193-210. 2011.

[6.263] Van de Vijver, B., Jarlman, A., de Haan, M. & Ector, L. New and interesting diatom species (Bacillariophyceae) from Swedish rivers. *Nova Hedw. Beih.* 141, 237–254. 2012.

[6.264] Van de Vijver, B. & Lange-Bertalot, H. New and interesting *Navicula* taxa from western and northern Europe. *Diatom Res.* 24, 415-429. 2009.

[6.265] Van de Vijver, B., Mataloni, G., Stanish, L. & Spaulding, S.A. New and interesting species of the genus *Muelleria* (Bacillariophyta) from the Antarctic Region and South Africa. *Phycologia* 49, 22-41, 2010.

[6.266] Vanelslander, B., Créach, V., Vanormelingen, P., Ernst, A., Chepurnov, V.A., Sahan, E., Muyzer, G., Stal, L.J., Vyverman, W. & Sabbe, K. Ecological differentiation between sympatric pseudocryptic spacies in the estuarine benthic diatom *Navicula phyllepta* (Bacillariophyceae). *J. Phycol.*, 45: 1278-1289. 2009.

[6.267] Vanormelingen, P., Chepurnov, V.A., Mann, D.G., Cousin, S., Sabbe, K. & Vyverman, W. Congruence of morphological, reproductive and ITS rDNA sequence data in some Australasian *Eunotia bilunaris*. *Eur. J. Phycol.*, 42, 61-79. 2007.

[6.268] Vanormelingen, P., Chepurnov, V.A., Mann, D.G., Sabbe, K. & Vyverman, W. Genetic divergence and reproductive barriers among morphologically heterogeneous sympatric clones of *Eunotia bilunaris* sensu lato (Bacillariophyta). *Protist*, 159, 73–90. 2008.

[6.269] Vanormelingen, P., Evans, K.M., Mann, D.G., Lance, S., DeBeer, A.-E., D'Hondt, S., Verstraete, T., De Meester, L & Vyverman, W. Genotypic diversity and differentiation among populations of two benthic freshwater diatoms as revealed by microsatellites. *Molecular Ecology* 24, 4433-4448. 2015.

[6.270] Watanabe, T., Asai, K. & Houki, A. Numerical estimation to organic pollution of flowing water by using the epilithic diatom assemblage. *Sci. Total Environ.* 55, 209-218. 1986.

[6.271] Williams, D.M. *Synedra, Ulnaria*: definitions and descriptions – a partial resolution. *Diatom Res.* 26, 149-153. 2011.

[6.272] Williams, D.M. Names, taxa and databases: some aspects of diatom nomenclature. *Diatom Res.* 36, 101-106. 2021.

[6.273] Wilson, C.J. & Holmes, R.W. The ecological importance of distinguishing between living and dead diatoms in estuarine sediments. *Br. Phycol. J.* 16, 345-349. 1981.

[6.274] Zimmermann, J., Abarca, N., Enk, N., Skibbe, O., Kusber, W.-H. & Jahn, R. Taxonomic reference libraries for environmental barcoding: a best practice example from diatom research. *PLos One* 9, e108793. 2014.

[6.275] Zingone, A., Percopo, I., Sims, P.A. & Sarno, D. Diversity in the genus *Skeletonema* (Bacillariophyceae). I. A reexamination of the type material of *S. costatum* with the description of *S. grevillei* sp. nov. *J. Phycol.*, 41, 140-150. 2005.

7

Diatom Biodiversity in Urban Freshwater Ecosystems: Opportunities and Challenges for Conservation

Majoi de Novaes Nascimento[1]* and Luciane Fontana[2]

[1]*Department of Ecosystem and Landscape Dynamics, Institute for Biodiversity and Ecosystem Dynamics, University of Amsterdam, Amsterdam, Netherlands*
[2]*Department of Engineering, Modeling and Applied Social Sciences, Universidade Federal do ABC (UFABC), São Paulo, Brazil*

Abstract
Artificial and natural urban water bodies could play a crucial role in preserving species, functions, habitats, and ecosystem services that may be unique to these systems or threatened elsewhere. Such ecosystems can contain a significant portion of local and regional biodiversity and have the potential to act as refugia in regions surrounded by sterile areas. A range of species, including diatoms, benefit from these islands of diversity. In this chapter we suggest that urban water systems could be important for maintaining diatom biodiversity and discuss the importance of diatoms for the monitoring and conservation of these systems. Our scientometric map showed that in urban systems, long-term research and rare species are understudied. The map also revealed a gap in research on benthic and periphytic habitats and the necessity for more integrated taxonomic and ecological approaches. We suggest that conservation and management of urban systems are crucial for preserving ecosystem services in areas where water bodies have been historically degraded by human activities. The understanding of the mechanisms driving diatom occurrence and biodiversity can provide insights into ecosystem resilience and restoration, which could guide conservation efforts and management decisions, and provide target conditions for policymakers and environmental managers.

**Corresponding author*: m.denovaesnascimento@uva.nl

Nora I. Maidana, Magdalena Licursi and Eduardo Morales (eds.) Diatom Ecology: From Molecules to Metacommunities, (203–232)

Keywords: Diatom biodiversity, biodiversity conservation, eutrophication, novel ecosystems

7.1 Contextualizing Urban and Novel Ecosystems

Humans have exerted influence on natural systems for thousands of years, but the magnitude of these impacts has increased exponentially over the last centuries [7.68]. Today most of the world's ecosystems are somehow impacted by humans [7.120] [7.138], leaving few ecosystems untouched. These impacts have had significant effects on the organization of species, leading to high rates of extinction and the introduction of non-native species as well as habitat fragmentation, and the loss of species pools, all of which can prevent the re-establishment of pre-existing assemblages [7.3] [7.37] [7.80] [7.109] [7.135] [7.141]. The concept of novel ecosystem was originated in response to the increased occurrence of rapidly transformed configurations of species and the idea that they need to be considered for conservation and restoration [7.21] [7.50] [7.86]. Novel ecosystems lie in between natural and intensively managed systems and have non-analog (that is compositionally different) configurations of species composition, relative abundance, and possibly ecosystem functions, all altered by human-induced conditions [7.50] [7.51] [7.120]. Although human agency mediates the formation of novel ecosystems, human intervention is not needed for the continued maintenance of these systems [7.51].

Urban freshwater systems have faced considerable biotic and abiotic reorganization, mediated by human activity, and as such, can be considered novel ecosystems [7.51]. Even when completely modified and intensively managed, urban water bodies provide an extensive range of services, including carbon sequestration, microclimate stabilization, water and energy consumption, food provision, and hold cultural and spiritual significance that is of an intangible nature [7.8] [7.29] [7.107] [7.136]. These novel ecosystems can also be important for conservation purposes, as they frequently function as refugial islands for resting, reproducing, or feeding birds, mammals, amphibians, fish, insects, plants and algae [7.1] [7.2] [7.27] [7.64] [7.96] [7.97]. In some cases, these urban water bodies are the only landscape locally available for a stopover, during a species' migration [7.1] [7.79] [7.82] [7.115] particularly in areas where historic freshwater systems are limited or absent. It is therefore fundamental to recognize that urban ecosystems can sometimes represent a unique and valuable alternative for species survival in urban areas [7.110].

Urban water systems could maintain high levels of biodiversity at local (alpha diversity) and regional (beta and gamma diversities) scales due to their productivity and environmental heterogeneity [7.77] [7.89] [7.132]. However, this biodiversity is threatened by processes resulting from population growth in urban areas, erosion, intensification of untreated sewage discharge, and changes in climatic patterns. The main driver of community change in urban aquatic systems is the increase in nutrient levels and eutrophication [7.15] [7.129] [7.147] [7.149], leading to a decrease in alpha and beta diversity, and an increase in community homogeneity and dominance of generalist diatoms [7.77] [7.144] [7.152]. This process results in the loss of resilience (Figure 7.1), reducing the system's ability to respond to environmental disturbances [7.30] [7.123] [7.134] [7.152]. Resilience can be defined as either the capacity to absorb changes and keep similar functioning and structure (ecological resilience) before shifting to an alternative state [7.52] [7.122] [7.123] or to rebound after change (engineering resilience) [7.106]. Resilience may reflect ecological redundancy (presence of species with equivalent functions in the ecosystem) within food webs [7.105] [7.116], which tends to be higher when biodiversity is also high, or natural buffering offered by the landscape.

Eutrophication is a well-documented environmental issue that affects almost all continental aquatic ecosystems, including non-urban ones [7.7] [7.42] [7.44] [7.60] [7.83] [7.144]. This phenomenon occurs as a biological response to an increase in nutrient concentration, mainly nitrogen and phosphorus, from domestic, agricultural and industrial effluents. Even after reducing external nutrient inputs, the effects of eutrophication can persist over long periods, considering anthropic time scales [7.18] [7.44] [7.150] [7.151]. Eutrophication not only diminishes the multiple uses of water services but is also associated with the proliferation and spread of diseases in humans. As species diversity decreases, the possibility of tolerant and toxic species dominating the community increases, resulting in the decline of natural controls for pests and diseases in urban areas [7.85] [7.90] [7.146]. Since all trophic levels of an ecosystem are interconnected, the loss of biodiversity at one level (e.g., autotrophs) can rapidly propagate to the next level of the trophic chain, causing ecological shifts and disrupting the historical balance and functioning of the entire ecosystem [7.99] [7.127]. Therefore, maintaining biodiversity of a system is crucial not only for its functioning but also for preventing trophic cascades [7.13] [7.36] [7.53]. This is a global problem that is far from being solved, and alternatives should be explored to decrease the velocity and intensity of eutrophication pressures on both non-urban and novel ecosystems.

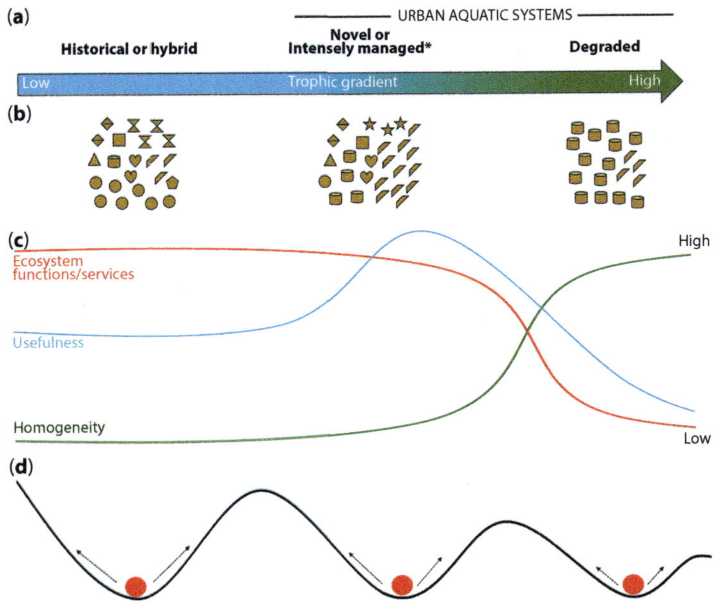

Figure 7.1 Conceptual figure showing urban water bodies as novel ecosystems [7.51] in relation to a progressive change in trophic gradient (a). Changes in trophic gradients will result in alterations in diatom biodiversity, where each golden geometric shape represents one diatom taxon (b). Shapes that are repeated represent taxa occurring at high relative abundances. Trophic gradients also affect ecosystem function and services, system usefulness for humans, species homogeneity (c, modified from Zorzal-Almeida, Bartozek [7.152], and ecosystem resilience (d), shown by valleys representing different stable equilibria, hilltops representing unstable equilibria (also called tipping point), red circles representing the system itself, and movement (arrows) representing environmental impacts (modified from Scheffer and Carpenter [7.123]). The lower the hilltop, the less resilient is the system. This shift to an alternative state occurs when a threshold in conditions is reached, leading to a strong and abrupt change in the state of a system. When an urban aquatic system presents low trophic states, diatom diversity can be high even if the relative abundances, or taxa occurrences change in comparison to the historical or hybrid system. *In this scenario some intensely managed aquatic systems (e.g., reservoirs used for water consumption or generation of electricity) can hold high diatom biodiversity. Ecosystem functions, services and species homogeneity can present levels that are similar to the ones found on historical systems. Usefulness for humans (blue line) can increase if the system is managed to supply populational demands that are not possible in natural, not managed systems (such as generation of electricity). The high biodiversity found in urban systems with low trophic gradients allows for high resilience to environmental impacts. As the trophic gradient increases, urban aquatic systems lose biodiversity, ecosystem functions and services, become homogenized, resulting in loss of resilience.

The consequences of eutrophication highlight the need for investment in restoration actions that are informed by scientific studies and science-based solutions for eutrophication-related problems. By doing so, environmental managers and policymakers can make informed decisions and prioritize the conservation of biodiversity [7.38] [7.45] [7.84]. This is especially important in urban areas, where increased absorption and storage of heat by artificial materials (such as buildings and pavements) result in higher local temperatures, also known as the "urban heat island effect" [7.48] [7.76] [7.92], which may accelerate biological processes. Climate change exacerbates the effects of urban heat islands and leads to extreme rainfall and drought anomalies, particularly in regions with poor land management [7.126] The environmental, economic, social, and cultural significance of urban freshwater ecosystems, often referred to as "blue spaces" in these urban heat islands, cannot be underestimated [7.19] [7.38] [7.46] [7.90] [7.103] [7.140].

Diatoms (Bacillariophyceae) are one of the groups that are abundant in urban waters, and are usually used as bioindicators of water quality [7.12] [7.25] [7.31] [7.41] [7.49] [7.91] [7.142]. On the other hand, urban water systems may also hold part of the local and regional diatom biodiversity, host endemic species, and serve as source of species dispersal in scenarios of future uncertainty such as the one posed by climate change. Diatoms occur in virtually all aquatic habitats, often dominating primary production of freshwater systems [7.117] [7.130]. One of the most widespread groups on Earth, diatoms are incredibly biodiverse and occupy a wide range of ecological niches [7.11] [7.14] [7.61].

In recent decades, the use of diatoms as bioindicators of limnological changes has gained significant attention and constantly demonstrated the sensitivity of this group to various chemical and physical environmental changes [7.24] [7.32] [7.43] [7.49] [7.62] [7.71] [7.87] [7.89]. Some of these approaches involve understanding the role of diatom biodiversity to obtain insights into the effects of multiple anthropogenic activities on these ecosystems, and the intensity and velocity at which ecosystems respond [7.65] [7.101] [7.102] [7.132]. In this chapter, we discuss the significance of urban aquatic systems in relation to diatom diversity as a tool to assess the impacts of human activities on freshwater ecosystems. We present an overview of the current state of the art in biodiversity studies carried out in urban regions, and compare it with those conducted in non-urban and rural freshwater ecosystems. Our aim is to provide a panoramic view of the current state of research in this field, highlighting potential trends, new promising approaches, and recommendations for future directions.

7.2 Diatom Diversity in Novel Freshwater Ecosystems: Insights and Future Directions from a Systematic Literature Review

To evaluate the state of art in diatom research in urban systems, we conducted a systematic bibliographic search utilizing the Scopus database and applied a combination of two specific keywords linked by the conjunction "and" — "urban freshwater diatoms" and "urban freshwater ecosystems" — associated with the overarching keyword "diatom diversity in freshwater ecosystems"; and only peer reviewed papers published in English language were included. This search aimed to retrieve studies related to diatom diversity in both novel freshwater ecosystems in urban settings and in non-urban or rural freshwater ecosystems. The search was limited to articles published from 1947 to 2023. In total, 9998 articles were found, of which 415 articles presented the word "urban" in the title (Figure 7.2), were published in English language, reviewed by pairs and contained the specific keywords previously described, and therefore, were the only ones used to discuss the geographic distribution of research in urban systems

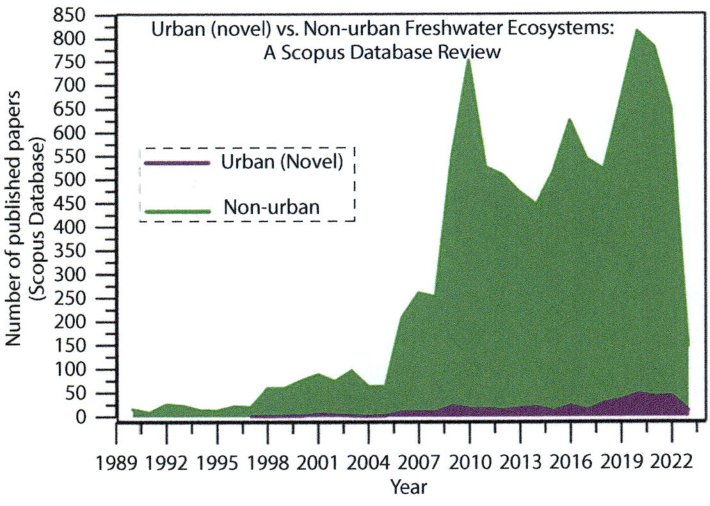

Figure 7.2 Graph showing trends in the number of published papers on novel and non-urban freshwater ecosystems worldwide, using data from the Scopus database. The green shade represents the number of studies conducted in non-urban or rural freshwater ecosystems per year and the purple line the number of studies conducted in novel ecosystems (urban freshwater ecosystems) per year. For better visualization we only included the articles published from 1986 to 2023.

(Figure 7.3) and the Scientometric analysis (Figure 7.4). Our analysis aimed to identify common themes, trends, and areas that require further research. Our goal was to gain insights into the current state of knowledge on this topic and to identify gaps in our understanding that can be addressed in future studies.

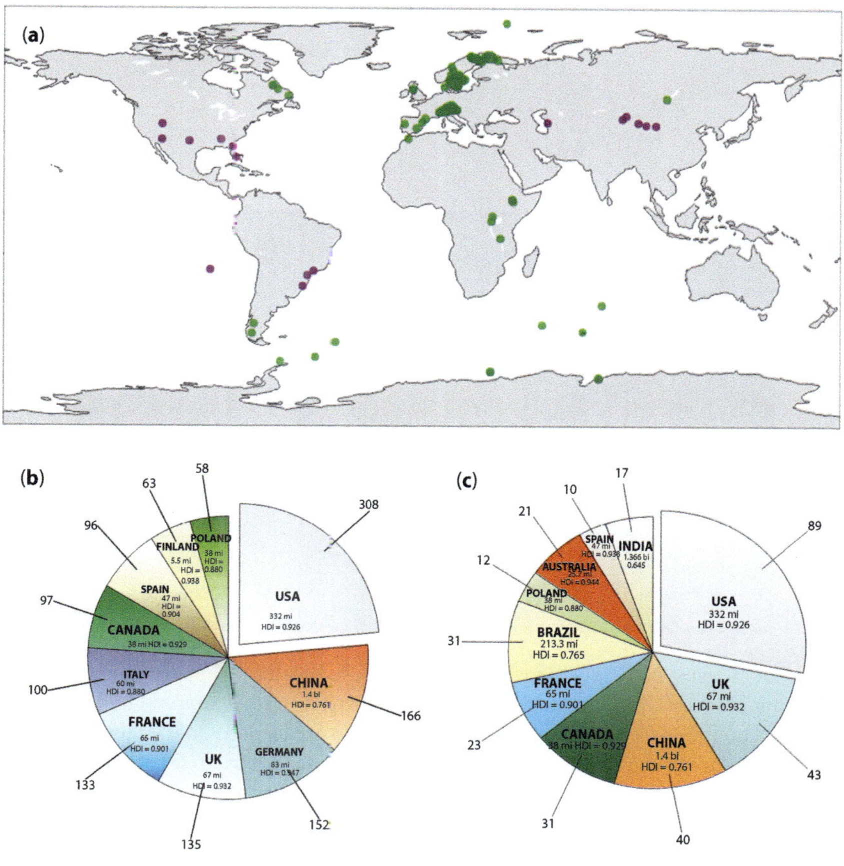

Figure 7.3 Distribution of studies retrieved from our search using the Scopus database (a). The search displays the global distribution of the 9998 studies related to the topic "biodiversity of diatoms in freshwater ecosystems" that were published in English language and peer reviewed (green) and of the 415 studies related to the topic "biodiversity of diatoms in urban freshwater ecosystems" (purple). The pie charts show the ten countries with the highest number of published papers on "biodiversity of diatoms in freshwater ecosystems" (b) and on "biodiversity of diatoms in urban freshwater ecosystems" (c), with population size and human development index (HDI) of each country included within the chart. The number of published papers for each country is displayed outside of the chart.

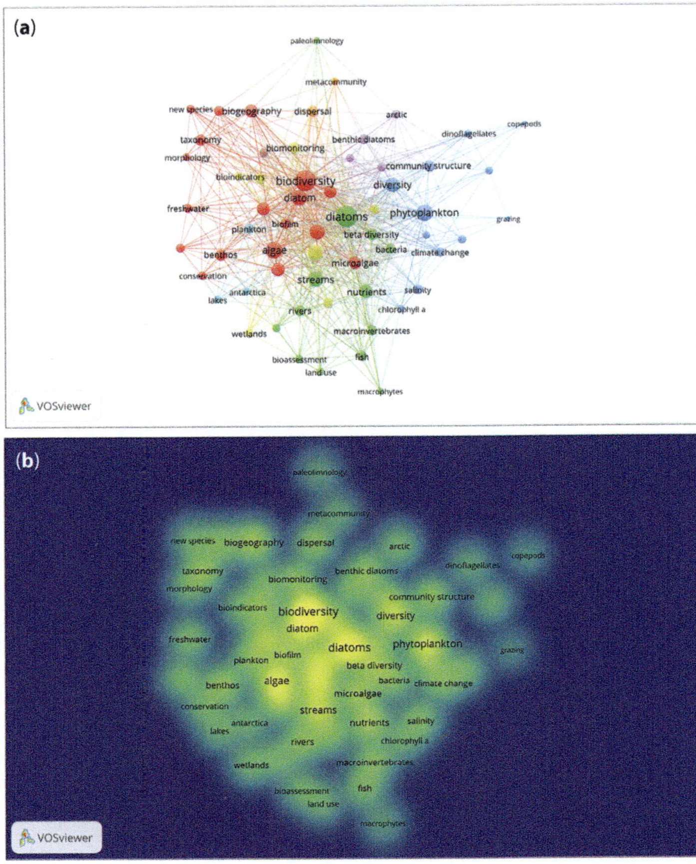

Figure 7.4 Scientometric VosViewer map created based on the 415 papers found on the Scopus database. The top panel (a) displays the map using balloons and highlights the connections between different topics and areas that are focused on the biodiversity of diatoms in urban ecosystems, ranging from earlier topics related to taxonomic and floristic surveys, to more recent experimental approaches using phytoplanktonic diatoms, and the current focus of ecological studies aimed at informing conservation and policy-making decisions. The bottom panel (b) shows a density color map that represents highly cited and more recent topics in warm colors (red) with larger font size, while earlier contributions and less impactful topics are represented by cooler colors (green).

We found a disparity in the total number of published articles between urban and non-urban freshwater ecosystems around the world (Figure 7.2). A temporal view of total number of publications addressing the topic “biodiversity of diatoms in freshwater ecosystems” in urban areas and non-urban or rural environments shows that the difference between both is hundreds of papers per year. From 1986 until 2005, the number of research papers in historical and non-urban systems varied from 20 to 100 publications per year, whilst in urban systems the number of papers per year never reached 20. After 2005, papers published in historical and non-urban areas increased greatly with more than 800 studies being published per year in some years, whilst in urban systems the number never exceeded 50 studies per year. There are several reasons why there are more published papers on diatom biodiversity in non-urban or rural freshwater ecosystems than in urban freshwater ecosystems. One is that urban freshwater ecosystems are subject to higher levels of anthropogenic stressors, such as pollution and habitat modification, which can have negative impacts on diatom communities and thus their diversity. This makes it more difficult to understand the biodiversity data information without access to reference conditions of these ecosystems. Another possible reason is that there may be a bias in research funding and priorities toward non-urban or rural freshwater ecosystems due to their importance for conservation and natural resource management [7.108]. Researchers may be more likely to receive funding or may be more interested in studies in non-urban or rural areas due to the perception that these ecosystems are more pristine, acting as a reference site and, thus, in need of protection. However, it is important to note that the trend in studies being conducted on diatom biodiversity in urban freshwater ecosystems is increasing (Figure 7.2), and this field of research has grown over the last 10 years. As urbanization continues to expand on the margins of freshwater resources, understanding the effects of human activity based on diversity will become increasingly important for effective urban ecosystem management [7.72].

The results of our search also reveal a significant disparity in the number of published articles on diatom diversity in novel freshwater ecosystems between countries with high and low human development indices (HDI; Fig. 7.3B and C). Specifically, our results indicate that countries with high HDI and high population sizes, such as the United States of America and China, have contributed with the highest numbers of papers, while other countries with comparable HDI, such as the United Kingdom, France, Italy, and Canada, have made modest contributions. However, countries with high HDI are between the top ten with more contributions on the topic, regardless of population sizes. The only exception is Germany, that has one-fourth of the USA's population and one-sixteenth of China's population, and has a paper production comparable to both countries. A similar disparity is observed when each country is considered. Even in countries where contribution is high, these contributions are concentrated in well-developed regions of the country, and are not well distributed (Figure 7.3a).

Notably, the global South displayed the lowest contribution to the field, with none of the countries featuring among the top ten contributors. This finding emphasizes the urgent need for research efforts and collaboration between developed and underdeveloped regions of the world to enhance our understanding of diatom diversity in freshwater ecosystems on a global scale. Given that conservation science requires an interdisciplinary approach that integrates social, economic, and political factors to effectively address biodiversity loss worldwide [7.39], it is crucial to consider the inadequacy of knowledge in several countries, as indicated by Figure 7.3.

The geographical discrepancy in the number of articles published on diatom diversity in urban freshwater ecosystems between countries with high and low HDI raises concerns about the global representation of scientific knowledge and is seen across different disciplines [7.131]. Therefore, we urge more scientific communication and collaboration between researchers, scientific funding institutions, and universities in underrepresented regions. Providing resources and training for local researchers and fostering international collaborations can aid in promoting research efforts and increasing the participation of underrepresented regions in scientific research.

The high contributions from the United States, China, and Germany may also reflect their larger economies, investments in research and development, and robust scientific infrastructure [7.94]. These countries are also known for having well-established research institutions, universities, and a strong scientific community. However, despite being developed countries, the United Kingdom, France, Italy, and Canada made more modest contributions, indicating the need for greater research efforts in these countries. The low contributions from the Global South, despite being home to a significant portion of the world's freshwater resources and, in some cases, with the largest urban areas (e.g., São Paulo State in Brazil), are particularly concerning. For instance, regions such as the Amazon experience high rates of environmental degradation, threatening their freshwater ecosystems [7.20]. Research efforts and collaboration in underrepresented regions are then crucial to enhance our understanding of diatom diversity and freshwater ecosystem functioning [7.131].

Regarding regions worldwide, such as São Paulo (Brazil), Buenos Aires (Argentina), Johannesburg (South Africa), Sydney (Australia), and Auckland (New Zealand), the number of published papers is generally lower compared to the top ten countries listed in this research (Figure 7.3). This aligns with the general trend observed in our analysis, where countries with higher HDI tend to publish more papers in this field, regardless of their population size. For instance, China, with the world's highest population, is the second-largest contributor to the field, behind only the United States of America. In contrast, India, which ranks second globally in terms of population, is considered underrepresented on the topic. It is important to consider that regional studies may not be included in our analysis due to limitations in the search database or language of publications may exist.

In Brazil, few studies have focused on diatom biodiversity, mainly in the southern regions of Paraná, Santa Catarina, and Rio Grande do Sul [7.10] [7.34] [7.73] [7.77] [7.81] [7.128]. In São Paulo City, one of the largest metropolitan areas of the world, there are even fewer studies on the topic [7.91] [7.144] [7.152] [7.153]. In addition to research efforts, adopting an interdisciplinary approach to conservation science [7.39], is essential. This approach can bridge the gap between scientific knowledge and practical solutions to protect freshwater ecosystems and the biodiversity they support. Promoting collaborations between developed and underdeveloped regions through scientific communication and collaboration can help increase the participation of underrepresented regions in scientific research. This can be achieved by supporting scientific institutions, universities, and research funding in underrepresented regions, providing resources and training for local researchers, and fostering international collaborations.

Although significant research efforts have been made to study diatom diversity in novel freshwater ecosystems, there is still a significant disparity in the number of articles published between developed and underdeveloped regions and between urban and non-urban regions. It is crucial to conduct further research to investigate patterns of diatom biodiversity in urban freshwater ecosystems in different regions of the world. Particularly in the most urbanized areas of the Global South so we can gain a more complete understanding of the topic's global status.

7.3 Mapping Gaps and Opportunities

In regions with high levels of urbanization near freshwater ecosystems, diatoms play a crucial role as an indicator of water quality. They are used in various studies because of their ability to contribute significantly to primary productivity in eutrophic waters. Some diatom species are well adapted to polluted conditions and widely recognized as a bioindicator of this impact (e.g., *Nitzschia palea*; *Aulacoseira granulata*), while other soft algae are not. Although urban systems are not usually used to access patterns of biodiversity, research shows urban freshwater ecosystems exhibit high diatom diversity, reaching up to 200 species in a single sample in some cases [7.72]. And even the species that are widely recognized as bioindicators, such as the ones cited above, can be erroneously identified, further underestimating the biodiversity of these systems.

Several factors can impact diatom diversity in urban freshwater ecosystems, such as eutrophication, water chemistry, land use, spatial heterogeneity and the presence of potentially toxic elements [7.16] [7.49] [7.144] [7.153]. Specifically, water pollution, whether from industrial or domestic sources, and heavy metal contamination due to anthropogenic activities, remain the main issues addressed today in novel ecosystems [7.70] [7.95] [7.113] [7.124]. However, to fully understand the main scientific approaches used by scientists to address diatom diversity, and the hot topics that emerged from analyzing Figure 7.3, we create a scientometric map from literature (Scopus database) by using VosVievwer® software (Figure 7.4). In the map analysis, the size and color of the balloons represent different characteristics of the papers included in the analysis. The size of the balloon indicates the number of citations received by the topic, with larger balloons representing more highly cited topics. The color of the balloon indicates the publication year of the paper, with earlier years represented by cooler colors such as blue and green, and more recent years represented by warmer colors such as red.

By combining these two characteristics, it is possible to quickly identify important and influential papers in the field, as well as trends in the research over time. The larger, more highly cited topics tend to be located towards the center of the map, indicating their importance in the field. Meanwhile, the cooler colors towards the edges of the map indicate older papers that may be less relevant to current research, while the warmer colors towards the center and in the other extreme represent more recent and potentially more relevant work.

The size and color of the balloons in the map provide a useful visual representation of the research landscape and can aid in identifying important papers and trends in the field. The generated map based on the Scopus database retrieved by this study reveals 3 main clusters differentiated by their balloon colors and sizes. The red cluster connects the subfield associated with the use of the diversity of diatoms to assess environmental changes, conservation, and monitoring of water quality (represented by the larger balloons). The blue cluster (in the middle) connects the subfields more associated with floristic surveys and taxonomical studies that also improve our knowledge about diatom diversity. It is important to emphasize that this cluster also interconnects with the other two clusters on the map, demonstrating that taxonomical and floristic surveys are closely related to ecological and experimental approaches. The last group connects all the papers related to experimental studies, mainly with planktonic diatoms that are closely related to productivity changes in urban freshwater ecosystems. These results provide insights into the different scientific approaches used to study the importance of diatom diversity in urban freshwater ecosystems. The identification of different clusters and their interconnectivity highlights the importance of interdisciplinary approaches to address the gaps and opportunities in the current literature on this topic.

Based on the results of the map, it is clear that diatoms play a critical role in understanding the conservation and management of freshwater ecosystems. The red cluster (recent trends), represented by larger balloons, connects the subfield associated with the use of the diversity of diatoms to assess environmental changes, conservation, and monitoring of water quality. This cluster highlights the importance of diatoms as bioindicators of water quality, and the need to use their diversity to assess changes in freshwater ecosystems and improve their conservation and management. The blue cluster, in the middle, connects the subfields more associated with floristic surveys and taxonomical studies that also incorporate our knowledge about diatom diversity. This cluster underscores the importance of taxonomic and floristic surveys to identify and describe diatom species, including rare or endemic species that can provide valuable information about ecosystem quality. However, since the group is in the middle of the map and with cooler color (blue), suggesting that these topics represent the earlier stage of the diatom diversity use in the literature. The last group (green) connects all the papers related to experimental studies, mainly planktonic diatoms, which are closely related to accessing productivity and seasonal changes in urban freshwater ecosystems. This cluster highlights the need for more research on sediment or periphyton diatoms to better understand the role of different habitats as a whole picture in the biodiversity in novel ecosystems. For example, benthic and periphytic diatoms can serve as important indicators of changes in productivity and other ecological processes.

Overall, the analysis highlights the importance of further research on diatom diversity to improve conservation and management efforts. By incorporating this information into the decision-making process, we can better protect and restore these valuable ecosystems for future generations. Finally, the studies using long-term monitoring, such as paleoenvironmental studies and paleolimnological approaches that incorporate a multiproxy approach (chemical, physical and genetic markers of the past) to better understand the causes, time, frequency and intensity of the biodiversity changes over time, and those studies related to forecasting and modeling future scenarios, both absent in the map, may represent the gaps that need to be focused on in future research on the topic.

7.4 Ecological Importance of Diatoms in Monitoring and Managing Urban Freshwater Ecosystems

Urban freshwater ecosystems are becoming increasingly important for conservation and management efforts due to their ability to support a range of species, functions, habitats, and ecosystem services that may be endemic to these systems or threatened elsewhere. Diatoms are one group of microalgae that have significant ecological importance and high capacity for monitoring the water quality of these ecosystems. Research on diatom diversity in urban freshwater ecosystems has the potential to provide valuable insights into the functioning and effective management of these systems. For example, studies have shown that diatom communities in urban freshwater ecosystems can vary greatly from those in non-urban systems, with the former often displaying lower diversity and different community composition due to the impacts of human activities such as pollution and eutrophication [7.126] [7.133] [7.142] [7.149]. By studying diatom diversity in urban freshwater ecosystems, researchers can identify key taxa that are adapted to these highly disturbed systems, which can provide insights into the potential for these systems to recover from human impacts. For example, combining diatom diversity data with paleolimnological approaches can reveal long-term changes in the ecosystem and help establish reference conditions for restoration efforts [7.111] [7.143] [7.166].

Studies that relate diatoms to certain environmental conditions can also provide much ecological status information useful for advising management decisions related to conservation [7.125] [7.137] [7.181]. Another approach that has been used is the development of a regional diatom trophic index [7.57] [7.133] to help environmental managers to delineate effective actions to reduce pollution or to recognize environmental thresholds. Researchers that focus on diatom diversity can inform management strategies, including helping managers to design and implement restoration plans, monitoring programs, and the establishment of conservation targets. Overall, research on diatom diversity in urban freshwater ecosystems represents a smart investment for biodiversity conservation and human well-being, since novel ecosystems in urban areas represent in several cases a unique way to obtain water for human consumption, recreation, hydroelectricity generation and other ecosystem services.

Diatoms have been used as water quality monitoring tools in many US state and federal agencies and the European Union. They have been employed to assess the biotic integrity of surface waters and as indicator taxa in federal programs such as the US Environmental Protection Agency's (US EPA) National Aquatic Resource Surveys, National Science Foundation's National Ecological Observatory Network [7.54], and the European Water Framework Directive [7.5] [7.78]. The US EPA has maintained a diatom sampling and monitoring program on the Laurentian Great Lakes for more than 20 years to assess annual variability, ecological changes, and long-term impacts [7.63] [7.112]. States, such as Kentucky, Minnesota, Wisconsin and Montana, use diatoms to assess stream quality and reference conditions of lakes, and as supporting evidence for developing standards, and site-specific management plans under the federal Clean Water Act criteria [7.47] [7.48] [7.59].

By understanding the ecological importance of diatoms as monitors and sentinels of multiple environmental changes in these historically highly disturbed systems, we can work towards a better understanding of how to conserve and manage urban freshwater ecosystems to ensure their ecosystem services are maintained and sustained over time. Therefore, research on diatom diversity can inform management strategies and help to protect non-urban ecosystems by learning from the negative effects that these highly disturbed systems demonstrated for several centuries of human activities.

7.5 Climate Change and Conservation of Diatom Biodiversity in Urban Systems

Ongoing and future anthropogenic climate change are expected to be one of the greatest threats to biodiversity, affecting species distributions and ecological interactions. The changes in climate already show the disruption of temperature and precipitation patterns and, of particular concern, the increase of the intensity and frequency of droughts [7.40] [7.56] [7.75] [7.100]. In aquatic systems, increase in temperatures and in drought events will result in reduced river runoff and water deficit, and will increase the risk and velocity in which eutrophication happens [7.28]. Aquatic systems in urban areas seem even more sensitive to the effects of climate change, warming at faster rates [7.58].

For diatoms, the community response to climate change can be similar to the response to eutrophication, with decrease in species richness, in structure and the functional biodiversity, in overall cell size, and change in species composition, with the increase in dominance of generalists [7.17] [7.23] [7.93] [7.104] [7.121] [7.139]. Diatom response to extreme drought seems to be less intense in larger water bodies [7.104], and recovery faster in environments when water quality and initial biodiversity is high [7.17]. These responses, however, may vary depending on the frequency and intensity of these droughts [7.93].

In a scenario where climatic patterns are changing, to prioritize areas for long-term conservation has become a difficult task. Populations of many species are likely to occur in microrefugia and with large risk of extinction [7.88]. Microrefugia are defined as isolated populations surviving in unusual microclimates that differ from the surroundings landscape [7.118] [7.119]. If these areas are large enough to support viable populations the species can survive in the microrefugia landscape, until more favorable conditions return to the larger, surrounding landscape. The duration of climatic isolation, the capacity for mutation, the size of the isolated population and the rate of gene flow to the nearest "mainland" population, all determine the genetic fate of the microrefugium [7.22] [7.88]. Climatic variables, however, operate at different scales and so differ in the degree to which they decouple from regional climate. Additionally, distributions of species are influenced individualistically by different climatic variables, which suggests that not all species from a given community will be favored by the same microrefugium [7.55]. Single-species conservation objectives, even associated with microrefugia, will then be less likely to be viable than planning on protecting a broad range of species with similar ecological tolerances.

Although less well studied than historic systems (Figure 7.2), urban systems can retain high diatom biodiversity, especially when water quality and spatial heterogeneity are high, and a well-preserved riparian zone is maintained [7.35] [7.89] [7.153]. Additionally, diatoms also present biogeographical patterns [7.9] [7.145], which means that by increasing the geographical range of protection we similarly increase the likelihood of preserving more species. This way urban systems could contribute by broadening this range of species/habitat types that can be protected.

To deposit our hopes in the possibility of high quality serving as microrefugia for some diatom species is not the purpose of this chapter. Instead, we believe that urban systems could play a role alongside historical systems in preserving a wider range of species in a world that is rapidly changing, and that diatoms are a valuable tool to evaluate the quality of these urban systems. More than that, investments in conservation and restoration of urban systems can increase their lifespan and optimize their multiple uses and their disproportionate importance to people, which can help diminish the pressure on historical systems.

7.6 New Approaches and Perspectives for the Future

Urban systems are characterized by multiple physical and chemical sources of pressure [7.15] [7.98], including domestic and industrial wastewater, surface runoff containing organic contaminants, fertilizers, pesticides, and metals that come from point and diffuse sources [7.74]. These systems are constantly under pressure from human activity and their restoration has become a complex challenge that requires a multifaceted approach [7.6]. One of the most important steps is to strengthen the global biodiversity framework, which includes a variety of strategies to reduce pollution, restore habitat and promote sustainable water use for human needs. However, there is no single solution to restore and protect urban freshwater ecosystems. Rather, a combination of approaches should increase the quality of the data and consequently generate a wider variety of solutions. Techniques such as barcoding, metagenomics, stable isotopes of frustules, and paleolimnology can be used as part of a multifaceted approach to restore and protect urban freshwater ecosystems. Barcoding and metagenomics can be used to identify and monitor aquatic organisms, including those that play important roles in freshwater ecosystem functioning [7.4] [7.69]. Stable isotopes of frustules can provide insights into the sources and cycling of nutrients in freshwater ecosystems [7.67]. Paleolimnology can be used to reconstruct past environmental conditions and track long-term changes in freshwater ecosystems, providing valuable information for ecosystem management and restoration [7.12] [7.41] [7.144], especially by providing the frequency, the causes and the intensity of changes. Besides, long-term studies can access the reference conditions of these systems in the absence of long-term monitoring available data. By applying these diversified methods, we are investing in understanding the long-term sustainability and adaptability of urban freshwater ecosystems and the important services they offer.

In this chapter we identified that gaps in research and investments are rooted in social economic disparities between and within countries and in the way our society and scientific community perceive the importance and value of natural vs. urban systems. Although we suggest that urban systems could serve as repositories and refugia for diatom diversity, the present body of work on these systems does not allow for certainty. We also emphasize that our point here is not to focus on urban systems instead of natural ones for conservation, but instead to include urban systems in the conservation discussion to optimize their uses, services, life span, and maybe their use as islands for recolonization. Finally, we advocate that conservation efforts should be placed on urban aquatic systems as a way to decrease pressure on historic "natural" systems, by optimizing their uses and ecological status. In this scenario diatoms offer a valuable tool in evaluating and controlling water quality and in offering reference conditions.

References

[7.1] Adams H, McGuire LP. 2022. Island biogeography theory and the urban landscape: Stopover site selection by the silver-haired bat (lasionycteris noctivagans). Canadian Journal of Zoology. 100(4):243-250

[7.2] Adams LW. 1994. Urban wildlife habitats: A landscape perspective. U of Minnesota Press.

[7.3] Aguilar R, Cristóbal-Pérez EJ, Balvino-Olvera FJ, de Jesús Aguilar-Aguilar M, Aguirre-Acosta N, Ashworth L, Lobo JA, Martén-Rodríguez S, Fuchs EJ, Sanchez-Montoya G. 2019. Habitat fragmentation reduces plant progeny quality: A global synthesis. Ecology Letters. 22(7):1163-1173.

[7.4] Alindonosi A, Baeshen M, Elsharawy N. 2021. Prospects for diatoms identification using metagenomics: A review. Appl Ecol Environ Res. 19(6):4281-4298.

[7.5] Allan IJ, Vrana B, Greenwood R, Mills GA, Knutsson J, Holmberg A, Guigues N, Fouillac A-M, Laschi S. 2006. Strategic monitoring for the european water framework directive. TrAC Trends in Analytical Chemistry. 25(7):704-715.

[7.6] Arthington AH. 2021. Grand challenges to support the freshwater biodiversity emergency recovery plan. Frontiers in Environmental Science. 9:118.

[7.7] Ayala-Borda P, Lovejoy C, Power M, Rautio M. 2021. Evidence of eutrophication in arctic lakes. Arctic Science. 7(4):859-871.

[7.8] Aylward B, Bandyopadhyay J, Belausteguigotia J-C, Borkey P, Cassar A, Meadors L, Saade L, Siebentritt M, Stein R, Tognetti S. 2005. Freshwater ecosystem services. Ecosystems and human well-being: policy responses. 3:213-256.

[7.9] Bartozek EC, da Silva-Lehmkuhl AM, Gregory-Eaves I, Bicudo DC. 2019. Environmental and spatial drivers of diatom assemblages in the water column and surface sediment of tropical reservoirs. Journal of Paleolimnology. 62:245-257.

[7.10] Bartozek ECR, Bueno NC, Ludwig TAV, Tremarin PI, Nardelli MS, Rocha ACRd. 2013. Diatoms (Bacillariophyceae) of Iguaçu National Park, Foz Do Iguaçu, Brazil. Acta Botanica Brasilica. 27:108-123.

[7.11] Behrenfeld MJ, Halsey KH, Boss E, Karp-Boss L, Milligan AJ, Peers G. 2021. Thoughts on the evolution and ecological niche of diatoms. Ecological Monographs. 91(3):e01457.

[7.12] Bennion H, Fluin J, Simpson GL. 2004. Assessing eutrophication and reference conditions for scottish freshwater lochs using subfossil diatoms. Journal of applied Ecology. 41(1):124-138.
[7.13] Biodiversity as a tool for environmental assessment. AIP Conference Proceedings; 2020: AIP Publishing LLC.
[7.14] Biodiversity, biogeography and conservation of diatoms. Biogeography of Freshwater Algae: Proceedings of the Workshop on Biogeography of Freshwater Algae, Held During the Fifth International Phycological Congress, Qingdao, China, June 1994; 1996: Springer.
[7.15] Birk S, Chapman D, Carvalho L, Spears BM, Andersen HE, Argillier C, Auer S, Baattrup-Pedersen A, Banin L, Beklioğlu M. 2020. Impacts of multiple stressors on freshwater biota across spatial scales and ecosystems. Nature Ecology & Evolution. 4(8):1060-1068.
[7.16] Blinn DW, Bailey PC. 2001. Land-use influence on stream water quality and diatom communities in victoria, australia: A response to secondary salinization. Hydrobiologia. 466:231-244.
[7.17] Calapez AR, Elias CL, Almeida SF, Feio MJ. 2014. Extreme drought effects and recovery patterns in the benthic communities of temperate streams. Limnetica. 33(2):281-296.
[7.18] Carpenter SR. 2005. Eutrophication of aquatic ecosystems: Bistability and soil phosphorus. Proceedings of the National Academy of Sciences. 102(29):10002-10005.
[7.19] Carrier-Belleau C, Pascal L, Nozais C, Archambault P. 2022. Tipping points and multiple drivers in changing aquatic ecosystems: A review of experimental studies. Limnology and Oceanography. 67:S312-S330.
[7.20] Castello L, Macedo MN. 2016. Large-scale degradation of amazonian freshwater ecosystems. Global change biology. 22(3):990-1007.
[7.21] Chapin III FS, Starfield AM. 1997. Time lags and novel ecosystems in response to transient climatic change in arctic alaska. Climatic change. 35(4):449-461.
[7.22] Cheddadi R, Birks HJB, Tarroso P, Liepelt S, Gömöry D, Dullinger S, Meier ES, Hülber K, Maiorano L, Laborde H. 2014. Revisiting tree-migration rates: *Abies alba* (mill.), a case study. Vegetation history and archaeobotany. 23(2):113-122.
[7.23] Chen C, Zhao L, Zhu C, Wang J, Jiang J, Yang S. 2014. Response of diatom community in Lugu Lake (Yunnan–Guizhou Plateau, China) to climate change over the past century. Journal of paleolimnology. 51:357-373.
[7.24] Chen X, Liang J, Zeng L, Cao Y, Stevenson MA. 2022. Heterogeneity in diatom diversity response to decadal scale eutrophication in floodplain lakes of the middle Yangtze reaches. Journal of Environmental Management. 322:116164.
[7.25] Chen X, Zhou W, Luo G, Luo P, Chen Z. 2020. Spatial and temporal variations of the diatom communities in megacity streams and its

implications for biological monitoring. Environmental Science and Pollution Research. 27:37581-37591.

[7.26] Chen X, Zhou W, Pickett ST, Li W, Han L, Ren Y. 2016. Diatoms are better indicators of urban stream conditions: A case study in Beijing, China. Ecological Indicators. 60:265-274.

[7.27] Chester E, Robson B. 2013. Anthropogenic refuges for freshwater bio diversity: Their ecological characteristics and management. Biological Conservation. 166:64-75.

[7.28] Christensen JH, Hewitson B, Busuioc A. 2007. Regional climate projections. In climate change 2007: The physical science basis. In: Solomon S, Qin D, Manning M, Chen Z, Marquis M, Averyt K, Tignor M, Miller H, editors. Contribution of working group i to the fourth assessment report of the intergovernmental panel on climate change. Cambridge, UK; New York, NY: Cambridge University Press. p. 847–940.

[7.29] Chung MG, Frank KA, Pokhrel Y, Dietz T, Liu J. 2021. Natural infrastructure in sustaining global urban freshwater ecosystem services. Nature Sustainability. 4(12):1068-1075.

[7.30] Clavel J, Julliard R, Devictor V. 2011. Worldwide decline of specialist species: Toward a global functional homogenization? Frontiers in Ecology and the Environment. 9(4):222-228.

[7.31] Costa-Böddeker S, Bennion H, de Jesus TA, Albuquerque ALS, Figueira RC, de C. Bicudo D. 2012. Paleolimnologically inferred eutrophication of a shallow, tropical, urban reservoir in southeast brazil. Journal of Paleolimnology. 48:751-766.

[7.32] Cremer H, Bunnik FP, Kirilova EP, Lammens EH, Lotter AF. 2009. Diatom-inferred trophic history of Ijsselmeer (The Netherlands). Palaeolimnological proxies as tools of environmental reconstruction in fresh water.279-287.

[7.33] da Silva-Lehmkuhl AM, Blanco S, Bicudo DC. 2022. A new trophic diatom index for tropical reservoirs (tdir). Hydrobiologia. 849(21):4651-4666.

[7.34] da Silva-Lehmkuhl AM, Tremarin PI, Vercellino IS, Ludwig TAV. 2019. Periphytic diatoms from an oligotrophic lentic system, Piraquara I Reservoir, Paraná State, Brazil. Biota Neotropica. 19.

[7.35] Donar CM, Neely RK, Stoermer EF. 1996. Diatom succession in an urban reservoir system. Journal of Paleolimnology. 15:237-243.

[7.36] Dudgeon D, Arthington AH, Gessner MO, Kawabata Z-I, Knowler DJ, Lévêque C, Naiman RJ, Prieur-Richard A-H, Soto D, Stiassny ML. 2006. Freshwater biodiversity: Importance, threats, status and conservation challenges. Biological reviews. 81(2):163-182.

[7.37] Duenas M-A, Hemming DJ, Roberts A, Diaz-Soltero H. 2021. The threat of invasive species to iucn-listed critically endangered species: A systematic review. Global Ecology and Conservation. 26:e01476.

[7.38] Elmqvist T, Setälä H, Handel S, van der Ploeg S, Aronson J, Blignaut JN, Gomez-Baggethun E, Nowak D, Kronenberg J, de Groot R. 2015. Benefits of restoring ecosystem services in urban areas. Current opinion in environmental sustainability. 14:101-108.

[7.39] Evans MC. 2021. Re-conceptualizing the role (s) of science in biodiversity conservation. Environmental Conservation. 48(3):151-160.

[7.40] Field C, Barros V, Dokken D, Mach K, Mastrandrea M, Bilir T, Chatterjee M, Ebi K, Estrada Y, Genova R. 2014. IPCC, 2014: Climate change 2014: Impacts, adaptation, and vulnerability. Part a: Global and sectoral aspects. Contribution of working group ii to the fifth assessment report of the intergovernmental panel on climate change. Cambridge University Press, Cambridge, United Kingdom and New York, NY, USA.

[7.41] Fontana L, Albuquerque ALS, Brenner M, Bonotto DM, Sabaris TP, Pires MA, Cotrim ME, Bicudo DC. 2014. The eutrophication history of a tropical water supply reservoir in brazil. Journal of Paleolimnology. 51:29-43.

[7.42] Gilarranz LJ, Narwani A, Odermatt D, Siber R, Dakos V. 2022. Regime shifts, trends, and variability of lake productivity at a global scale. Proceedings of the National Academy of Sciences. 119(35):e2116413119.

[7.43] Gregersen R, Howarth JD, Atalah J, Pearman JK, Waters S, Li X, Vandergoes MJ, Wood SA. 2023. Paleo-diatom records reveal ecological change not detected using traditional measures of lake eutrophication. Science of The Total Environment.161414.

[7.44] Harper DM. 1992. Eutrophication of freshwaters. London: Chapman & Hall: Springer.

[7.45] Hassall C. 2014. The ecology and biodiversity of urban ponds. Wiley Interdisciplinary Reviews: Water. 1(2):187-206.

[7.46] Heino J, Virkkala R, Toivonen H. 2009. Climate change and freshwater biodiversity: Detected patterns, future trends and adaptations in northern regions. Biological Reviews. 84(1):39-54.

[7.47] Heiskary SA, Swain EB, Edlund MB. 2004. Reconstructing historical water quality in Minnesota lakes from fossil diatoms. Environmental Bulletin. 4.

[7.48] Heiskary SA. 2005. Minnesota lake water quality assessment report: Developing nutrient criteria. Water Assessment & Environmental Information Section, Environmental Analysis & Outcomes Division, Minnesota Pollution Control Agency.

[7.49] Heramza K, Barour C, Djabourabi A, Khati W, Bouallag C. 2021. Environmental parameters and diversity of diatoms in the Aïn Dalia Dam, northeast of Algeria. Biodiversitas Journal of Biological Diversity. 22(9).

[7.50] Hobbs RJ, Arico S, Aronson J, Baron JS, Bridgewater P, Cramer VA, Epstein PR, Ewel JJ, Klink CA, Lugo AE *et al.* 2006. Novel ecosystems: Theoretical and management aspects of the new ecological world order. Global Ecology and Biogeography. 15(1):1-7.

[7.51] Hobbs RJ, Higgs ES, Hall CM. 2013. Defining novel ecosystems. Novel ecosystems: Intervening in the new ecological world order.58-60.

[7.52] Holling CS. 1973. Resilience and stability of ecological systems. Annual review of ecology and systematics. 4(1):1-23.

[7.53] Hong P, Schmid B, De Laender F, Eisenhauer N, Zhang X, Chen H, Craven D, De Boeck HJ, Hautier Y, Petchey OL. 2022. Biodiversity promotes ecosystem functioning despite environmental change. Ecology letters. 25(2):555-569.

[7.54] Hunsaker CT, Carpenter DE. 1990. Environmental monitoring and assessment program: Ecological indicators. Atmospheric Research and Exposure Assessment Laboratory, Office of Research ….

[7.55] Hylander K, Ehrlén J, Luoto M, Meineri E. 2015. Microrefugia: Not for everyone. Ambio. 44(1):60-68.

[7.56] Jiménez-Muñoz JC, Mattar C, Barichivich J, Santamaría-Artigas A, Takahashi K, Malhi Y, Sobrino JA, Van Der Schrier G. 2016. Recordbreaking warming and extreme drought in the Amazon rainforest during the course of el niño 2015–2016. Scientific reports. 6:33130.

[7.57] Kahlert M, Fölster J, Tapolczai K. 2023. No lukewarm diatom communities-the response of freshwater benthic diatoms to phosphorus as basis for a new phosphorus diatom index (pdise). INCOMPLETE reference

[7.58] Kaushal SS, Likens GE, Jaworski NA, Pace ML, Sides AM, Seekell D, Belt KT, Secor DH, Wingate RL. 2010. Rising stream and river temperatures in the United States. Frontiers in Ecology and the Environment. 8(9):461-466.

[7.59] Kentucky D. 2002. Methods for assessing biological integrity of surface waters in kentucky. In: Protection DoE, editor.: Frankfort, Kentucky.

[7.60] Khan FA, Ansari AA. 2005. Eutrophication: An ecological vision. The botanical review. 71(4):449-482.

[7.61] Kléparski L, Beaugrand G, Edwards M, Schmitt FG, Kirby RR, Breton E, Gevaert F, Maniez E. 2022. Morphological traits, niche-environment interaction and temporal changes in diatoms. Progress in Oceanography. 201:102747.

[7.62] Kock A, Taylor J, Malherbe W. 2019. Diatom community structure and relationship with water quality in Lake Sibaya, Kwazulu-Natal, South Africa. South African Journal of Botany. 123:161-169.

[7.63] Kovalenko KE, Reavie ED, Barbiero RP, Burlakova LE, Karatayev AY, Rudstam LG, Watkins JM. 2018. Patterns of long-term dynamics of aquatic communities and water quality parameters in the great lakes: Are they synchronized? Journal of Great Lakes Research. 44(4):660-669.

[7.64] Lafont M, Marsalek J, Breil P. 2014. Urban aquatic habitats: Characteristics and functioning. Aquatic habitats in sustainable urban water management. CRC Press. p. 9-24.

[7.65] Lai GG, Padedda BM, Ector L, Wetzel CE, Lugliè A, Cantonati M. 2020. Mediterranean karst springs: Diatom biodiversity hotspots under the pressure of hydrological fluctuation and nutrient enrichment. Plant

Biosystems-An International Journal Dealing with all Aspects of Plant Biology. 154(5):673-684.
[7.66] Larocque-Tobler I, Lamentowicz M, Luoto TP, Perga M-E, Perga M, Frossard V, Jenny J, Alric B, Arnaud F, Berthon V. 2015. High-resolution paleolimnology opens new management perspectives for lakes adaptation to climate warming. Frontiers in Ecology and Evolution. 3(72).
[7.67] Leng MJ, Swann GE. 2010. Stable isotopes in diatom silica.
[7.68] Lewis SL, Maslin MA. 2015. Defining the anthropocene. Nature. 519(7542):171-180.
[7.69] Lim NK, Tay YC, Srivathsan A, Tan JW, Kwik JT, Baloğlu B, Meier R, Yeo DC. 2016. Next-generation freshwater bioassessment: Edna metabarcoding with a conserved metazoan primer reveals species-rich and reservoir-specific communities. Royal Society open science. 3(11):160635.
[7.70] Liu B, Hou W, Kattel G, Li D, Yan X. 2023. Advancing potential of diatom indicators for stream and lake water quality assessment against nutrient-induced pollution. Journal of Applied Phycology. 35(2):721-737.
[7.71] Liu J, Lin Z, Zhang H, Han B-P. 2012. Hydrodynamic change recorded by diatoms in sediments of liuxihe reservoir, southern china. Journal of paleolimnology. 47:17-27.
[7.72] Liu P, Xu S, Lin J, Li H, Lin Q, Han B-P. 2020. Urbanization increases biotic homogenization of zooplankton communities in tropical reservoirs. Ecological Indicators. 110:105899.
[7.73] Lobo EA, Heinrich CG, Schuch M, Wetzel CE, Ector L. 2016. Diatoms as bioindicators in rivers. River algae.245-271.
[7.74] Mallin MA, Johnson VL, Ensign SH. 2009. Comparative impacts of stormwater runoff on water quality of an urban, a suburban, and a rural stream. Environmental monitoring and assessment. 159:475-491.
[7.75] Marengo JA, Williams ER, Alves LM, Soares WR, Rodriguez DA. 2016. Extreme seasonal climate variations in the amazon basin: Droughts and floods. Interactions between biosphere, atmosphere and human land use in the Amazon basin. Springer. p. 55-76.
[7.76] Markovic D, Carrizo SF, Kärcher O, Walz A, David JN. 2017. Vulnerability of european freshwater catchments to climate change. Global Change Biology. 23(9):3567-3580.
[7.77] Marra RC, Algarte VM, Ludwig TA, Padial AA. 2018. Diatom diversity at multiple scales in urban reservoirs in southern brazil reveals the likely role of trophic state. Limnologica. 70:49-57.
[7.78] Masouras A, Karaouzas I, Dimitriou E, Tsirtsis G, Smeti E. 2021. Benthic diatoms in river biomonitoring—present and future perspectives within the water framework directive. Water. 13(4):478.
[7.79] Matthews SN, Rodewald PG. 2010. Urban forest patches and stopover duration of migratory swainson's thrushes. The Condor. 112(1):96-104.

[7.80] McKinney ML, Lockwood JL. 1999. Biotic homogenization: A few winners replacing many losers in the next mass extinction. Trends in Ecology & Evolution. 14(11):450-453.

[7.81] Medeiros G, Padial AA, Amaral MWW, Ludwig TAV, Bueno NC. 2020. Environmental variables likely influence the periphytic diatom community in a subtropical lotic environment. Limnologica. 80:125718.

[7.82] Mehlman DW, Mabey SE, Ewert DN, Duncan C, Abel B, Cimprich D, Sutter RD, Woodrey M. 2005. Conserving stopover sites for forest-dwelling migratory landbirds. The Auk. 122(4):1281-1290.

[7.83] Mendiondo EM. 2009. Global review of lake and reservoir eutrophication and associated management challenges.

[7.84] Meyer J. 2009. Urban aquatic ecosystems. In: Tockner K, Likens G, editors. Encyclopedia of inland waters. Academic Press: Oxford. p. pp 367–377.

[7.85] Miksic JN. 1999. Water, urbanization, and disease in ancient indonesia. Archeological Papers of the American Anthropological Association. 9(1):167-184.

[7.86] Milton SJ. 2003. 'Emerging ecosystems'-a washing-stone for ecologists, economists and sociologists?: News and views. South African Journal of Science. 99(9):404-406.

[7.87] Minaoui F, Hakkoum Z, Douma M, Mouhri K, Loudiki M. 2021. Diatom communities as bioindicators of human disturbances on suburban soil quality in arid marrakesh area (morocco). Water, Air, & Soil Pollution. 232:1-19.

[7.88] Mosblech NA, Bush MB, Van Woesik R. 2011. On metapopulations and microrefugia: Palaeoecological insights. Journal of Biogeography. 38:419-429.

[7.89] Mutinova PT, Kahlert M, Kupilas B, McKie BG, Friberg N, Burdon FJ. 2020. Benthic diatom communities in urban streams and the role of riparian buffers. Water. 12(10):2799.

[7.90] Muyodi FJ, Hecky RE, Kitamirike JM, Odong R. 2009. Trends in health risks from water-related diseases and cyanotoxins in ugandan portion of lake victoria basin. Lakes & Reservoirs: Research & Management. 14(3):247-257.

[7.91] Nascimento MdN, Bush M, Bicudo DdC. 2021. Water quality and spatial and seasonal dynamics in the largest water supply reservoir in brazil and implications for diatom assemblages. Acta Limnologica Brasiliensia. 33.

[7.92] Nazari-Sharabian M, Ahmad S, Karakouzian M. 2018. Climate change and eutrophication: A short review. Engineering, Technology and Applied Science Research. 8(6):3668.

[7.93] Nemes-Kókai Z, Kovács K, Borics G, Mayer R, Novák Z, Robotka ÁG, József J, Érczes K, Lukács Á, Viktória B. 2023. Continuous precipitation loss induced more pronounced compositional and diversity changes in the lotic phytobenthos than one-off drought events. Ecological Indicators. 148:110051.

[7.94] Nkem J, Kalame FB, Idinoba M, Somorin OA, Ndoye O, Awono A. 2010. Shaping forest safety nets with markets: Adaptation to climate change under changing roles of tropical forests in congo basin. environmental science & policy. 13(6):498-508.

[7.95] Ochieng H, Magezi G, Gandhi WP, Okot-Okumu J, Odong R. 2022. Epipelic diatom diversity as a bioindicator in River Aturukuku, Eastern Uganda. Plant Ecology & Diversity. 15(1-2):77-92.

[7.96] Ofori BY, Garshong RA, Gbogbo F, Owusu EH, Attuquayefio DK. 2018. Urban green area provides refuge for native small mammal biodiversity in a rapidly expanding city in ghana. Environmental Monitoring and Assessment. 190:1-11.

[7.97] Oliver AJ, Hong-Wa C, Devonshire J, Olea KR, Rivas GF, Gahl MK. 2011. Avifauna richness enhanced in large, isolated urban parks. Landscape and Urban Planning. 102(4):215-225.

[7.98] Ormerod SJ, Dobson M, Hildrew AG, Townsend C. 2010. Multiple stressors in freshwater ecosystems. Wiley Online Library. p. 1-4.

[7.99] Pace ML, Cole JJ, Carpenter SR, Kitchell JF. 1999. Trophic cascades revealed in diverse ecosystems. Trends in ecology & evolution. 14(12):483-488.

[7.100] Parsons L, LeRoy S, Overpeck J, Bush M, Cárdenes-Sandí G, Saleska S. 2018. The threat of multi-year drought in western amazonia. Water Resources Research. 54(9):5890-5904.

[7.101] Passy SI. 2008. Continental diatom biodiversity in stream benthos declines as more nutrients become limiting. Proceedings of the National Academy of Sciences. 105(28):9663-9667.

[7.102] Passy SI. 2010 A distinct latitudinal gradient of diatom diversity is linked to resource supply. Ecology. 91(1):36-41.

[7.103] Pauchard A, Barbosa O. 2013. Regional assessment of Latin America: Rapid urban development and social economic inequity threaten biodiversity hotspots. Urbanization, biodiversity and ecosystem services: Challenges and opportunities: A global assessment.589-608

[7.104] Peszek Ł, Zgrundo A, Noga T, Kochman-Kędziora N, Poradowska A, Rybak M, Puchalski C, Lee J. 2021. The influence of drought on diatom assemblages in a temperate climate zone: A case study from the Carpathian Mountains, Poland. Ecological Indicators. 125:107579.

[7.105] Peterson G, Allen CR, Holling CS. 1998. Ecological resilience, biodiversity, and scale. Ecosystems. 1(1):6-18.

[7.106] Pimm SL. 1984. The complexity and stability of ecosystems. Nature. 307(5949):321-326.

[7.107] Postel S, Carpenter S. 1997. Freshwater ecosystem services. Nature's services: Societal dependence on natural ecosystems. 195.

[7.108] Potapova M, Charles DF. 2003. Distribution of benthic diatoms in us rivers in relation to conductivity and ionic composition. Freshwater biology. 48(8):1311-1328.

[7.109] Püttker T, Crouzeilles R, Almeida-Gomes M, Schmoeller M, Maurenza D, Alves-Pinto H, Pardini R, Vieira MV, Banks-Leite C, Fonseca CR. 2020. Indirect effects of habitat loss via habitat fragmentation: A crosstaxa analysis of forest-dependent species. Biological Conservation. 241:108368.

[7.110] Qadir M, Sharma BR, Bruggeman A, Choukr-Allah R, Karajeh F. 2007. Non-conventional water resources and opportunities for water augmentation to achieve food security in water scarce countries. Agricultural water management. 87(1):2-22.

[7.111] Qin B, Kong X, Wang R, Zhao Y, Yang X. 2022. Lake restoration time of lake taibai (china): A case study based on paleolimnology and ecosystem modeling. Journal of Paleolimnology.1-14.

[7.112] Reavie ED, Barbiero RP, Allinger LE, Warren GJ. 2014. Phytoplankton trends in the Great Lakes, 2001–2011. Journal of Great Lakes Research. 40(3):618-639.

[7.113] Relationship of diatom phytoplankton (bacillariophyceae) diversity with water quality and heavy metal in Angke and Grogol Rivers, Jakarta City. IOP Conference Series: Earth and Environmental Science; 2022: IOP Publishing.

[7.114] Rivas E, Pérez GR, Tundisi J, Vammen K, Örmeci B, Forde M. 2020. Eutrophication: A growing problem in the americas and the caribbean. Brazilian Journal of Biology. 80:688-689.

[7.115] Rodewald PG, Matthews SN. 2005. Landbird use of riparian and upland forest stopover habitats in an urban landscape. The Condor. 107(2):259-268.

[7.116] Rosenfeld JS. 2002. Functional redundancy in ecology and conservation. Oikos. 98(1):156-162.

[7.117] Round FE, Crawford RM, Mann DG. 1990. Diatoms: Biology and morphology of the genera. Cambridge university press.

[7.118] Rull V, Schubert C, Aravena R. 1988. Palynological studies in the Venezuelan Guayana shield: Preliminary results. Current Research in the Pleistocene. 5:54-56.

[7.119] Rull V. 2009. Microrefugia. Journal of Biogeography. 36(3):481-484.

[7.120] Sanderson EW, Jaiteh M, Levy MA, Redford KH, Wannebo AV, Woolmer G. 2002. The human footprint and the last of the wild the human footprint is a global map of human influence on the land surface, which suggests that human beings are stewards of nature, whether we like it or not. BioScience. 52(10):891-904.

[7.121] Saros J, Anderson N. 2015. The ecology of the planktonic diatom *Cyclotella* and its implications for global environmental change studies. Biological reviews. 90(2):522-541.

[7.122] Scheffer M, Carpenter S, Foley JA, Folke C, Walker B. 2001. Catastrophic shifts in ecosystems. Nature. 413:591-596.
[7.123] Scheffer M, Carpenter SR. 2003. Catastrophic regime shifts in ecosystems: Linking theory to observation. Trends in Ecology and Evolution. 18(12):648-656.
[7.124] Selimović S, Huseinović S, Kamberović J, Bektić S, Đozić A. 2022. Epilithic diatoms in the river spreča affected by urban and industrial pollution, bosnia and herzegovina. Current Journal of Applied Science and Technology. 41(48):59-69.
[7.125] Shibabaw T, Beyene A, Awoke A, Tirfie M, Azage M, Triest L. 2021. Diatom community structure in relation to environmental factors in human influenced rivers and streams in tropical Africa. PLoS One. 16(2):e0246043.
[7.126] Shukla PR, Skea J, Calvo Buendia E, Masson-Delmotte V, Pörtner HO, Roberts D, Zhai P, Slade R, Connors S, Van Diemen R. 2019. Ipcc, 2019: Climate change and land: An IPCC special report on climate change, desertification, land degradation, sustainable land management, food security, and greenhouse gas fluxes in terrestrial ecosystems.
[7.127] Shurin JB, Gruner DS, Hillebrand H. 2006. All wet or dried up? Real differences between aquatic and terrestrial food webs. Proceedings of the Royal Society B: Biological Sciences. 273(1582):1-9.
[7.128] Silva AMd, Ludwig TAV, Tremarin PI, Vercellino IS. 2010. Diatomáceas perifíticas em um sistema eutrófico brasileiro (reservatório do iraí, estado do paraná). Acta Botanica Brasilica. 24:997-1016.
[7.129] Silva SCAd, Cervi AC, Bona C, Padial AA. 2014. Aquatic macrophyte community varies in urban reservoirs with different degrees of eutrophication. Acta Limnologica Brasiliensia. 26:129-142.
[7.130] Smol JP, Stoermer EF. 2010. The diatoms: Applications for the environmental and earth sciences. Cambridge University Press.
[7.131] Sonne C, Dietz R, Alstrup AK. 2020. Factors affecting global flow of scientific knowledge in environmental sciences. Science of The Total Environment. 701:135012.
[7.132] Stevenson RJ, Novoveska L, Riseng CM, Wiley MJ. 2009. Comparing responses of diatom species composition to natural and anthropogenic factors in streams of glaciated ecoregions. Nova Hedwigia. 135:1-13.
[7.133] Teittinen A, Taka M, Ruth O, Soininen J. 2015. Variation in stream diatom communities in relation to water quality and catchment variables in a boreal, urbanized region. Science of the Total Environment. 530:279-289.
[7.134] Thrush SF, Halliday J, Hewitt JE, Lohrer AM. 2008. The effects of habitat loss, fragmentation, and community homogenization on resilience in estuaries. Ecological applications. 18(1):12-21.
[7.135] Tilman D, May RM, Lehman CL, Nowak MA. 1994. Habitat destruction and the extinction debt. Nature. 371:65-66.

[7.136] UNESCO. 2001. Convention concerning the protection of the world cultural and natural heritage, world heritage committee 24th session, cairnes, australia 27 november - 2 december 2000. Paris: UNESCO, Distribution Limited.
[7.137] Urrea G, Sabater S. 2009. Epilithic diatom assemblages and their relationship to environmental characteristics in an agricultural watershed (Guadiana River, SW Spain). Ecological indicators. 9(4):693-703.
[7.138] Venter O, Sanderson EW, Magrach A, Allan JR, Beher J, Jones KR, Possingham HP, Laurance WF, Wood P, Fekete BM. 2016. Global terrestrial human footprint maps for 1993 and 2009. Scientific data. 3(1):1-10.
[7.139] Viktória B, Tóthmérész B, Bácsi I, Borics G, Abonyi A, Tapolczai K, Rimet F, Bouchez Á, Várbíró G, Török P. 2019. Autumn drought drives functional diversity of benthic diatom assemblages of continental intermittent streams. Advances in Water Resources. 126:129-136.
[7.140] Vörösmarty CJ, McIntyre PB, Gessner MO, Dudgeon D, Prusevich A, Green P, Glidden S, Bunn SE, Sullivan CA, Liermann CR. 2010. Global threats to human water security and river biodiversity. nature. 467(7315):555-561.
[7.141] Wake DB, Vredenburg VT. 2008. Are we in the midst of the sixth mass extinction? A view from the world of amphibians. Proceedings of the National Academy of Sciences. 105(Supplement 1):11466-11473.
[7.142] Walker CE, Pan Y. 2006. Using diatom assemblages to assess urban stream conditions. Advances in Algal Biology: A Commemoration of the Work of Rex Lowe.179-189.
[7.143] Watson VT, Kehler D, Medeiros AS. 2022. A paleolimnological context of ecological vulnerability for the freshwater ecosystems of Sable Island National Park Reserve, Canada. Science Progress. 105(4):00368504221126865.
[7.144] Wengrat S, Padial AA, Jeppesen E, Davidson TA, Fontana L, CostaBöddeker S, Bicudo DC. 2018. Paleolimnological records reveal biotic homogenization driven by eutrophication in tropical reservoirs. Journal of Paleolimnology. 60:299-309.
[7.145] Wetzel CE, Bicudo DdC, Ector L, Lobo EA, Soininen J, Landeiro VL, Bini LM. 2012. Distance decay of similarity in neotropical diatom communities.
[7.146] Williams PC, Bartlett AW, Howard-Jones A, McMullan B, Khatami A, Britton PN, Marais BJ. 2021. Impact of climate change and biodiversity collapse on the global emergence and spread of infectious diseases. Journal of paediatrics and child health. 57(11):1811-1818.
[7.147] Yang J, Yu X, Liu L, Zhang W, Guo P. 2012. Algae community and trophic state of subtropical reservoirs in southeast fujian, china. Environmental Science and Pollution Research. 19:1432-1442.

[7.148] Yang L, Yang Y, Villarini G, Li X, Hu H, Wang L, Blöschl G, Tian F. 2021. Climate more important for chinese flood changes than reservoirs and land use. Geophysical Research Letters. 48(11):e2021GL093061.
[7.149] Yang Y, Cao J-X, Pei G-F, Liu G-X. 2015. Using benthic diatom assemblages to assess human impacts on streams across a rural to urban gradient. Environmental Science and Pollution Research. 22:18093-18106.
[7.150] Yu C, Li Z, Xu Z, Yang Z. 2020. Lake recovery from eutrophication: Quantitative response of trophic states to anthropogenic influences. Ecological Engineering. 143:105697.
[7.151] Zhao J, Gao Q, Liu Q, Fu G. 2020. Lake eutrophication recovery trajectories: Some recent findings and challenges ahead. Ecological Indicators. 110:105878.
[7.152] Zorzal-Almeida S, Bartozek ECR, Bicudo DC. 2021. Homogenization of diatom assemblages is driven by eutrophication in tropical reservoirs. Environmental Pollution. 288:117778.
[7.153] Zorzal-Almeida S, Bini LM, Bicudo DC. 2017. Beta diversity of diatoms is driven by environmental heterogeneity, spatial extent and productivity. Hydrobiologia. 800:7-16.

8

Guilds for Diatoms: History and Future Prospects

Csilla Stenger-Kovács[1,2]* and Viktória B-Béres[3]

[1]*University of Pannonia, Center for Natural Science, Limnology Research Group, Veszprém, Hungary*

[2]*HUN-REN Limnoecology Research Group, Veszprém, Hungary*

[3]*HUN-REN Centre for Ecological Research, Institute of Aquatic Ecology, Department of Tisza Research, Functional Algology Research Group, Debrecen, Hungary*

Abstract

The term "guild" as a fundamental unit of communities had already been used in terrestrial research from the 1900s, but its definition was first provided in 1967. In diatom studies, however, guild-based approaches became familiar only in the last 15 years, beginning with Passy's diatom guild concept, which induced a number of studies dealing with diatom guilds in the recent years. These papers mainly focus on, *inter alia*, the relationship between guilds and environmental, spatial and biotic factors highlighting the significant effects of processes induced by anthropogenic pressures on diatom guilds from local to global scale. Climate change and the accelerating decline of diversity have drawn the attention to the perceptible changes in the distribution and dispersal of guilds, and gave great impetus to the conservation researches and protection of diatoms in the future. Furthermore, guild-based approaches with their practical application have also opened up new possibilities in paleolimnological studies and highlighted the ecological role of their appearance in the phytoplankton. In conclusion, guild-based analyses can fill knowledge gaps and provide deeper insights into the structural characteristics of

**Corresponding author*: stenger-kovacs.csilla@mk.uni-pannon.hu
Csilla Stenger-Kovács: https://scholar.google.com/citations?user=-UzL8QMAAAAJ&hl=hu&oi=ao, https://www.researchgate.net/profile/Csilla-Stenger-Kovacs
Viktória B-Béres: https://scholar.google.com/citations?user=k2eFEPIAAAAJ&hl=hu&oi=ao, https://www.researchgate.net/profile/B-Beres-Viktoria

Nora I. Maidana, Magdalena Licursi and Eduardo Morales (eds.) Diatom Ecology: From Molecules to Metacommunities, (233–280)

benthic diatom communities, leading to a clear and general understanding of the operation of ecosystems.

Keywords: High profile, low profile, motile, planktic, guild distribution, metacommunities, conservation, environmental extremes

8.1 The Origin and History of the Guild Concept

The term "guild" or "Genossenschaften" was already used in terrestrial vegetation research in the 1900s as a fundamental unit of communities [8.1]. Richard B. Root was the zoologist and ecologist who first provided the definition of guild in 1967 as "a group of species that exploit the same class of environmental resources in a similar way" (Figure 8.1). According to Root [8.2], the members of the defined groups have similar niche requirements, therefore, there is an intense competition among them. This means the original guild concept focused on the similarities in resource sharing within a group of species without considering their ecological and evolutionary relatedness and consequences in contrast to the functional group concept [8.3], which was developed shortly after the guild concepts. The members of a guild can originate from different taxonomic categories; moreover, a species can be the member of even more guilds. As "super-species," guilds can more accurately indicate the changes in resources, therefore their response is more predictable than that of individual species with a variety of responses [8.4]. However, the response of the members (different species) of a given guild can be different to other environmental factors, which is consistent with the individualistic hypothesis [8.5]. Therefore, a guild member is not necessarily able to represent the response of the whole guild [8.6].

It is clear that in the guild concept the subject is the species [8.7] and the type of guilds can be determined based on the resources. In most cases, the resources are food/nutrients, but habitat guilds [8.8], nesting guilds [8.9] and reproductive guilds [8.10] can also be found in the literature. Sometimes guild definition [8.11] and separation is restricted to only closely related species called "taxon guild" [8.12], derived mostly by experts on particular taxa (Figure 8.1). However, there are more examples where guild separations were not based on the resource at all, and the original definition of guilds was clearly ignored and violated, as in the case of mobility, body mass guilds [8.13] or microhabitat (different heights of vegetation [8.14]). Furthermore, in a short time the guild approach was matched with the idea of the indicator species. Verner [8.15] provided the definition of management guild (Figure 8.1) as those species with similar response to environmental changes,

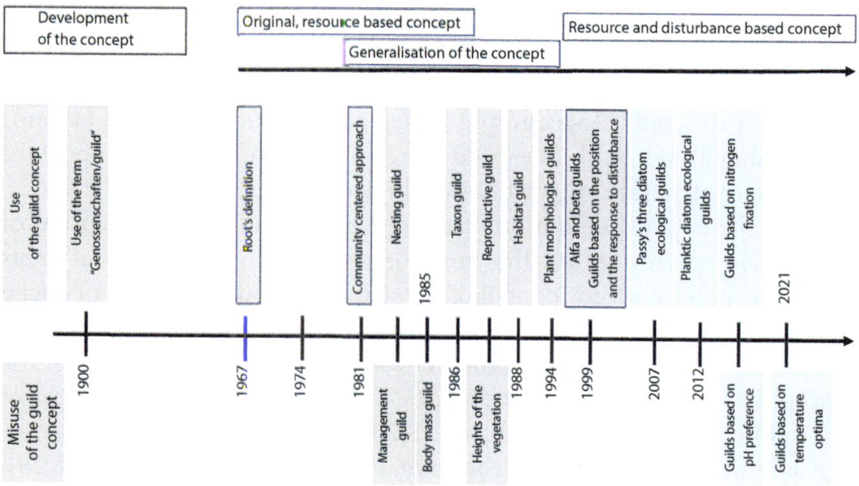

Figure 8.1 An overview of the history and development of the guild concept with the relevant adaptations and misinterpretations. (Light blue background: Specific diatom guilds. Dark blue frames: Guild definitions).

which accelerates the establishment of further "guilds" [8.16]. It is obvious that the guild concept was misinterpreted and used in an incoherent way, leading to several connotations. The original narrow meaning of the term has been neglected and has been applied in a broader view in community ecology [8.6]. The words trait, functional group and guild are used as synonyms [8.17] and they were applied to distinguish different groups of species from an ecological perspective, and it was too late to control the term and its usage further. Therefore, Wilson [8.18] made the "classification of classifications," defining alpha and beta guilds and their subdivisions (Figure 8.1). Several studies suggested the exclusion of the "similar way" from the definition and led to a community centered approach established by MacMahon *et al.* [8.19], which can be applicable for all organisms "which use the same investigator-defined resource" (Figure 8.1). According to Landres [8.20], guilds are artificial creations of the researchers and with their help we can answer specific questions. Several studies support the usefulness of guilds, emphasizing that quantitative methods should be used to create these groups [8.6] and their validity should be justified with verifying the existence of different guilds [8.20].

In fact, there was a universal need [8.6] to reduce the number of members and their variabilities in the community beside retaining the important ecological information in the basic blocks of a building [8.21], which makes possible the easier understanding of the organization of communities [8.22], food webs and diversity [8.23]. Using guild as an operational unit or macro descriptor [8.24] provides cheap and simple community analyses, clear community structures, and comparable datasets of different biogeographical regions in contrast to the taxonomy-based analyses [8.4]. The guild concept is one of the most important "thinking in structure" theories of ecology of the last decades [8.25].

Using the original definition of guild, we can study the coevolution of competitive species within the same or overlapping geographical areas [8.6]. The guild concept contributes to further understanding of niche theory and to discovering the structuring processes of communities and ecosystems [8.20]. A variety of theoretical and practical application of the concept has emerged in recent decades. Guild structure can be indicative for trophic structure and food web [8.26]. Guilds are applicable as ecological indicators [8.27] in environmental impact assessment [8.28] [8.29], in forest service, wildlife and ecosystem management (e.g., [8.30] [8.31]); moreover, in conservation and restoration management [8.20]. Guilds can be applied for tracing population and evolutionary dynamics to understand species diversity and community stability [8.32]. These ecological categories are useful in comparative community studies [8.2] because of their region independent usage [8.29].

8.2 Passy's Idea: Use of the Guild Concept in Diatom Research

The guild approach had already become popular in the sixties and seventies among the community ecologists, especially in animal studies (for birds and mammals) [8.7]. For plants in relation to nutrients, this classification was not obvious [8.6]. Perhaps this could be the reason why in plant [8.41] and phytoplankton ecology (e.g., [8.42] [8.43]), the use of functional groups gained widespread acceptance similar to other aquatic organisms, like functional feeding groups for aquatic invertebrates [8.44]. The first guild-based approach in diatom ecology was introduced by Passy [8.45] (Figure 8.1). In Passy's concept several approaches were amalgamated such as the combination of food and habitat resources [8.46], the role of positioning on the substrate, i.e., (i) horizontally or vertically [8.47] [8.48], (ii) the stalk [8.49] of the diatom species, and (iii) the strength of the attachment and cohesive physiognomy [8.47].

For benthic diatoms the main resources are N and P forms as nutrients, light for photosynthesis, and the substrate as habitat for their attachments. The community compositional response of diatoms to the various temporal

Box 8.1: Definitions

Guild: A group of taxa which use the same investigator-defined resource [8.33].

Functional group: A group of species which have a similar function in the ecosystem processes [8.7].

Resource: All things that are consumed by organisms, or as quantities that can be reduced by the activity of the organisms [8.34].

Disturbance: An event with a given frequency which allows the community to reach the equilibrium state [8.35].

Stress: An event with higher frequency which does not allow the community to reach the original dynamics and shift it to a new path [8.35].

Growth forms: A set of adaptation evolved by different taxa for a certain way of life [8.36].

Fundamental niche: Sum of all physical and biological variables required for a species to persist [8.37].

Facilitation: A positive interaction among the species, in which the presence of a species changes the environment enhancing the growth, survival or reproduction of another species [8.38].

Eurheic condition: Water discharge is high enough to maintain all types of the aquatic habitats and allow the optimal hydraulic connectivity between them [8.39].

Nestedness: The biotas of local communities with a smaller number of species are the subsets of the biotas of the larger, species rich communities [8.40].

and spatial changes in resources is well-known, similar to the changes in physical structure of the benthic layer built by a high diversity of algal growth forms such as (e.g., [8.50]): prostrate, adnate, erect, filamentous, branched, chain-forming, tube-forming, colonial, stalked and motile, which are evolutionary adaptation forms [8.51] to the different environments (different habitats, substrates, grazing pressures, current velocities, light, nutrients (e.g., [8.52] [8.53] [8.54]). The niche overlap of these growth forms allows grouping them into guilds [8.45], in which members utilize the same range of resources along the resource gradient, although implying that they adapt differently to the abiotic factors. Therefore, the relationship between the defined guilds and environmental variables should be stronger than that for species alone [8.51].

Originally, three diatom ecological guilds were defined by Passy [8.45]: low profile, high profile and motile. The species of the low-profile guilds are in the bottom layer of the biofilm (colored brown in Figure 8.2), because they have short stature. This guild consists of the adnate, prostrate, erect, solitary centrics and slow-moving species, which adapted to the resource (both nutrient and light) limitation, because the diffusion of nutrients into the biofilm is restricted and only a reduced amount can reach the lower layer, which is shaded by the upper one. Species with tall stature form the upper benthic layer and comprise the branched, chain-forming, erect, filamentous, tube-forming, stalked and colonial centrics that can easily reach the dissolved nutrients (PO_4^{3-}, NO_3^-) and light. This second group—called high-profile guild (colored yellow in Figure 8.2)—benefits from the availability of these variables (they are in direct connection), which is not reduced by another canopy. The third guild is the motile guild (colored grey in Figure 8.2) that collects relatively fast-moving species, which are able to select the most suitable habitats in the phytobenthos. These features can predict their distribution: the low-profile guild will dominate in resource-poor environments, the high-profile guild will characterize resource-rich habitats, while the taxa of the motile guild especially prefer nutrient-rich ones.

This approach recalls plant morphological guilds created by their light harvesting strategy at lawn communities [8.55], or by their position in the forest canopy, where the resources, especially the light and space usage, depends on the stratification of the plant species [8.18]. This plant classification was criticized by the scientific community because this approach determined only the main "building blocks" of a community, and did not specify the single resource [8.6]. Similar to Fagerstrom's classes [8.56] [8.57] [8.58], considering the morphology and skeleton of coral reefs was completely rejected by the scientific community [8.59]. Therefore, to avoid the problem with the resource specification [8.6], the term "guild" was extended to the response of a group to disturbances [8.18], as was also applied to diatom guilds by Passy based on their position in the biofilm: the lower layer (low-profile guild) is more resistant to physical disturbance (higher water flow, floods, grazing) than the upper layer (high-profile guild). The motile guild with its physical capability can avoid the marginal disturbances. Thus, in the diatom guilds, characteristics of morphology-based "functional

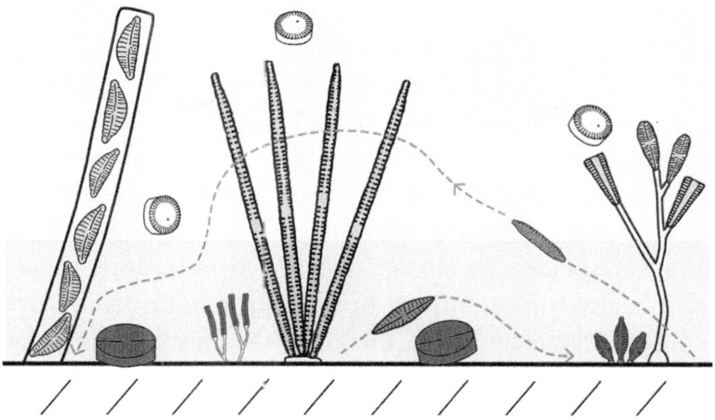

Figure 8.2 Schematic structure of the biofilm built by diatom ecological guilds (low-profile guild: brown, high-profile guild: yellow, motile guild: grey, planktic guild: black and white). (Figure drawn by Edina Lengyel).

groups" and environmental preference-based "response groups" seem to be merged. However, Soininen *et al.* [8.60] note that the response of the high-profile guild to resources and disturbances is not as unambiguous as that for motile and low-profile guilds.

The habitat/substrate preference of these three guilds and their distribution along rivers is also characteristic: the dominance of the low-profile guild is obvious on the psammon where the short stature and strong attachment allow them to live on moving sand grains. The high-profile guild dominates on more or less stable substrates like stones and macrophytes, while the motile guild is characteristic for the unstable sludge which can cover most of the algae species [8.45]. The roughness of the substrate can further support the dominance of the motile guild [8.61]. On wood or macrophyte substrates, the endogenous nutrient loading can also modify the guild composition; on these substrate types, the low- and high-profile guild have a competitive advantage over the motile guild [8.45] [8.62] [8.63], where the macrophyte leaves even provide refuge for the high-profile guild against water turbulence or flow [8.45] [8.63]. In algal succession a neutral coexistence of the different guilds can be observed. In the first phase of the succession processes, the tolerant low-profile guild is characteristic, followed by overgrowth of the motile and high-profile guilds controlled by the environment [8.64].

A high proportion of the low-profile groups can be found upstream with high current velocity and low nutrient concentrations. Motile species are most abundant downstream, where the waters are characterized by lower current velocity, higher nutrient levels, high siltation and large water mass [8.65] [8.66], in addition to in floodplains, which are seasonally inundated

[8.67]. High-profile species mostly prevail midstream with moderate velocities and siltation (e.g., [8.66] [8.68]).

Passy provided the major genera of the three guilds; however, she indicated that the selection of species from a heterogeneous genus into guilds can be species specific, since these genera (e.g., *Gomphonema* with short stalks or long-branched stalks and motile or non-motile secreting *Navicula* spp.) contain several different growth forms which might belong to distinct guilds. Moreover, species can change their life-form in time (*Cymbella* spp.) [8.51], and it is also important that unusual growth forms of a species can appear in different habitats [8.45], therefore, the check of the growth forms of the live or preserved samples can be necessary. Rimet and Bouchez [8.51] further refined the species/genera classifications of guilds, e.g., species with larger size (>1500 μm^3) from the low profile guild (like *Cymbella lanceolata*, *Eucocconeis flexella*) were put into the high-profile guild and they completed the list of the high-profile guild with further large-sized (e.g., *Cymbopleura*) and tube-forming genus (e.g., *Amphipleura*, *Berkeleya*) and extended the list of the genera in the motile guild (e.g., Mastogloiales) (Table 8.1).

However, Rimet and Bouchez [8.51] suggested the establishment of a new guild, the planktic guild (Figure 8.1) and proposed to move the solitary centrics from the low profile (e.g., *Cyclotella* spp.) and filamentous centrics (*Aulacoseira* spp.) from the high profile guild to this group, because these species sedimented from the plankton, especially downstream [8.65] [8.66] and they cannot resist strong physical disturbances. The relevance of the planktic guild was supported by subsequent studies [8.69] [8.70]. Furthermore, the study of Rimet and Bouchez [8.51] provided detailed, species-specific information about guild classification. This database allows avoiding the confusions if a genus can be found in two different guilds. Later, other guilds were also introduced, such as the nitrogen-fixing guild (*Denticula, Epithemia, Rhopaloida*) (Table 8.1) and non-nitrogen fixers as well as the acid-tolerant (acidophilus and acidobiontic species based on [8.71] and alkaliphilus guild [8.60]) (Figure 8.1). Moreover, recently, three thermal guilds were defined [8.72] using a weighted average method: cold-water diatoms (temperature optima <20 °C; majority of low- and high-profile diatoms), cool-water and warm-water diatoms (temperature optima >20 °C; most of the motile species) were separated (Figure 8.1). However, it should be noted that these guilds are not related to common resources, species groups with similar pH preference and temperature optima are not guilds [8.73] according to the original guild definition [8.2]. However, it can be connected to an alternative definition found in the literature as a group of sympatric species with similar response to autecological pressures [8.74]. However, in this case the term of

Table 8.1 Assignment of taxa to diatom guilds according to Passy [8.45], Rimet and Bouchez [8.51] and Soininen *et al.* [8.60]. Taxa moved to another guild during the revision by Rimet and Bouchez [8.51] are marked in bold.

Diatom guilds	Taxa
Low profile	*Achnanthes, Achnanthidium, Amphora, Brachysira, Cocconeis,* **medium and small sized *Cymbella***, *Cymbopleura,* **medium and small sized *Diploneis***, *Encyonopsis,* **small and medium sized *Eucocconeis***, *Karayevia, Kobayasiella, Kolbesia, Lemnicola, Meridion, Nupela, Planothidium, Platessa, Psammothidium, Reimeria, Rhoicosphenia, Rossithidium*
High profile	*Achnanthidium catenatum, Amphipleura, Ctenophora,* **large sized *Cymbella***, *Cymbopleura rhomboidea,* ***Delicata***, *Diadesmis, Diatoma, Didymosphenia,* ***large sized Diploneis***, *Ellerbeckia,* ***Encyonema***, **large sized *Eucocconeis***, *Eunotia,* ***Fragilaria***, *Frustulia, Gomphoneis, Gomphonema, Gomphosphenia, Melosira, Muelleria, Parlibellus, Pleurosira, Pseudostaurosira, Stauroforma, Staurosira, Staurosirella, Tabellaria, Tabularia,* **large sized *Ulnaria***
Motile	*Adlafia, Aneumastus, Anomoeoneis, Bacillaria, Caloneis, Campylodiscus, Cavinula, Chamaepinnularia, Craticula, Cymatopleura, Denticula, Entomoneis, Eolimna, Epithemia, Fallacia, Fistulifera, Geissleria, Gyrosigma, Hantzschia, Hippodonta, Luticola, Mastogloia, Mayamaea, Microcostatus, Navicula, Naviculadicta, Neidium, Nitzschia, Nupela, Pinnularia, Placoneis, Pulchella, Rhopalodia, Sellaphora, Simonsenia, Stauroneis, Surirella, Tryblionella*
Planktic	***Actinocyclus, Asterionella, Aulacoseira, Cyclostephanos, Cyclotella, Diatoma tenuis, Discostella, Fragilaria, Nitzschia, Puncticulata, Skeletonema, Stephanodiscus, Thalassiosira***, **medium sized *Ulnaria***
Nitrogen-fixers	*Denticula, Epithemia, Rhopalodia*

functional response group is more appropriate, where the species has a given functional role in the ecosystem [8.6].

These separations can be connected to well-known ecological theories such as resource partitioning. In the case of a shared preference, species invest in good quality sources, and they dominate in optimal conditions like the motile guild in high resource availability. But under high stress, composition of assemblages shifts to the dominance of tolerant taxa, like the low-profile guild at low resource availability. When different species have unique optima but shared tolerance along the environmental gradients, they try to avoid the competition by occupying only a part of their fundamental niche (to reach their highest fitness); this can be well exemplified by the acidophilus (pH <7) and alkaliphilus guilds (pH >7). In the case of facilitation as a part of the resource partitioning, when one species benefits from the presence of another under stress conditions, its realized niche will be extended to the original, fundamental niche. Under nitrogen limitation, nitrogen-fixing diatoms provide the benefits and non-nitrogen fixing, eutrophic diatoms receive the benefits [8.75].

8.3 Relationship Between Guilds and Environmental Variables

The growth, resilience and resistance of algal species living in biofilms, as well as the structure and composition of benthic assemblages, are altered by a number of environmental factors (e.g., nutrient supply, flow conditions, light, substrate, pH, water temperature, biotic interaction) [8.76]. Obviously, these factors display spatial and temporal gradients and often interact. When Passy introduced the diatom guild concept in 2007, she focused on availability and disturbance pressure as key environmental variables controlling the distribution of algal growth forms, and she suggested using these groups in ecological status assessment. Though, not immediately, she set off an avalanche with her revolutionary proposal in diatom research. All of the results mentioned below support the idea that diatom guilds respond strongly to local and regional environmental factors and this observation was also supported on a global scale [8.60] (Figure 8.3).

8.3.1 Nutrients

Eutrophication, which can be traced mainly to human activity, was considered to be the most significant environmental impact on freshwaters in the last century [8.77]. Therefore, the first guild-based diatom studies, especially from European regions, paid particular attention to testing guild distribution in space [8.68] [8.78] [8.79] [8.80] [8.81] [8.82] [8.83] and time along the nutrient gradient [8.84] [8.85], and they tested it even in a field experiment [8.86]. Initially, different results were obtained regarding the trophic state indicated by the guilds. A clear correlation was revealed

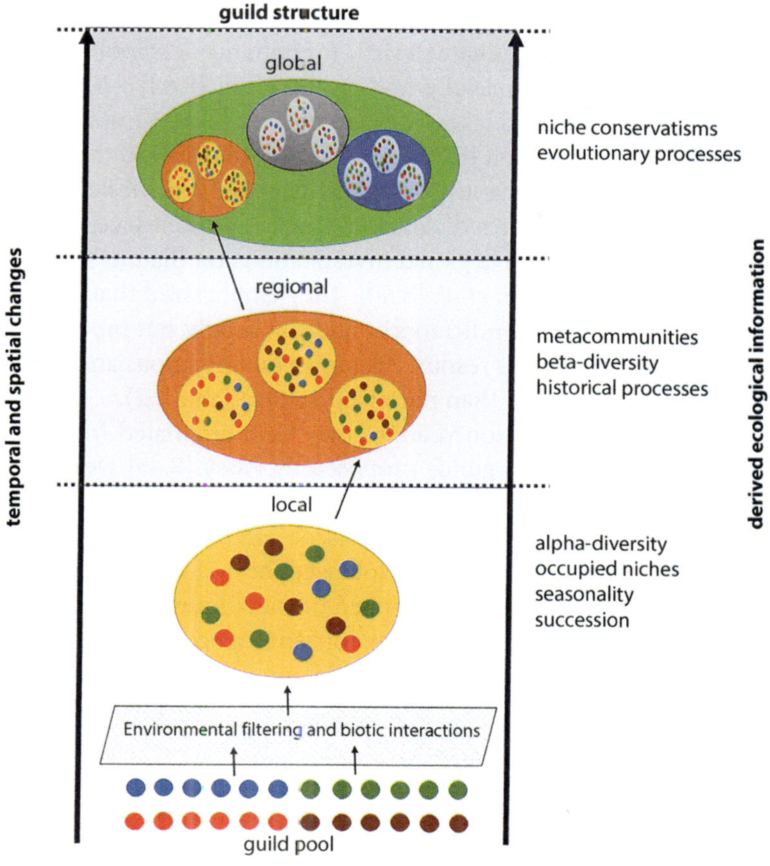

Figure 8.3 Overview of guilds in diatom ecological studies—from local to global perspectives.

between guild composition and nutrient load in French watercourses [8.78] [8.86], highlighting a strong negative selection pressure of a nutrient-rich environment on the low profile guild. Increase in trophic level of waters and organic pollution, however, favor the motile guild [8.78]. Goldenberg Vilar *et al.* [8.68] also found significant differences in the distribution of guilds along the eutrophication gradient of lakes in the Netherlands. Similar to Berthon *et al.* [8.78] and Rimet *et al.* [8.86], waters with lower trophic level were characterized by the low-profile guild. In addition, the planktic guild was strongly related to the hypertrophic state. High-profile and motile guilds showed hump-shaped distribution along the nutrient gradient [8.68]. Other studies, however, found no correlation between the guild distribution and nutrient supply of waters, or the results were contradictory. Low-profile and motile guilds, for example, were dominant in both nutrient-poor and nutrient-rich Swedish lakes [8.79], while in a Hungarian watercourse [8.84]

[8.85], low- and high-profile guilds either did not relate to nutrient supply of water [8.85], or they were characteristic for both nutrient-poor and nutrient-rich environments (especially low-profile guild [8.84]). Moreover, in Fennoscandian streams and lakes, nutrient content had no impact on guild composition and distribution [8.80] [8.81] [8.83]. In contrast, the high-profile guild proliferated in oligotrophic streams in the US, while the motile guild dominated with greater diversity at higher nutrient levels [8.87]. To date the most comprehensive global investigations on diatom distribution have been done by Soininen *et al.* [8.60]. They emphasized that low-profile and motile guilds are good indicators of nutrient supply, but the response of the high-profile guild to this resource is not so unambiguous and it is probably driven by factors other than purely nutrients (Table 8.2).

The contradictions mentioned above may have originated from the fact that only the original three guilds proposed by Passy [8.45] were involved in the analyses of the earliest studies (e.g., [8.78] [8.79]). In the early 2010s a revision and extension of the guild concept was proposed by Rimet and Bouchez [8.51], but some authors still followed the original description (e.g., [8.80] [8.85]), while others already applied the new, four guild systems (e.g., [8.81]). The other side of the coin is that nutrient supply is important but not the only factor controlling the distribution of microorganisms [8.76].

8.3.2 Physical Disturbances

Physical disturbances, e.g., shear forces due to high water velocity or grazing, are also pivotal in guild distribution [8.45], since flow-impeded immigration or even nutrient uptake can be negatively impacted by them [8.64]. The effect of a physical disturbance on low- and high-profile guilds seems to be clear (low profile: tolerant; high profile: sensitive [8.66] [8.88] [8.89]). As for the motile guild, the relationship between its distribution and current velocity or grazing is a complex phenomenon that leads to contradictory results (no effects [8.80] [8.83] [8.90]; positive effects [8.88] [8.89]; negative effects [8.66]). An explanation for this contradiction could be that the effect of current velocity and grazing on guild composition is regulated by nutrient supply [8.64] [8.88]; while high velocity and low nutrient content resulted in lower algal density in all guilds in the early phase of the colonization, high flow conditions may reduce nutrient depletions in the matured biofilm late in succession [8.64]. Furthermore, strong physical disturbances may have a positive effect on algal density in nutrient-rich environments, especially in low-profile and motile guilds, stressing that faster flow or grazing pressure can provide an opportunity for these guilds, which are less competitive for resources than the high-profile guild [8.64] [8.88]. Living in the open water (in the plankton) clearly requires different adaptation strategies than in biofilms [8.51]; therefore, with few exceptions, the proportion of planktic guild is low in phytobenthos [8.69]. While Wu *et al.* [8.89] found a strong negative relationship between high flow conditions and planktic guild, grazing had no significant effect on this guild [8.83]. Cell size and life forms, e.g., filamentous or unicellular, or even the shape of cells, e.g. round-valved or needle-like [8.91], may be additional characteristics that may determine how physical disturbances regulate this guild (Table 8.2).

Table 8.2 The effects of environmental variables on guilds' abundance.

	Low profile	High profile	Motile	Planktic	Studied waters	References
Nutrient supply	negative		positive		French watercourses	[8.86]
	negative		positive		French watercourses	[8.78]
	negative	hump shaped	hump shaped	positive	Dutch lakes	[8.68]
			contradictory and opposite		Swedish lakes	[8.79]
	none	none			Hungarian stream	[8.85]
	contradictory and opposite				Hungarian stream	[8.84]
	none	none	none		Swedish streams	[8.80]
	none	none	none		Finnish lakes	[8.83]
	none	none	none		Fennoscandian streams and lakes	[8.81]
		negative	positive		American streams	[8.87]

(*Continued*)

Table 8.2 The effects of environmental variables on guilds' abundance. (*Continued*)

	Low profile	High profile	Motile	Planktic	Studied waters	References
Disturbance	positive	negative			Laboratory experiment and American river	[8.88]
	positive	negative			Russian river	[8.66]
	positive	negative			Danish streams	[8.89]
			none		Swedish streams	[8.80]
			none		Hungarian stream	[8.84]
			none		Finnish lakes	[8.83]
			positive		Laboratory experiment and American river	[8.88]
			positive		Danish streams	[8.89]
			negative		Russian river	[8.66]
				negative	Danish streams	[8.89]
				none	Finnish lakes	[8.83]
Light	negative	positive	positive		New Zealand stream	[8.98]
	negative	positive	negative		Argentine streams	[8.96]
	negative				Swedish streams	[8.97]
	negative	positive	none		Argentine streams	[8.94]
	negative	positive			Chinese streams	[8.99]

(*Continued*)

Table 8.2 The effects of environmental variables on guilds' abundance. (*Continued*)

	Low profile	High profile	Motile	Planktic	Studied waters	References
	negative		negative		Hungarian stream	[8.85]
	positive				Hungarian lake	[8.100]
	positive				Italian lake	[8.101]
	positive	positive	negative		Brazilian lakes	[8.102]
	positive				Hungarian watercourses	[8.103]
		positive	negative		Argentine streams	[8.104]
			negative		Mayotte island streams	[8.107]
Alkalinity and pH	none	none	none		Swedish lakes	[8.79]
	positive	negative	negative		global dataset	[8.60]
	positive	negative	negative		French watercourses	[8.108]
	positive	negative	negative		South African wetlands	[8.109]

8.3.3 Light

Since diatoms are primarily photosynthetic organisms [8.92], light is a crucial environmental factor that structures the benthic assemblages through growth forms and guilds (e.g., stalked diatoms), thus guild-based studies can provide similar results as species-based analyses along the canopy cover [8.92]. By controlling guild composition and distribution, light availability also determines colonization success, and finally, community complexity [8.94]. Low-profile guild taxa are considered as *a priori* shade-tolerant ones [8.85] [8.95]. The strong negative relationship between this guild and light availability has been widely demonstrated [8.94] [8.96] [8.97] [8.98] [8.99]. The members of this guild are mainly r-strategist taxa, which are able to tolerate extreme conditions, even extreme light availability caused by high transparency [8.100] or rising light intensity [8.101]. However, under certain conditions, such as an early stage of colonization, low concentration of total suspended solids in waters, low proportion of tree canopy, or even rapid water level fluctuations, can result in high light environments dominated by a low-profile guild [8.100] [8.102] [8.103]. In contrast, the high-profile guild prefers well-illuminated waters and is negatively correlated with shading [8.96] [8.98] [8.99] [8.102] [8.104]. Furthermore, this guild has a strategic advantage in accessing light due to its location in thick biofilms [8.45], which enable this guild to tolerate even high turbidity [8.105] [8.106]. Motility allows taxa to change their position within the biofilm, thus the motile guild can avoid resource-stressed environments [8.45]. This would assume that a motile guild avoids shaded habitats and prefers transparent freshwaters. Although Lange *et al.* [8.98] revealed a strong positive relationship between high light conditions and the motile guild, others found the opposite [8.85] [8.96] [8.102] [8.104] [8.107], or have not found any correlation [8.94] [8.105]. However, environmental factors like nutrient supply (enrichment [8.104] [8.107]), hydrological conditions (low velocity [8.85] [8.96] [8.102]) or organic pollution (high concentration [8.107]) seem to override or cover the group's relationship to light. Finally, in freshwaters, the planktic guild was found to be negatively correlated with shading [8.99] and could attain high proportions in turbid waters [8.105] (Table 8.2).

8.3.4 Alkalinity and pH

Alkalinity and pH of waters also appear to be important environmental factors controlling guild composition and distribution [8.60] [8.79] [8.108] [8.109]. At a local scale, however, nutrient supply and water transparency seem to be key factors influencing guild composition [8.79] and clear pH-related guild distribution was not found in Swedish lakes despite the wide pH range (4.9–9.2). Nutrient rich, alkaline lakes promoted the dominance of a motile guild, but low light conditions favored the low-profile guild. In acid lakes, however, high light level, low grazing pressure and nutrient level are characteristic. These oligotrophic, undisturbed environments are preferred by the high-profile guild, which can be explained by the high light conditions. Global and regional studies revealed the negative effects of alkalinity and pH on the high-profile and motile guilds, while a strong positive effect on the low-profile guild [8.60] [8.108] [8.109] (Table 8.2).

8.3.5 Extreme Environments

Not only in freshwaters, but in inland saline waters, the guild composition reflects the extreme conditions well. The dominance of the motile guild [8.70] [8.110] and reduced guild diversity [8.111] is characteristic for natural soda pans with naturally high conductivity, pH, turbidity, nutrient content, highly daily temperature fluctuation and ephemeral features [8.110]. Among these conditions the amount of the motile group is significantly higher than that of low- and high-profile guilds, since this guild is capable of surviving the multiple high stresses. Their abundance is further increasing with the increased turbidity and decreased light availability [8.85] [8.110], due to its active habitat selection [8.45] among the sediment particles and "light trap function" of the more elongated, smaller cells of the *Nitzschia* and *Navicula* species representing the motile guild [8.70]. With the reduction of these harsh conditions (in degraded status), the low- and high-profile guilds prevail [8.112] [8.113], and together with planktic guild, they indicate the decreasing conductivity as well [8.70] (Table 8.2).

8.4 Effect of Global Threatening Processes on Guilds

8.4.1 Direct Anthropogenic Pressures: Land Use, Urbanization, Mining and Salinization

Agricultural land use and urbanization are key anthropogenic pressures contributing to increases in nutrients, ion concentrations, and conductivity values in freshwaters. In parallel with the increasing pressures the guild distribution changes in waters, mainly shifting to a higher proportion of motile guild within the assemblages [8.78] [8.114] [8.115] [8.116]. Furthermore, the areas exposed to intense land use are often located at the lower altitudes [8.114] [8.116], where the residence time of the water is longer, contributing to increased turbidity and siltation. In addition, the riparian shading is usually lower in these lowland, mainly agricultural or urban areas than in pristine, often forested areas, resulting in a change in light supply. All of these factors support the development of a more complex biofilm [8.116]. In contrast, the low-profile guild, which tolerates high physical disturbances and even lower nutrient content and poor light availability, characterizes areas less impacted by urbanization and agricultural land use [8.78] [8.114] [8.115] [8.116]. But it must be emphasized that the low-profile guild can also be common in nutrient-rich lowland areas, simply because its members spread quickly [8.115]. The relationship between the land use and distribution of the high-profile guild is not so clear. While some studies suggest negative correlation with waters less exposed to land use [8.78] [8.114], others did not find clear correlations [8.115] [8.116]. One possible reason could be that distribution of this guild in nutrient-rich areas is strongly determined by the regional species pool [8.64], but to verify this suggestion, further studies are needed.

Beside the problem of nutrient load by agricultural land use and urbanization, intensive farming also uses pesticides and antibiotics, which remain permanently in aquatic environments [8.117] [8.118] [8.119]. Single-diatom-species bioassays or ecotoxicological studies based on community composition at the species level are regularly used to determine individual sensitivities or identify tolerant/sensitive taxa [8.117] [8.120] [8.121]. Although the practical benefit of these assays is unquestionable, general trends are difficult to find in the field because processes within natural communities are complex (unlike single species bioassay) and species pools are different. However, the effect of toxic agents used in agriculture on guilds can be detected in freshwater ecosystems. The influence of different types of herbicides and fungicides on the guild composition were tested at environmental concentrations of chronic and acute levels [8.118]. Motile and low-profile (including tube-forming *Encyonema* taxa; *confer* [8.45]) guilds tolerated the exposure of pesticides, while high-profile guilds were sensitive to them. The ecophysiological reason behind their findings could be that the high-profile guild, especially the pedunculate forms, exploits dissolved nutrients, while the motile guild utilizes nutrients adsorbed on the substrate easier. Therefore, the high-profile guild is more exposed to dissolved pesticides than the motile guild. Tolerance of the low-profile guild might be due to tube-forming species, because mucous tubules might serve as shields against the dissolved toxic agents, which might explain their resistance. Other low-profile guild forms (e.g., pioneers), however, were not significantly affected by pesticides. Veterinary antibiotics, such as monensin, also have significant effects on guild composition in stream ecosystems with different availability of light [8.119]. The most impacted guilds were the high-profile and motile guilds, whose dominance decreased with increasing antibiotic concentration in an ambient light environment. In contrast, these guilds were more tolerant to monensin in reduced light conditions. This study also highlighted that a short contamination pulse did not have an irreversible impact on guild composition [8.119].

Beside urbanization and land use, mining is another human activity that has a big and complex impact on freshwaters; although heavy metals alone or in combination with macronutrients, as well as high alkalinity and salinity of acid mine drainage from coal mining act as selective factors for tolerant species with specific growth [8.122] [8.123]; however, we have very limited information on how these stressors affect guild distribution. High-profile guilds seems to be characteristic in environments affected by mining [8.123] [8.124]. Heine-Fuster *et al.* [8.124] pointed out that only a few well-adapted and tolerant taxa resulted in the dominance of a high-profile guild in extreme areas such as the very high altitude salt flats in Andean Altiplano. In contrast, motile and low-profile guilds were less abundant but more diverse and strongly connected to unimpacted areas [8.124].

Freshwater salinization is an increasingly common phenomenon worldwide [8.125] [8.126]. Recent research has pointed out that more than 30% of the freshwater ecosystems can be affected by salinization [8.127], which strongly alters community compositions and threatens biodiversity [8.126] of aquatic organisms Although the effect of the secondary salinization of diatom guilds can be predicted [8.128] based on their strong relationships with conductivity [8.115] [8.116] [8.123] [8.124], this research area is still in its infancy, and the number of papers on this topic is still limited.

8.4.2 Climatic Extremes, Intermittence, Floods

Hydrological regimes of streams and rivers worldwide are progressively modified by climate change, which manifests itself in historically unusual extreme weather events such as droughts and heavy rains [8.129] [8.130]. Flash floods and drying events, as a direct consequence of these climate extremes, are harsh disturbances and work selective pressures that can significantly alter the composition and structure of aquatic communities, thereby affecting their diversity and functioning [8.130] [8.131]. Since disturbance is one of the most important factors for the separation of the four diatom guilds, the effect of water regime changes, or even intermittency on lotic diatom assemblages based on guilds, could be expected.

However, early studies revealed only slight, non-significant water discharge, or intermittence-dependent variability in guild composition both in the Mediterranean [8 132] [8.133] and in the continental region [8.84] [8.85] [8.90], and the results were contradictory. The low-profile and motile guilds mainly characterized the permanent streams in Italy [8.132], but increasing water discharge and drastic water-level drop were also preferred by motile guilds in Hungarian rivers [8.84] [8.85]. In contrast, no flow preference of high-profile guilds was found by B-Béres *et al.* [8.84], while Stenger-Kovács *et al.* [8.85] and Falasco *et al.* [8.132] revealed a slight positive relationship between this guild and flow reduction. Low-profile guilds did not show any flow or water discharge preference either in Portuguese streams [8.133] or Hungarian watercourses [8.84] [8.85]. One possible reason that helps to understand and resolve this contradiction is that nutrients were the key factor influencing guild composition within the biofilm and flow plays only a secondary role in shaping the guild distribution. Another explanation could be that the applied guild system was not the same for the different studies, since some studies followed Passy's guild classification [8.85] [8.133], while others followed the four-guild classification [8.84] [8.132].

The same guild categorization based on the four guilds proposed by Rimet and Bouchez in 2012 [8.51], however, highlighted strong guild-water regime relationships. Although there is some variation in results depending on the study type (sampling from permanent and intermittent streams or only from intermittent streams during dry and flowing phases), sampling design and sampling effort, general conclusions can be drawn. The distribution of a low-profile guild is strongly related to its high reproductive rates, its tolerance against high flow velocity during floods and its ability to spread rapidly under eurheic conditions [39] even in permanent [8.134] and intermittent streams [8.103] [8.135] [8.136] [8.137] [8.138] [8.139]. However, it is very important to emphasize that a low-profile guild may also be dominant in pools during drying [8.139] and may be indicative of intermittent streams in their flowing phase [8.135] [8.136]. Although no clear water scarcity dependent distribution of high-profile guild was found [8.134] [8.137] [8.139], this guild can indicate both the permanent [8.103] [8.138] and even dry phase of intermittent streams [8.136], which is consistent with this guild preferring stable, non-disturbed conditions [8.45]. Taxa belonging to this guild usually formed long colonies or filaments, therefore, they are threatened by physical disturbances (e.g., floods), which explains why they can prefer permanent streams. Moreover, many high-profile guild taxa live in mucous tubes or attach to the substrate with long stalks, which provides an advantage by retaining water during drying [8.140], explaining their dominance during the dry phase in intermittent streams. The motile guild in intermittent streams is under strong environmental filtering [8.103], but members of this guild, which did not filter out from the community, can successfully prevail in drying periods [8.134] [8.137] [8.139], since motile guild taxa can tolerate increasing siltation [8.132] and/or increasing nutrient content [8.135] during decreasing water level and currents. In addition, they are able to move to the deeper layer of the mud, where they can survive the dry phase [8.141].

8.5 Diatom Guilds in Plankton and Their Ecological Role

Planktic habitats can be considered as "extreme" environments for benthic diatoms, since surviving in the water column would require different adaptation strategies from benthic species [8.51], therefore their presence in plankton is usually temporary. However, phytoplankton can be dominated by benthic elements in the upstream river sections [8.142] [8.143] or in springs [8.144], since there, the environmental conditions, such as short water residence time, water flow, or even slight seasonality, could favor them. On the other hand, open water in mid- and downstream usually does not provide suitable conditions for benthic diatoms [8.142] [8.143]; therefore, the increase in proportion of benthic taxa here has great ecological meaning: alteration in water conditions caused by climate change have resulted in extreme water flows (floods and highly reduced water level) in the Danube River. These extreme water discharges lead to a clear compositional shift in the phytoplankton community: the dominance of planktic taxa decreased in the plankton, while the relative abundance of benthic diatoms increased [8.145]. Benthic algae can also be abundant in the phytoplankton after floods [8.143] [8.146], especially those which adapt better to lentic habitats, because diatoms attached to macrophytes or living on the surface of psammon are more easily detached from substrates during floods than forms inhabiting stones [8.143]. Moreover, along the river, the distribution of guilds in the plankton is not random, it is mainly driven by seasonality and nutrient supply, and their abundance is strongly related to physical disturbances [8.144] [8.146]. High flow and/or low conductivity favor the low-profile guild, which conditions are usually characteristic during spring and early summer. The presence of the high-profile guild is strongly related to stable environments; thus, this guild often characterizes diatom assemblages of plankton in autumn and winter. Finally, the dominance/appearance of a motile guild in the plankton is driven mainly by the elevated suspended solids in summer [8.144] [8.146].

8.6 Guild Dispersal, Guild-Based Metacommunity Analysis and B-Diversity

In the above sections we discussed the importance of local factors in the development of diatom guilds. However, spatial processes also have an influence on the composition of local assemblages. An increasing amount of evidence suggests that the grouping of diatom species into ecological guilds can enhance our understanding of the organization of metacommunities [8.147] and it has an important role in discovering dispersal mechanisms and spatial distribution of diatoms [8.148].

8.6.1 Dispersal

Since the determination of the dispersal rate of different diatom species is very difficult, and in similar cases dispersal proxies are commonly applied [8.149], diatom ecological guilds can be indirect indicators (depending on their attachment they can be weak or strong dispensers) of dispersal processes in these studies [8.147]. Uncertainties and contradictions arising in dispersal and metacommunity studies can be resolved by considering the distinctive and special characteristics of guilds. The strong attachment of low-profile taxa to substrates determines their dispersal ability and makes them weak dispersers. However, most species belonging to this guild are small [8.51] r-strategists, characteristics which may contribute to their spread in certain environments (e.g., during/after flood, [8.135] [8.139]). In contrast, the high-profile guild can spread easier along streams passively facilitated by rushing water and a steep slope, but topographic barriers (high mountains) limit the spread of taxa [8.150]. The abundance of high-profile guilds is highly controlled by hydrological conditions (shear effect) and environmental variables, resulting in species sorting and their moderate dispersal in these aquatic ecosystems. The motile guild has a similar phenomenon; it has a clear relationship with the environmental variables, but no spatial structure due to the active movements, but weak resistance to high current velocity and grazer pressure [8.83] [8.147] [8.150]. However, the spatial process plays the highest role on the planktic guild, which is that it easily drifts in water [8.83] [8.147]. In highly connected waters there are no barriers to spread [8.151], the intensity of dispersion depends only on the dispersal potential of the different guilds [8.83]. In high mountain, oligotrophic streams the low-profile species prevailed [8.150], but at lower altitudes and less steep slopes the high-profile guild became dominant and metacommunities are structured according to the species sorting paradigm [8.80].

8.6.2 Metacommunity Studies

By definition, "a metacommunity is a set of local communities linked by the dispersal of multiple potentially interacting species" [8.152]. Processes, such as dispersion and selection, determine the composition and diversity within local communities [8.153], and can also be extended to ecological guilds. At a local scale, environmental filtering first determines which species belonging to a certain guild are able to survive in the given environment, which results in convergence within the community. Next, biotic interactions provide sufficient diversity for the utilization of resources [8.154]. These selecting forces finally regulate the emigration and immigration, and influence the dispersal and the colonization, determining the abundance and occupancy of guilds at regional and global scales (Figure 8.2).

Diatom metacommunity works to date are still in their infancy, but it is obvious that metacommunities across diatom ecological guilds are affected by environmental, spatial and biotic factors (grazers) both in lotic and lentic ecosystems [8.60] [8.75] [8.83] [8.80] [8.148] [8.150] [8.155]. However, the environmental filter is usually bigger on diatom metacommunities (e.g., [8.80]) but spatial effects can be scale dependent [8.75] [8.80]: the spatial influence on diatom guilds can increase from small scale [8.156] [8.150] to continental scales [8.60], which is consistent with that general view in metacommunities [8.157] that the spatial control is characteristic for generalists (here tolerant guild; low profile), and environmental for specialist (dominant guilds; motile). In highly connected running waters, the spatial processes (dispersion) can override the effect of the local environment (e.g., [8.158]), especially in the case of small microorganisms like diatoms as small passive, but effective dispersers. Moreover, the relative role of these fundamental constraints can be different depending on the selected groups (e.g., three or four guilds or other ecological groups) of metacommunities [8.155] and they are specific to the different guilds [8.80].

There are more guild-based metacommunity analyses primarily focused on subarctic and tropical streams and lakes and high mountain streams, where primarily the spatial and environmental factors with different ratios explained the guild and inside guild structure of the metacommunities [8.80] [8.83] [8.147] [8.150]. In Finnish oligotrophic lake systems, spatial factors, wave actions, and slope have clear effects on the diversity within the guilds [8.83]: richness in the planktic, low-profile and high-profile guilds are explained by the spatial processes, while on the motile species' richness, the physical habitat (slope, particle size of the sediment) had greater influence, similar to the diversity of the motile and low-profile group [8.83]. Quite similar phenomena were found in lotic ecosystems: the distribution of a low-profile guild was explained by both environmental and spatial variables [8.150], while Lindholm *et al.* [8.147] revealed the greater importance of local variables in this guild's dispersion, but spatial and catchment factors were also important. The abundance of the high-profile guild was driven by local and spatial variables in the arctic region [8.147], while in high mountain streams, local environment and topography were the main factors controlling the guild distribution [8.150]. Local factors were the most important in distribution of motile guild and spatial factors that drove the pattern of planktic guilds. In tropical lakes, dispersal mechanisms were the major factor which organized the metacommunity structure [8.148], however, the results were latitude dependent: in the case of the weaker dispersers, high- and low-profile guilds, the spatial processes overrode the environmental ones, especially climate stability around the equator. But in high latitudes, where the climatic conditions are more variable, the regional variable environment forms the diatom metacommunities [8.148]. However, for the strong dispersers, such as the motile and planktic guilds, the importance of these two factors were equal. In high-latitude lakes the physical distance, and at low latitudes terrain heterogeneity, were the most important part of the spatial processes [8.148]. In tropical lowland, floodplain lakes, the environmental effect on the motile guild was significant, because of the high environmental heterogeneity, similar to tropical reservoirs, where the pure effect of the environmental variables on the guild composition was detected [8.155]. In contrast to this, high elevational gradients and steep valleys make aerial dispersion in mountain lakes possible [8.65] [8.150], resulting in a pure spatial effect on the planktic guilds [8.148].

Along with the environmental, spatial and biotic gradient, large dataset based on guilds (655 stream sites, 16 years) were analyzed from the US. The motile guild richness negatively affected the ratio of the low-profile guild (tolerant guild), highlighting the importance of the competition between these guilds, which have shared preference. The spatial (low latitudes and middle longitudes) and environmental factors (primarily the phosphate concentration) had less pronounced but similar negative significant effects on the low-profile guild. The proportion of the motile guild (sensitive, dominant guild) was explained mainly by the environmental variables (N and P concentration). The competition between the guilds as biotic impact has a key role in the variability of the guild structure of benthic diatoms, which must be taken into account in metacommunity studies [8.75].

The pH-tolerance is also a significant characteristic that regulates the distribution and abundance of guilds, but while the acidophilus guild abundance is controlled by pH, the alkaliphilus guild was determined by the acidophilus guild richness (environmental derived competitor), pH and spatial factor (latitude) in order of importance. Which means that with moving away from the optimal pH, the alkaliphilus guild is stressed by physiological and biotic effects. The acidophilus guild has more diverse growth morphology than the alkaliphilus (only with low-profile taxa) guild including low, high and motile guilds leading to more efficient resource utilization, which may result in a significantly weaker biotic effect on this guild [8.75].

The ability of nitrogen-fixing may enhance the guild's success in spreading and colonizing nitrogen-limited areas. However, Passy [8.75] highlighted that the donor nitrogen-fixers guild was negatively affected by the recipient motile guild richness, and was not affected by nitrogen concentration of the water, which increased the abundance of the motile guild, besides the positive effect of the nitrogen-fixer richness. Consequently, the strong environmental impact and the facilitation together determine the abundance of the motile guild. Along the nitrogen gradients, there is a shift from facilitation to competition between the two guilds, which affects their relative abundance [8.75]. The abundance of nitrogen fixers is primarily determined by temperature, since nitrogen fixation is strongly influenced by this parameter [8.159]. In addition, other environmental variables, such as the concentration of inorganic, easily absorbed forms of nitrogen, also play a key role in the proportion of nitrogen-fixing diatoms found in the community. The eutrophic feature of the motile guild [8.45] was further supported by its phosphate and conductivity dependence [8.75].

Besides the inter-guild interactions (competition/facilitation), the environmental and spatial factors can also influence the niche realized by the guilds; however, the first two impacts override the dispersal and historic mechanisms [8.75]. The unexplained fraction of the guild composition is generally notable because of the factors not tested (other environmental factors, disturbance, grazing, etc.).

8.6.3 Beta Diversity

Metacommunity and guild dispersal studies create an opportunity to interpret biodiversity at a regional level. At a local scale, alpha diversity provides information about the structural and resource-sharing diversity in the biofilm [8.160], which can greatly contribute to the successful and cost-effective implementation of water management and nature conservation plans. This is, however, only a part of the whole; responsible environmental protection and nature conservation can only be realized at the regional level, for which beta diversity, i.e., the partitioning of habitats by guilds, provides an important clue. But to date, the number of guild-based beta-diversity studies are limited [8.65] [8.124]; thus, there is a gap in our knowledge on the topic.

In French streams, the motile and the low-profile species were the most abundant and their contribution to the overall beta diversity was the highest. The beta diversity of the guilds differed along the watercourse (from upstream to downstream), explained by both environmental and spatial processes. The dominance of the motile (that tolerate the siltation) and subdominance of planktic species (that favor bigger water volumes) in mid- and downstream sections of the stream resulted in the lowest beta diversity and turnover, and the highest nestedness. The mass effect influences these two guilds, based on the relationship between the beta diversity and spatial factors (stream sections) [8.65]. In contrast to the low-profile guild, whose proportion decreases along the stream, its beta diversity and turnover increase. Any obvious pattern was observed for high-profile guilds [8.65].

In salt lakes with different conditions (undisturbed nor affected by mining), beta diversity of all guilds was mainly explained by species turnover, especially in the case of low and motile guilds, which means that the species of these groups are more sensitive to mining, and species replacement is characteristic, while species composition of the high guild was more similar among the lakes [8.124].

8.7 Guilds in Conservation Biology Studies

Recently, an increasing number of studies are highlighting the role and importance of the diatom ecological guilds in nature conservation of aquatic habitats. In high mountain lakes, diatom guilds can indicate non-native fish production and their grazing pressure; moreover, they are sensitive to global warming, and therefore can be used as an indicator of undesirable processes. Thus, they may contribute to planning and tracing the actions aimed at sustainable management of these ecosystems [8.161]. Furthermore, in small rivers, hydropower dams have significant longitudinal effects on benthic diatom guilds, resulting in a shift from the strongly attached low-profile guild dominating in stream segments with high current velocity, to planktic guilds below dams and outlets and high-profile taxa below dams. By monitoring these guilds and adjusting the flow velocity and water depth accordingly, we can ensure the adequate connectivity and the health of stream ecosystems [8.162]. Maintaining the instream flows (as the flow regime and natural baseflow) [8.89] ensures the primary dispersal pathways of diatoms, in addition to preserving various habitats [8.150] that are essential in diatom biodiversity conservation of small rivers [8.89].

In tropical regions, deforestation is one of the most threatening anthropogenic activities to biodiversity. Interestingly, diatom guilds were good indicators of this phenomenon, especially the low- and high-profile guilds [8.163]. In deforested regions, where resource (light and nutrient) availability is not limited, habitat heterogeneity is characteristically large (with the appearance of filamentous algae, the number of the available microhabitats increase), and high-profile species in tubes (*Encyonema* spp.) or with long stalks (*Gomphonema* spp.) prevail. Contrarily, in undisturbed forested areas, low-profile taxa with short talks and strong attachment (*Achnanthidium* spp.) are dominant [8.164]. Land use can also modify the guild composition. For example, in streams surrounded by agricultural areas, the motile guild takes control; therefore, the preservation of guild assemblages in natural, non-degraded streams characterized by low profile guilds is fundamental from a conservational point of view [8.116].

The usefulness of guilds in saline ecosystems are also demonstrated from conservation perspectives. Under natural conditions, soda pans are dominated by the motile guild [8.110], while in the reconstructed and degraded pans, low- and high-profile taxa prevail. Preserving this benthic diatom guild structure can indicate the success of conservation management [8.113]. In salt flats (naturally salt or affected by mining) the species richness of the motile guilds was also the highest; however, the motile guild richness and composition were different in the two types of lakes. The unimpacted area was more valuable for all parameters examined, which has conservation significance since preservation of this key guild helps maintain the characteristic diatom community [8.124]. Since species turnover is the main force of the diatom ecological guild structure (especially motile and low-profile guilds), to maintain the diverse microflora in natural areas, a regional and integrated conservation plan is needed that focuses on more sampling sites of the different salt flats [8.124].

The guild-based approach is well applicable to the discovery of biodiversity-ecosystem function. Biomass production (density and total biovolume) of benthic diatoms mostly depends on the guild composition and the number of added macro- and micronutrients (like nitrogen, phosphorus, iron, manganese), and the species richness and evenness have less influence on it, indicating high functional redundancy among the species. The added nutrients more effectively enhance the richness of the high-profile and motile guilds than the low-profile guild. Grazing decreases the abundance of the high-profile guilds, and at the same time the low-profile and motile guild can increase, indicating their complementary resource utilization. In the absence of grazing, the high-profile guild proliferates, the biomass increases, and evenness decreases, which means that few productive species dominate and contribute most of the biomass [8.88].

In conclusion, benthic diatom guilds can help to understand the ecosystem processes and provide guidance in biodiversity conservation [8.162]. The occupancy-abundance relationship connected to guilds can also provide further knowledge for relevant biodiversity conservation [8.165]. Guild composition can also support the selection of the sampling sites for conservation purposes [8.166]. However, in conservation planning of guilds, not only the environmental factors should be taken into account, but the intra-guild relationships, such as facilitation and competition, in which case the applied conservation management strategies should be changed [8.75].

8.8 Guilds in the Past, Present and Future

8.8.1 Guilds in Paleolimnological Studies

Species composition of diatoms is often used as a paleoecological proxy [8.167]. Moreover, recent results proved that diatom guilds can be successfully involved in paleoreconstruction studies: guilds are considered good indicators of past (e.g., Holocene) or subfossil (i.e., last 200-1000 years) changes in nutrient concentrations and in water level fluctuations; they also indicate anthropogenic disturbances or even geomorphological alteration of floodplains [8.101] [8.168] [8.169] [8.170]. Resistance to high scour seem to be one of the most important strengths of low-profile guilds [8.101], which allows a guild to indicate hydrodynamically energetic environments such as sandy habitats [8.101] [8.169], littoral rock tops [8.169], water flow disturbances [8.168] [8.170], and productivity transitions [8.168]. In contrast, a high-profile guild indicates stable conditions in the past, e.g., changes of floodplains from semi-lotic to lentic environments [8.170], or epilithic or epiphytic habitats, which also provide shelter from consumers [8.101]. Paleolimnological studies also indicate that light conditions can modify the adaptation of a high-profile guild to turbid environments, which allows them to dominate in physically disturbed habitats, resulting in strong biotic interaction between high- and low-profile guilds [8.101]. Since motile guild members can move into fine sediment grains, they are good indicators in epipelic [8.101] and epilithic [8.169] habitats, but avoid disturbed conditions [8.170]. Open water environments with deep sediments are mainly inhabited by planktic guilds [8.169] in the fossil records.

8.8.2 Recent Trends of Guild Distributions

At a regional scale, there is a considerable overlap between metacommunity theories and biogeography; therefore, in view of recent global processes, such as climate change, habitat fragmentation and pollution, a combination of these scientific fields is required [8.171]. Using ecological guilds, we can model the dispersal ability of diatoms [8.148] and we can predict the diatom occupancy and abundance in lakes and streams [8.165]. Many studies revealed that the different guilds show similar positive occupancy and abundance relationships in aquatic ecosystems [8.165]: the broader the niche, the wider the distribution and greater the local abundance [8.172]. The niche of the well-known four guilds (high, low, motile and planktic) of continental lakes has non-marginal positions, broad niche and wide dominance, except for planktic guilds, which is common [8.172]. It was also shown that attached diatoms have higher occupancy than non-attached ones in boreal streams [8.173], especially low-profile diatoms compared to high-profile and motile guilds in US streams [8.174]. Niche conservatism, i.e., maintaining niche characteristics over time and space, is pivotal in large-scale distribution of species [8.175] and thereby guilds, which strongly highlights regional or even global changes in guild diversity patterns during, i.e., climate change or under the pressure of anthropogenic activities. On a global scale, there is weak niche conservatism among guilds based on environmental and climate models [8.175] which urges the calibration of regional diatom data to specify the niche of diatom guilds. However, there are few examples where niche conservatism was characteristic for guilds, especially in the case of the acid-tolerant group [8.175].

At a regional scale, ecological guilds have individual responses to contemporary and historical drivers (the adaptation of guilds at an evolutionary scale is possible because of the more or less permanent environment in space and in time), especially in the Neotropics [8.176]. The temporal changes (seasonal) of the diatom ecological guilds is obvious in a temperate, small stream [8.85]: the high-profile guild prevails in spring, the low profile in summer, while the motile group reaches its maximum abundance in winter. Seasonality drives guild diversity too: the highest guild diversity can be found in summer and the lowest in winter [8.85]. However, in Finnish boreal streams, the guild type does not influence their temporal occurrence, but the attachment ability of a species supports its temporal occurrence through permanent populations in changing environments [8.177]. Low-profile taxa are characteristic of temporary streams in eurheic conditions [8.103] [8.139] and high-profile taxa for perennial streams of the Mediterranean region [8.103], while motile taxa for continental streams are characteristic independent of their hydrological condition [8.103]. Furthermore, a recent study also revealed that a motile guild has the highest abundance in moderately heterogeneous communities in alpine lakes; at lower and higher levels of heterogeneity its abundance decreases [8.178].

It seems that lake size is also an important factor determining the guild diversity in lacustrine ecosystems at a regional scale [8.179]: the species richness of diatom ecological guilds increased with size (surface area: 10^2 m^2 - 10^{10} m^2) of the lakes, but while this positive relationship is clearly linear in the case of planktic, low- and high-profile guilds, it is not true for the motile guild, which is almost constant (the highest species richness) among the size categories, but greatly increased in the largest lakes (10^{10} m^2), revealing the small island effect as a characteristic of the benthic diatoms species-area relationship. However, the relative contribution of the motile guild among the guilds is the highest in small-sized lakes (10^{-2} m^2 - 10^3 m^2) [8.179].

At a continental scale, nitrogen fixers are controlled by temperature affecting nitrogen fixation, and the presence of this guild further supports the appearance of the motile guild (phenomenon of facilitation) [8.75]. The global pattern of guild distribution was also studied [8.60], where high regional overlap of the guilds was highlighted (~no regional differences), except for acid-tolerant species restricted mostly to the Finnish region. Moreover, the planktic taxa had a positive relationship with the latitude. At the local environmental level, spatial and climatic variables together form the guild composition, which was better explained by the environmental variables than the species composition and less dependent on historic factors. This feature of the diatom ecological guilds will be very applicable in studies dealing with global environmental changes (e.g., eutrophication, climate change, salinization, acidification; [8.60].

The Water Framework Directive (WFD EU 2000) requires member states to monitor the quality of surface waters. Since diatoms are key elements in assessing trophic conditions both in lakes and rivers, long-term datasets were compiled in recent years, which provide a good opportunity to study long-term dynamics in the composition of benthic diatoms. These studies revealed a general improvement in the quality of rivers [8.180] [8.181]. The decreasing nutrient concentration and biological oxygen demand induce a rearrangement and a shift in diatom ecological guilds from motile (generally eutrophic species) and planktic dominance to the increased significance of the low-profile guild in French rivers [8.182], which can co-tolerate low nutrient availability and even high disturbances, as a consequence of different hydrological regimes caused by climate change [8.139].

8.8.3 Future Distribution of Diatom Guilds

Climate change will reconstruct the structure of benthic diatoms. Model projections for the following decades predict the expansion of the warm-water guild (mostly motile guilds), which replace the cold-water taxa (low- and high-profile taxa) [8.72]. These changes in composition threaten the building of the benthic layer, since the low-profile guild is the base and the high-profile guild creates the upper biofilm layer, and together they provide the major part of the biofilm. Besides the enhancement of motile warm-water taxa induced by the increasing water temperature, microalgae-induced biofilm-formation will be limited by nutrients, which are less and less available in streams due to oligotrophication [8.180] [8.181] [8.182]. Their consequence is the low primary productivity and its negative effect on the next trophic level (herbivores). The removal of climate barriers supports the spread of the warm-water guild, leading to reduced beta diversity and biotic homogenization. However, the decline of cold-water taxa cannot offset the faster expansion of the warm-water taxa, which will presumably increase the species richness at a subcontinental scale [8.72].

8.9 Conclusion and Future Directions of the Guild Concept in Diatom Community Ecology

Recently, freshwaters are threatened by several anthropogenic pressures (e.g., eutrophication, climate change, salinization) from local to global scales [8.126]. Guild-based studies contribute to explaining the organization of diatom communities under these pressures and predict their reorganization due to environmental changes [8.60]. Despite some limitations of the diatom guild concept, i.e., their distribution among resources in Scandinavian freshwaters is not clear enough [8.79] [8.81] [8.82], the use of diatom guilds has indisputable advantages over taxonomic methods: (i) due to the extraordinary diversity of the group [8.183] and the large number of taxonomically problematic classes and cryptic species [8.184], classifying taxa into guilds is often much easier than species-level identification. (ii) For accurate taxonomic identification, training experts is time-consuming and expensive, but using guilds is more cost-effective and results in fewer scientific errors. (iii) In traditional species-based analyses rare species are often excluded from the statistical analyses, therefore their ecological information content is lost, while guild-based approaches preserve it. (iv) The variance between the environmental variables and guild structure is higher than in the case of species-based approaches; furthermore, (v) this approach can also provide complementary ecological information. (vi) Species pools can differ significantly between ecoregions or even between habitats; guild-based approaches, however, can facilitate the comparison of results independent of regions and improve understanding of wide-scale processes on the structure and formation of phytobenthos.

With the appearance of the guild concepts, published ecological studies led to several misused terminologies for guilds like *traits* or *functional groups*. The consistent use of these completely different terms is strongly recommended [8.7]. Guilds are defined based on similarity of resource sharing, while functional groups contain species with similar morphological and physiological traits and have a similar role in ecosystem functioning. The use and development of new guilds should follow the right definition to draw appropriate conclusions using guild-based studies in diatom research. Moreover, the application of three guilds in parallel [8.45], four guilds and refined guild systems [8.51], can lead to different results; therefore, the use of the last system with validated guilds (low, high, motile, planktic, nitrogen fixer and non-nitrogen fixers) is suggested so that the results are comparable. For further development (refined guilds, new guilds) of the guild system, it would be essential to standardize and harmonize the grouping of taxa from species to life forms, based on live benthic diatom samples and from life forms to guilds, which can be supported by a database, such as that of Rimet and Bouchez [8.51], kept up to date.

Recent studies highlighted that combining guilds with ecologically meaningful, easy-to-measure diatom traits (e.g., cell size, length and width ratio [8.70] [8.90]) can further enhance the robustness of Passy's guilds and provide opportunities to deepen our knowledge of the diatoms-environment relationship, which may lead to involving guilds and guild-based systems in economically and ecologically significant research with great practical benefit. The framework of guilds based on ecological concepts like resource partitioning, the stress gradient hypothesis proposed by Passy [8.45], opened the way for new research towards guild-based biogeography and discovery of biodiversity-ecosystem functioning relationships [8.75], where the role of competition and facilitation in forming high biodiversity and biomass is an unknown field [8.45].

Acknowledgment

This work was supported by the National Research, Development and Innovation Office–NKFIH-K137950 (CS-K), NKFIH-FK132142 (VB-B), by the János Bolyai Research Scholarship of the Hungarian Academy of Sciences BO-00458-20-8 (VB-B), by the ÚNKP-21-5 (VB-B) New National Excellence Program of the Ministry for Innovation and Technology from the Source of the National Research, Development and Innovation Fund, by the Centre for Ecological Research - BA4902/21 (Riverine Project) (VBB), and by the National Laboratory for Water Science and Water Safety program (RRF-2.3.1-21-2022-00008) from the National Research, Development and Innovation Office (CS-K). The authors thank Edina Lengyel for her contribution in the preparation of Figure 8.1.

References

[8.1] Schimper, A.F.W. (1903) Plant-geography upon a physiological basis. Translated by WR Fisher; rev. and edited by Percy Groom and IB Balfour.
[8.2] Root, R.B. (1967) The niche exploitation pattern of the blue-gray gnatcatcher. *Ecological Monographs* 37(4), 317–350.
[8.3] Cummins, K.W. (1974) Structure and function of stream ecosystems. *BioScience* 24(11), 631–641.
[8.4] Austen, D.J., Bayley, P.B. and Menzel, B.W. (1994) Importance of the guild concept to fisheries research and management. *Fisheries* 19 (6), 12–20.
[8.5] Gleason, H.A. (1926) The individualistic concept of the plant association. *Bulletin of the Torrey Botanical Club* 53(1), 7–26.
[8.6] Simberloff, D., and Dayan, T. (1991) The guild concept and the structure of ecological communities. *Annual Review of Ecology and Systematics* 22, 115–143.
[8.7] Blondel, J. (2003) Guilds or functional groups: does it matter? *Oikos* 100, 223–231.
[8.8] Gorman, O.T. (1988) The dynamics of habitat use in a guild of Ozark minnows. *Ecological Monographs* 58 (1), 1–18.
[8.9] Knopf, F.L. and Olson, T.E. (1984) Naturalization of Russian-olive: implications to Rocky Mountain wildlife. *Wildlife Society Bulletin* 12 (3), 289–298.
[8.10] Berkman, H.E. and Rabeni, C.F. (1987) Effect of siltation on stream fish communities. *Environmental Biology of Fishes* 18 (4), 285–294.
[8.11] Mac Nally, R.C. and Doolan, J.M. (1986) An empirical approach to guild structure: habitat relationships in nine species of eastern-Australian cicadas. *Oikos* 47, 33–46.
[8.12] Schoener, T.W. (1986) Overview: kinds of ecological communities-ecology becomes pluralistic. In *Community Ecology*. J. Diamond and T.J. Case (eds) NY: Harper and Row, New York, USA: 467–479.
[8.13] Guillet, A. and Crowe, T.M. (1985) Patterns of distribution, species richness, endemism and guild composition of water-birds in Africa. *African Journal of Ecology* 23(2), 89–120.

[8.14] Sedgwick, J.A. and Knopf, F.L. (1987) Breeding bird response to cattle grazing of a cottonwood bottomland. *The Journal of Wildlife Management* 51(1), 230–237.

[8.15] Verner, J. (1984) The guild concept applied to management of bird populations. *Environmental Management* 8(1), 1–13.

[8.16] Skagen, S.K., Knight, R.L. and Orians, G.H. (1991) Human disturbance of an avian scavenging guild. *Ecological Applications* 1(2), 215–225.

[8.17] Boutin, C. and Keddy, P.A. (1993) A functional classification of wetland plants. *Journal of Vegetation Science* 4(5), 591–600.

[8.18] Wilson, J.B. (1999) Guilds, functional types and ecological groups. *Oikos* 86(3), 507–522.

[8.19] MacMahon, J.A., Schimpf, D.J., Andersen, D.C., Smith, K.G. and Bayn Jr, R.L. (1981) An organism-centered approach to some community and ecosystem concepts. *Journal of Theoretical Biology* 88(2), 287–307.

[8.20] Landres, P.B. (1983) Use of the guild concept in environmental impact assessment. *Environmental Management* 7(5), 393–397.

[8.21] Hawkins, C.P. and MacMahon, J.A. (1989) Guilds: the multiple meanings of a concept. *Annual Review of Entomology* 34, 423–451.

[8.22] Hawkins, C.P. and Sedell, J.R. (1981) Longitudinal and seasonal changes in functional organization of macroinvertebrate communities in four Oregon streams. *Ecology* 62(2), 387–397.

[8.23] Adams, J. (1985) The definition and interpretation of guild structure in ecological communities. *The Journal of Animal Ecology* 54(1), 43–59.

[8.24] Orians, G.H. (1980) Micro and macro in ecological theory. *Bioscience* 30(2), 79–79.

[8.25] Koran, M. and Kropil, R. (2014) What are ecological guilds? Dilemma of guild concepts. *Russian Journal of Ecology* 45(5), 445–447.

[8.26] Paine, R.T. (1980) Food webs: linkage, interaction strength and community infrastructure. *Journal of Animal Ecology* 49(3), 667–685.

[8.27] Niemi, G.J. and McDonald, M.E. (2004) Application of ecological indicators. *Annual Review of Ecology, Evolution, and Systematics* 35, 89–111.

[8.28] Johnson, R.A. (1981) Application of the guild concept to environmental impact analysis of terrestrial vegetation. *Journal of Environmental Management* 13(3), 205–222.

[8.29] Severinghaus, W. (1981) Guild theory development as a mechanism for assessing environmental impact. *Environmental Management* 5, 187–190.

[8.30] Smith, K.G. (1980) Nongame birds of the Rocky Mountain spruce-fir forests and their management. In: *Management of western forest and grassland for nongame birds.* R. DeGraaf (tech. coord.) USDA Forest Service General Technical Report INT-89: 258–279.

[8.31] O'Connell, T.J., Jackson, L.E. and Brooks, R.P. (2000) Bird guilds as indicators of ecological condition in the central Appalachians. *Ecological Applications* 10(6), 1706–1721.

[8.32] Colwell, R.K. (1979) Toward a unified approach to the study of species diversity. *Ecological diversity in theory and practice.* International Cooperative Publishing House, Fairland, Maryland, USA, 75–91.
[8.33] MacMahon, J.A., Schimpf, D.J., Andersen, D.C., Smith, K.G. and Bayn Jr, R.L. (1981) An organism-centered approach to some community and ecosystem concepts. *Journal of Theoretical Biology* 88(2), 287–307.
[8.34] Begon, M., Harper, J.L. and Townsend, C.R. (1986) *Ecology. Individuals, populations and communities.*, Blackwell Scientific Publications, Oxford, UK.
[8.35] Borics, G., Várbíró, G. and Padisák, J. (2013) Disturbance and stress: different meanings in ecological dynamics? *Hydrobiologia* 711(1), 1–7.
[8.36] Hutchinson, G.E. (1975) *A treatise on limnology:limnological botany.* Wiley, New York, USA.
[8.37] Hutchinson, G.E. (1957) Concluding remarks. population studies: animal ecology and demography. *Cold Spring Harbor Symposia on Quantitative Biology* 22, 415–427.
[8.38] Bronstein, J.L. (2009) The evolution of facilitation and mutualism. *Journal of Ecology* 97(6), 1160–1170.
[8.39] Gallart, F., Prat, N., García-Roger, E.M., Latron, J., Rieradevall, M., Llorens, P., Barberá, G.G., Brito, D., De Girolamo, A.M., Lo Porto, A., Buffagni, A., Erba, S., Neves, R., Nikolaidis, N.P., Perrin, J.L., Querner, E.P., Quiñonero, J.M., Tournoud, M.G., Tzoraki, O., Skoulikidis, N., Gómez, R., Sánchez-Montoya, M.M. and Froebrich, J. (2012) A novel approach to analysing the regimes of temporary streams in relation to their controls on the composition and structure of aquatic biota. *Hydrology and Earth System Sciences* 16(9), 3165–3182.
[8.40] Wright, D.H. and Reeves, J.H. (1992) On the meaning and measurement of nestedness of species assemblages. *Oecologia*, 92, 416–428.
[8.41] Lozanovska, I., Ferreira, M.T. and Aguiar, F.C. (2018) Functional diversity assessment in riparian forests – Multiple approaches and trends: A review. *Ecological Indicators* 95, 781–793.
[8.42] Reynolds, C.S. (1988) The concept of ecological succession applied to seasonal periodicity of freshwater phytoplankton: With 4 figures in the text. *SIL Proceedings, 1922-2010* 23(2), 683–691.
[8.43] Reynolds, C.S., Huszar, V., Kruk, C., Naselli-Flores, L. and Melo, S. (2002) Towards a functional classification of the freshwater phytoplankton. *Journal of Plankton Research* 24(5), 417–428.
[8.44] Cummins, K.W. and Klug, M.J. (1979) Feeding ecology of stream invertebrates. *Annual Review of Ecology and Systematics* 10, 147–172.
[8.45] Passy, S.I. (2007) Diatom ecological guilds display distinct and predictable behavior along nutrient and disturbance gradients in running waters. *Aquatic Botany* 86(2), 171–178.
[8.46] Toda, M.J. (1984) Guild structure and its comparison between two local drosophilid communities. *Physiology and Ecology Japan* 21, 131–172.

[8.47] Hoagland, K.D., Roemer, S.C. and Rosowski, J.R. (1982) Colonization and community structure of two periphyton assemblages, with emphasis on the diatoms (Bacillariophyceae). *American Journal of Botany* 69(2), 188–213.

[8.48] Kutka, F.J. and Richards, C. (1996) Relating diatom assemblage structure to stream habitat quality. *Journal of the North American Benthological Society* 15(4), 469–480.

[8.49] Molloy, J.M. (1992) Diatom communities along stream longitudinal gradients. *Freshwater Biology* 28(1), 59–69.

[8.50] Fore, L.S. and Grafe, C. (2002) Using diatoms to assess the biological condition of large rivers in Idaho (USA). *Freshwater Biology* 47(10), 2015–2037.

[8.51] Rimet, F. and Bouchez, A. (2012) Life-forms, cell-sizes and ecological guilds of diatoms in European rivers. *Knowledge and Management of Aquatic Ecosystems* (406), 01.

[8.52] Julius, M.L. and Theriot, E.C. (2010) The diatoms: a primer. In: *The diatoms: applications for the environmental and earth sciences.* J.P. Smol and E.F. Stoermer Cambridge University Press, Cambridge, UK: 8–22.

[8.53] Round, F.E., Crawford, R.M. and Mann, D.G. (1990) *Diatoms: biology and morphology of the genera*, Cambridge University Press, Cambridge, UK.

[8.54] Tuji, A. (2000) Observation of developmental processes in loosely attached diatom (Bacillariophyceae) communities. *Phycological Research* 48(2), 75–84.

[8.55] Wilson, J.B. and Roxburgh, S.H. (1994) A demonstration of guild-based assembly rules for a plant community, and determination of intrinsic guilds. *Oikos* 69(2), 267–276.

[8.56] Fagerstrom, J.A. (1987) *The evolution of reef communities.* Wiley, New York, USA.

[8.57] Fagerstrom, J.A. (1988) A structural model for reef communities. *Palaios* 3(2), 217–220.

[8.58] Fagerstrom, J.A. (1991) Reef-building guilds and a checklist for determining guild membership. *Coral Reefs* 10(1), 47–52.

[8.59] Precht, W.F. (1994) The use of the term guild in coral reef ecology and paleoecology: a critical evaluation. *Coral Reefs* 13(3), 135–136.

[8.60] Soininen, J., Jamoneau, A., Rosebery, J. and Passy, S.I. (2016) Global patterns and drivers of species and trait composition in diatoms. *Global Ecology and Biogeography* 25(8), 940–950.

[8.61] Schneck, F., Schwarzbold, A. and Melo, A.S. (2011) Substrate roughness affects stream benthic algal diversity, assemblage composition, and nestedness. *Journal of the North American Benthological Society* 30(4), 1049–1056.

[8.62] Trbojević, I., Jovanović, J., Kostić, D., Popović, S., Krizmanić, J., Karadžić, V. and Subakov Simić, G. (2017) Structure and succession of periphyton in an urban reservoir: artificial substrate specificity. *Oceanological and Hydrobiological Studies* 46(4), 379–392.

[8.63] Nemes-Kókai, Zs., Borics, G., Csépes, E., Lukács, Á., Török, P., T-Krasznai, E., Bácsi, I. and B-Béres, V. (2023) Role of microhabitats in shaping diversity of periphytic diatom assemblages. *Hydrobiologia*. https://doi.org/10.1007/s10750-023-05336-x.

[8.64] Passy, S.I. and Larson, C.A. (2011) Succession in Stream Biofilms is an Environmentally Driven Gradient of Stress Tolerance. *Microbial Ecology* 62(2), 414–424.

[8.65] Jamoneau, A., Passy, S.I., Soininen, J., Leboucher, T. and Tison-Rosebery, J. (2018) Beta diversity of diatom species and ecological guilds: Response to environmental and spatial mechanisms along the stream watercourse. *Freshwater Biology* 63(1), 62–73.

[8.66] Rusanov, A.G. and Khromov, V.M. (2016) Longitudinal distribution of periphyton algae in the Moskva river under eutrophication. *Water Resources* 43(3), 513–521.

[8.67] Mackay, A.W., Davidson, T., Wolski, P., Woodward, S., Mazebedi, R., Masamba, W.R. and Todd, M. (2012) Diatom sensitivity to hydrological and nutrient variability in a subtropical, flood-pulse wetland. *Ecohydrology* 5(4), 491–502.

[8.68] Goldenberg Vilar, A., van Dam, H., van Loon, E.E., Vonk, J.A., van Der Geest, H.G. and Admiraal, W. (2014) Eutrophication decreases distance decay of similarity in diatom communities. *Freshwater Biology* 59(7), 1522–1531.

[8.69] B-Béres, V., Török, P., Kókai, Z., Lukács, Á., T-Krasznai, E., Tóthmérész, B. and Bácsi, I. (2017) Ecological background of diatom functional groups: Comparability of classification systems. *Ecological Indicators* 82, 183–188.

[8.70] Stenger-Kovács, C., Körmendi, K., Lengyel, E., Abonyi, A., Hajnal, É., Szabó, B., Buczkó, K. and Padisák, J. (2018) Expanding the trait-based concept of benthic diatoms: Development of trait-and species-based indices for conductivity as the master variable of ecological status in continental saline lakes. *Ecological Indicators* 95, 63–74.

[8.71] Van Dam, H., Mertens, A. and Sinkeldam, J. (1994) A coded checklist and ecological indicator values of freshwater diatoms from The Netherlands. *Netherlands Journal of Aquatic Ecology* 28(1), 117–133.

[8.72] Pound, K.L., Larson, C.A. and Passy, S.I. (2021) Current distributions and future climate-driven changes in diatoms, insects and fish in US streams. *Global Ecology and Biogeography* 30(1), 63–78.

[8.73] de Kroon, H. and Olff, H. (1995) On the use of the guild concept in plant ecology. *Folia Geobotanica* 30(4), 519–528.

[8.74] Walter, G.H. (1988) Competitive exclusion, coexistence and community structure. *Acta Biotheoretica* 37(3), 281–313.

[8.75] Passy, S.I. (2017) Framework for community functioning: synthesis of stress gradient and resource partitioning concepts. *PeerJ*, 5, e3885.
[8.76] Law, R.J. (2011) A Review of the Function and uses of, and Factors Affecting, Stream Phytobenthos. *Freshwater Reviews* 4(2), 135–166.
[8.77] Scholten, M.C.T., Foekema, E.M., Dokkum, H.P., Kaag, N.H.B.M., Jak, R.G. and Scholten, M.C.T. (2005) *Eutrophication Management and Ecotoxicology*, Springer-Verlag Berlin Heidelberg, Berlin, Heidelberg.
[8.78] Berthon, V., Bouchez, A. and Rimet, F. (2011) Using diatom life-forms and ecological guilds to assess organic pollution and trophic level in rivers: a case study of rivers in south-eastern France. *Hydrobiologia* 673(1), 259–271.
[8.79] Gottschalk, S. and Kahlert, M. (2012) Shifts in taxonomical and guild composition of littoral diatom assemblages along environmental gradients. *Hydrobiologia* 694(1), 41–56.
[8.80] Göthe, E., Angeler, D.G., Gottschalk, S., Löfgren, S. and Sandin, L. (2013) The Influence of Environmental, Biotic and Spatial Factors on Diatom Metacommunity Structure in Swedish Headwater Streams. *PLoS ONE* 8(8), e72237.
[8.81] Kahlert, M. and Gottschalk, S. (2014) Differences in benthic diatom assemblages between streams and lakes in Sweden and implications for ecological assessment. *Freshwater Science* 33(2), 655–669.
[8.82] Kahlert, M., Bailet, B., Chonova, T., Karjalainen, S.M., Schneider, S.C. and Tapolczai, K. (2021) Same same, but different: The response of diatoms to environmental gradients in Fennoscandian streams and lakes – barcodes, traits and microscope data compared. *Ecological Indicators* 130, e108088.
[8.83] Vilmi, A., Tolonen, K.T., Karjalainen, S.M. and Heino, J. (2017) Metacommunity structuring in a highly-connected aquatic system: effects of dispersal, abiotic environment and grazing pressure on microalgal guilds. *Hydrobiologia* 790(1), 125–140.
[8.84] B-Béres, V., Török, P., Kókai, Z., Krasznai, E.T., Tóthmérész, B. and Bácsi, I. (2014) Ecological diatom guilds are useful but not sensitive enough as indicators of extremely changing water regimes. *Hydrobiologia* 738(1), 191–204.
[8.85] Stenger-Kovács, C., Lengyel, E., Crossetti, L.O., Üveges, V. and Padisák, J. (2013) Diatom ecological guilds as indicators of temporally changing stressors and disturbances in the small Torna-stream, Hungary. *Ecological Indicators* 24, 138–147.
[8.86] Rimet, F., Ector, L., Cauchie, H.-M. and Hoffmann, L. (2009) Changes in diatom-dominated biofilms during simulated improvements in water quality: implications for diatom-based monitoring in rivers. *European Journal of Phycology* 44(4), 567–577.
[8.87] Budnick, W.R., Mruzek, J.L., Larson, C.A. and Passy, S.I. (2021) The impacts of nutrient supply and imbalance on subcontinental

co-occurrence networks and metacommunity composition of stream algae. *Ecography* 44(7), 1109–1120.

[8.88] Passy, S.I. and Larson, C.A. (2019) Niche dimensionality and herbivory control stream algal biomass via shifts in guild composition, richness, and evenness. *Ecology* 100 (11), e02831.

[8.89] Wu, N., Thodsen, H., Andersen, H.E., Tornbjerg, H., Baattrup-Pedersen, A. and Riis, T. (2019) Flow regimes filter species traits of benthic diatom communities and modify the functional features of lowland streams-a nationwide scale study. *Science of the Total Environment* 651, 357–366.

[8.90] B-Béres, V., Lukács, Á., Török, P., Kókai, Z., Novák, Z., T-Krasznai, E., Tóthmérész, B. and Bácsi, I. (2016) Combined eco-morphological functional groups are reliable indicators of colonisation processes of benthic diatom assemblages in a lowland stream. *Ecological Indicators* 64, 31–38.

[8.91] Gordon, R., Björklund, N.K., Robinson, G.G.C. and Kling, H.J. (1996) Sheared drops and pennate diatoms. *Nova Hedwigia* 112, 287-297.

[8.92] Villanova, V. and Spetea, C. (2021) Mixotrophy in diatoms: Molecular mechanism and industrial potential. *Physiologia Plantarum* 173(2), 603–611.

[8.93] Leland, H.V. and Porter, S.D. (2000) Distribution of benthic algae in the upper Illinois River basin in relation to geology and land use. *Freshwater Biology* 44(2), 279–301.

[8.94] Cibils-Martina, L., Principe, R.E., Márquez, J.A., Gari, E.N. and Albariño, R.J. (2017) Succession of algal communities in headwaters: a comparison of pine afforested and natural grassland streams. *Ecological Research* 32(3), 423–434.

[8.95] Tapolczai, K., Bouchez, A., Stenger-Kovács, C., Padisák, J. and Rimet, F. (2016) Trait-based ecological classifications for benthic algae: review and perspectives. *Hydrobiologia* 776(1), 1–17.

[8.96] Cibils, L., Principe, R., Márquez, J., Gari, N. and Albariño, R. (2015) Functional diversity of algal communities from headwater grassland streams: How does it change following afforestation? *Aquatic Ecology* 49(4), 453–466.

[8.97] Johnson, R.K. and Almlöf, K. (2016) Adapting boreal streams to climate change: effects of riparian vegetation on water temperature and biological assemblages. *Freshwater Science* 35(3), 984–997.

[8.98] Lange, K., Liess, A., Piggott, J.J., Townsend, C.R. and Matthaei, C.D. (2011) Light, nutrients and grazing interact to determine stream diatom community composition and functional group structure: Diatom responses to light, nutrients and grazing. *Freshwater Biology* 56(2), 264–278.

[8.99] Zhang, J., Shu, X., Zhang, Y., Tan, X. and Zhang, Q. (2020) The responses of epilithic algal community structure and function to light and nutrients and their linkages in subtropical rivers. *Hydrobiologia* 847(3), 841–855.

[8.100] Kókai, Z., Borics, G., Bácsi, I., Lukács, Á., Tóthmérész, B., Csépes, E., Török, P. and B-Béres, V. (2019) Water usage and seasonality as primary drivers of benthic diatom assemblages in a lowland reservoir. *Ecological Indicators*, 106, e105443.

[8.101] Leira, M., Filippi, M.L. and Cantonati, M. (2015) Diatom community response to extreme water-level fluctuations in two Alpine lakes: a core case study. *Journal of Paleolimnology* 53(3), 289–307.

[8.102] Algarte, V.M., Dunck, B., Leandrini, J.A. and Rodrigues, L. (2016) Periphytic diatom ecological guilds in floodplain: Ten years after dam. *Ecological Indicators* 69, 407–414.

[8.103] Várbíró, G., Borics, G., Novais, M.H., Morais, M.M., Rimet, F., Bouchez, A., Tapolczai, K., Bácsi, I., Usseglio-Polatera, P. and Viktória, B. (2020) Environmental filtering and limiting similarity as main forces driving diatom community structure in Mediterranean and continental temporary and perennial streams. *Science of the Total Environment* 741, e140459.

[8.104] Nicolosi Gelis, M.M., Cochero, J., Donadelli, J. and Gómez, N. (2020) Exploring the use of nuclear alterations, motility and ecological guilds in epipelic diatoms as biomonitoring tools for water quality improvement in urban impacted lowland streams. *Ecological Indicators* 110, e105951.

[8.105] Goldenberg Vilar, A., Vonk, J.A., Bichebois, S., van Dam, H., Admiraal, W. and van der Geest, H.G. (2015) Suspended organic particles drive the development of attached algal communities in degraded peatlands. *Hydrobiologia* 744(1), 211–221.

[8.106] Rimet, F., Bouchez, A. and Tapolczai, K. (2016) Spatial heterogeneity of littoral benthic diatoms in a large lake: monitoring implications. *Hydrobiologia* 771(1), 179–193.

[8.107] Tapolczai, K., Bouchez, A., Stenger-Kovács, C., Padisák, J. and Rimet, F. (2017) Taxonomy- or trait-based ecological assessment for tropical rivers? Case study on benthic diatoms in Mayotte island (France, Indian Ocean). *Science of The Total Environment* 607–608, 1293–1303.

[8.108] Marcel, R., Berthon, V., Castets, V., Rimet, F., Thiers, A., Labat, F. and Fontan, B. (2017) Modelling diatom life forms and ecological guilds for river biomonitoring. *Knowledge and Management of Aquatic Ecosystems* (418), e1.

[8.109] Riato, L., Della Bella, V., Leira, M., Taylor, J.C. and Oberholster, P.J. (2017) A diatom functional-based approach to assess changing environmental conditions in temporary depressional wetlands. *Ecological Indicators* 78, 205–213.

[8.110] Stenger-Kovács, C., Lengye, E., Buczkó, K., Tóth, F.M., Crossetti, L.O., Pellinger, A., Doma, Z.Z. and Padisák, J. (2014) Vanishing world: alkaline, saline lakes in Central Europe and their diatom assemblages. *Inland Waters* 4(4), 383–396.

[8.111] Ács, É., Földi, A., Vad, C.F., Trábert, Z., Kiss, K.T., Duleba, M., Borics, G., Grigorszky, I. and Botta-Dukát, Z. (2019) Trait-based community assembly of epiphytic diatoms in saline astatic ponds: a test of the stress-dominance hypothesis. *Scientific Reports* 9(1), e15749.

[8.112] Ćirić, M., Gavrilović, B., Krizmanić, J., Dojčinović, B.P. and Vidaković, D. (2021) Can a benthic diatom community complement chemical analyses and discriminate between disturbed and undisturbed saline wetland habitats? A study of seven soda pans in Serbia. *Wetlands Ecology and Management* 29(3), 451–466.

[8.113] Lengyel, E., Padisák, J., Hajnal, É., Szabó, B., Pellinger, A. and Stenger-Kovács, C. (2016) Application of benthic diatoms to assess efficiency of conservation management: a case study on the example of three reconstructed soda pans, Hungary. *Hydrobiologia* 777(1), 95–110.

[8.114] Bona, F., La Morgia, V., Fenoglio, S., Morandi, L. and Falasco, E. (2021) Diatom communities and ecological status classification in the upper Po River basin. *Journal of Limnology* 80(3), e2025.

[8.115] Hlúbiková, D., Novais, M.H., Dohet, A., Hoffmann, L. and Ector, L. (2014) Effect of riparian vegetation on diatom assemblages in headwater streams under different land uses. *Science of the Total Environment* 475, 234–247.

[8.116] Stenger-Kovács, C., Lengyel, E., Sebestyén, V. and Szabó, B. (2020) Effects of land use on streams: traditional and functional analyses of benthic diatoms. *Hydrobiologia* 847, 2933–2946.

[8.117] Larras, F., Bouchez, A., Rimet, F. and Montuelle, B. (2012) Using Bioassays and Species Sensitivity Distributions to Assess Herbicide Toxicity towards Benthic Diatoms. *PLoS ONE* 7(8), e44458.

[8.118] Rimet, F. and Bouchez, A. (2011) Use of diatom life-forms and ecological guilds to assess pesticide contamination in rivers: Lotic mesocosm approaches. *Ecological Indicators* 11(2), 489–499.

[8.119] Winkworth, C.L., Salis, R.K. and Matthaei, C.D. (2015) Interactive multiple-stressor effects of the antibiotic monensin, cattle effluent and light on stream periphyton. *Freshwater Biology* 60(11), 2410–2423.

[8.120] Dorigo, U., Bourrain, X., Bérard, A. and Leboulanger, C. (2004) Seasonal changes in the sensitivity of river microalgae to atrazine and isoproturon along a contamination gradient. *Science of The Total Environment* 318(1–3), 101–114.

[8.121] Pérès, F., Florin, D., Grollier, T., Feurtet-Mazel, A., Coste, M., Ribeyre, F., Ricard, M. and Boudou, A. (1996) Effects of the phenylurea herbicide isoproturon on periphytic diatom communities in freshwater indoor microcosms. *Environmental Pollution* 94(2), 141–152.

[8.122] Belando, M.D., Marín, A., Aboal, M., García-Fernández, A.J. and Marín-Guirao, L. (2017) Combined in situ effects of metals and nutrients on marine biofilms: Shifts in the diatom assemblage structure and biological traits. *Science of The Total Environment* 574, 381–389.

[8.123] Riato, L., Leira, M., Della Bella, V. and Oberholster, P.J. (2018) Development of a diatom-based multimetric index for acid mine drainage impacted depressional wetlands. *Science of The Total Environment* 612, 214–222.

[8.124] Heine-Fuster, I., López-Allendes, C., Aránguiz-Acuña, A. and Véliz, D. (2021) Differentiation of Diatom Guilds in Extreme Environments in the Andean Altiplano. *Frontiers in Environmental Science* 9, e701970.

[8.125] Porter-Goff, E.R., Frost, P.C. and Xenopoulos, M.A. (2013) Changes in riverine benthic diatom community structure along a chloride gradient. *Ecological Indicators* 32, 97–106.

[8.126] Reid, A.J., Carlson, A.K., Creed, I.F., Eliason, E.J., Gell, P.A., Johnson, P.T.J., Kidd, K.A., MacCormack, T.J., Olden, J.D., Ormerod, S.J., Smol, J.P., Taylor, W.W., Tockner, K., Vermaire, J.C., Dudgeon, D. and Cooke, S.J. (2019) Emerging threats and persistent conservation challenges for freshwater biodiversity. *Biological Reviews* 94(3), 849–873.

[8.127] Cañedo-Argüelles, M. (2020) A review of recent advances and future challenges in freshwater salinization. *Limnetica* 39, 185–211.

[8.128] Pandey, L.K., Bergey, E.A., Lyu, J., Park, J., Choi, S., Lee, H., Depuydt, S., Oh, Y.-T., Lee, S.-M. and Han, T. (2017) The use of diatoms in ecotoxicology and bioassessment: Insights, advances and challenges. *Water Research* 118, 39–58.

[8.129] Messager, M.L., Lehner, B., Cockburn, C., Lamouroux, N., Pella, H., Snelder, T., Tockner, K., Trautmann, T., Watt, C. and Datry, T. (2021) Global prevalence of non-perennial rivers and streams. *Nature* 594(7863), 391–397.

[8.130] Soria, M., Leigh, C., Datry, T., Bini, L.M. and Bonada, N. (2017) Biodiversity in perennial and intermittent rivers: a meta-analysis. *Oikos* 126(8), 1078–1089.

[8.131] Stubbington, R., England, J., Wood, P.J. and Sefton, C.E.M. (2017) Temporary streams in temperate zones: recognizing, monitoring and restoring transitional aquatic-terrestrial ecosystems. *WIREs Water* 4(4), e1223.

[8.132] Falasco, E., Piano, E. and Bona, F. (2016) Diatom flora in Mediterranean streams: flow intermittency threatens endangered species. *Biodiversity and Conservation* 25(14), 2965–2986.

[8.133] Novais, M.H., Morais, M.M., Rosado, J., Dias, L.S., Hoffmann, L. and Ector, L. (2014) Diatoms Of Temporary And Permanent Watercourses In Southern Europe (Portugal). *River Research and Applications* 30(10), 1216–1232.

[8.134] Peszek, Ł., Zgrundo, A., Noga, T., Kochman-Kędziora, N., Poradowska, A., Rybak, M., Puchalski, C. and Lee, J. (2021) The influence of drought on diatom assemblages in a temperate climate zone: A case study from the Carpathian Mountains, Poland. *Ecological Indicators* 125, e107579.

[8.135] B-Béres, V., Tóthmérész, B., Bácsi, I., Borics, G., Abonyi, A., Tapolczai, K., Rimet, F., Bouchez, Á., Várbíró, G. and Török, P. (2019) Autumn drought drives functional diversity of benthic diatom assemblages of continental intermittent streams. *Advances in Water Resources* 126, 129–136.

[8.136] B-Béres, V., Kókai, Z., Várbíró, G., Mustazhapova, G., Csabai, Z., Pernecker, B., Borics, G., Bácsi, I. and Boda, P. (2022) Flow Intermittence Drives the Benthic Algal Composition, Biodiversity and Diatom-Based Quality of Small Hilly Streams in the Pannonian Ecoregion, Hungary. *Frontiers in Ecology and Evolution* 10, e834548.

[8.137] Cantonati, M., Kelly, M.G., Demartini, D., Angeli, N., Dörflinger, G., Papatheodoulou, A. and Armanini, D.G. (2020) Overwhelming role of hydrology-related variables and river types in driving diatom species distribution and community assemblage in streams in Cyprus. *Ecological Indicators* 117, e106690.

[8.138] Falasco, E., Bona, F., Risso, A.M. and Piano, E. (2021) Hydrological intermittency drives diversity decline and functional homogenization in benthic diatom communities. *Science of The Total Environment* 762, e143090.

[8.139] Novais, M.H., Morales, E.A., Penha, A.M., Potes, M., Bouchez, A., Barthès, A., Costa, M.J., Salgado, R., Santos, J. and Morais, M. (2020) Benthic diatom community dynamics in Mediterranean intermittent streams: Effects of water availability and their potential as indicators of dry-phase ecological status. *Science of The Total Environment* 719, e137462.

[8.140] Sabater, S., Timoner, X., Bornette, G., De Wilde, M., Stromberg, J.C. and Stella, J.C. (2017) Chapter 4.2 - The Biota of Intermittent Rivers and Ephemeral Streams: Algae and Vascular Plants, in *Intermittent Rivers and Ephemeral Streams,* Datry, T., Bonada, N., and Boulton, A.(eds.), Academic Press, pp. 189–216.

[8.141] McKew, B.A., Taylor, J.D., McGenity, T.J. and Underwood, G.J.C. (2011) Resistance and resilience of benthic biofilm communities from a temperate saltmarsh to desiccation and rewetting. *The ISME Journal* 5(1), 30–41.

[8.142] Abonyi, A., Leitão, M., Lançon, A.M. and Padisák, J. (2012) Phytoplankton functional groups as indicators of human impacts along the River Loire (France). *Hydrobiologia* 698(1), 233–249.

[8.143] Bolgovics, Á., Várbíró, G., Ács, É., Trábert, Z., Kiss, K.T., Pozderka, V., Görgényi, J., Boda, P., Lukács, B.-A., Nagy-László, Z., Abonyi, A. and Borics, G. (2017) Phytoplankton of rhithral rivers: Its origin, diversity and possible use for quality-assessment. *Ecological Indicators* 81, 587–596.

[8.144] Zelnik, I., Balanč, T. and Toman, M. (2018) Diversity and Structure of the Tychoplankton Diatom Community in the Limnocrene Spring Zelenci (Slovenia) in Relation to Environmental Factors. *Water* 10(4), e361.

[8.145] Abonyi, A., Ács, É., Hidas, A., Grigorszky, I., Várbíró, G., Borics, G. and Kiss, K.T. (2018) Functional diversity of phytoplankton highlights long-term gradual regime shift in the middle section of the Danube River due to global warming, human impacts and oligotrophication. *Freshwater Biology* 63(5), 456–472.

[8.146] Wang, C., B-Béres, V., Stenger-Kovács, C., Li, X. and Abonyi, A. (2018) Enhanced ecological indication based on combined planktic and benthic functional approaches in large river phytoplankton ecology. *Hydrobiologia* 818(1), 163–175.

[8.147] Lindholm, M., Grönroos, M., Hjort, J., Karjalainen, S.M., Tokola, L. and Heino, J. (2018) Different species trait groups of stream diatoms show divergent responses to spatial and environmental factors in a subarctic drainage basin. *Hydrobiologia* 816(1), 213–230.

[8.148] Benito, X., Fritz, S.C., Steinitz-Kannan, M., Vélez, M.I. and McGlue, M.M. (2018) Lake regionalization and diatom metacommunity structuring in tropical South America. *Ecology and Evolution* 8(16), 7865–7878.

[8.149] Grönroos, M., Heino, J., Siqueira, T., Landeiro, V.L., Kotanen, J. and Bini, L.M. (2013) Metacommunity structuring in stream networks: roles of dispersal mode, distance type, and regional environmental context. *Ecology and Evolution* 3(13), 4473–4487.

[8.150] Dong, X., Li, B., He, F., Gu, Y., Sun, M., Zhang, H., Tan, L., Xiao, W., Liu, S. and Cai, Q. (2016) Flow directionality, mountain barriers and functional traits determine diatom metacommunity structuring of high mountain streams. *Scientific Reports* 6(1), 1–11.

[8.151] Kristiansen, J. (1996) 16. Dispersal of freshwater algae — a review. *Hydrobiologia* 336(1–3), 151–157.

[8.152] Leibold, M.A., Holyoak, M., Mouquet, N., Amarasekare, P., Chase, J.M., Hoopes, M.F., Holt, R.D., Shurin, J.B., Law, R., Tilman, D., Loreau, M. and Gonzalez, A. (2004) The metacommunity concept: a framework for multi-scale community ecology: The metacommunity concept. *Ecology Letters* 7(7), 601–613.

[8.153] Vellend, M. (2010) Conceptual Synthesis in Community Ecology. *The Quarterly Review of Biology* 85(2), 183–206.

[8.154] Weiher, E. and Keddy, P. (eds.) (2001) *Ecological assembly rules: perspectives, advances, retreats*, Cambridge Univ. Press, Cambridge, 2nd.

[8.155] Zorzal-Almeida, S., Soininen, J., Bini, L.M. and Bicudo, D.C. (2017) Local environment and connectivity are the main drivers of diatom species composition and trait variation in a set of tropical reservoirs. *Freshwater Biology* 62(9), 1551–1563.

[8.156] Liu, J., Soininen, J., Han, B.-P. and Declerck, S.A.J. (2013) Effects of connectivity, dispersal directionality and functional traits on the metacommunity structure of river benthic diatoms. *Journal of Biogeography* 40(12), 2238–2248.

[8.157] Pandit, S.N., Kolasa, J. and Cottenie, K. (2009) Contrasts between habitat generalists and specialists: an empirical extension to the basic metacommunity framework. *Ecology* 90(8), 2253–2262.

[8.158] Tonkin, J.D., Altermatt, F., Finn, D.S., Heino, J., Olden, J.D., Pauls, S.U. and Lytle, D.A. (2018) The role of dispersal in river network metacommunities: Patterns, processes, and pathways. *Freshwater Biology* 63(1), 141–163.

[8.159] Marcarelli, A.M. and Wurtsbaugh, W.A. (2006) Temperature and nutrient supply interact to control nitrogen fixation in oligotrophic streams: an experimental examination. *Limnology and Oceanography* 51(5), 2278–2289.

[8.160] Cornell, H.V. and Lawton, J.H. (1992) Species Interactions, Local and Regional Processes, and Limits to the Richness of Ecological Communities: A Theoretical Perspective. *The Journal of Animal Ecology* 61(1), 1–12.

[8.161] Cantonati, M., Zorza, R., Bertoli, M., Pastorino, P., Salvi, G., Platania, G., Prearo, M. and Pizzul, E. (2021) Recent and subfossil diatom assemblages as indicators of environmental change (including fish introduction) in a high-mountain lake. *Ecological Indicators* 125, e107603.

[8.162] Wang, Y., Wu, N., Tang, T., Wang, Y. and Cai, Q. (2022) Small run-of-river hydropower dams and associated water regulation filter benthic diatom traits and affect functional diversity. *Science of The Total Environment* 813, e152566.

[8.163] Barlow, J., Lennox, G.D., Ferreira, J., Berenguer, E., Lees, A.C., Nally, R.M., Thomson, J.R., Ferraz, S.F. de B., Louzada, J. and Oliveira, V.H.F. (2016) Anthropogenic disturbance in tropical forests can double biodiversity loss from deforestation. *Nature* 535(7610), 144–147.

[8.164] González-Trujillo, J.D., Petsch, D.K., Córdoba-Ariza, G., Rincón-Palau, K., Donato-Rondon, J.C., Castro-Rebolledo, M.I. and Sabater, S. (2019) Upstream refugia and dispersal ability may override benthic-community responses to high-Andean streams deforestation. *Biodiversity and Conservation* 28(6), 1513–1531.

[8.165] Vilmi, A., Karjalainen, S.M., Wang, J. and Heino, J. (2019) Using traits to explain interspecific variation in diatom occupancy and abundance across lakes and streams. *Journal of Biogeography* 46(7), 1419–1428.

[8.166] Vilar, A.G., van Dam, H., Vonk, J.A., van der Geest, H.G. and Admiraal, W. (2014) Typology of diatom communities in the Dutch delta: Recognizing patterns of environmental drivers in nutrient rich ditches. *Ecological indicators* 45, 561–569.

[8.167] Moser, K.A., MacDonald, G.M. and Smol, J.P. (1996) Applications of freshwater diatoms to geographical research. *Progress in Physical Geography: Earth and Environment* 20(1), 21–52.

[8.168] Hubay, K., Braun, M., Buczkó, K., Pál, I., Veres, D., Túri, M., Biró, T. and Magyari, E. (2018) Holocene environmental changes as recorded in the geochemistry of glacial lake sediments from Retezat Mountains, South Carpathians. *Quaternary International* 477, 19–39.

[8.169] Pla-Rabés, S. and Catalan, J. (2018) Diatom species variation between lake habitats: implications for interpretation of paleolimnological records. *Journal of Paleolimnology* 60(2), 169–187.
[8.170] Ruwer, D.T., Bernardes, M.C. and Rodrigues, L. (2018) Diatom responses to environmental changes in the Upper Paraná River floodplain (Brazil) during the last ~ 1000 years. *Journal of Paleolimnology* 60(4), 543–551.
[8.171] Jenkins, D.G. and Ricklefs, R.E. (2011) Biogeography and ecology: two views of one world. *Philosophical Transactions of the Royal Society B: Biological Sciences* 366(1576), 2331–2335.
[8.172] Lengyel, E., Szabó, B. and Stenger-Kovács, C. (2020) Realized ecological niche-based occupancy–abundance patterns of benthic diatom traits. *Hydrobiologia* 847(14), 3115–3127.
[8.173] Heino, J. and Soininen, J. (2006) Regional occupancy in unicellular eukaryotes: a reflection of niche breadth, habitat availability or size-related dispersal capacity? *Freshwater Biology* 51(4), 672–685.
[8.174] Passy, S.I. (2016) Abundance inequality in freshwater communities has an ecological origin. *The American Naturalist* 187(4), 502–516.
[8.175] Soininen, J., Jamoneau, A., Rosebery, J., Leboucher, T., Wang, J., Kokociński, M. and Passy, S.I. (2019) Stream diatoms exhibit weak niche conservation along global environmental and climatic gradients. *Ecography* 42(2), 346–353.
[8.176] González-Trujillo, J.D., Pedraza-Garzón, E., Donato-Rondon, J.C. and Sabater, S. (2020) Ecoregional characteristics drive the distribution patterns of neotropical stream diatoms. *Journal of Phycology* 56(4), 1053–1065.
[8.177] Virtanen, L.K., Köngäs, P., Aitto-Oja, S. and Soininen, J. (2011) Is temporal occurrence of diatoms related to species traits, local abundance, and regional distribution? 1. *Journal of Phycology* 47(6), 1445–1453.
[8.178] Rimet, F., Feret, L., Bouchez, A., Dorioz, J.-M. and Dambrine, E. (2019) Factors influencing the heterogeneity of benthic diatom communities along the shoreline of natural alpine lakes. *Hydrobiologia* 839(1), 103–118.
[8.179] Bolgovics, Á., Ács, É., Várbíró, G., Görgényi, J. and Borics, G. (2016) Species area relationship (SAR) for benthic diatoms: a study on aquatic islands. *Hydrobiologia* 764(1), 91–102.
[8.180] Minaudo, C., Meybeck, M., Moatar, F., Gassama, N. and Curie, F. (2015) Eutrophication mitigation in rivers: 30 years of trends in spatial and seasonal patterns of biogeochemistry of the Loire River (1980–2012). *Biogeosciences* 12(8), 2549–2563.
[8.181] Verbeek, L., Gall, A., Hillebrand, H. and Striebel, M. (2018) Warming and oligotrophication cause shifts in freshwater phytoplankton communities. *Global Change Biology* 24(10), 4532–4543.
[8.182] Tison-Rosebery, J., Leboucher, T., Archaimbault, V., Belliard, J., Carayon, D., Ferréol, M., Floury, M., Jeliazkov, A., Tales, E. and Villeneuve, B.

(2022) Decadal biodiversity trends in rivers reveal recent community rearrangements. *Science of the Total Environment* 823, e153431.
[8.183] Falkowski, P.G., Katz, M.E., Knoll, A.H., Quigg, A., Raven, J.A., Schofield, O. and Taylor, F.J.R. (2004) The Evolution of Modern Eukaryotic Phytoplankton. *Science* 305(5682), 354–360.
[8.184] Besse-Lototskaya, A., Verdonschot, P.F.M., Coste, M. and Van de Vijver, B. (2011) Evaluation of European diatom trophic indices. *Ecological Indicators* 11(2), 456–467.

9

Soil Diatoms and Their Use in Bioindication

Jasper Foets[1]* and Eveline Pinseel[2,3]

[1]*Mol, Belgium*
[2]*Department of Biological Sciences, University of Arkansas, Fayetteville, Arkansas, USA*
[3]*Laboratory of Protistology & Aquatic Ecology, Department of Biology, Ghent University, Ghent, Belgium*

Abstract
Soils are complex, multifunctional ecosystems. Safeguarding soil quality is therefore critical for ensuring the sustainability and biodiversity of terrestrial ecosystems worldwide. For this reason, biological indicators of soil quality are considered useful tools for the monitoring of soil health. Within this context, soil diatoms have been tested as potential bioindicators since they are a key component of the soil protist community. They are also studied as a potential tracer for water flow path connectivity and surface runoff processes and as a potential source of information in forensic investigations. The potential of soil diatoms as bioindicators is clear, but there are still several hurdles to overcome before they can be applied on a routine basis.

Development of DNA metabarcoding for the characterization of soil diatom communities is seen as a crucial next step for using diatoms as bioindicators of soil quality. This approach can improve diatom identification and species delimitation, especially when combined with molecular phylogenetics. For example, molecular techniques have already revealed high levels of (pseudo)cryptic species-level diversity in soil diatom species complexes surrounding *Pinnularia borealis* and *Hantzschia amphioxys*. However, the use of metabarcoding for soil diatom biomonitoring is still in its infancy.

Keywords: Terrestrial diatom, environmental marker, hydrological tracer, soil algae

**Corresponding author*: jasper.foets1@gmail.com
Jasper Foets: https://www.researchgate.net/profile/Jasper-Foets
Eveline Pinseel: https://evelinepinseel.weebly.com; https://www.researchgate.net/profile/Eveline-Pinseel; https://scholar.google.com/citations?user=5miI2aQAAAAJ&hl=nl&oi=ao

Nora I. Maidana, Magdalena Licursi and Eduardo Morales (eds.) Diatom Ecology: From Molecules to Metacommunities, (281–328) © 2024 Scrivener Publishing LLC

9.1 Introduction

Diatoms are generally regarded as inhabitants of water bodies, both freshwater and marine. However, numerous taxa can colonize a variety of non-aquatic habitats, including soils, mosses, and rocks. The first of these terrestrial diatoms were already described in the 19th century: *Pinnularia borealis* Ehrenb. and *Hantzschia amphioxys* Grunow [9.22]. Also in the 19th century, Graebner [9.38] was the first researcher to compile an algae species list that included diatom taxa occurring on soils, namely *H. amphioxys* and *Achnanthes coarctata* Bréb. Despite this long-term awareness that diatoms inhabit terrestrial environments, there is little ecological information available on terrestrial diatoms. This stands in stark contrast with the vast array of knowledge that has been gathered on freshwater and marine diatoms. Besides the large difference in the total number of studies on aquatic and terrestrial diatoms, there is also a noticeable difference in the covered research topics. Specifically, whereas research on freshwater diatoms comprised a wide range of different fields (e.g., taxonomy, (paleo-)ecology, physiology...), the work on terrestrial diatoms has until recently been strongly restricted to floristics (see Smol and Stoermer [9.94] and references therein), which generally provides little information on habitat preferences and ecology. However, the last decade saw a substantial increase in the number of publications on terrestrial diatoms, a higher proportion of which focuses on their ecological aspects and potential applications (Figure 9.1).

This chapter will focus on soil diatoms, which are terrestrial diatoms that live or grow on soils moistened solely by direct atmospheric precipitation. Typical soil diatom taxa are amongst others *H. amphioxys*, *Humidophila contenta* (Grunow) R.L.Lowe *et al.*, *P. borealis*, *Pinnularia neorabenhorstii* Gogorev, *Luticola ventricosa* (Kütz.) D.G.Mann, *Mayamaea atomus* (Kütz.) Lange-Bert., and *Mayamaea permitis* (Hust.) K.Bruder & L.K.Medlin. The occurrence of these species, and soil diatoms in general, is not restricted to soils alone. Apart from other terrestrial environments such as rock surfaces and terrestrial mosses, most soil diatoms are also found in freshwater environments, although their relative abundances in these habitats is often significantly smaller compared to soils. Exceptions to this rule are, amongst others, *M. permitis* and *M. atomus*, since both species can reach high numbers in freshwater and terrestrial environments. In contrast, many freshwater diatoms cannot survive in soils. Ecological characteristics of soil diatoms have

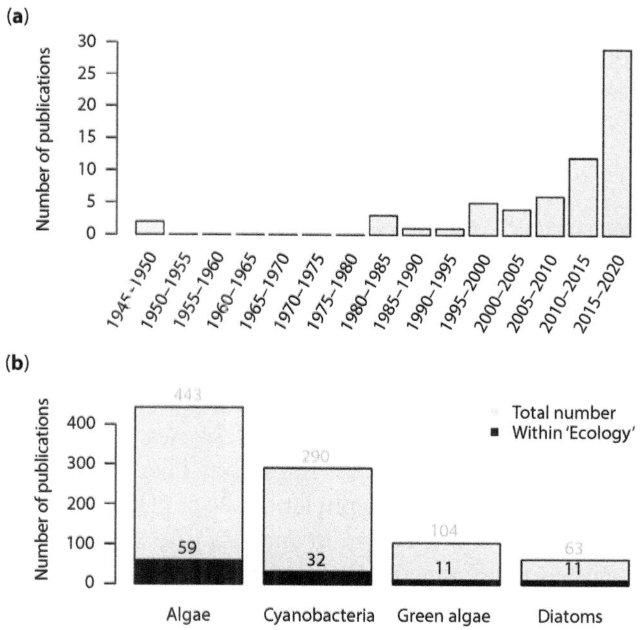

Figure 9.1 (a) Number of publications on terrestrial diatoms between 1945 and 2020. (b) Algae, cyanobacteria, and green algae. In black, the number of publications assigned to the Web of Science category 'Ecology'. The search queries used for retrieving data were constructed as follows: Title = (algae), (green algae OR Chlorophyta OR Chlorophyceae), (cyanobacteria OR blue-green algae), (diatom OR Bacillariophyta OR Bacillariophyceae) for algae, green algae, cyanobacteria and diatoms respectively, each time in combination with a title = (soil OR terrestrial OR aerial OR subaerial). Afterwards, the articles were checked for their correspondence with our search query. It is important to note that papers written in another language than English were not included in Figure 9.1a.

been described in several publications [9.57] [9.111]. Of these, the early research of Lund [9.57] is still an important guide for both soil diatom taxonomy and ecology in contemporary research.

Compared to many other biotas, diatoms are more specific in their preferences of, and tolerances to, environmental conditions [9.62]. They are, therefore, widely regarded as excellent bioindicators of the ecological status of water bodies, including rivers, lakes, and wetlands. Like freshwater diatoms, it is believed that soil diatoms could be useful bioindicators for soil quality. Although extensive research is needed before soil diatoms can be launched as effective bioindicators, current research shows that there is potential in several applications. In this chapter, we explain how and where soil diatoms can be used as bioindicators, and describe the current methods and future perspectives for soil diatom sampling and analyses for bioindication. In addition, we give an overview of the taxonomical and ecological aspects of soil diatom communities, highlighting the environmental variables that affect soil diatom distributions and adaptations.

9.2 Taxonomy

In 1995, Ettl and Gärtner [9.24] mentioned the presence of 1291 different diatom taxa in Central Europe, whereof 224 (17%) were regarded as terrestrial forms, and four years later 340 diatom taxa were reported from soils by Johansen [9.47]. Since then, this number substantially increased mainly due to three reasons [9.47]. First, many new genera were erected or resurrected to accommodate the numerous taxa belonging to genera which are now believed to be polyphyletic (i.e., containing species that are derived from more than one common ancestor). Second, species concepts in diatoms shifted from a coarse-grained taxonomy ('lumpers') to a fine-grained taxonomy ('splitters'). Today, species are much more narrowly defined, leading to the fragmentation of broadly defined taxa into several species (e.g., [9.122]). Third, there has been an increased interest in investigating soil diatoms [9.17] [9.27] [9.29]. This was facilitated by the design and application of novel or improved approaches for the taxonomic characterization of diatoms. For example, the increased resolution of scanning electron microscopes (SEM), development of morphometric techniques, and increasingly common use of molecular phylogenetics have helped diatom researchers significantly with the discovery and description of new genera and taxa [9.51] [9.54].

Whenever soils are studied in detail, they have proven to be a source of new and interesting diatom species. Several publications show the wealth of diatom diversity encountered on different soils, especially in Antarctica. For example, Van de Vijver *et al.* [9.117] described six new species of the genus *Muelleria* (Freng.) Freng. from Antarctic soils, whereas Zidarova *et al.* [9.128] discovered five new taxa belonging to the genus *Hantzschia* Grunow from Livingston Island (Antarctica). Van de Vijver and Le Cohu [9.113] found two new *Geissleria* Lange-Bert. & Metzeltin species on dry soils from the Crozet Archipelago, while *Fallacia emmae* Van de Vijver & E.J.Cox was described from the Sub-Antarctic soils by Van de Vijver and Cox [9.112]. New soil diatoms species have also been discovered in regions outside Antarctica. For example, three soil-inhabiting species were described from South Africa [9.104], and Stanek-Tarkowska *et al.* [9.101] found two new *Microcostatus* J.R.Johans. & Sray species in Central

European soils, a genus previously circumscribed as typically terrestrial [9.48]. In addition, several new taxa and genera from soils have been described from around the world by various authors [9.4] [9.37] [9.101] [9.114] [9.116] [9.123] [9.124]. For illustration, some common soil diatoms are shown in Figure 9.2 and Figure 9.3. From the above studies, it became clear that soil diatoms typically are raphid pennates, with most of the species being biraphid. However, some centric diatoms have adapted to a terrestrial lifestyle, *Orthoseira* Thwaites for instance [9.33]. Even though it has previously been reported that large diatoms of more than 50 μm are rare in soils, with the size of species generally ranging between 10 and 40 μm [9.57], this could also be a sampling artifact given that large diatoms are easily overlooked due to their generally low abundances [9.77] [9.83]. For example, by using culture approaches Pinseel *et al.* [9.75] commonly found large-celled *Pinnularia* Ehrenb. species in soils around the world (Figure 9.3). The most species-rich diatom genera in soils are generally *Pinnularia*, *Navicula* Bory, *Nitzschia* Hassall, *Mayamaea* Lange-Bert., *Stauroneis* Ehrenb., and *Sellaphora* Mereschk. Many of the common soil diatoms shown in Figure 9.2 also occur in freshwater ecosystems, and there generally are no morphological differences between conspecific freshwater and terrestrial populations.

The contribution of molecular techniques to the taxonomy and systematics of diatoms has greatly increased in the last decade. Molecular techniques allow us to unravel whether common cosmopolitan diatom taxa, which were described based on the morphology of their frustules, are indeed one species or, alternatively, a species complex (i.e., a group of closely related species with highly similar morphologies). The two most notable diatom species complexes that inhabit soils are *P. borealis* and *H. amphioxys*. Souffreau *et al.* [9.96] investigated several strains of both complexes from polar and temperate environments, and found that they both consist of multiple lineages, which are likely to be different species. This included distinct Antarctic lineages which have diverged in their temperature tolerance from strains of temperate regions. However, most of these lineages did not express clearly distinct valve morphologies and correspond only partially to identifications based on valve morphology alone. In fact, application of morphometric techniques on phylogenetically distinct lineages in the *P. borealis* complex revealed that, although these lineages show subtle morphological differences, high levels of intraspecific morphological variation preclude their conclusive identification using light microscopy (LM) and SEM techniques, even when morphometric analyses are used [9.76] (Figure 9.3). This study also pointed out that previously described varieties of *P. borealis*, such as *P. borealis* var. *scalaris* A.Mayer,

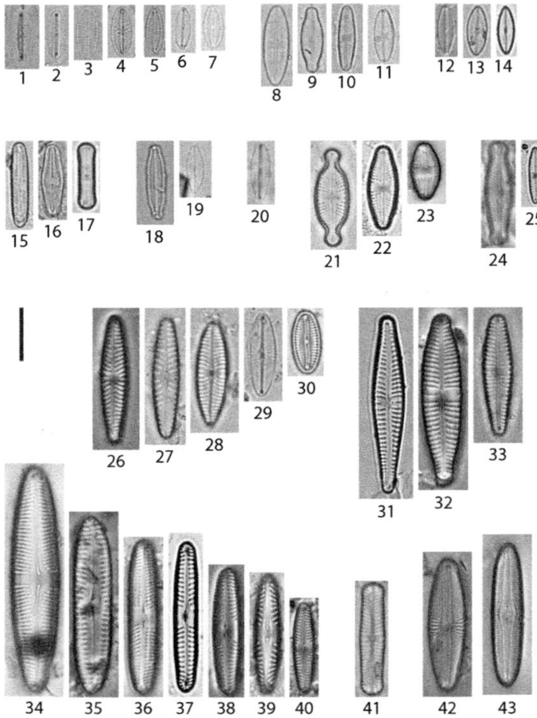

Figure 9.2 Light microscopy images of some common soil diatoms. 1, *Mayamaea excelsa* (Krasske) Lange-Bert.; 2, *Mayamaea permitis* (Hust.) Bruder & L.K.Medlin; 3, *Mayamaea fossalis* (Krasske) Lange-Bert.; 4, *Mayamaea atomus* (Kütz.) Lange-Bert.; 5, *Mayamaea agrestis* (Hust.) Lange-Bert.; 6, *Mayamaea lacunolaciniata* (Lange-Bert. & K.Bonik) Lange-Bert.; 7, *Mayamaea aff. arida* (Bock) Lange-Bert.; 8, *Sellaphora nana* (Hust.) Lange-Bert. *et al.*; 9, *Sellaphora lundii* C.E.Wetzel *et al.*; 10, *Sellaphora atomoides* (Grunow) C.E.Wetzel & Van de Vijver; 11, *Sellaphora subseminulum* (Hust.) C.E.Wetzel; 12, *Microcostatus* sp.1; 13, *Microcostatus* sp.2; 14, *Microcostatus aerophilus* Stanek-Tarkowska *et al.*; 15, *Humidophila irata* (Grunow) R.L.Lowe *et al.*; 16, *Humidophila perpussila* (Grunow) R.L.Lowe *et al.*; 17, *Humidophila contenta* (Grunow) R.L.Lowe *et al.*; 18, *Adlafia bryophila* (J.B.Petersen) Gerd Moser *et al.*; 19, *Adlafia minuscula* var. *muralis* (Grunow) Lange-Bert.; 20, *Craticula minuscoloides* (Hust.) Lange-Bert.; 21, *Luticola ventricosa* (Kütz.) D.G.Mann; 22, *Luticola imbricata* (Bock) Levkov *et al.*; 23, *Luticola acidoclinata* Lange-Bert.; 24, *Chamaepinnularia krookiformis* (Krammer) Lange-Bert. & Krammer; 25, *Chamaepinnularia obsoleta* (Hust.) C.E.Wetzel & Ector; 26, *Navicula microcari* Lange-Bert.; 27, *Navicula cincta* var. *heufleri* f. *curta* J.B.Petersen; 28, *Navicula recens* (Lange-Bert.) Lange-Bert.; 29, *Navicula vaucheriae* var. *densistriata* J.W.G.Lund; 30, *Navicula fritschii* J.W.G.Lund; 31, *Gomphonema* cf. *angustatum* (Kütz.) Rabenh.; 32, *Gomphonema micropus* Kütz.; 33, *Gomphonema minutum* f. *minutum* (C.Agardh) C.Agardh; 34, *Pinnularia microstauron* (Ehrenb.) Cleve; 35, *Pinnularia subrupestris* Krammer; 36, *Pinnularia schoenfelderii* Krammer; 37, *Pinnularia subcapitata* f. *typica* J.B.Petersen; 38, *Pinnularia microstauron* var. *angusta* Krammer; 39, *Pinnularia obscura* Krasske; 40, *Pinnularia perirrorata* Krammer; 41, *Navicula ignota* var. *palustris* (Hust.) J.W.G.Lund; 42, *Muelleria gibbula* (Cleve) S.A.Spauld. & Stoermer; 43, *Muelleria terrestris* (J.B.Petersen) S.A.Spauld. & Stoermer. The scale bar is 10 μm.

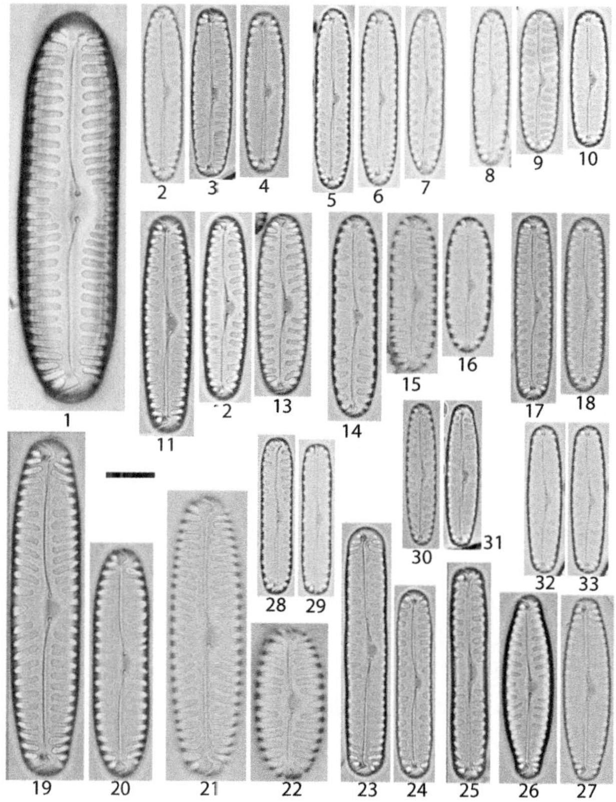

Figure 9.3 Light microscopy images of *Pinnularia* section Distantes, a clade of diatoms that commonly inhabits soils worldwide. The plate includes (cryptic) taxa belonging to the *P. borealis* species complex (2-18, 23-33) and the *P. neorabenhorstii* species complex (19-22), as well as other large-celled Distantes taxa (1). All figures represent cultures obtained by Pinseel *et al.* [9.75], and were collected from terrestrial environments (soils and mosses). All taxa on this plate have been flagged as different species using automated molecular species delimitation methods and molecular phylogenetic inference [9.75]. Most species in the complex are not yet formally described, and therefore, taxon names in this figure header correspond with the taxon identifiers from Pinseel *et al.* [9.75]. If multiple micrographs per taxon are shown, they were obtained from different cultures. The localities from which the cultures were obtained are indicated in the figure header. 1, lineage NY15_22c_02a from Svalbard; 2-4, lineage JRI15_10_06 from Chile and Antarctica; 5-7, lineage BE14_E03_03 from Belgium; 8-10, lineage JRI15_18b_09 from Antarctica; 11-13, lineage JRI15_32_13 from Antarctica; 14-16, lineage TAS17_41_03 from Tasmania; 17-18, lineage LU15_04_08 from Switzerland and Belgium; 19-20, lineage ED15_E33_03 from Svalbard; 21-22, lineage MI16_20b_08 from sub-Antarctica; 23-24, lineage MA15_11_04 from Mallorca; 25, lineage MAU16_03_03 from Mauritius; 26-27, lineage JRI15_50b_10 from Antarctica; 28-29, lineage IT03_15 from Italy; 30-31, lineage LU15_09_07 from Switzerland and Italy; 32-33, lineage OT16_01_11 from Canada. The scale bar is 10 μm.

var. *sublinearis* Krammer, var. *islandica* Krammer, and var. *subislandica* Krammer, which are based on subtle, often overlapping differences in valve dimensions and shape, might not correspond to biological meaningful units (i.e., taxa). Similarly, phylogenetically distinct lineages in *H. amphioxys* show subtle morphological differences [9.59]. Detailed LM and SEM analyses have already led to the description of seven new *Hantzschia* species from 25 strains formerly assigned to "*Hantzschia amphioxys* sensu lato" [9.59]. As molecular techniques are becoming more routinely used, it will just be a matter of time before new species complexes, and yet unrecognized diversity within known complexes, will be discovered and unraveled.

9.2.1 Methodological Protocols

Morphological identification of diatoms hinges on detailed observations of their silicified cell wall with a light microscope. This is generally performed after removal of the protoplasm (i.e., cell content) with, for example, hydrogen peroxide. Contrary to aquatic diatom samples, soil samples generally contain much more inorganic material (e.g., silt and clay) and other acid-unsolvable particles. This results in microscopy slides that are less clean [9.40] [9.55]. As those particles obscure diatom valves, their presence makes identifications more difficult and limits concentrating sample material on the slides. Together with the generally lower diatom abundances in terrestrial habitats relative to aquatic ones, this means that analysis of microscopy slides is often more time-consuming, and that it is often difficult to reach the generally accepted criterion of 400 to 500 counted valves per sample [9.5]. Hence, researchers developed several sampling techniques and protocols to make the investigation of terrestrial diatoms more feasible. However, no standard sampling protocol is widely accepted so far. To date, four different kinds of methods have been used: (*i*) liquid culture, (*ii*) moist plate culture, (*iii*) buried slide technique, and (*iv*) direct microscopic examination after acid cleaning. Combining several of these methods, such as cultures and oxidized material, can be especially powerful to build a comprehensive idea of soil diatom diversity [9.57] [9.75] [9.76].

Liquid cultures were particularly common in the early days of terrestrial diatom research. Until the early 90s, researchers predominantly focused on characterizing the overall soil algal and cyanobacteria community, but not diatoms specifically [9.13] [9.31] [9.38] [9.49] [9.66]. Contrary to diatom identifications, the identifications of, for instance, green algae are based on their life cycle, which is most easily observed in culture. During liquid culturing, inorganic material is removed and the concentration of algae is increased, which benefits taxonomic identifications. It is also possible to count the number of algal cells and calculate the number of algae per unit of soil [9.50]. However, the main disadvantage of this method is that algae are exposed to artificial conditions, which may favor certain taxa. Hence, ecological studies are not possible. Additionally, many algae, particularly diatoms, have shown a poor growth rate on most culture media [9.7] [9.14] [9.55] [9.57] [9.89], further challenging the usefulness of liquid culturing to characterize soil diatom communities.

The moist plate method was used in research by Lund, John, Stanek-Tarkowska and Noga, and Heger *et al.* [9.41] [9.49] [9.57] [9.98]. The method is similar to liquid culturing, with the difference being that the soil is periodically moistened instead of using a nutrient mixture to stimulate the development of diatoms. The sample is generally kept in natural conditions, but sometimes also incubated in warmer conditions with a longer light-dark regime to further stimulate growth and reproduction. In addition, a glass slide is often placed on the soil surface during incubation to allow diatoms to attach and grow on it, enabling researchers to obtain clean diatom slides [9.45] [9.57] [9.58] [9.60] [9.61]. However, the addition of the slide assumes that all diatoms become equally attached, which, in practice, is not the case [9.78]. Nevertheless, the moist plate culture method achieves a better representation of the soil diatom community than liquid culturing, since the soil habitat is not completely altered, and the moist cultures are easier to set up, maintain, and study for a longer period. As a result, more species are observed in general [9.14] [9.125].

The buried slide or contact slide technique is another diatom sampling method which, until 1980, was only used for the examination of aquatic algae, soil bacteria, and soil fungi [9.78]. The procedure is simple. Specifically, a glass slide is implanted vertically into the soil, with a small part protruding above the soil surface. This permits light to pass downwards from the upper exposed and illuminated portion of the glass slide. This illumination encourages algal development on the slide surface that is in contact with the soil. After incubation, the slide is removed from the soil and can directly be examined under a microscope [9.77]. This technique is convenient for the direct counting and identification of soil algae under *in-situ* conditions. For instance, it has been used to study the motility of common diatom species, such as *H. amphioxys,* including the effect of chemical compounds or pesticides on cell movement [9.79]. Moreover, Elster *et al.* [9.23] and Evans *et al.* [9.25] used the buried slide method to examine the vertical distribution of algae in soils, and the micro-stratification of diatoms in dry pond margins, respectively. Despite the wide range of possibilities, there are two major drawbacks. First, the extra light that channels down into the lower parts of the soil also allows for algal development on the glass surface at a level that is lower than they naturally occur in soils (± 1 cm), possibly altering conclusions on the composition and behavior of natural soil diatom communities [9.23]. Second, the technique excludes less motile species and consequently does not reflect a fully representative community [9.78].

At last, researchers have used acid cleaning of soil samples for direct, detailed microscopy analyses, following the same procedures for removal of organic material that are widely used for investigating aquatic diatoms [9.118]. To this end, different sampling and preparation methods have been developed and applied to extract diatoms from soils, although these were not as successful in alleviating issues with acid-resistant soil particles as the previously discussed cultivation methods. Some researchers only collect the first 1-2 mm of soil surface, thus limiting the amount of soil material analyzed [9.57] [9.70], whereas others collect the topsoil layer (e.g., 0–4 cm), followed by mixing and acid cleaning [9.100]. Mechanical fractionation, where soil samples were sieved prior to examination, has also been used [9.91]. The most recently developed method for sampling and extracting diatoms from soils was described in Coles *et al.* [9.18] and adapted by Barragán *et al.* [9.5]. Those authors used metal rings, comparable to 'Kopecky' rings, to obtain an undisturbed soil layer. Subsequently, distilled or sparkling water was used to extract the diatoms from the topsoil layer, after which sample preparation followed the same (acid) cleaning procedures as is routine for aquatic samples. The communities that are collected and processed this way have been found to be representative of their surrounding area (>75 m^2) [9.5]. Several other techniques that have been developed for challenging diatom samples have not yet been applied to soil diatoms, as far as we are aware, but could be promising avenues to explore in the future. For instance, in paleo-ecological research, diatoms are extracted from sediment samples using heavy liquid separation [9.86].

Today, direct examination using microscopy slides (see previous paragraph) is preferred for investigating soil diatoms, since this method enables researchers to study diatom communities and morphology without the influence of other effects [9.50]. Using this method, the impact of environmental variables on the species composition of soil diatom communities can be investigated, enabling ecological studies. For instance, direct examination of microscopy slides showed that although overall soil diatom community composition was similar before and after a 14-day experiment where soils were moisturized, soil moisture did impact the abundance of some soil diatom species [9.98]. Another advantage of direct examination is that most diatom taxa are observed, thus resulting in the most complete diatom community compared to the other methods discussed previously

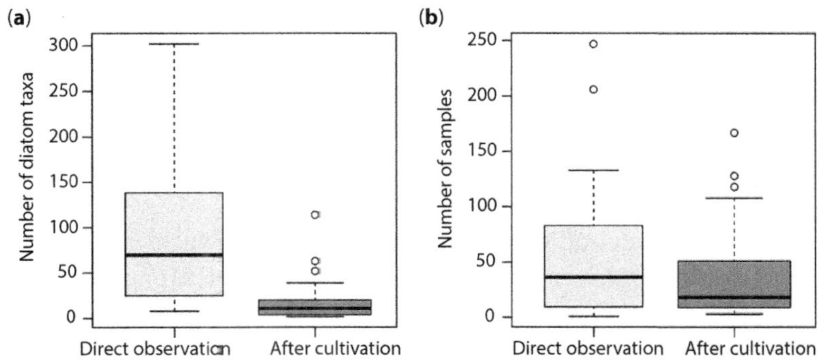

Figure 9.4 Comparison between diatom preparation techniques. Direct microscope observation of dead cells (n = 27) and diatom identification after culturing (liquid and moist) (n = 36). (a) Number of diatom taxa; (b) Number of samples analyzed. The list of publications used for this analysis can be found in the Appendix (Table 9.1).

[9.14] [9.90]. Indeed, when comparing a series of studies that used these different methods, more species are observed with direct observation compared to cultivation methods (i.e., liquid culturing, moist plate method, buried slide method) (two-sample Wilcoxon test, $P < 0.001$), despite no significant difference in the number of examined samples ($P > 0.05$) (Figure 9.4). This is at least partially due to the fact that researchers who used the cultivation methods generally focused on the overall algal community, and did not treat diatoms with much detail and taxonomic accuracy [9.31] [9.35] [9.36] [9.67] [9.105] [9.125]. Hence, many smaller and/or rare diatom taxa were overlooked. However, the main disadvantage of the direct observation technique is the presence of large amounts of mineral soil particles, since they may obscure the small number of diatom cells present in the slides [9.40]. As a result, microscopy analysis of soil diatom slides can be highly time-consuming [9.5]. Furthermore, it has to be noted that, similar to aquatic environments, lowly abundant large-celled species might be overlooked using direct examination of microscopy slides [9.82], but will be more easily detected using culturing methods [9.75].

9.3 Ecology

9.3.1 Adaptations to Non-Aquatic Conditions

Terrestrial environments present a much more adverse habitat for diatoms than aquatic environments. Physical variables such as moisture and temperature can vary significantly between consecutive days or even during the course of one day (e.g., day and night rhythms), especially in polar and temperate regions [9.15]. Consequently, soil diatoms can experience frequent and prolonged periods of desiccation or suboptimal temperatures, including prolonged heating or freezing of their habitats. Terrestrial diatom species have developed morphological and physiological mechanisms to cope with variable and extreme soil habitats.

Morphological adaptations retain raphe formation, valve size, and ultrastructure. For example, most terrestrial diatoms have a raphe, allowing for active moment away from adverse conditions [9.9] [9.40] [9.57], such as vertical migration through soils to avoid strong ultra-violet radiation or low moisture availability [9.97]. In addition, it has been suggested that alterations in cell size could provide an adaptation to moisture availability, allowing diatoms to retain and sequester moisture as much as possible, in addition to limiting moisture loss. For example, aquatic diatoms have been found to have significantly shorter cells than terrestrial ones, both on a community level [9.83], and for several species that inhabit both environments: *Sellaphora nana* (Hust.) Lange-Bert. *et al.*, *Stauroneis borrichii* (J.B.Petersen) J.W.G.Lund, and *Stauroneis thermicola* (J.B.Petersen) J.W.G.Lund [9.100]. However, statistical evidence showed that the valve lengths of e.g., *Humidophila contenta*, *Achnanthidium minutissimum* (Kütz.) Czarn. and the species complex *P. borealis,* significantly decreased with shrinking moisture levels [9.110]. How cell size aids in sequestering moisture is not yet known, but these examples demonstrate that a decrease or increase in cell length is a taxon-dependent feature, rather than a general response to lower moisture levels. It is possible that it infers adaptive advantages that are correlated with moisture availability, rather than an adaptation to moisture itself. There are several other examples of taxon- or genus-dependent responses which have been linked to moisture availability. For example, in drier environments, *Humidophila* (Lange-Bert. & Werum) R.L.Lowe *et al.* species reduce the number of areolae (i.e., openings in the cell wall) and/or occlude them with siliceous lamina. The latter is also observed in species of the genera *Nupela* W.Vyverman & P.Compère, *Chamaepinnularia* Lange-Bert. & Krammer, and *Eunotia* Ehrenb. In contrast, in *Hantzschia* spp. and *Melosira dickiei* (Thwaites) Kütz., internal valves are produced and the open area on the valve face is decreased [9.56] [9.83]. All these modifications are believed to aid in the retention of internal moisture levels as osmotic pressure increases.

Physiological adaptations to soil environments have been observed through laboratory experiments [9.95]. Specifically, soil diatoms show higher tolerances to desiccation and temperature extremes, such as heating and freezing, than their aquatic counterparts, and resting cell formation increases this tolerance [9.42] [9.95] [9.97]. The most extreme examples come from the terrestrial *P. borealis* complex, as well as the intertidal mudflat diatom *Opephora guenter-grassi* (Witkowski & Lange-Bert.) K.Sabbe & W.Vyverman, which can both survive temperatures as low as -180 °C [9.42] [9.103]. The genes and molecular pathways that enable these adaptations are unknown, but it is likely that they show similarities to adaptations evolved by other diatoms that inhabit extreme environments, such as the antifreeze proteins of the sea-ice diatom *Fragilariopsis cylindrus* (Grunow) Krieger [9.6]. Transcriptomics in combination with targeted candidate gene approaches, such as CRISPR/Cas9, is needed to unravel the genomic basis of soil adaptations in diatoms.

9.3.2 Functionality on Soils

Soil protists constitute a global biomass that is higher than that of all humans combined [9.34]. Soil diatoms, which are a key component of the soil protist community, play an important role in the functioning of the soil ecosystem. As primary producers they live on top of the soil surface [9.30]. Through movement or attachment to the soil, soil diatoms produce an extracellular mucopolysaccharide matrix, which has been suggested to bind soil particles and stabilize the soil surface [9.11] [9.46] [9.69] [9.106]. It is thought that this aggregation will in turn reduce soil erosion and water loss by evaporation, and improve water infiltration, thereby providing favorable conditions for seed germination [9.11] [9.44], yet more research is necessary to confirm this. Additionally, diatoms stimulate the release of nutrients from insoluble compounds and the weathering of silicates by creating a slightly acidic environment [9.44] [9.126]. Soil diatoms might even be an important carbon input source for decomposing soil animals [9.88], and their death and decay provides organic matter that other microorganisms and plants can utilize [9.32] [9.93] [9.102].

Given that soils are less buffered against environmental extremes, soil diatoms are affected more directly by climatological changes than their aquatic counterparts [9.43] [9.85]. For this reason, we expect soil diatoms to respond more quickly to environmental change [9.83] [9.85] [9.97]. So far, barely any consideration has been given to terrestrial diatoms in relation to environmental change, which stands in stark contrast to the vast body of work done on aquatic diatoms. The nature of aquatic environments, where sediments can accumulate in a relatively undisturbed manner for hundreds of years, providing an archive of past environmental change, naturally renders these habitats more suitable for examining long-term climate impacts through diatom communities. However, biomonitoring of soil diatoms in the face of ongoing environmental change is possible, in parallel with the already widely used biomonitoring programs of freshwater diatoms. Within this context, it would be valuable to experimentally investigate soil diatom responses and recovery to environmental extremes, focusing both on physiological and morphological modifications.

9.3.3 The Influence of Environmental Factors

9.3.3.1 Vitality and Absolute Abundances

Thanks to their morphological and physiological adaptions, diatoms can thrive in soils. For instance, during periods of high soil moisture availability in temperate regions, diatom productivity reaches peak abundances that are higher than green algae and cyanobacteria combined [9.30]. However, because of the significant relationship with soil moisture, absolute abundances of soil diatoms follow a strong seasonal gradient resulting in low minimum values during summer. Hence, diatoms comprise only a small part of the algal biomass in arid ecosystems despite being found in almost every sample [9.3], as most soil diatoms prefer a minimum soil moisture content of around 20% [9.27]. In contrast to diatoms, abundances of green algae remain high throughout the year, suggesting that their adaptations are more effective in sequestering and retaining moisture than diatoms. Another major environmental factor that influences diatom productivity is pH. Diatoms generally prefer soils with a pH between 5 and 8 [9.30] [9.44], while green algae seem to tolerate a wider pH range and generally dominate the algal floras of acidic soils [9.44]. Furthermore, vegetation coverage also seems to have a positive effect on diatom abundances [9.30], probably due to its capacity to retain moisture and protect against UV radiation. Overall, absolute abundances of soil diatoms are mainly determined by pH and soil moisture availability and follow a seasonal gradient [9.30].

As diatom community composition is generally determined by counting cleaned samples (i.e., empty cells), it is important to understand to what extent these fixated samples reflect the living communities in the field. For instance, recent work showed that whereas on average 92% of the diatoms were alive on (agricultural) grasslands, this was only 50% for samples taken in a forest at a riparian zone [9.5], and other studies replicated this result in similar locations [9.18]. This low percentage is probably due to the close proximity of a waterway which introduces cells of aquatic diatoms into the soils by various dispersal routes (e.g., wind, flooding, animals) [9.53]. It is likely that many of these aquatic diatoms do not survive long term in soils, contributing to the large number of dead cells in riparian zones. Indeed, riparian soil communities are often more diverse and comprise several typically planktonic and benthic freshwater species, such as *Fragilaria* Lyngb. spp. and *Aulacoseira* Thwaites spp. [9.108], which are not well adapted to changing moisture conditions and desiccation. We thus expect that forests generally have a similar percentage of viable cells as grasslands since they likely create a more stable and humid terrestrial environment [9.29]. Despite the recent insights gathered on this topic, a knowledge gap still exists regarding the (changing) viability of terrestrial diatom communities throughout the year.

9.3.3.2 Community Composition

Research on the impact of environmental factors on soil diatom communities found that anthropogenic disturbances play a key role in determining species compositions. Differences are even visible within farming systems. For example, soil diatom community structure differs between organic and conventional agricultural systems [9.41], whereas communities under reduced tillage showed a higher diversity than under traditional tillage [9.98] [9.99]. Similarly, studies focusing on a wide range of habitat types observed significantly more species in undisturbed soils (i.e., forest) than in anthropogenically disturbed areas, and noted very different communities between the habitat types [9.5] [9.27] [9.29]. Increased sampling has the potential to further uncover key differences between soil diatom communities. For example, Foets *et al.* [9.29], who sampled 16 sites monthly for a period of 14 months, assigned indicator species for five different habitat types. They found that species, such as *Humidophila irata* (Krasske) R.L.Lowe, *Sellaphora harderi* (Hust.) J.Foets & C.E.Wetzel, *Sellaphora nana*, and *Mayamaea fossalis* (Bock) Lange-Bert., are typical for undisturbed soils (forest and grassland), while amongst others *H. amphioxys*, *Stauroneis borichii*, and *Luticola ventricosa* (Kütz.) D.G.Mann indicate disturbances (pasture, farmland, meadow). Besides assigning indicator species, the authors also followed the community composition after a drastic change in the landscape occurred (i.e., ploughing, grazing), and observed that *P. borealis*, *H. amphioxys*, and *H. abundans* Lange-Bert. serve as important pioneer species. In addition, they noticed that soil diatom communities need one to two months to colonize a disturbed habitat. Altogether, while anthropogenic disturbances do not seem to influence diatom absolute abundances, it is perhaps the most important factor defining the species composition.

Farming practices not only affect communities directly, but also indirectly by influencing, amongst others, organic matter, nitrogen, and carbon content. For instance, lower organic matter concentrations have been observed in conventional tillage systems compared to reduced tillage [9.98]. In turn, this leads to a lower moisture content, since organic matter has a positive effect on the moisture holding capacity [9.12]. In addition to impacting the absolute abundances of soil diatoms, several studies regarded soil moisture to be an important driver of species composition [9.81] [9.115] [9.119]. However, its impact depends on other variables, such as anthropogenic activity. For instance, soil moisture is not a major driver of soil diatom community composition in anthropogenically disturbed areas [9.27] [9.29], which is probably due to the presence of more desiccation-tolerant species. Besides soil moisture, pH is also considered as an important factor [9.2] [9.57] [9.119], but contrary to the former, the effect of disturbance is less expressed [9.27]. Furthermore, nutrients such as carbon, nitrogen, phosphorous, and magnesium have all been documented to influence soil diatom communities [9.2] [9.17] [9.65] [9.108] [9.115] [9.127]. However, there is a need for more research on these variables, since it is not yet possible to establish autecological values for them and many researchers used different methods to extract and analyze these variables leading to different, non-comparable results [9.27]. Thus, similar to freshwater diatoms, terrestrial diatoms have shown to be sensitive to several environmental variables, which could pave the way forward for their use as environmental markers.

9.3.3.3 The Impact of Herbicides and Oil Pollution on Soil Diatoms

Several studies explored the impact of different herbicides on soil diatoms, observing that herbicides intervene at different levels. At the individual level, herbicides can inhibit cell movement leading to a decrease in colonization rates [9.78] [9.79] [9.129], and can cause significant reductions in carbon fixation and chlorophyll concentrations, with the latter resulting in a decreased photosynthetic activity [9.8] [9.80]. Eventually, these effects can be lethal, even when sub-lethal herbicide doses are used. Furthermore, different species seem to have different sensitivity levels to herbicides. For instance, *Pinnularia obscura* Krasske was found to be more abundant on soils treated with herbicides, whereas *Stauroneis obtusa* Lagerst. and *M. atomus* were more numerous on untreated soils [9.8]. Similarly, *H. amphioxys* is known to be one of the most pesticide-tolerant terrestrial diatom taxa [9.79]. Clearly, the use of herbicides impacts soil diatom communities. In general, researchers noted a lower diatom diversity after the application of herbicides, and the resulting communities consisted of the most tolerant species [9.8] [9.19]. Overall, the impact of herbicides on soil diatoms is substantial, and more long-lasting than the immediate impact of soil disturbances caused by ploughing and mowing.

Industrial oil production is another anthropogenic activity that is impacting soil diatoms. Generally, soils in oil production areas are polluted with petroleum products, drilling muds, crude oil, chemical treatment, etc. These pollutants cause drastic changes in soil and vegetation properties and soil biota assemblages. Research of polluted soils in Russia demonstrated that soil diatoms are highly vulnerable to the impact of oil, and are generally more sensitive to oil pollution than other algal groups [9.21] [9.26]. For instance, Dorokhova [9.21] note that at low oil concentrations, a persistent, yet reversible change in the dominant diatom species was noted. However, at high oil concentrations, or in soils with a low self-purification ability, the entire diatom species composition is altered in an irreversible way following oil pollution. Communities impacted by oil pollution are characterized by the presence of only the most tolerant species such as *H. abundans*, *H. amphioxys*, and *P. obscura*. Due to this responsivity, soil diatoms are considered good bioindicators of soil conditions in oil production regions and could potentially be used for biological monitoring in such areas in the future.

9.4 Diatoms as Environmental Markers

9.4.1 Soil Quality Biomonitoring with Diatoms

Water quality assessments usually involve the analysis of the degree of pollution that directly impacts human and animal consumption and health, or the functioning of natural ecosystems [9.16] [9.20]. In contrast, soil quality is more site-specific and complex, since soil constitutes solid, liquid, and gaseous phases and can be used for a larger variety of purposes. In addition, there are numerous linkages between soil functions and soil-based ecosystem services [9.68]. Due to this complexity and multifunctionality, biological indicators are considered highly useful tools as they simultaneously incorporate different parameters. In this regard, diatoms have been tested as soil quality indicators to assess, amongst others, the degree of anthropogenic disturbances.

Like freshwater diatoms, the functionality of soil diatoms in ecological applications lies in their sensitivity to environmental variables. Recently, researchers defined autecological values for pH and soil moisture for the most widespread soil diatoms [9.27]. This involved the merging of several datasets from Luxembourg, Poland, Greenland, and the sub-Antarctic island Crozet in order to obtain more robust indicator values. Prior to that study, only Lund [9.57] and Van de Vijver *et al.* [9.115] defined ecological ranges for pH and moisture availability. Lund [9.57] assigned pH ranges for only 24 common soil diatoms, and Van de Vijver *et al.* [9.115] applied transfer functions for soil moisture to ~90 sub-Antarctic soil diatom taxa. In addition, van Dam *et al.* [9.109] developed an ecological checklist containing indicator values for pH, saprobity, trophic state, and moisture for 102 terrestrial species. However, the assigned values were based on aquatic samples and consequently may not fully reflect the behavior of soil diatoms in true terrestrial conditions. Indeed, the pH optimum of soil diatoms appears to be 0.5 lower than in freshwater conditions, whereas there is no indication that typical "terrestrial" species generally occur on drier soils compared to species [9.27]. Therefore, the assigned values of Foets *et al.* [9.27] provide essential progress to the field, and together with the defined indicator species for some types of land use (see Foets *et al.* [9.29]), they provide a good basis for tracing purposes.

There are, however, little to no data available on the impact of many environmental variables or anthropogenic activities on soil diatoms. In such cases, indices developed for aquatic diatoms can still be useful. For instance, researchers have applied the IPS index, a quality index developed for aquatic diatoms to measure organic pollution, to soil diatom communities [9.2]. Compared to other diatom-based indices, the IPS index includes the largest number of taxa and is the most up-to-date in taxonomy. Results showed that the IPS index gave an indication of the degree of anthropogenic disturbance of soils, allowing to distinguish four different land uses (farmland, pastures, undisturbed grassland, and forests) [9.2] [9.29]. However, due to a lack of ecological information, some important diatom species (sometimes up to 40% of the counted cells) could not be taken into account as there were no indicator values assigned for them [9.5] [9.29], further underscoring the need to expand soil diatom research.

9.4.2 Tracing Hydrological Connectivity with Soil Diatoms

Researchers explored the potential for terrestrial diatoms as a tracer for water source and hydrological connectivity between different landscape positions in catchments since no other hydrological tracer could do this unambiguously. The concept is based on the hypothesis that drift diatoms, sampled in the stream during rainfall-runoff events, could record their geographical origins at the watershed scale thanks to their sensitivity to several abiotic parameters [9.71]. The first experiment dates back to 2009, when two distinct rainfall-runoff events were sampled by automatic samplers in the Attert River basin (Luxembourg). This study used the moisture classification of van Dam *et al.* [9.109] to allocate drift diatom species to either aquatic (categories 1, 2, and 3) or terrestrial (categories 4 and 5) conditions (Figure 9.5). Their results showed strong increases in the abundance of terrestrial species during peak flows, indicating the onset of surface runoff connected to the stream. Several years later, the experiment was repeated in three contrasted catchments located in Luxembourg, Slovakia, and Oregon (USA) [9.73]. Similar results were obtained showing clear responses in the drift diatom community composition to the precipitation pulses in all three areas. The encouraging results of both studies paved the way for several other studies, which confirmed the potential for terrestrial diatoms to serve as a tracer at the hillslope–riparian–stream scale [9.18] [9.52] [9.63] [9.64]. Specifically, Martínez-Carreras *et al.* [9.64] demonstrated in laboratory experiments that diatoms do not percolate through the soil and thus suggested that surface runoff is the most important pathway to the stream. Furthermore, riparian zones have been found to be the largest terrestrial diatom reservoirs and the main source for terrestrial diatoms in streams [9.63], whereas diatom reservoir exhaustion seems unlikely even during heavy rainfall events [9.18]. Furthermore, the work by Klaus *et al.* [9.52] extended the use of terrestrial diatoms to trace hydrological connectivity at large spatial scales. Within a set of seven nested catchments (0.45–249 km^2), they showed that different communities of species could be attributed to sub-catchments with different physiographic characteristics. As a result, the occurrence of certain species at the 249 km^2-catchment outlet during precipitation events provided information on both the occurrence

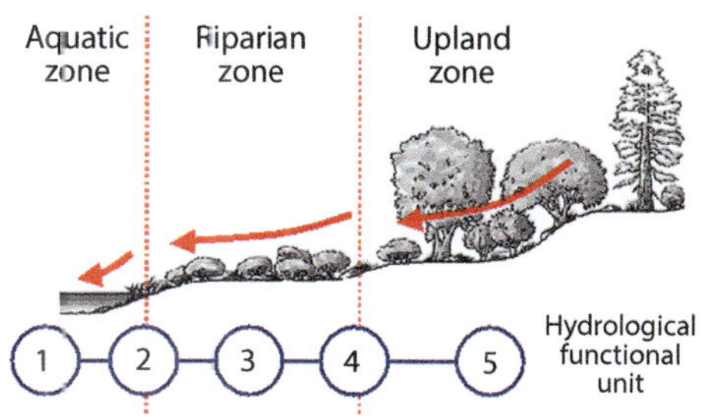

Figure 9.5 Figure explaining the concept of terrestrial diatoms as hydrological tracers. The hydrological functional units correspond to the moisture classification designed by van Dam *et al.* [9.109].

of riparian-stream connectivity in the different sub-catchments and the source of stream water to the outlet.

Although exploratory work has shown the potential for terrestrial diatoms to serve as complementary tracers in studies on hydrological processes, more research is needed to overcome current limitations. The first limitation relates to the large number of samples that need to be investigated to assess the variability of the diatom community composition during a rainfall-runoff event. For example, Klaus *et al.* [9.52] investigated 155 stream samples to analyze a single rainfall-runoff event. This limitation has recently been tackled by Foets *et al.* [9.28], who explored the potential of the Philips sampler, a time-integrated mass-flux sampler, to sample representative diatom communities. The Philips sampler, which consists of a PVC pipe (98 mm × 1000 mm) with a very small inlet and outlet (4 mm), was originally developed to trap fine-grained (<62.5 μm) suspended sediment for the assessment of geochemical, physical, and magnetic features of transported material. This sampler permits the collection of a time-integrated composite suspended sediment sample from an entire storm runoff event or series of events. The study found that in certain conditions the usage of the sampler could substantially reduce the number of aquatic samples that need to be collected, could complement automatic sampling, and could be applied in areas where automatic sampling is not possible due to logistical reasons. The second limitation relates to the large number of soil samples that need to be collected from a catchment area to accurately assess from which area the soil diatom species enter a stream during runoff events. We propose that to reduce the number of terrestrial samples to be collected, diatom taxa should be mapped according to their defined ecological preferences. Thus far, preferences are only available for pH, soil moisture, and habitat type for the most common species, meaning that preliminary mapping could already be achieved. However, many species are not yet ecologically characterized or even (formally) discovered, indicating the need for more research.

9.4.3 Soil Diatoms as Environmental Markers: Major Challenges and Solutions

One of the major issues with applying soil diatoms for biomonitoring and hydrology research is the high cost of microscopy analysis, due to its time-consuming nature. This cost are perhaps even more profound for soil diatoms compared to their aquatic counterparts because microscopy slides of soils are more challenging to analyze due to high concentrations of inorganic material present in the slides. Because molecular techniques (i.e., DNA metabarcoding) are less expensive and do not require expert knowledge on diatom taxonomy, they can significantly decrease the time spent on diatom identifications. As such, metabarcoding may finally unlock the large potential of soil diatoms for biomonitoring purposes. To date, only a handful of studies used metabarcoding to investigate soil diatoms, with a specific focus on *Pinnularia* section Distantes (Cleve) Patrick, including the *P. borealis* complex [9.74] [9.75]. These case studies, however, underscore the feasibility of metabarcoding to reveal soil diatom biodiversity, and show that the technique can even be used to detect (pseudo)cryptic diversity in environmental samples [9.74] [9.84]. Nevertheless, there is still much work to be done before it can rival the quality of microscopic analyses in routine surveys [9.87] [9.120].

A second prerequisite for the routine application of soil diatoms in biomonitoring research is the development of more comprehensive ecological classifications of soil diatoms that go beyond soil moisture and pH, but also include other environmental variables such as soil organic matter [9.27]. A strong starting point for this endeavor represents the ecological check list of soil diatoms which was recently developed by Foets *et al.* [9.27], and which only includes species that frequently inhabit soils. Although that list does not contain all soil diatom taxa, it currently represents the most complete overview and provides a significant leap forward from the widely used van Dam *et al.* [9.109] classification, which was not specifically developed for soil diatoms and might introduce bias when used for soil diatom analysis. The presence of (pseudo)cryptic diversity will, however, complicate the development of ecological classifications given that (pseudo)cryptic diatoms, including the common soil diatom complexes *P. borealis* and *H. amphioxys*, can have different growth optima, and thus ecological niches [9.96]. If we are to leverage these challenges to our advantage, soil diatoms will one day likely become a routine aspect of soil quality biomonitoring.

9.5 Soil Diatoms in Forensic Sciences

Environmental analysis of water, plants, and soils plays an important role in the process of criminal investigation, as it not only provides valuable information about the course of the crime, but it can also clarify the connection between the victim, the suspect, and the environment. A valuable way of linking the suspect and victim to a specific environment is by analyzing their clothing for the presence of diatoms [9.10] [9.39]. Studies showed that if clothes are submersed in aquatic environments, even for a very short period of time (i.e., 30 sec – 3 min), a distinctive diatom species composition can be obtained. However, data on soil samples revealed that the transferability of soil diatoms to clothing was generally lower, and that the communities were not sufficiently specific [9.92]. This lower transferability and specificity are due to the lower abundance and diversity of diatoms on soils than in aquatic environments, and the fact that dry soil particles attach less well to clothes than wet particles. Also, terrestrial environments, such as a forests, can consist of many different micro-habitats, each of which contain a different soil diatom community [9.29]. Therefore, the most optimal opportunity for obtaining viable information from soil-contaminated clothing is when the crime scene is located in a riparian area. There, the soil is generally moist or wet and contains a highly diverse diatom community which includes aquatic diatoms from the nearby stream or river [9.5] [9.107]. This is evidenced by a case study in the Hunan province of China [9.39], which demonstrates that soil diatoms could play an important role in linking the suspect and/or victim to a specific environment.

9.6 Future Perspectives

Despite the substantial progress that has been made on the ecology of soil diatoms during the last decade, many questions remain unanswered. In particular, many geographic regions and climate zones remain understudied. So far, most of the ecological studies on soil diatoms were conducted in temperate Europe and some parts of North America and Antarctica, whereas other continents were barely investigated, suggesting that many species remain undiscovered. Therefore, future studies should aim to increase sampling in understudied regions, especially with a focus towards anthropogenic undisturbed areas since they generally harbor a higher diversity and a higher percentage of rare species than disturbed areas [9.27] [9.29]. Another advantage of sampling in different climate zones is that the morphology and physiology of terrestrial diatom species can be compared between environments, which will give more information on soil life-style adaptations as well as environmental tolerances useful for biomonitoring. Overall, there is a need to conduct more research in understudied areas, since it will provide us with valuable information regarding soil diatom taxonomy, ecology, and evolution.

The introduction of metabarcoding will be an essential next step for using terrestrial diatoms in ecological applications [9.72]. We believe that the exploration of soil diatom diversity with metabarcoding can be an important stepping-stone to building larger, standardized datasets of soil diatom ecological preferences. In turn, this will give valuable ecological information that would provide the basis for calculations of well-defined autecological values of several environmental factors for specific regions of interest. So far, indicator values have only been calculated from data-limited datasets [9.57] [9.115] or from a combination of (relatively small) datasets encompassing different climate zones [9.27]. Although the latter should give a decent basis, the values could give unsatisfactory results in certain areas, since environmental variables are context-dependent and may vary across regions and time [9.1] [9.121]. Besides, these studies only reported autecological values for pH and soil moisture content. If autecological values become more widely available, we will be able to map diatom taxa on different soils and perform species distribution modeling. Overall, we believe that molecular techniques will play a key role in future terrestrial diatom research and their quantitative use of (hydro-)ecological applications.

A quick look at the number of published articles reveals that the interest in soil diatoms has grown significantly over the last two decades. Many new taxa have been described during this period, and given that these new taxa have been reported from a relatively small part of the world, many more taxa are awaiting discovery. The use of molecular techniques will only enhance interest in soil diatom biodiversity, which in turn will enable the routine application of soil diatoms as a bioindicator in various applications. Soil diatoms have already been successfully tested in forensic science, as a tracer for water source and hydrological connectivity, and as an indicator for soil pollution and soil quality. In an era where it is increasingly recognized that maintaining good soil quality is of fundamental importance for economic and societal purposes (e.g., food production), there has never been a more exciting time to investigate soil diatom biodiversity and ecology.

Appendix

Table 9.1 Articles used for data analysis in Figure 9.1. The first author(s), year of publication, title, the approach (direct or culturing), the number of samples collected, and the number of diatom species found during the study are provided.

Reference	Year	Title	Approach	No. diatom taxa	No. samples
Antonelli *et al.* [9.2]	2017	On the potential for terrestrial diatom communities and diatom indices to identify anthropic disturbance in soils	Direct	251	114
Ashley *et al.* [9.3]	1985	Soil algae of cryptogamic crusts from the Uintah basin, Utah, USA	Moist Culture	24	18
Bakieva *et al.* [9.130]	2012	Ecological-floristic analysis of soil algae and cyanobacteria on the Tra-Tau and Yurak-Tau Mounts, Bashkiria	Culturing	35	12
Barragán *et al.* [9.5]	2018	A standard method for the routine sampling of terrestrial diatom communities for soil quality assessment	Direct	129	40
Bataeva *et al.* [9.131]	2017	Composition of phototrophs in different soil types of Astrakhan Oblast	Culturing	4	5
Brendemühl [9.132]	1949	Über die Verbreitung der Erddiatomeen	Culturing	114	167

(*Continued*)

Table 9.1 Articles used for data analysis in Figure 9.1. The first author(s), year of publication, title, the approach (direct or culturing), the number of samples collected, and the number of diatom species found during the study are provided. (*Continued*)

Reference	Year	Title	Approach	No. diatom taxa	No. samples
Broady [9.133]	2005	The distribution of terrestrial and hydro-terrestrial algal associations at three contrasting locations in southern Victoria Land, Antarctica	Direct	15	247
Broady [9.14]	1978	The terrestrial algae of Signy Island, South Orkney Islands	Culturing	39	24
Chattová and Uher [9.134]	2013	A preliminary study on diversity and ecology of soil diatoms in urban habitats	Direct	111	84
Dorokhova *et al.* [9.135]	2015	Algae and cyanobacteria in soils of Moscow	Culturing	18	27
Elster *et al.* [9.23]	1999	Diversity and abundance of soil algae in the polar desert, Sverdrup Pass, central Ellesmere Island.	Culturing	10	54
Fairchild and Willson [9.136]	1967	The algal flora of two Washington soils	Culturing	4	60
Fathi and Zaki [9.137]	2003	Preliminary survey of edaphic algae in El-Minia region, Nile Valley, Egypt	Culturing	25	6

(*Continued*)

Table 9.1 Articles used for data analysis in Figure 9.1. The first author(s), year of publication, title, the approach (direct or culturing), the number of samples collected, and the number of diatom species found during the study are provided. (*Continued*)

Reference	Year	Title	Approach	No. diatom taxa	No. samples
Fazlutdinova & Sukhanova [9.26]	2014	Composition of soil diatoms in zones of impact from oil production complexes	Direct	64	10
Fehér [9.138]	1936	Untersuchungen über die regionale Verbreitung der Bodenalgen	Culturing	63	118
Fermani *et al.* [9.139]	2007	Soil microalgal communities on an Antarctic active volcano (Deception Island, South Shetlands)	Direct	18	74
Ferreira da Silva *et al.* [9.140]	2019	Distribution of aerophilous diatom communities associated with terrestrial green macroalgae in the South Shetland Islands, Maritime Antarctica	Direct	23	6
Flechtner *et al.* [9.141]	1998	Algal composition of microbiotic crusts from the Central Desert of Raja California, Mexico	Direct	10	10

(*Continued*)

Table 9.1 Articles used for data analysis in Figure 9.1. The first author(s), year of publication, title, the approach (direct or culturing), the number of samples collected, and the number of diatom species found during the study are provided. (*Continued*)

Reference	Year	Title	Approach	No. diatom taxa	No. samples
Flechtner *et al.* [9.142]	2008	The biological soil crusts of the San Nicolas Island: Enigmatic algae from a geographically isolated ecosystem	Direct	8	7
Foets *et al.* [9.29]	2020	Temporal and spatial variability of terrestrial diatoms at the catchment scale: controls on communities	Direct	302	206
Al-Fredan & Fathi [9.143]	2007	Preliminary survey of edaphic algae in Al-Hasa Region, Saudi Arabia	Culturing	16	5
Gremmen *et al.* [9.144]	2007	Distribution of moss-inhabiting diatoms along an altitudinal gradient at sub-Antarctic Iles Kerguelen	Direct	130	39
Grondin & Johansen [9.145]	1995	Seasonal succession in a soil algal community associated with a beech-maple forest in northeastern Ohio, USA	Culturing	5	10

(*Continued*)

Table 9.1 Articles used for data analysis in Figure 9.1. The first author(s), year of publication, title, the approach (direct or culturing), the number of samples collected, and the number of diatom species found during the study are provided. (*Continued*)

Reference	Year	Title	Approach	No. diatom taxa	No. samples
Hahn & Neuhaus [9.146]	1997	Boden-diatomeen einer landwirtschaftlichen nutzfläche bei Potsdam, Deutschland	Culturing	5	12
Hayek & Hulbary [9.40]	1956	A survey of soil diatoms	Direct	15	1
Heger [9.41]	2011	Impact of farming practices on soil diatoms and testate amoebae: a pilot study in the DOK-trial at Therwil, Switzerland	Culturing	16	3
Hoffmann *et al.* [9.45]	2007	Algal flora from limed and unlimed forest soils in the Ardenne (Belgium)	Culturing	3	24
Hunt *et al.* [9.147]	1979	Soil algae in field and forest environments	Culturing	19	51
Ibraheem [9.148]	2003	Preliminary survey of microalgal soil crusts in a xeric habitats (Wadi-Araba, Eastern Desert, Egypt)	Culturing	24	12

(*Continued*)

Table 9.1 Articles used for data analysis in Figure 9.1. The first author(s), year of publication, title, the approach (direct or culturing), the number of samples collected, and the number of diatom species found during the study are provided. (*Continued*)

Reference	Year	Title	Approach	No. diatom taxa	No. samples
Johansen *et al.* [9.149]	1982	Effects of burning on the algal communities of a high desert soil near Wallsburg, Utah	Culturing	19	30
Johansen *et al.* [9.150]	1984	Recovery patterns of cryptogamic soil crusts in desert rangelands following fire disturbance	Culturing	31	32
John [9.49]	1942	An ecological and taxonomic study of the algae of British soils: I. the distribution of the surface-growing algae	Culturing	11	54
Lin *et al.* [9.151]	2013	Biodiversity of soil algae in the farmlands of mid-Taiwan	Direct	27	15
Loescher [9.55]	1981	Diatoms from a native Iowa prairie	Direct	74	133
Lukešová & Hoffmann [9.152]	1996	Soil algae from acid rain impacted forest areas of the Krusňé Hory Mts. 1. Algal communities	Culturing	10	70

(*Continued*)

Table 9.1 Articles used for data analysis in Figure 9.1. The first author(s), year of publication, title, the approach (direct or culturing), the number of samples collected, and the number of diatom species found during the study are provided. (*Continued*)

Reference	Year	Title	Approach	No. diatom taxa	No. samples
Lukešová & Komárek [9.153]	2011	Succession of soil algae on dumps from strip coal-mining in the most region (Czechoslovakia)	Culturing	10	10
Lund [9.57]	1945	Observations on Soil Algae. I. The ecology, size and taxonomy of British soil diatoms	Culturing	52	66
Maltsev & Maltseva [9.58]	2018	The influence of forest-forming tree species on diversity and spatial distribution of algae in forest litter	Culturing	20	108
Maltsev *et al.* [9.61]	2017	Use of soil biota in the assessment of the ecological potential of urban soils	Culturing	2	9
Maltsev *et al.* [9.154]	2017	Specific features of algal communities in forest litter of forest biogeocenoses of the steppe zone	Culturing	5	12
Maxwell [9.155]	1991	Floristic changes in soil algae and cyanobacteria in reclaimed metal-contaminated land at Sudbury, Canada	Culturing	2	6

(*Continued*)

Table 9.1 Articles used for data analysis in Figure 9.1. The first author(s), year of publication, title, the approach (direct or culturing), the number of samples collected, and the number of diatom species found during the study are provided. (*Continued*)

Reference	Year	Title	Approach	No. diatom taxa	No. samples
Moravcová *et al.* [9.156]	2010	Diatom communities in soils influenced by the wandering albatross (*Diomedea exulans*)	Direct	163	109
Neustupa [9.67]	2001	Soil algae from marlstone-substratum based biotopes in the Nature park Džbán (Central Bohemia, Czech Republic) with special attention to the natural treeless localities	Culturing	8	128
Noga *et al.* [9.157]	2014	Diatoms (Bacillariophyceae) in rivers and streams and on cultivated soils of the Podkarpcie region in the years 2007-2011	Direct	147	82
Novakovskaya *et al.* [9.158]	2020	Influence of ecological factors on soil algae in different types of mountain tundra and sparse forests in the Northern Urals	Culturing	3	10
Phillipson [9.159]	1935	Some algae of Victorian soils	Culturing	2	3

(*Continued*)

Table 9.1 Articles used for data analysis in Figure 9.1. The first author(s), year of publication, title, the approach (direct or culturing), the number of samples collected, and the number of diatom species found during the study are provided. (*Continued*)

Reference	Year	Title	Approach	No. diatom taxa	No. samples
Rosa & Lhotský [9.160]	1971	Edaphische Algen und Protozoen im Isergebirge, Tschechoslowakei	Culturing	14	18
Round [9.161]	1958	Observations on the diatom flora of Braunton Burrows, N. Devon	Direct	20	9
Shcherbyna *et al.* [9.162]	2017	Post-pyrogenic changes in vegetation cover and biological soil crust in steppe ecosystems	Culturing	11	28
Stanek-Tarkowska *et al.* [9.163]	2016	Diatom assemblages growing on cropping soil in Pogórska Wola near Tarnów	Direct	47	8
Stanek-Tarkowska & Noga [9.98]	2012	Diversity of diatoms (Bacillariophyceae) in the soil under traditional and reduced tillage	Direct	62	4
Stanek-Tarkowska *et al.* [9.164]	2015	The diversity of diatom assemblages developed on fallow soil in Pogórska Wola near Tarnów (southern Poland)	Direct	57	8

(*Continued*)

Table 9.1 Articles used for data analysis in Figure 9.1. The first author(s), year of publication, title, the approach (direct or culturing), the number of samples collected, and the number of diatom species found during the study are provided. (*Continued*)

Reference	Year	Title	Approach	No. diatom taxa	No. samples
Stanek-Tarkowska *et al.* [9.99]	2018	Effects of reduced and traditional tillage on soil properties and diversity of diatoms under winter wheat	Direct	197	60
Stanek-Tarkowska *et al.* [9.165]	2017	Physicochemical properties of silt loamy soil and diversity of diatom species under winter wheat and oats	Direct	91	24
Terlizzi & Karlander [9.166]	1979	Soil algae from a Maryland serpentine formation	Culturing	3	7
Tittor [9.105]	1956	Untersuchungen über die Bodenmikroflora von Gatersleben und Umgebung	Culturing	12	17
Vacht *et al.* [9.108]	2014	Variability of riparian soil diatom communities and their potential as indicators of anthropogenic disturbances	Direct	105	11

(*Continued*)

Table 9.1 Articles used for data analysis in Figure 9.1. The first author(s), year of publication, title, the approach (direct or culturing), the number of samples collected, and the number of diatom species found during the study are provided. (*Continued*)

Reference	Year	Title	Approach	No. diatom taxa	No. samples
Vacht *et al.* [9.107]	2018	Diatom and microarthropod communities of three airfields in Estonia – their differences and similarities and possible linkages to airfield properties	Direct	49	48
Van de Vijver & Beyens [9.111]	1998	A preliminary study on the soil diatom assemblages from Iles de la Possession (Crozet, Sub Antarctica)	Direct	165	48
Van de Vijver *et al.* [9.115]	2002	Soil diatom communities from Ile de la Possession (Crozet, sub-Antarctica)	Direct	230	104
Van Kerckvoorde *et al.* [9.119]	2000	Terrestrial soil diatom assemblages from different vegetation types in Zackenberg (Northeast Greenland)	Direct	81	30
Willson & Forest [9.125]	1957	An exploratory study on soil algae	Culturing	3	6

(*Continued*)

Table 9.1 Articles used for data analysis in Figure 9.1. The first author(s), year of publication, title, the approach (direct or culturing), the number of samples collected, and the number of diatom species found during the study are provided. (*Continued*)

Reference	Year	Title	Approach	No. diatom taxa	No. samples
Zancan *et al.* [9.167]	2005	Soil algae composition under different agro-ecosystems in North-Eastern Italy	Culturing	10	8
Zhang *et al.* [9.127]	2020	Soil diatom communities and their relation to environmental factors in three types of soil from four cities in central-west China	Direct	66	34
Zurek *et al.* [9.129]	1981	The influence of the herbicides of the Lenacil and Pyrazon on the soil algae	Culturing	4	48

References

[9.1] Álvarez-Blanco, I., Cejudo-Figueiras, C., Bécares, E., and Blanco, S. 2011. Spatiotemporal changes in diatom ecological profiles: Implications for biomonitoring. Limnology (Tokyo), 12 (2), 157–168.

[9.2] Antonelli, M., Wetzel, C.E., Ector, L., Teuling, A.J., and Pfister, L. 2017. On the potential for terrestrial diatom communities and diatom indices to identify anthropic disturbance in soils. *Ecological Indicators*, 75, 73–81.

[9.3] Ashley, J., Rushforth, S.R., and Johansen, J.R. 1985. Soil algae of cryptogamic crusts from the Uintah basin, Utah, U.S.A. *The Great Basin Naturalist*, 45 (3), 432–442.

[9.4] Barragán, C., Ector, L., and Wetzel, C.E. 2017. *Mayamaea petersenii* sp. nov., a new diatom from European aerial habitats and a brief appraisal on the morphological diversity of the genus. *Algological Studies*, 153, 71–87.

[9.5] Barragán, C., Wetzel, C.E., and Ector, L. 2018. A standard method for the routine sampling of terrestrial diatom communities for soil quality assessment. *Journal of Applied Phycology.*

[9.6] Bayer-Giraldi, M., Weikusat, I., Besir, H., and Dieckmann, G. 2011. Characterization of an antifreeze protein from the polar diatom *Fragilariopsis cylindrus* and its relevance in sea ice. *Cryobiology*, 63 (3), 210–219.

[9.7] Bérard, A., Dorigo, U., Humbert, J.F., and Martin-Laurent, F. 2005. Microalgae community structure analysis based on 18S rDNA amplification from DNA extracted directly from soil as a potential soil bioindicator. *Agronomy for Sustainable Development*, 25 (2), 285–291.

[9.8] Bérard, A., Rimet, F., Capowiez, Y., and Leboulanger, C. 2004. Procedures for Determining the Pesticide Sensitivity of Indigenous Soil Algae: A Possible Bioindicator of Soil Contamination? *Archives of Environmental Contamination and Toxicology*, 46 (1), 24–31.

[9.9] Bock, W. (1963) Diatomeen extrem trockener Standorte. *Nova Hedwigia*, 5, 199–254.

[9.10] Bogusz, I., Bogusz, M., and Żelazna-Wieczorek, J. 2022. Diatoms from inland aquatic and soil habitats as indestructible and nonremovable forensic environmental evidence. *Journal of Forensic Sciences*, 67, 1490-1504.

[9.11] Booth, W.E. 1941. Algae as pioneers in plant succession and their importance in erosion control. *Ecology*, 22 (1), 38–46.

[9.12] Bot, A., and Benites, J. 2005. The importance of soil organic matter: Key to drought-resistant soil and sustained food production, *FAO Soils Bulletin.*

[9.13] Bristol, M.B. 1919. On the retention of vitality by algae from old stored soils. *New Phytologist*, 18 (3), 92–107.

[9.14] Broady, P.A. 1979. The terrestrial algae of Signy Island, South Orkney Islands, *British Antarctic Survey.*

[9.15] Campbell, G.S., and Norman, J.M. 1998. Introduction, in *An Introduction to Environmental Biophysics* (eds. Campbell, G.S., and Norman, J.M.), Springer New York, New York, NY, pp. 1–13.

[9.16] Carter, M.R., Gregorich, E.G., Anderson, D.W., Doran, J.W., Janzen, H.H., and Pierce, F.J. 1997. Concepts of soil quality and their significance, in *Developments in Soil Science*, vol. 25, Elsevier, pp. 1–19.

[9.17] Chattová, B., Cahová, T., Pinseel, E., Kopalová, K., Kohler, T.J., Hrbáček, F., Van de Vijver, B., and Nývlt, D. 2022. Diversity, ecology, and community structure of the terrestrial diatom flora from Ulu Peninsula (James Ross Island, NE Antarctic Peninsula). *Polar Biology*, 45 (5), 873–894.

[9.18] Coles, A.E., Wetzel, C.E., Martínez-Carreras, N., Ector, L., McDonnell, J.J., Frentress, J., Klaus, J., Hoffmann, L., and Pfister, L. 2016. Diatoms as a tracer of hydrological connectivity: are they supply limited? *Ecohydrology*, 9 (4), 631-645.

[9.19] Cullimore, D.R., and McCann, A.E. 1977. Influence of four herbicides on the algal flora of a prairie soil. *Plant and Soil*, 46 (3), 499–510.

[9.20] Davidson, D.A. 2000. Soil quality assessment: recent advances and controversies. Progress in Environmental Science, 2 (4), 342–350.

[9.21] Dorokhova, M.F. 2007. Diatoms as indicators of soil conditions in oil production regions. *Oceanological and Hydrobiological Studies*, 36 (1), 129–135.

[9.22] Ehrenberg, C.G. 1843. Verbreitung und Einfluss des mikroskopischen Lebens in Süd-und Nord-Amerika, Königlichen Akademie der Wissenschaften.

[9.23] Elster, J., Lukesová, A., Svoboda, J., Kopecky, J., and Kanda, H. 1999. Diversity and abundance of soil algae in the polar desert, Sverdrup Pass, central Ellesmere Island. *Polar Record*, 35 (194), 231–254.

[9.24] Ettl, H., and Gärtner, G. 1995. Syllabus der Boden-, Luft- und Flechtenalgen, Springer Spektrum.

[9.25] Evans, J.H. 1958. The survival of freshwater algae during dry periods: part I . An investigation of the algae of five small ponds. *Journal of Ecology*, 46 (1), 149–167.

[9.26] Fazlutdinova, A.I., and Sukhanova, N. v. 2014. Composition of soil diatoms in zones of impact from oil production complexes. *Russian Journal of Ecology*, 45 (3), 188–193.

[9.27] Foets, J., Stanek-Tarkowska, J., Teuling, A.J., Van de Vijver, B., Wetzel, C.E., and Pfister, L. 2021. Autecology of terrestrial diatoms under anthropic disturbance and across climate zones. *Ecological Indicators*, 122.

[9.28] Foets, J., Wetzel, C.E., Carreras, N.M., Teuling, A.J., Iffly, J., and Pfister, L. 2020. Technical note: A time-integrated sediment trap to sample diatoms for hydrological tracing. *Hydrology and Earth System Sciences*, 24, 4709-4725.

[9.29] Foets, J., Wetzel, C.E., Teuling, A.J., and Pfister, L. 2020. Temporal and spatial variability of terrestrial diatoms at the catchment scale: controls on communities. *PeerJ*. (8:e8296), 1–19.

[9.30] Foets, J., Wetzel, C.E., Teuling, A.J., and Pfister, L. 2020. Temporal and spatial variability of terrestrial diatoms at the catchment scale: controls on productivity and comparison with other soil algae. *PeerJ*, 8 (2), 1–21.

[9.31] Forest, H.S. 1962. Analysis of the soil algal community. *Transactions of the American Microscopical Society*, 81 (2), 189–198.

[9.32] Fritsch, F.E. 1907. The role of algal growth in the colonization of new ground and in the determination of scenery. *The Geographical Journal*, 30 (5), 531–548.

[9.33] Gargas, C.B., Theriot, E.C., Ashworth, M.P., and Johansen, J.R. 2018. Phylogenetic Analysis Reveals that the 'Radial Centric' Diatom *Orthoseira* Thwaites (Orthoseiraceae, Bacillariophyta) is a Member of a 'Multipolar' Diatom Lineage. *Protist*, 169 (6), 803–825.

[9.34] Geisen, S., Briones, M.J.I., Gan, H., Behan-Pelletier, V.M., Friman, V.P., de Groot, G.A., Hannula, S.E., Lindo, Z., Philippot, L., Tiunov, A.V., and Wall, D.H. 2019. A methodological framework to embrace soil biodiversity. *Soil Biology and Biochemistry*, 136, 107536.

[9.35] Gistl, R. 1932. Zur Kenntnis der Erdalgen. *Archiv für Mikrobiologie*, 3, 634–649.

[9.36] Gistl, R. 1933. Erdalgen und Düngung. Erdalgen und Anionen. *Archiv für Mikrobiologie*, 4, 348–378.

[9.37] Glushchenko, A., Kezlya, E., Maltsev, Y., Genkal, S., Kociolek, J.P., Kulikovskiy, M., and Timiryazev, K.A. 2022. Description of the Soil Diatom *Sellaphora terrestris* sp. nov. (Bacillariophyceae, Sellaphoraceae) from Vietnam, with Remarks on the Phylogeny and Taxonomy of *Sellaphora* and Systematic Position of *Microcostatus. Plants*, 11 (16).

[9.38] Graebner, P. 1895. *Studien über die Norddeutsche Heide*, Engelmann.

[9.39] Guo, H., Yao, Y., Li, Y., Wang, P., Hu, C., Yuan, M., Mei, H., and Zhu, J. 2022. A case study in forensic soil comparison. *Journal of Forensic Sciences*, 67 (2), 766–774.

[9.40] Hayek, J.M.W., and Hulbary, R.L. 1956. A survey of soil diatoms. *Proceedings of the Iowa Academy of Science*, 63, 327–338.

[9.41] Heger, T.J., Straub, F., and Mitchell, E.A.D. 2012. Impact of farming practices on soil diatoms and testate amoebae: a pilot study in the DOK-trial at Therwil, Switzerland. *European Journal of Soil Biology*, 49, 31–36.

[9.42] Hejduková, E., Pinseel, E., Vanormelingen, P., Nedbalová, L., Elster, J., Vyverman, W., and Sabbe, K. 2019. Tolerance of pennate diatoms (Bacillariophyceae) to experimental freezing: comparison of polar and temperate strains. *Phycologia*, 58 (4), 382–392.

[9.43] Hinder, S.L., Hays, G.C., Edwards, M., Roberts, E.C., Walne, A.W., and Gravenor, M.B. 2012. Changes in marine dinoflagellate and diatom abundance under climate change. *Nature Climate Change*, 2 (4), 271–275.

[9.44] Hoffmann, L. 1989. Algae of terrestrial habitats. *Botanical Review*, 55 (2), 77–105.

[9.45] Hoffmann, L., Ector, L., and Kostikov, I. 2007. Algal flora from limed and unlimed forest soils in the Ardenne (Belgium). *Systematics and Geography of Plants*, 77 (1), 15–90.

[9.46] Jewson, D.H., Lowry, S.F., and Bowen, R. 2006. Co-existence and survival of diatoms on sand grains. *European Journal of Phycology*, 41 (2), 131–146.

[9.47] Johansen, J.R. 2010. Diatoms of aerial habitats, in *The Diatoms*, Cambridge University Press, pp. 465–472.

[9.48] Johansen, J.R., and Sray, J.C. 1998. *Microcostatus* gen. nov., a new aerophilic diatom genus based on *Navicula* krasskei Hustedt. Diatom Research, 13 (1), 93–101.

[9.49] John, R.P. 1942. An ecological and taxonomic study of the algae of British soils : I. the distribution of the surface-growing algae. *Annals of Botany*, 6 (22), 323–349.

[9.50] Kabirov, R.R., and Gaisina, L.A. 2009. Parameters of the productivity of soil algae in terrestrial ecosystems. Eurasian Soil Science, 42 (12), 1374–1379.

[9.51] Kezlya, E., Glushchenko, A., Kociolek, J.P., Maltsev, Y., Martynenko, N., Genkal, S., and Kulikovskiy, M. 2020. *Mayamaea vietnamica* sp. Nov.: A new terrestrial diatom (bacillariophyceae) species from Vietnam. *Algae*, 35 (4), 325–335.

[9.52] Klaus, J., Wetzel, C.E., Martínez-Carreras, N., Ector, L., and Pfister, L. 2015. A tracer to bridge the scales: on the value of diatoms for tracing fast flow path connectivity from headwaters to meso-scale catchments. *Hydrological Processes*, 29 (25), 5275–5289.

[9.53] Kristiansen, J. 1996. Dispersal of freshwater algae -- a review. *Hydrobiologia*, 336 (1–3), 151–157.

[9.54] Kulikovskiy, M., Maltsev, Y., Glushchenko, A., Kuznetsova, I., Kapustin, D., Gusev, E., Lange-Bertalot, H., Genkal, S., and Kociolek, J.P. 2020. *Gogorevia*, a New Monoraphid Diatom Genus for *Achnanthes exigua* and Allied Taxa (Achnanthidiaceae) Described on the Basis of an Integrated Molecular and Morphological Approach. *Journal of Phycology*, 56 (6), 1601–1613.

[9.55] Loescher, J.E.H. 1972. Diatoms from a native Iowa prairie. Iowa State University

[9.56] Lowe, R.L., Furey, P.C., Ress, J.A., and Johansen, J.R. 2007. Diatom biodiversity and distribution on wetwalls in Great Smoky Mountains National Park. *Southeastern Naturalist*, 6 (2), 135–153.

[9.57] Lund, J.W.G. 1945. Observations on soil algae. I. The ecology, size and taxonomy of British soil diatoms (part 1). *New Phytologist*, 44 (2), 196–219.

[9.58] Maltsev, Y., and Maltseva, I. 2018. The influence of forest-forming tree species on diversity and spatial distribution of algae in forest litter. *Folia Oecologica*, 45 (2), 72–91.

[9.59] Maltsev, Y., Maltseva, S., Kociolek, J.P., Jahn, R., and Kulikovskiy, M. 2021. Biogeography of the cosmopolitan terrestrial diatom *Hantzschia amphioxys* sensu lato based on molecular and morphological data. *Scientific Reports*, 11 (1).

[9.60] Maltsev, Y.I., Didovich, S. v., and Maltseva, I.A. 2017. Seasonal changes in the communities of microorganisms and algae in the litters of tree plantations in the steppe zone. *Eurasian Soil Science*, 50 (8), 935–942.

[9.61] Maltsev, Y.I., Maltseva, I.A., Solonenko, A.N., and Bren, A.G. 2017. Use of soil biota in the assessment of the ecological potential of urban soils. *Biosystems Diversity*, 25 (4), 257–262.

[9.62] Mangadze, T., Bere, T., and Mwedzi, T. 2016. Choice of biota in stream assessment and monitoring programs in tropical streams: A comparison of diatoms, macroinvertebrates and fish. *Ecological Indicators*, 63, 128–143.

[9.63] Martínez-Carreras, N., Wetzel, C.E., Frentress, J., Ector, L., McDonnell, J.J., Hoffmann, L., and Pfister, L. 2015. Hydrological connectivity inferred from diatom transport through the riparian-stream system. *Hydrology and Earth System Sciences*, 19 (7), 3133–3151.

[9.64] Martínez-Carreras, N., Wetzel, C.E., Frentress, J., Ector, L., McDonnell, J.J., Hoffmann, L., Pfister, L., Wetzel, C.E., Martínez-Carreras, N., Iffly, J.F., Klaus, J., Holko, L., McDonnell, J.J., Wetzel, C.E., Martínez-Carreras, N., Ector, L., Pfister, L., Abbott, B.W., Baranov, V., Mendoza-Lera, C., Nikolakopoulou, M., Harjung, A., Kolbe, T., Balasubramanian, M.N., Vaessen, T.N., Ciocca, F., Campeau, A., Wallin, M.B., Romeijn, P., Antonelli, M., Gonçalves, J., Datry, T., Laverman, A.M., de Dreuzy, J.-R., Hannah, D.M., Krause, S., Oldham, C., Pinay, G., Pfister, L., Wetzel, C.E., Klaus, J., Martínez-Carreras, N., Antonelli, M., Teuling, A.J., McDonnell, J.J., Tauro, F., Martínez-Carreras, N., Barnich, F., Juilleret, J., Wetzel, C.E., Ector, L., Hissler, C., Pfister, L., McDonnell, J.J., Wrede, S., Hlúbiková, D., Matgen, P., Fenicia, F., Ector, L., and Hoffmann, L. 2015. Diatom percolation through soils: a proof of concept laboratory experiment. *Hydrological Processes*, 23 (3), 2841–2845.

[9.65] Minaoui, F., Hakkoum, Z., Douma, M., Mouhri, K., and Loudiki, M. 2021. Diatom Communities as Bioindicators of Human Disturbances on Suburban Soil Quality in Arid Marrakesh Area (Morocco). *Water, Air, & Soil Pollution*, 232 (4), 146.

[9.66] Moore, G.T., and Karrer, J.L. 1919. A Subterranean Algal Flora. *Annals of the Missouri Botanical Garden*, 6 (4), 281.

[9.67] Neustupa, J. 2001. Soil algae from marlstone-substratum based biotopes in the Nature park Džbán (Central Bohemia, Czech Republic) with special attention to the natural treeless localities. *Algological Studies*, 101, 109–120.

[9.68] Nortcliff, S. 2002. Standardisation of soil quality attributes. *Agriculture, Ecosystems and Environment*, 88 (2), 161–168.

[9.69] Paterson, D.M. 1986. The migratory behaviour of diatom assemblages in a laboratory tidal scanning electron microscopy micro-ecosystem examined by low temperature. *Diatom Research*, 1 (2), 227–239.

[9.70] Petersen, J.B. 1928. The aerial algae of Iceland, in *The botany of Iceland*, vol. 2, Wheldon and Wesley, Copenhagen & London, pp. 326–447.

[9.71] Pfister, L., McDonnell, J.J., Wrede, S., Hlúbiková, D., Matgen, P., Fenicia, F., Ector, L., and Hoffmann, L. 2009. The rivers are alive: on the potential for diatoms as a tracer of water source and hydrological connectivity. *Hydrological Processes*, 23 (19), 2841–2845.

[9.72] Pfister, L., Wetzel, C.E., Klaus, J., Martínez-Carreras, N., Antonelli, M., Teuling, A.J., and McDonnell, J.J. 2017. Terrestrial diatoms as tracers in catchment hydrology: a review. *Wiley Interdisciplinary Reviews: Water.*

[9.73] Pfister, L., Wetzel, C.E., Martínez-Carreras, N., Iffly, J.F., Klaus, J., Holko, L., and McDonnell, J.J. 2015. Examination of aerial diatom flushing across watersheds in Luxembourg, Oregon and Slovakia for tracing episodic hydrological connectivity. *Journal of Hydrology and Hydromechanics*, 63 (3).

[9.74] Pinseel, E. 2019. Diversity, Biogeography and Evolutionary History of Diatoms: *Pinnularia Borealis* as a Case Study, Ghent University.

[9.75] Pinseel, E., Janssens, S.B., Verleyen, E., Vanormelingen, P., Kohler, T.J., Biersma, E.M., Sabbe, K., Van de Vijver, B., and Vyverman, W. 2020. Global radiation in a rare biosphere soil diatom. *Nature Communications*, 11 (1).

[9.76] Pinseel, E., Kulichová, J., Scharfen, V., Urbánková, P., Van de Vijver, B., and Vyverman, W. 2019. Extensive cryptic diversity in the terrestrial diatom *Pinnularia borealis* (Bacillariophyceae). *Protist*, 170 (2), 121–140.

[9.77] Pipe, A.E. 1992. Pesticide effects on soil algae and cyanobacteria, in *Reviews of Environmental Contamination and Toxicology*, vol. 127, Springer-Verlag, pp. 95–171.

[9.78] Pipe, A.E., and Cullimore, D.R. 1980. An implanted slide technique for examining the effects of the herbicide diuron on soil algae. *Bulletin of Environmental Contamination and Toxicology*, 24 (1), 306–312.

[9.79] Pipe, A.E., and Cullimore, R.D. 1984. Influence of five phenylurea herbicides on the diatom *Hantzschia* in a sandy loam soil. *Bulletin of Environmental Contamination and Toxicology*, 33, 439–443.

[9.80] Plumley, G.F., and Davis, D.E. 1980. The effects of a photosynthesis inhibitor atrazine, on salt marsh edaphic algae, in culture, microecosystems, and in the field. *Estuaries*, 3 (4), 271–277.

[9.81] Poradowska, A. 2020. Diatoms (Bacillariophyta) from the genus *Eunotia* and *Pinnularia* developing on soils in the open landscape of the low beskids. *Journal of Ecological Engineering*, 21 (7), 257–270.

[9.82] Potapova, M.G., Kersey, M.H., and Aycock, L.L. 2023. Diversity and distribution of spine-bearing species of *Pinnularia* in eastern North America. *Diatom Research*, 38 (1), 33–53.

[9.83] Ress, J.A. 2012. The ecology of aerial algae, Bowling Green State University.

[9.84] Rimet, F., Pinseel, E., Bouchez, A., Japoshvili, B., and Mumladze, L. 2023. Diatom endemism and taxonomic turnover: Assessment in high-altitude alpine lakes covering a large geographical range. *Science of The Total Environment*, 871, 161970.

[9.85] Rindi, F. 2011 Terrestrial green algae: Systematics, biogeography and expected responses to climate change, in *Climate Change, Ecology and*

Systematics (eds. Hodkinson, T.R., Jones, M.B., Waldren, S., and Parnell, J.A.N.), Cambridge University Press, pp. 201–228.
[9.86] Rings, A., Lücke, A., and Schleser, G.H. 2004. A new method for the quantitative separation of diatom frustules from lake sediments. *Limnology and Oceanography: Methods*, 2 (1), 25–34.
[9.87] Rivera, S.F., Vasselon, V., Bouchez, A., and Rimet, F. 2020. Diatom metabarcoding applied to large scale monitoring networks: Optimization of bioinformatics strategies using Mothur software. *Ecological Indicators*, 109 (105775), 1–13.
[9.88] Schmidt, O., Dyckmans, J., and Schrader, S. 2016. Photoautotrophic microorganisms as a carbon source for temperate soil invertebrates. *Biology Letters*, 12 (1).
[9.89] Schramm, J.R. 1914. Some pure culture methods in the algae. *Annals of the Missouri Botanical Garden*, 1 (1), 23–45.
[9.90] Schulz, K., Mikhailyuk, T., Dreßler, M., Leinweber, P., and Karsten, U. 2016. Biological soil crusts from coastal dunes at the Baltic Sea: Cyanobacterial and algal biodiversity and related soil properties. *Microbial Ecology*, 71 (1), 178–193.
[9.91] Schuttler, P.L., and Weaver, T. 1986. Concentrating soil diatoms for assemblage description. *Soil Biology and Biochemistry*, 18 (4), 389–394.
[9.92] Scott, K.R., Morgan, R.M., Jones, V.J., and Cameron, N.G. 2014. The transferability of diatoms to clothing and the methods appropriate for their collection and analysis in forensic geoscience. *Forensic Science International*, 241, 127–137.
[9.93] Shubert, L.E., and Starks, T.L. 1979. Algal succession on orphaned coal mine spoils. *Ecology and coal resource development*, 2 (2), 661–669.
[9.94] Smol, J.P., and Stoermer, E.F. 2010. *The diatoms: Applications for the environmental and earth sciences*, Cambridge University Press.
[9.95] Souffreau, C., Vanormelingen, P., Sabbe, K., and Vyverman, W. 2013. Tolerance of resting cells of freshwater and terrestrial benthic diatoms to experimental desiccation and freezing is habitat-dependent. *Phycologia*, 52 (3), 246–255.
[9.96] Souffreau, C., Vanormelingen, P., Van de Vijver, B., Isheva, T., Verleyen, E., Sabbe, K., and Vyverman, W. 2013. Molecular evidence for distinct Antarctic lineages in the cosmopolitan terrestrial diatoms *Pinnularia borealis* and *Hantzschia amphioxys*. *Protist*, 164 (1), 101–115.
[9.97] Souffreau, C., Vanormelingen, P., Verleyen, E., Sabbe, K., and Vyverman, W. 2010. Tolerance of benthic diatoms from temperate aquatic and terrestrial habitats to experimental desiccation and temperature stress. *Phycologia*, 49 (4), 309–324.
[9.98] Stanek-Tarkowska, J., and Noga, T. 2012. Diversity of diatoms (Bacillariophyceae) in the soil under traditional tillage and reduced tillage. *Inżynieria Ekologiczna*, (30), 287–296.

[9.99] Stanek-Tarkowska, J., Czyz, E.A., Dexter, A.R., and Sławiński, C. 2018. Effects of reduced and traditional tillage on soil properties and diversity of diatoms under winter wheat. *International Agrophysics*, 32 (3), 403–409.

[9.100] Stanek-Tarkowska, J., Noga, T., Pajączek, A., and Peszek, U. 2013. The occurrence of *Sellaphora nana* (Hust.) Lange-Bert. Cavacini, Tagliaventi & Alfinito, *Stauroneis borichii* (J.B. Petersen) J.W.G. Lund, *S. parathermicola* (J.B. Petersen) J.W.G. Lund on agricultural soils. *Algological Studies*, 142 (1), 109–119.

[9.101] Stanek-Tarkowska, J., Wetzel, C.E., Noga, T., and Ector, L. 2016. Study of the type material of *Navicula egregia* Hustedt and descriptions of two new aerial *Microcostatus* (Bacillariophyta) species from Central Europe. *Phytotaxa*, 280 (2), 163–172.

[9.102] Starks, T.L., Shubert, L.E., and Trainor, F.R. 1981. Ecology of soil algae: a review. *Phycologia*, 20 (1), 65–80.

[9.103] Stock, W., Pinseel, E., De Decker, S., Sefbom, J., Blommaert, L., Chepurnova, O., Sabbe, K., and Vyverman, W. 2018. Expanding the toolbox for cryopreservation of marine and freshwater diatoms. *Scientific Reports*, 8 (1), 4279.

[9.104] Taylor, J.C., Levanets, A., Blanco, S., and Ector, L. 2010. *Microcostatus schoemanii* sp. nov., *M. cholnokyi* sp. nov. and *M. angloensis* sp. nov. three new terrestrial diatoms (Bacillariophyceae) from South Africa. *Phycological Research*, 58 (3), 177–187.

[9.105] Tittor, I. 1956. Untersuchungen über die Bodenmikroflora von Gatersleben und Umgebung. *Die Kulturpflanze*, 4 (1), 195–207.

[9.106] Tolhurst, T.J., Gust, G., and Paterson, D.M. 2002. The influence of an extracellular polymeric substance (EPS) on cohesive sediment stability. *Proceedings in Marine Science*, 5 (C), 409–425.

[9.107] Vacht, P., Kuu, A., Puusepp, L., Koff, T., Kutti, S., Raamets, J., and Küttim, L. 2018. Diatom and microarthropod communities of three airfields in Estonia-their differences and similarities and possible linkages to airfield properties. *European Journal of Ecology*, 4 (2), 1–14.

[9.108] Vacht, P., Puusepp, L., Koff, T., and Reitalu, T. 2014. Variability of riparian soil diatom communities and their potential as indicators of anthropogenic disturbances. *Estonian Journal of Ecology*, 63 (3), 168–184.

[9.109] van Dam, H., Mertens, A., and Sinkeldam, J. 1994. A coded checklist and ecological indicator values of freshwater diatoms from the Netherlands. *Netherland Journal of Aquatic Ecology*, 28 (1), 117–133.

[9.110] Van de Vijver, B., and Beyens, L. 1997. The epiphytic diatom flora of mosses from Stromness Bay area, South Georgia. *Polar Biology*, 17 (6), 492–501.

[9.111] Van de Vijver, B., and Beyens, L. 1998. A preliminary study on the soil diatom assemblages from Ile de la Possession (Crozet, Subantarctica). *European Journal of Soil Biology*, 34 (3), 133–141.

[9.112] Van de Vijver, B., and Cox, E.J. 2015. *Fallacia emmae* sp. nov., (Bacillariophyta) a New Soil-Inhabiting Diatom Species from the Sub-Antarctic Region. *Cryptogamie Algologie*, 36 (3), 245–254.

[9.113] Van de Vijver, B., and le Cohu, R. 2003. Two new species of the genus *Geissleria* Lange-Bertalot & Metzeltin (Bacillariophyceae) from the Kerguelen and Crozet archipelagos (TAAF, Subantarctica). *Nova Hedwigia*, 77, 341–349.

[9.114] Van de Vijver, B., Frenot, Y., Beyens, L., and Lange-Bertalot, H. 2005. *Labellicula*, a new diatom genus (Bacillariophyta) from Ile de la Possession (Crozet Archipelago, Subantarctica). *Cryptogamie, Algologie*, 26 (2), 125-133.

[9.115] Van de Vijver, B., Ledeganck, P., and Beyens, L. 2002. Soil diatom communities from Ile de la Possession (Crozet, sub-Antarctica). *Polar Biology*, 25, 721–729.

[9.116] Van de Vijver, B., Ledeganck, P., and Beyens, L. 2002. Three new species of *Diadesmis* from soils of Ile de la Possession (Crozet Archipelago, Subantarctic). *Cryptogamie, Algologie*, 23 (4), 333-341.

[9.117] Van de Vijver, B., Mataloni, G., Stanish, L., and Spaulding, S.A. 2010. New and interesting species of the genus *Muelleria* (Bacillariophyta) from the Antarctic region and South Africa. *Phycologia*, 49 (1), 22–41.

[9.118] van der Werff, A. 1953. A new method of concentrating and cleaning diatoms and other organisms. *SIL Proceedings*, 1922-2010, 12 (1), 276–277.

[9.119] van Kerckvoorde, A., Trappeniers, K., Nijs, I., and Beyens, L. 2000. Terrestrial soil diatom assemblages from different vegetation types in Zackenberg (Northeast Greenland). *Polar Biology*, 23 (6), 392–400.

[9.120] Vasselon, V., Rimet, F., Tapolczai, K., and Bouchez, A. 2017. Assessing ecological status with diatoms DNA metabarcoding: Scaling-up on a WFD monitoring network (Mayotte island, France). *Ecological Indicators*, 82, 1–12.

[9.121] Weilhoefer, C.L., and Pan, Y. 2008. Using change-point analysis and weighted averaging approaches to explore the relationships between common benthic diatoms and in-stream environmental variables in mid-Atlantic highlands streams, USA. *Hydrobiologia*, 614 (1), 259–274.

[9.122] Wetzel, C.E., Ector, L., de Vijver, B., Compere, P., and Mann, D.G. 2015. Morphology, typification and critical analysis of some ecologically important small naviculoid species (Bacillariophyta). *Fottea*, 15 (2), 203–234.

[9.123] Wetzel, C.E., Martínez-Carreras, N., Hlúbiková, D., Hoffmann, L., Laurent, P., and Ector, L. 2013. New combinations and type analysis of *Chamaepinnularia* species (Bacillariophyceae) from aerial habitats. *Cryptogamie, Algologie*, 34 (2), 149–168.

[9.124] Wetzel, C.E., van de Vijver, B., Kopalová, K., Hoffmann, L., Pfister, L., and Ector, L. 2014. Type analysis of the South American diatom *Achnanthes haynaldii* (Bacillariophyta) and description of *Planothidium amphibium* sp. nov., from aerial and aquatic environments in Oregon (USA). *Plant Ecology and Evolution*, 147 (3), 439–454.

[9.125] Willson, D., and Forest, H.S. 1957. An exploratory study on soil algae. *Ecology*, 38 (2), 309–313.

[9.126] Wu, Y., Rao, B., Wu, P., Liu, Y., Li, G., and Li, D. 2013. Development of artificially induced biological soil crusts in fields and their effects on top soil. *Plant and Soil*, 370 (1–2), 115–124.

[9.127] Zhang, Y., Ouyang, S., Nie, L., and Chen, X. 2020. Soil diatom communities and their relation to environmental factors in three types of soil from four cities in central-west China. *European Journal of Soil Biology*, 98 (March), 103175.

[9.128] Zidarova, R., van de Vijver, B., Quesada, A., and de Haan, M. 2010. Revision of the genus *Hantzschia* (Bacillariophyceae) on Livingston Island (South Shetland Islands, Southern Atlantic Ocean). *Plant Ecology and Evolution*, 143 (3), 318–333.

[9.129] Zurek, L. 1981. The influence of the herbicides Lenacil and Pyrazon on the soil algae. *Polish Journal of Ecology*, 29 (3), 327–342.

[9.130] Bakieva, G. R., Khaibullina, L. S., Gaisina, L. A., Kabirov, R. R. 2012. Ecological-floristic analysis of soil algae and cyanobacteria on the Tra-Tau and Yurak-Tau mounts, Bashkiria. *Eurasian Soil Science*, 45 (9), 873-881.

[9.131] Bataeva Y. V., Dzerzhinskaya, I. S., and Yakovleva, L. V. 2017. Composition of phototrophs in different soil types of Astrakhan oblast. *Eurasian Soil Science*, 50 (8), 943-951.

[9.132] Brendemühl, I. 1949. Über die Verbreitung der Erddiatomeen. *Archiv für Mikrobiologie*, 14 (3), 407-449.

[9.133] Broady, P. 2005. The distribution of terrestrial and hydro-terrestrial algal associations at three contrasting locations in southern Victoria Land, Antarctica. *Algological studies*, 118, 95-112.

[9.134] Chattová, B., and Uher, B. 2013. A preliminary study on diversity and ecology of soil diatoms in urban habitats. *Študenská Vedecká Konferencia*, 105-110.

[9.135] Dorokhova, Marina F., Kosheleva, Natalia E., and Terskaya, Elena V. 2015. Algae and cyanobacteria in soils of Moscow. *American Journal of Plant Sciences*, 6, 2461-2471.

[9.136] Fairchild, E., and Wilson, D. 1967. The algal flora of two Washington soils. *Ecology*, 48 (6), 1053-1055.

[9.137] Fathi, A., and Zaki, F. 2003. Preliminary survey of edaphic algae in El-Minia region, Nile valley, Egypt. *Egyptian Journal of Phycology*, 4 (2), 131-148.

[9.138] Fehér, D. 1936. Untersuchungen über die regionale Verbreitung der Bodenalgen. *Archiv für Mikrobiologie*, 439-476.

[9.139] Fermani, P., Mataloni, G., and Van de Vijver, B. 2007. Soil microalgal communities on an Antarctic active volcano (Deception Island, South Shetlands). 30 (11), 1381-1393.

[9.140] Ferreira da Silva, J., Linton, M. A. O., Ribeiro da Anunciação, R., Pereira da Silva, E., Paidano Alves, R. Luis Schünemann, A., de Carvalho Victoria, F., Pereira de Albuquerque, M., and Batista Pereira, A. 2019. Distribution of aerophilous diatom communities associated with terrestrial green macroalgae in the South Shetland Islands, Maritime Antarctica. *PLoS ONE*, 14 (12), 1-15.

[9.141] Flechtner, V., Johansen, J. R., and Clark, W. 1998. Algal composition of microbiotic crusts from the Central Desert of Raja California, Mexico. *Great Basin Naturalist*, 58 (4), 295-311.

[9.142] Flechtner, V., Johansen, J. R., and Belnap, J. 2008. The biological soil crusts of the San Nicolas Island: Enigmatic algae from a geographically isolated ecosystem. Western North American Naturalist, 68 (4), 405-436.

[9.143] Al-Fredan, M. A., and Fathi, A. A. 2007. Preliminary survey of edaphic algae in Al-Hasa region, Saudi Arabia. *Pakistan Journal of Biological Sciences*, 10 (18), 3210-3214.

[9.144] Gremmen, N., Van de Vijver, B., Frenot, Y., and Lebouvier, M. 2007. Distribution of moss-inhabiting diatoms along an altitudinal gradient at sub-Antarctic Îles Kerguelen. *Antarctic Science*, 19 (1), 17-24.

[9.145] Grondin, A., and Johansen, J. R. 1995. Seasonal succession in a soil algal community associated with a beech-maple forest in northeastern Ohio, USA. Nova Hedwigia, 60 (1-2), 1-12.

[9.146] Hahn, A., and Neuhaus, W. 1997. Boden-diatomeen einer landwirtschaftlichen nutzflache bei Potsdam, Deutschland. Nova Hedwigia, 65 (1-4), 285-298.

[9.147] Hunt, M., Floyd, G., and Stout, B. 1979. Soil algae in field and forest environments. *Ecology*, 60 (2), 362-375.

[9.148] Ibraheem, B. 2003. Preliminary survey of microalgal soil crusts in a xeric habitats (Wadi-Araba, Eastern Desert, Egypt). *Egyptian Journal of Phycology*, 4 (1), 19-35.

[9.149] Johansen J. R., Javakul, A., and Rushforth S. R. 1982. Effects of Burning on the Algal Communities of a High Desert Soil near Wallsburg, Utah. *Journal of Range Management*, 35 (5), 598-600.

[9.150] Johansen, J. R., St. Clair, L. L., Webb, B. L., and Nebeker, G. T. 1984. Recovery patterns of cryptogamic soil crusts in desert rangelands following fire disturbance. *The Bryologist*, 87 (3), 238-243.

[9.151] Lin, C., Chou, T., and Wu, J. 2013. Biodiversity of soil algae in the farmlands of mid-Taiwan. Botanical Studies, 54 (1), 1-12.

[9.152] Lukešová, A., and Hoffmann, L. 1996. Soil algae from acid rain impacted forest areas of the Krušné hory Mts. 1. Algal communities. *Vegetatio*, 125 (2), 123-136.

[9.153] Lukešová, A., and Komárek, J. 2011. Succession of soil algae on dumps from strip coal-mining in the most region (Czechoslovakia). *Folia Geobotanica and Phytotaxonomica*, 22 (4), 355-362.

[9.154] Maltsev, Y. I., Pakhomov, A. Y., and Maltseva, I. A. 2017. Specific features of algal communities in forest litter of forest biogeocenoses of the steppe zone. *Contemporary Problems of Ecology*, 10 (1), 71-76.
[9.155] Maxwell, C. 1991. Floristic changes in soil algae and cyanobacteria in reclaimed metal-contaminated land at Sudbury, Canada. *Water, Air, and Soil Pollution*, 60, 381-393.
[9.156] Moravcová, A., Beyens, L., and Van de Vijver, B. 2010. Diatom communities in soils influenced by the wandering albatross (Diomedea exulans). *Polar Biology*, 33 (2),241-255.
[9.157] Noga, T., Kochman, N., Peszek, Ł., Stanek-Tarkowska, J., and Pająjczek, A. 2014. Diatoms (Bacillariophyceae) in rivers and streams and on cultivated soils of the Podkarpacie region in the years 2007-2011. *Journal of Ecological Engineering*, 15 (1), 6-25.
[9.158] Novakovskaya, I. V., Dubrovskiy, Y. A., Patova, E. N., Novakovskiy, A. B., and Sterlyagova, I. N. 2020. Influence of ecological factors on soil algae in different types of mountain tundra and sparse forests in the Northern Urals. *Phycologia*, 59 (4), 320-329.
[9.159] Phillipson, J. 1935. Some algae of Victorian soils. *Proceedings of the Royal Society of Victoria*, 47, 262-287.
[9.160] Rosa, K., and Lhotský, O. 1971. Edaphische Algen und Protozoen im Isergebirge, Tschechoslowakei. *Oikos*, 22 (1), 21-29.
[9.161] Round, F. 1958. Observations on the diatom flora of Braunton Burrows, N. Devon. *Hydrobiologia*, 11 (2),119-127.
[9.162] Shcherbyna, V. V., Maltseva, I. A., Maltsev, Y. I., and Solonenko, A. N. 2017. Post-pyrogenic changes in vegetation cover and biological soil crust in steppe ecosystems. *Regulatory Mechanisms in Biosystems*, 8 (4), 633-638.
[9.163] Stanek-Tarkowska, J., Noga, T., Kochman-Kędziora, N., and Rybak, M. 2016. Diatom Assemblages Growing on Cropping Soil in Pogórska Wola Near Tarnów. *Ecological Engeneering*, 46 (46), 128-134.
[9.164] Stanek-Tarkowska, J., Noga, T., Kochman-Kędziora, N., Peszek, Ł., Pajączek, A., and Kozak, A. 2015. The diversity of diatom assemblages developed on fallow soil in Pogórska Wola near Tarnów (southern Poland). Acta Agrobotanica, 68 (1),33-42.
[9.165] Stanek-Tarkowska, J., Czyz, E. A., Kaniuczak, J., and Poradowska, A. 2017. Physicochemical properties of silt loamy soil and diversity of diatom species under winter wheat and oats. *Journal of Ecological Engineering*, 19 (6), 142-151.
[9.166] Terlizzi, D., and Karlander, E. 1979. Soil algae from a maryland serpentine formation. *Soil Biology and Biochemistry*, 11 (2), 205-207.
[9.167] Zancan, S., Trevisan, R., and Paoletti, M. G. 2005. Soil algae composition under different agro-ecosystems in North-Eastern Italy. *Agriculture, Ecosystems and Environment*, 112 (1), 1-12.

10

Lacustrine Diatoms as Paleoclimate Proxies and Their Use in Climate Change Research

Gabriela S. Hassan

Instituto de Investigaciones Marinas y Costeras (IIMyC), Consejo Nacional de Investigaciones Científicas y Técnicas (CONICET)- Facultad de Ciencias Exactas y Naturales, Universidad Nacional de Mar del Plata (UNMdP), Mar del Plata, Argentina

Abstract

In this chapter, the potentialities and limitations of using diatoms to infer a range of climate-driven environmental variables are discussed. Given their sensibility and differential distribution along environmental gradients, diatoms have played a significant role in paleoclimatic research. Particularly since the development of statistical inference techniques, diatom-based paleoclimatic reconstructions shifted from strictly qualitative to a quantitative science based on the development of calibration sets and transfer functions. Studies have applied a series of diatom-inferred parameters, such as salinity, conductivity, water level, temperature, pH, and nutrients, to paleoclimatic reconstructions. As the links between salinity, conductivity and climate are more clearly understood, diatom responses to changes in these variables have been successfully linked to past climatic changes, producing a series of robust reconstructions of Holocene paleoclimates. The usefulness of pH, nutrients and temperature inferences is less clear and still under debate, as the links between these variables and climate are often complex and difficult to disentangle from non-climatic controls of lake dynamics. Local and regional conditions can also play a major role in structuring diatom assemblages, signaling the need for more localized studies covering a variety of timescales and waterbodies, particularly in aquatic systems of the Southern Hemisphere, which have received far less attention than their Northern Hemisphere counterparts.

Email: ghassan@mdp.edu.ar

Gabriela S. Hassan: https://www.researchgate.net/profile/Gabriela-Hassan; https://www.iimyc.gob.ar/iimyc/es/grupos-de-investigacion/grupo-epaac/

Nora I. Maidana, Magdalena Licursi and Eduardo Morales (eds.) Diatom Ecology: From Molecules to Metacommunities, (329–360) © 2024 Scrivener Publishing LLC

Keywords: Diatoms, salinity, pH, temperature, water level, nutrients, ice cover, holocene

10.1 Introduction

Since its origins in the 17th century, paleoclimatology has changed our view of the Earth's climate system, both by providing evidences of the occurrence of past climates that were significantly different from present ones, and by allowing to recognize past global temperature as a dynamic system, intricately linked to several factors, such as changes in solar radiation, continental configurations, oceanic patterns, and CO_2 concentrations [10.13] [10.14] [10.70]. In the centuries that have passed since then, our understanding of the natural climatic shifts that occurred in the history of the Earth has increased, providing valuable information to hypothesize about the impact that future climate changes can exert on ecological systems [10.13]. This knowledge is of central importance in the current context of climate change, as models based on different scenarios of CO_2 emission suggest a significant impact on the future climate system, with a predicted global temperature (GT) increase between 1.8 and 5.6 °C by 2100 [10.35]. These changes are expected to have differential impacts among regions, highlighting the importance of conducting studies addressing local scenarios. As an example, while the South American monsoon region and some mid-latitude and semi-arid zones are expected to present the highest increase in the maximum temperatures of hottest days (up to 1.5 to 2 GT), the Arctic is projected to experience the highest increase of temperature of the coldest days (3 GT) [10.35]. Additionally, the occurrence of extreme precipitation and drought events are also predicted, with precipitation increases projected over high latitudes, the equatorial Pacific and parts of the monsoon regions, and decreases over parts of the subtropics and in limited areas of the tropics [10.35]. In this context, the paleoclimatic record brings a natural baseline against which the magnitude and rate of the projected changes can be analyzed and compared among regions [10.60].

Recent studies suggest that modern rates of global warming are unprecedented when compared with natural fluctuations that occurred during the Holocene ([10.14] and references therein). According to quantitative comparisons of similarities between future projected climate scenarios and geohistorical data, analogous situations of 2030 CE projected climates have been identified in the Mid-Pliocene (3.3-3 Ma), while by 2150 CE, they would highly resemble Early Eocene (ca. 50 Ma) climates [10.13].

These advances of paleoclimatology have relied on the development of several lines of indirect research, as there are no direct indicators of the changes in temperature or precipitation. Robust climate change inferences are obtained by intersecting reconstructions from a variety of independent indicators (i.e., proxies) of geological, chemical and biological origin [10.54]. In all cases, the value of the proxy as indicator is based on its sensitivity to a certain environmental condition that is related, directly or indirectly, to climate. Assuming that their capability to reflect environmental changes has remained stable throughout time, paleoclimatologists can interpret their patterns of change in the fossil record to reconstruct the environmental conditions that originated them [10.69].

Among biological proxies, diatoms have played a major role in the reconstruction of past climate changes, being used as indicators of Holocene climate variability worldwide, especially in continental aquatic ecosystems [10.46]. As the distribution of diatoms in these environments has shown statistical responses of different strengths to environmental variables related both directly (i.e., surface-water and air temperature, ice cover) or indirectly (i.e., salinity, pH, nutrients, water level) to climate, paleoclimatologists have used modern ecological information to infer these environmental variables from diatom fossil assemblages. Most recent diatom-based paleoclimatological studies applied quantitative multivariate techniques to reconstruct past environmental variables; although qualitative studies of changes in diatom species habitats, life forms and autoecologies are also common. In both cases, an important assumption of these reconstructions is that "the environmental variable to be reconstructed is, or is linearly related to, an ecologically important variable in the system of interest" [10.12]. Although the links between salinity, nutrients, pH and diatom community responses are well known, there is a high degree of uncertainty about the physiological effects of temperature on diatoms [10.17]. There are multiple intercorrelated factors controlling diatom distribution in nature, and evidence suggests that temperature is far less important than other factors, posing doubts on the applicability of diatom-based temperature transfer functions in climate change research [10.3]. Moreover, as lakes clearly respond to local and regional climate, temporal and regional variability in diatom responses to climate can also impact on paleoclimatic reconstructions, signaling the need for more localized studies covering a variety of timescales [10.3]. This is particularly true for vast extensions of the Southern Hemisphere, where high-resolution paleoclimatology is in its infancy [10.51].

In this chapter, the potentialities and limitations of using diatoms to infer a range of climate-driven environmental variables are discussed, with an emphasis on lacustrine environments. Particularly, the rationale of applying diatom studies to reconstruct salinity, water level, pH, nutrients, temperature and ice cover is discussed, analyzing the links between these variables, climate, lake dynamics and diatom responses. The main objective is to highlight the contributions of diatom paleolimnology to the study of climate change and identify the most promising lines of research.

10.2 Lacustrine Diatoms as Archives of Climatic Signals

In order to use lacustrine diatoms as climate change proxies, significant and ecologically meaningful relationships between assemblage characteristics and climate-driven limnological variables are necessary. The sensitivity of lakes to climate has been extensively proven in the literature, as several examples of the responses of their physical, chemical and biological properties to climate changes have been published in the last decades ([10.1] and references therein). These studies demonstrated that lakes are good sentinels of climate change, both because of their sensitivity to environmental changes and their capability to integrate changes in their surrounding environment in the sedimentary record [10.43] [10.72]. The main mechanisms of climate forcing in lakes include changes in the solar radiation and water balances (Figure 10.1a). Radiation determines light and temperature regimes, which are modulated by winter ice, snow cover and wind, and influences indirectly the depth, duration and intensity of water-column stratification, subsequently controlling many chemical and biological processes (e.g., primary production, nutrient cycling, O_2 consumption, alkalinity and pH). Changes in precipitation/evaporation ratios, on the other hand, influence lake level and littoral habitats development, and can significantly alter the chemical characteristics of the water body (e.g., pH, nutrients, ionic composition and conductivity [10.5]). These changes, in turn, may affect the physiological and ecological responses of the lake's biota, including diatoms (Figure 10.1b).

Diatom assemblages respond to climate-driven environmental changes in several ways (Figure 10.1c): abundance and compositional changes include shifts in species composition, dominance structure, diversity, and variations in the proportional abundances of planktonic and benthic taxa

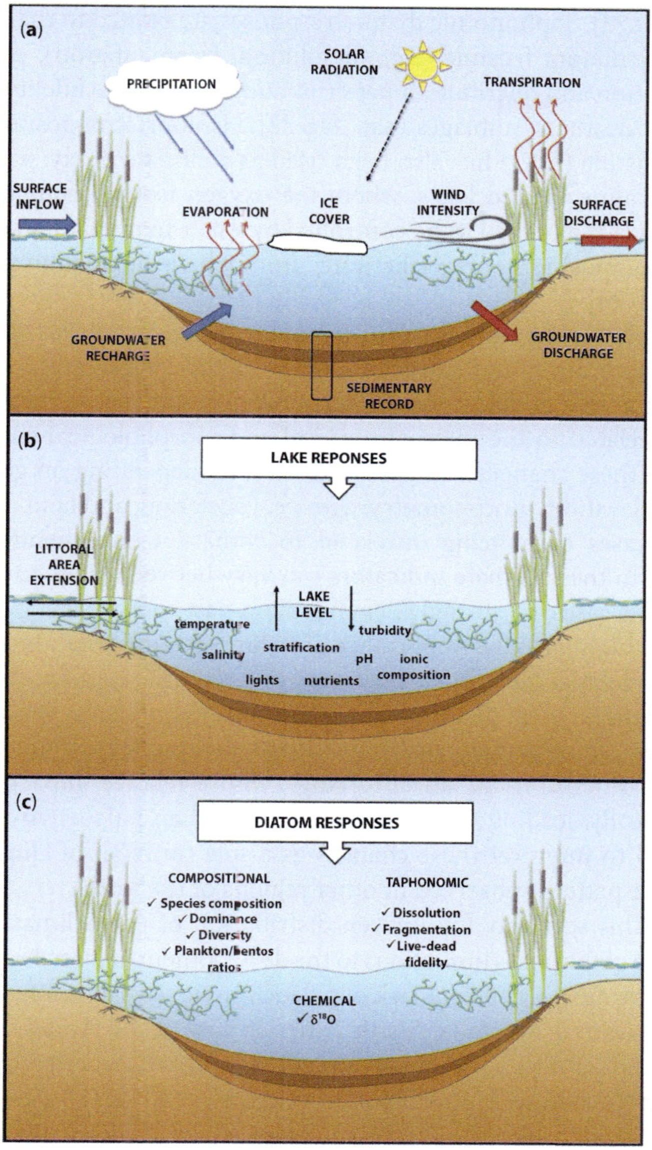

Figure 10.1 Diagram showing the potential effects of climatic forcings (a) and lake responses (b) on diatom assemblages (c). See Section 10.1 for details.

[10.25] [10.73]. Taphonomic diatom responses are related to the preservation of the diatom frustule (e.g., dissolution, fragmentation), changes in sedimentation and distribution patterns, and biases in the fidelity between living and death assemblages (e.g., [10.32]). Isotopic composition of the diatom frustule ($\delta^{18}O$) has also been used as a climate proxy, particularly in non-alkaline diluted lakes, where the oxygen isotope composition of diatom silica is assumed to be controlled by water temperature and by the isotopic composition of the lake water from which the diatom frustule is formed [10.44].

Although the literature is full of examples of successful uses of response variables as indirect indicators of climate change, these are not broadly applicable across all lakes. As the responses of indicator variables to climate are related to the internal physical and morphological structure of each lake, these changes will behave differently depending on geographic location, elevation, morphometry, climate, vegetation, and land use [10.1], in some cases even being unrelated to climate. Consequently, diatom responses to these climate indicators can vary between lakes and regions, and in some cases are strongly dependent on past and present local conditions. This highlights the importance of a complete understanding of internal lake dynamics and climatic forcings as a requisite to infer climatically driven changes from diatom signals. Even environmental variables that have proven to be influenced by climate to a certain extent are subjected to several other non-climatical forcings, whose relative impact can vary geographically, leading to misinterpretations when paleoenvironmentalists intend to interpret these changes as a sole function of climate or to extrapolate patterns observed in other regions of the world.

Under this scenario, the uneven distribution of paleoclimatic records around the globe poses limitations to the development of inter-hemispheric models. The most significant lack of paleoclimatic records is found in most of Africa, central Australia, South America east of the Andes, the south Atlantic and Southern Oceans. Among these, South America is the only region where various high-resolution records exist [10.51]. Paleoclimatic research has increased significantly in tropical and Andean areas of South America in recent decades, and diatom research has played an important role in this development (e.g., [10.47] [10.50] [10.55] [10.68]). In this chapter, some of these works are used to exemplify paleoclimatic uses of diatoms, as a way to highlight the potential development of the discipline in this largely unexplored area. In the following sections, the relationship between diatoms and a selected group of climate-driven environmental variables is analyzed, and the pros and cons of their application to paleoclimatic studies are discussed.

10.3 Diatoms, Limnology and Climate

10.3.1 Numerical Methods

Paleolimnology, as well other branches of paleoecology, started primarily as a qualitative science, with data from the different proxies used presented as qualitative categories (e.g., acid, brackish, cool, temperate, moist, dry, etc.) [10.11]. This has changed in the last decades, as the development of quantitative techniques (i.e., transfer functions) allowed numerically modeling of environmental variables through the development of training sets, which can then be applied to the fossil record [10.38]. In doing so, past environmental conditions can be quantitatively inferred from the species composition of the fossil assemblages, based on estimates of species responses derived from modern environments [10.11].

Quantitative reconstructions assume that knowledge of an organism's modern ecology and environmental preferences can be used to make inferences about past conditions, and involves three separate steps: (1) the development of a modern training set, (2) the development of a numerical model relating modern biological responses to contemporary physicochemical environmental predictors, and (3) the application of the model to the fossil record and the evaluation of the resulting quantitative reconstruction (see [10.11] for details on the statistical analyses involved). The transfer function approach has been extensively applied in diatom-based paleoenvironmental reconstructions since the 1980s, when the controversy about acid rain and the rise of acidification projects boosted this area of research [10.38]. In the following sections, examples of the application of these techniques to the reconstruction of the discussed environmental variables are given, illustrating the applicability of these methods in different areas of diatom-based paleoclimatic research.

10.3.2 Salinity and Conductivity

Salinity, conductivity and ionic composition are among the main environmental variables that drive diatom distribution in continental settings [10.23]. The hydrologic sensibility to climate change varies significantly among lakes, being closed-basin lakes without surface outflow more susceptible to show changes in level related to changes in the precipitation and evaporation (P-E) balance. The evaporative concentration of salts within these basins causes salinization, leading to climate-driven increases in salinity (and the related variable, conductivity), and altering the ionic composition (i.e., carbonates, sulfates, chlorides, and the major cations Ca,

Mg and Na) via precipitation of minerals during evaporative concentration [10.25]. This is of great importance for diatom-based paleoclimatic reconstructions, as diatoms vary in their tolerances to salinity, conductivity and brine composition, with some species being distributed along a wide range of conditions, and others restricted to narrow environmental gradients (e.g., [10.16] [10.24] [10.27]). While significant correlations between modern diatom assemblages and ionic concentrations are well documented around the world, the mechanisms responsible for these patterns are not completely understood. Salinity clearly poses osmotic pressures, with the subsequent energetic cost for diatoms [10.25]. Experimental and field studies have demonstrated the impact of salinity on nutrient requirements, availability and uptake (see [10.23] for a review). This differential distribution of taxa along salinity gradients allows the use of relative abundances of diatoms showing distinct salinity optima and tolerances as indicators of past salinity fluctuations. To do so, researchers have historically relied on qualitative techniques based on the salinity classifications of taxa (i.e., halobion system [10.39]). In the last decades, diatom transfer functions have been successfully applied to reconstruct salinity, conductivity, cation and anion compositions all over the world, showing high and significant correlation coefficients (e.g., [10.24] [10.26] [10.58]).

Whatever the technique used, the inferred salinity fluctuations can be further related to climatic changes, as ionic concentration and composition have proven to be very sensitive and easily measurable indicators of changes in water balance and weathering rate [10.1]. Consequently, the potential application of diatom-inferred salinity as a proxy of climate depends on the sensibility of lakes to changes in the P-E balance and the subsequent changes in the concentration and composition of total dissolved solids [10.25]. Ionic-sensitive lakes are mostly found in semiarid and arid regions [10.27], were the shifts in the water balance (precipitation/evaporation ratios) can be large enough to cause strong and noticeable fluctuations in salinity [10.1], particularly in closed-basin lakes [10.25]. As these lakes are widely distributed around the globe, they have been extensively used to reconstruct past climate variability based on diatom-inferred salinity and conductivity changes in arid and semiarid regions of all continents (see [10.25] for examples).

Salinity and brine composition have also been related to frustule preservation, both in experimental and field-based studies ([10.18] and references therein). Silica dissolution is positively correlated with salinity [10.59], although the chemical causes of this interaction are not clearly understood [10.56]. It has been demonstrated that the addition of electrolytes to reacting solutions at near neutral pH accelerates the rate of quartz dissolution, probably by strengthening the nucleophilic properties of water molecules at the silica surface [10.20]. This effect has also been recognized in diatom experiments [10.18] [10.59], and is likely to be a key factor for diatom preservation in saline lakes. In this context, taphofacies analyses (i.e., facies defined on the basis of diagnostic taphonomic traits [10.65]) have demonstrated to be an independent source of paleoenvironmental information, as it allows inferring past environmental fluctuations from diatom taphonomic signatures both at local and regional scales. Taphofacies analysis is a powerful methodology, as it allows underscoring the non-random relationship between taphonomic signatures and environmental conditions in modern settings, highlighting the feasibility of reconstructing paleoenvironments from taphonomic analyses [10.32]. In shallow lakes from the Argentinean Pampas, for instance, valve dissolution was found to be a more sensitive indicator of salinity differences among lakes than shifts in relative abundances of diatom taxa [10.30]. In that work, examination of dissolution marks in valves of the target taxa *Cyclotella meneghiniana* in surface sediments of 25 lakes located along a strong salinity gradient (0.4–13.2 ppt) showed that salinity explained a higher percentage of the variance in taphonomical (33% of total variance explained; $p<0.0001$) rather than in compositional data (11% of total variance explained; $p<0.01$). This allowed the use of Holocene taphonomic profiles as proxies of salinity fluctuations, highlighting the potentiality of combining diatom taphonomy and community paleoecology as a promising tool for paleoclimatic reconstruction, particularly in cases where assemblages are dominated by euryhaline taxa or when bad preservation prevents a complete analysis of compositional data.

10.3.3 Water-Level Fluctuations

Lake-level fluctuations may have a direct impact on many limnological variables, such as habitat availability, light penetration, and chemical conditions, that can affect diatom communities [10.73]. In non-regulated lakes, water level is a good indicator of climate change, because it reflects the dynamic balance between water input (e.g., precipitation, runoff) and water loss (e.g., evaporation [10.1]). This is particularly true in closed-basin lakes, where climate signals are often strong, with water levels rising during wet periods and falling during dry phases. The climatic signal can, however, be confounded by several other factors that can also impact lake levels, such as geological (e.g., isostatic rebound), and biological (e.g., vegetation changes) processes, as well as human impact (e.g., catchment deforestation and nutrient inputs). Hence, while diatoms can be used to reconstruct past water-level fluctuations, the paleoclimatic interpretation of this signal should be supported by additional and independent evidences [10.73].

Most studies using diatoms to infer water-level and depth fluctuations apply a qualitative approach based on the classification of diatom taxa according to their life forms and habitats. Lacustrine diatoms are abundant both in planktonic and benthic habitats, which are the source communities for the sediment record [10.6]. Planktonic diatoms are related to the water column, where they can spend their whole life cycle (euplankton). Some taxa can successfully inhabit the benthic zone, and also survive if swept into the open water (tychoplankton). Benthic taxa are associated with substrates, being able to be attached to stones (epilithon), plants (epiphyton), sand grains (epipsammon), or mud (epipelon [10.6]). Generally, when lake levels are low, the development of habitats available for taxa living attached to substrates is often higher, and decreases when the water body deepens, favoring planktonic taxa. Exceptions to this pattern can occur in steep-sided lakes or lakes with complex morphologies, where the relationship between water depth and habitat availability may show the opposite pattern. Also, in some shallow lakes, increased depth may lead to greater turbidity or productivity-driven light limitation, leading to a decrease of planktonic taxa [10.73].

Overall, relating lower water levels to higher availability of benthic habitats has allowed qualitative assessment of water levels in the fossil record, with peaks of planktonic diatoms usually associated with higher water levels, and dominance of benthic taxa associated with shallower habitats (e.g., [10.47] [10.67] [10.68] [10.74]). In an example of the use of diatom-inferred water levels and salinity in paleoclimatic research, Sylvestre [10.67] analyzed relative variations of planktonic and benthic taxa to infer climatic changes between 21,000 and 17,500 ^{14}C yr BP in Lake Coipasa (Bolivian Altiplano). Given this is a closed-basin lake, these variations in taxa were associated directly with lake-level changes, and the application of a diatom-based salinity transfer function allowed quantitatively reconstructing the salinity changes that accompanied these fluctuations (Figure 10.2). Diatom results suggest that the lake experienced short-term fluctuations of water depth and salinity, with shallow water conditions indicated by low proportions of planktonic diatoms and high values of inferred salinities (up to 76 $g \cdot l^{-1}$), and greater depths inferred from higher percentages of planktonic taxa and decreased salinity values (Figure 10.2). Hence, the diatom record suggests that Lake Coipasa underwent short-term fluctuations of water depth and salinity related to increases in precipitation that led to episodic freshwater inputs. Comparison of Lake Coipasa and Lake

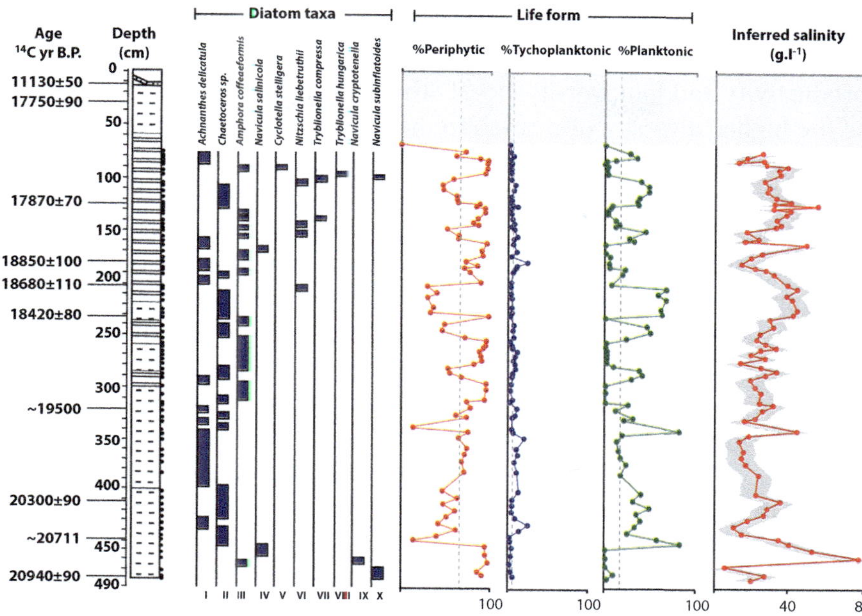

Figure 10.2 Paleoenvironmental analysis of a sediment core covering the period between 21,000 and 17,500 ^{14}C yr BP from Lake Coipasa (Bolivian Altiplano). See Section 10.3.3 for details. (Reprinted from [10.67] with permission from Springer Nature).

Titicaca records suggested that precipitation from Atlantic and Amazonia sources could have played an important role on the Altiplano during the Last Glacial Maximum [10.67].

While the plankton to benthic (P/B) ratio has been commonly used as a way to semi-quantitatively infer water-level changes from diatom assemblages [10.66], the use of transfer functions to quantitatively reconstruct water depth has been limited [10.28] [10.42], mainly as a consequence of the previously mentioned confounding factors that can obscure climatic signals [10.37]. Although the shifts between planktonic and benthic diatoms tend to reflect water depth in lakes with relatively simple bathymetry, they can be less useful in complex lakes where littoral areas can change nonlinearly, both increasing or decreasing its surface with lake depth. For example, Vélez and collaborators [10.71] found that epiphytic, aerophilous and planktonic diatoms at ca. 14,200 ^{14}C yr BP in Lago Fúquene (Colombia) indicated an increase (rather than a decrease) in lake level that extended the littoral area available for those diatoms to colonize, based on the increase in the planktonic *Aulacoseira ambigua* and on the fact that the epiphytic and aerophilous diatom species were also highly represented during high lake levels in the lake.

Increased diversity has also been signaled as an indicator of benthic habitat availability in lakes, as littoral zones are often hotspots of diatom productivity and biodiversity [10.2]. This has been related to factors such as the higher diversity of near-shore habitats that can host diatoms with different growth habits, a higher variation in light availability in the benthic zone of shallow areas, and the transfer of planktonic taxa though wind action into the littoral zone [10.42]. Accordingly, diversity profiles along sedimentary sequences have been interpreted as a function of water depth, with high diversity of benthic diatoms associated with increasing availability of shallow habitats with enough light availability and well-developed macrophyte growth (e.g., [10.50]). These profiles need to be interpreted with caution, however, as changing sedimentation rates over time can affect the time-averaging of individual diatom samples. In such cases, diversity variations may be artifacts of sedimentation rates, rather than a reflection of the ecology and biology of the species involved [10.61]. Once again, diatom studies should rely on independent sources of data and a complete knowledge of the sedimentary system in order to infer paleoclimatic changes from fossil data.

10.3.4 Temperature

The use of diatoms to reconstruct past temperature changes remains a controversial topic. Some authors have stated that diatoms are inappropriate for reconstructing lake water or air temperatures, mainly because of the limited influence that this variable exerts on diatom assemblages [10.3] [10.5]. While several studies suggest that temperature is an important aspect of algal ecology, as it affects key metabolic processes, such as respiration and photosynthesis (see [10.17] for a review), there are many potentially stronger influences on diatom assemblages. Temperature is more likely correlated with other limnological variables, such as salinity, pH, nutrients or water depth, which have a more direct and stronger impact on diatoms [10.5]. As discussed in the previous sections, salinity, conductivity and water depth are dependent on the precipitation/evaporation ratio, and consequently very likely to covariate with temperature (see Sections 10.3.2 and 10.3.3). Additionally, changes in vegetation cover, rates of weathering, and metabolic processes of lake biota that ultimately influence pH and nutrients, and consequently, diatom composition, have also been related to long-term changes in climate (see [10.3] and references therein). Hence, temperature is not always a dominant or independent explanatory variable in many calibration sets, and it is often not possible to demonstrate statistical independence of temperature as a controlling variable (e.g., [10.3]

[10.37]). As the models used to infer past changes for a given environmental variable also include the influence of other correlated variables, this covariance is included in the final reconstructions, resulting in inferred values that are not independent of other factors affecting the distribution in the original training set [10.3]. This may be the case of temperature, as any change in one of the mentioned correlated factors may change the diatom assemblage structure while temperature remains constant, producing biased temperature inferences. This may constitute a significant problem during long periods of time, as the variables controlling diatom distribution would probably not remain constant [10.3].

The success of diatom-based transfer functions to reconstruct other environmental variables, such as pH and salinity, lead to the inevitable expansion of this technique to include various climatic and hydrological variables, including temperature [10.37]. Besides the aforementioned limitations, temperature has been widely used to infer past climate changes, as several studies showed that it captures a significant proportion of the variance in the species data independently of lake chemistry ([10.34] and references therein). These studies argue that the indirect relationships between diatoms and temperature do not preclude the development of transfer functions, if temperature explains a significant and independent portion of the variance in the diatom data [10.52]. As an example, Bigler and Hall [10.10] found diatoms to be reliable indicators of summer temperature in northern Sweden, based on comparisons of inferred values and meteorological records. They also noticed, however, that lake-water pH was stronger in explaining assemblage composition than July temperature, and consequently, changes in lake-water pH over time would obscure diatom-based temperature inferences. Hence, the accuracy of diatom-inferred temperatures become reduced during periods of changing pH and should be interpreted with caution. As an alternative, they suggest that using a multi-proxy approach based on several independent lines of evidence (e.g., pollen, macrofossils, chironomids, stable isotopes) may make it possible to identify periods when diatom inferences become unreliable. Nonetheless, regardless of the statistical strength of these models and the potentiality of multi-proxy analyses to dodge their limitations, diatomists still doubt that they will ever be sufficiently robust and sensitive to be used as independent methods of temperature inference [10.5].

Another approach to the inference of temperature from diatom analyses has been the use of the oxygen isotopic composition of diatom frustules ($\delta^{18}O_{diatoms}$). In simple systems (non-alkaline, dilute, and/or open lakes, where carbonates may be rare or absent [10.44]), the $\delta^{18}O_{diatoms}$ is assumed to be controlled by water temperature and by the isotope composition of the lake water from which the frustule is formed (see Figure 1 in [10.44]). Accordingly, palaeotemperature equations based on $\delta^{18}O_{diatoms}$ can be used to infer past temperature variations, as long as the oxygen isotope composition of the lake water remains constant. In many lakes, however, the diatom oxygen isotope response to temperature will be dominated by changes in the isotope composition of precipitation (δ_p), leading to biases when $\delta^{18}O_{diatoms}$ are solely interpreted in terms of local temperature changes that ignore the potential effect of changes in δ_p driven by regional climatic processes. Consequently, in practice, deriving quantitative palaeotemperature data from $\delta^{18}O_{diatoms}$ may become very difficult, except over major climatic boundaries [10.44]. Despite these limitations, temperature reconstructions based on $\delta^{18}O_{diatoms}$ records have increased in number during the last decades (e.g., [10.15]). Alternatively, in areas where the $\delta^{18}O_{diatoms}$ registers changes in the $\delta^{18}O$ composition of the lake water, diatom studies have been successfully related to other aspects of climate rather than temperature, such as precipitation/evaporation balances [10.44]. For example, Hernández and collaborators [10.33] reconstructed a series of decadal-to-centennial ENSO-related dry–wet oscillations in Lake Chungará (Chilean Altiplano) based on oscillations of $\delta^{18}O_{diatoms}$, with dry periods inferred from relatively high isotope values and wet episodes indicated by lower values. These changes were also supported by compositional changes of assemblages [10.4], that showed decreasing plankton proportions during arid phases, as well as previous reconstruction of the effective moisture balance of the lake based on independent proxy data (Figure 10.3).

In a novel approach, Kuefner and collaborators [10.40] proposed the use of the diatom silicification values (SiVa) as proxies of temperature in mountain lakes of the Alps. As increasing lake temperatures cause higher productivity which lead to shorter time spans for valve silicification, it may support a shift in diatom assemblages towards species that produce lighter valves. In the studied lakes, the average SiVa of diatom communities on surface sediment samples decreased with temperature and total phosphorous concentration. Further application of this approach to the fossil record demonstrated its usefulness in reconstructing climate change impacts on mountain lakes, as it allowed the quantification of lake warming within the last decades validated against a 100-year temperature reconstruction from a varved sediment core [10.41]. Although the method is limited to shallow lakes of subpolar and temperate climates (as it is based on a formula applicable to a teoretical temperature range of 7.1 to 22.2 °C), it constitutes a promising technique whose applicability in other lakes and climatic situations needs to be futher investigated.

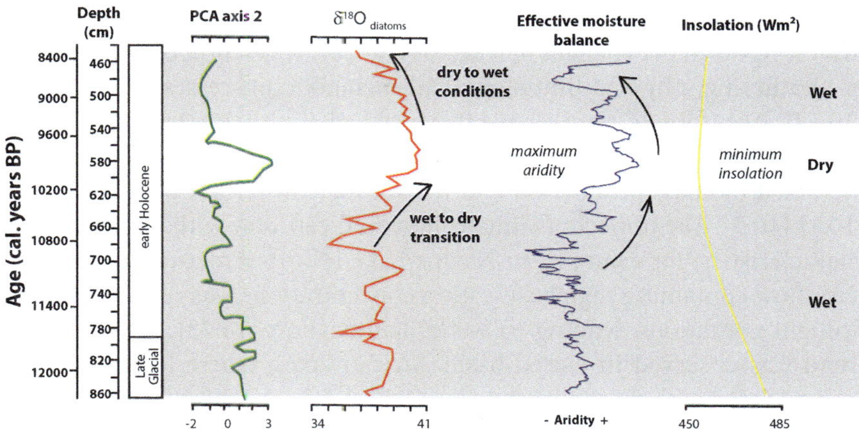

Figure 10.3 Paleoenvironmental reconstruction of a sedimentary sequence of the period 12,400–8300 cal. years BP in Lake Chungará (Chilean Altiplano). The figure shows a comparison between diatom assemblage composition (synthesized as sample scores of Principal Component Analysis (PCA) axis 2 [10.4]), diatom oxygen isotopes ($\delta^{18}O_{diatoms}$), effective moisture balance and insolation curve in austral summer at 18°S [10.33]. (Modified from [10.4] and [10.33]).

10.3.5 pH

The use of diatoms to study past pH fluctuations has a long history, although the recognition of its importance has significantly increased since the identification of the role of acid deposition in the acidification of lakes [10.7]. The significance of pH as one of the most important controlling variables of species composition in freshwater systems has long been documented in the diatom literature (see [10.6] [10.7] and references therein). However, the application of this relationship in climate change research is not straightforward, as the links between water pH and climate are not always clearly distinguished from natural and anthropogenic processes that control pH, such as acid deposition, bedrock composition, fluxes of humic acids to the lakes with vegetation succession and soil development, or the depletion over time of base cations due to chemical weathering of catchment soils, that can be more important regulators of pH than the climatic signal [10.22]. Despite these limitations, the use of pH in climate change research has progressed over the last decades, after Psenner and Schmidt [10.53] linked instrumental temperature records from the last ~ 200 years and diatom-inferred pH from two cores in the Austrian Tyrol. Since then, pH has been actively used as a proxy climate, particularly in pre-industrial records [10.7]. These studies hypothesized that climate can drive long-term pH changes in lakes, as it affects vegetation cover and rates of weathering, physical limnology and metabolic processes of lake biota [10.53]. Accordingly, they related increased alkalinity to warmer and drier periods, associated with longer water residence times, lower lake levels, increased biomass production and the subsequent release of base cations [10.1] [10.5]. The impact of climate, however, can vary with the lake basin characteristics: for example, in Nevins Lake (North America) groundwater inflow containing high base-cation concentrations was reduced during drought conditions, leading to water acidification [10.73]. The opposite trend was observed in closed-basin African lakes, where low water levels associated with drier climates result in cation concentrations increases, subsequently raising pH (e.g., [10.25]).

The significance of pH in controlling the composition of diatom assemblages in freshwater systems has been long recognized as a tool for paleoenvironmental research, from Hustedt's pioneer classification of diatoms according to their pH preferences in the 1930s to the semiquantitative and quantitative techniques developed in the recent decades [10.7]. Besides this importance, the mechanisms of physiological influence of pH on diatom growth are not fully comprehended. Evidence shows that pH controls a number of biogeochemical processes in lakes, such as the carbonate-bicarbonate balance, nutrient availability, metals solubility, and enzymatic activity [10.6]. The development of diatom-based pH transfer functions was supported by the strong and significant relationships between species distributions and pH gradients in many contemporary datasets, where pH explained most of the compositional variance, although the influence of covariance with other environmental factors needs to be considered (see [10.31] and references therein). Although quantitative models were mostly constructed for acid deposition studies, they have also been used in climate change reconstruction [10.5] [10.22]. Their application to reconstruct the decades following the industrial revolution, however, can be challenging, as interpretations can be further complicated by confounding factors, such as acid-rain acidification [10.1].

As previously mentioned, one of the earliest approaches to climate change reconstructions from diatom pH shifts was offered by Psenner and Schmidt [10.53]. In that work, a pH transfer function was applied to quantitatively reconstruct pH variations in two soft-water high-elevation lakes in the central Alps, and the results were compared with temperature and precipitation instrumental records. Results demonstrated that the acidity of both lakes was correlated with climate during the 19th century:

according to the authors, in warmer and drier years this can be explained by increased evaporation, longer residence times, more stable stratification, and higher biomass production that led to higher pH values in lake water. These processes reversed during cold periods, as the formation of snow fields contributed to the discharge of unbuffered or acidic meltwater during summer, consequently reducing the intensity of the biochemical processes that produce alkalinity. At the beginning of the 20th century, the pH-temperature relationship became obscured as a consequence of the onset of anthropogenically derived acid precipitation. Overall, this work demonstrated the impact that climate has in lacustrine pH and opened the door for further application of diatom-based pH transfer functions in paleoclimatic research.

Examples of the use of diatom-inferred pH models in paleoclimatic reconstructions of South America water bodies are scarce. In a quantitative approach, Vélez and collaborators [10.71] used diatom-based pH and conductivity transfer functions to reconstruct the late Pleistocene–Holocene paleolimnological and paleoclimatic history of Lake Fúquene (Colombian Andes). The reconstructed values of pH and conductivity were further related to water-depth fluctuations. Lower lake levels were associated with higher conductivity, as smaller volumes of water led to higher concentration of solutes, and higher pH values suggested that the water input from local aquifers was low. Periods of low pH values, on the other hand, were associated with a major influence of runoff and local aquifers carrying more acidic water into the lake, with higher lake levels and the subsequent decrease in conductivity. These results were supported by pollen analyses and the integration with other records from northern South America, that suggested that the changes in lake level were the result of precipitation variations driven by latitudinal shifts of the Intertropical Convergence Zone [10.71].

10.3.6 Nutrients

The concentrations of nutrients in lakes can be altered by numerous factors, such as external processes related to climatic influences on weathering rates, precipitation, runoff, fire frequency, terrestrial primary productivity, internal processes related to changes in thermal structure and/or primary productivity, and anthropogenic activity [10.1]. Bedrock composition is also significant, as some lakes lie in naturally fertile catchments or receive high natural loads of nutrients from groundwater and are naturally eutrophic [10.29]. As diatom individual species are sensitive to changes in concentrations, supply rates, and ratios of nutrients (e.g., Si, P), they have been widely used as proxies to reconstruct past changes in nutrient signals, particularly those associated with human impacts within lake basins [10.9] [10.29]. The use of diatom-inferred nutrients in climate change has been scarce, and some studies provided have interpreted nutrient enrichment as a result of low water levels related to warmer and drier periods [e.g., [10.49] and references therein]. This approach needs to be applied with caution, however, as the effects of climate on nutrient loadings are still debatable and difficult to predict. Field and laboratory studies suggest that warming could intensify many symptoms of eutrophication, but responses are complex and diverse depending on the initial conditions and location of the lake. Metabolic responses of ecosystems to warming are not clearly understood, but several lines of evidence hint at biological feedback mechanisms that may result in increased respiratory production of carbon dioxide [10.36]. Moreover, anthropogenic signals also need to be identified and separated in order to disentangle climatic signals [10.73]. For lakes situated in populated or highly impacted areas, the discrimination of both signals can be challenging for recent periods of time, where responses of diatom assemblages to climate change may become equivocal as a consequence of the confounding effect of human-induced eutrophication. In such cases, multi-proxy studies are needed in order to evaluate the relative impact of these processes though independent lines of evidence [10.8].

Diatoms are photoautotrophic organisms that are directly affected by changes in the availability of nutrients and light, and each taxon is usually characterized by an optimum and tolerance that can be quantified and modeled [10.29]. Several studies have documented differential distributions of diatom taxa along nutrient gradients, with silica, nitrogen and phosphorous the key nutrients for diatom growth [10.9]. Among these, phosphorous is usually the main nutrient limiting diatom productivity and driving species composition; therefore, this is the key variable reconstructed in lake eutrophication studies [10.6]. This ability to quantify the responses of diatom taxa to nutrient concentrations has provided diatomists with a powerful tool for quantifying eutrophication and recovery, particularly through the development of diatom-phosphorous transfer functions [10.29], and has also been applied to climate change studies [10.49]. Qualitative studies, on the other hand, mostly focused on the ratio of planktonic to non-planktonic diatoms as proxies of nutrient shifts (e.g., [10.48]). Although increasing the availability of phosphorous can affect both benthic and planktonic diatom communities, it usually causes most alteration to phytoplankton, being that this impact typically is reflected as an increase in the ratio of planktonic to non-planktonic diatoms in sediment cores [10.6].

As an example, Moos and collaborators [10.49] studied climate-induced changes in water quality in northwestern Ontario (Canada), based on the analysis of the species composition of diatom assemblages and associated quantitative inferences of total phosphorus (TP). Other independent lines of evidence, such as the proportion of chrysophyte scales to diatoms and changes to organic matter content, were also analyzed and compared to diatom results. The lake showed changes in water quality that were associated with lake levels at least 8 m lower than today, evidenced by the increases in planktonic nutrient-rich diatom assemblages (*Aulacoseira subarctica*, *Fragilaria crotonensis*), TP inferred values, and diatom accumulation, as well as the decrease in the chrysophytes relative to diatoms. This period of large decrease in water depth and inferred increase in nutrient concentrations was related to the warmer and drier climate prevailing in the mid-Holocene. Hence, results from this study provided evidence that changes in trophic status can be related to climate, mediated by its impact on lake levels. As the mid-Holocene warming was found to be consistent with some future warming scenarios, the results of this work proved to be useful analogues in a future scenario of anthropogenic warming [10.49].

The application of diatom-inferred nutrient shifts in climate research has been scarce in South America, being works were based on qualitative rather than quantitative criteria. For example, in the southern Patagonia of Argentina, Massaferro and collaborators [10.47] studied the paleoclimatic history of Laguna Potrok Aike, one of the few permanent lakes with a well-preserved continuous and high-resolution climatic record from the semiarid Patagonian steppe. Fluctuations in water levels during the last 16,000 years BP were inferred from the combined analysis of diatoms and chironomids. The qualitative analysis of diatom shifts along the core included the discrimination between planktonic and non-planktonic diatom taxa and the consideration of nutrient and salinity classifications of dominant species. Periods of lower lake levels were dominated by nutrient-rich waters (indicated by *Discostella stelligera*), accompanied by indicators of warmer temperatures and higher turbidity (signaled by *Aulacoseira granulata*). Overall, diatom results, combined with chironomid inferences, allowed inferring that humid conditions during the Late Glacial followed by drier conditions during the Holocene, resulted in large variations in lake level [10.47].

10.3.7 Ice Cover

Ice cover is a key element influencing lakes in cold climate. It plays a major role in isolating the water body from the atmosphere and sun radiation, consequently slowing down physical and biological lacustrine processes. Ice cover can be seasonal (mostly in the boreal zone, tundra, and mountain regions) or permanent (high elevations and high polar latitudes). The dynamics of freezing is controlled by local climate, as fluxes between the atmosphere and lakes depend on the precipitation, air pressure, cloud coverage, air temperature, humidity, and wind. Freezing lakes are found on all continents in zones of snow and polar climates where the winter mean air temperature is less than 0 °C (i.e., the "cold climate zone" [10.45]). In the Southern Hemisphere, only some mountain areas are included in the zone of freezing lakes, although in the high Andes, the cold zone reaches the Equator. In these regions, most shallow freshwater lakes freeze in normal years, although there are also some geothermal, hypersaline, and/or very deep lakes that can stay ice-free in cold winters [10.45].

In contrast to the relatively small variations recorded in water temperature, lakes in cold climate zones may exhibit large seasonal and/or interannual differences in the extent of ice cover, leading to noticeable changes in diatom communities. It has been pointed out that in lakes with perennial ice cover, diatom assemblages are mostly dominated by aerophilic or benthic taxa characteristic of shallow waters during very cold periods, that progressively include deeper water taxa as climate ameliorates [10.62]. This approach has been widely used to infer changes in the duration of ice cover in early paleolimnological studies on Arctic and Antarctic lakes [10.19] [10.64]. Later findings demonstrated more complex shifts in diatom life strategies associated with warming in Arctic and subarctic lakes, signaling the necessity of considering both the ecological and physiological characteristics of the local flora and the historical and limnological characteristics of the lake being investigated in order to interpret these changes [10.57]. Diatom absolute abundances have also been used as indicators of fluctuations in past ice coverage, as warmer and ice-free periods are characterized by higher productivity because of the higher light availability [10.63]. In the High Arctic, pH and conductivity transfer functions have also been used, as they tend to increase during warmer periods: cool periods were related to lower pH levels due to extended ice cover that traps respired CO_2 in the water column, whereas reduced ice cover during warm periods allows respired CO_2 to escape to the atmosphere, while greater photosynthetic activity increases the consumption of limnetic CO_2, thereby increasing pH [10.19].

Most of the paleoclimatic studies in cold, ice-covered lakes have been conducted in the Northern Hemisphere, particularly in the polar region, where lakes are highly sensitive to climate change (see [10.19] for a review). In a comprehensive study conducted in the Arctic region, Smol [10.62] analyzed 42 diatom records from high-altitude lakes located in remote circumpolar areas. Given the remoteness of the sites, this dataset allowed studying climate change in an area where anthropogenic land use was almost negligible. Several lakes showed increases in planktonic diatoms (genus *Cyclotella*) and a reduction in benthic and tychoplanktonic taxa (such as *Fragilaria* and *Aulacoseira*). This shift to plankton-dominated assemblages was related to a reduction of ice cover in the region in recent times and associated to climate warming. This finding agrees with regional trends observed in other lakes from central subarctic Canada, also showing recent shifts in diatom life strategy from benthic to planktonic species. Thus, the results of this work demonstrated that in a large part of the Arctic, shifts in the composition of diatom assemblages in lakes were recorded from the 19th to 20th century, consistent with expected responses to climate warming, in contrast to regions where no temperature increases were recorded through instrumental records, such as Labrador and northern Quebec (Canada [10.62]).

The use of diatoms to infer past ice cover fluctuations in South America is still scarce, although some studies have been conducted, demonstrating the usefulness of this approach for understanding climatic variability in the Southern Hemisphere. Among these, works conducted in Southern Patagonia are of key importance to understand Quaternary climates, as this is the southernmost continuous landmass outside Antarctica in which the ice balance and climatic development are closely related to the Antarctic climate ([10.55] and references therein). For example, the study of diatoms present in a sediment core from Laguna Las Vizcachas (Santa Cruz, Argentina) was used to estimate regional climate fluctuations in Southern Patagonia for the last ca. 1600 cal. yrs. BP [10.21]. The lake is situated at 1100 m a.s.l. and is ice-covered for several months of the year. The diatom record indicates temperature changes, evidenced by the highest proportions of benthic diatoms indicative of cold conditions during the 15th–17th centuries, followed by relatively warm conditions until the mid-18th century, suggested by a decrease of benthic taxa and the increase of planktonic species as a consequence of the decrease in the length of winter ice cover (Figure 10.4).

On the other hand, Laguna Potrok Aike, a subsaline lake located at a lower altitude (116 m a.s.l.) and approximately 150 km from Laguna Las Vizcachas, is strongly influenced by westerly winds, which prevents the

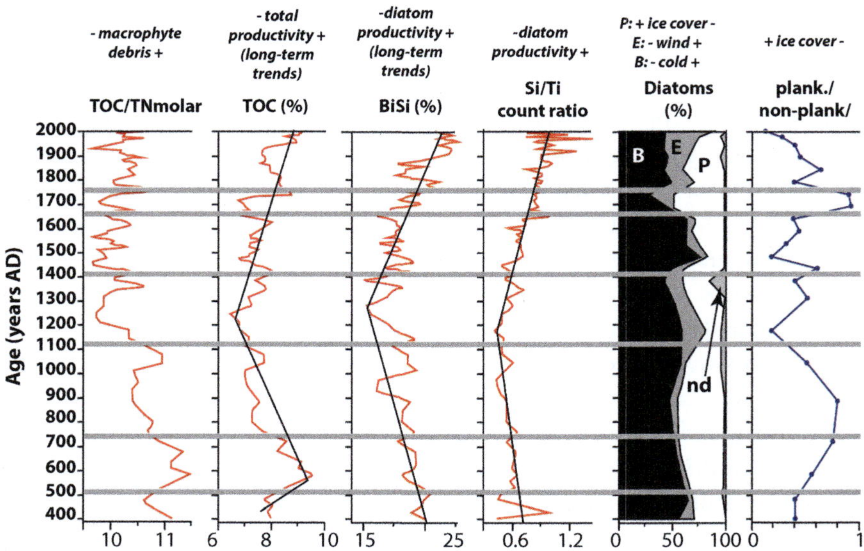

Figure 10.4 Selected parameters and paleoenvironmental inferences in a core covering the last ca. 1600 cal. yrs. BP from Laguna Las Vizcachas (Santa Cruz, Argentina). TOC: total organic carbon, TN: total nitrogen, BiSi: biogenic silica, B: benthic diatoms, E: epiphytic diatoms, P: planktonic diatoms, ND: no data, plankt./non-plankt.: %-ratio planktonic/non-planktonic diatoms, TDC: total diatom concentration. Si/Ti: silica to titanium ratio. Straight lines in TOC, BiSi and Si/Ti count ratio refer to long-term trends. (Reprinted from [10.21] with permission from Elsevier).

development of ice cover in winter [10.55]. The diatom record of this lake extends to ca. 50 cal. ka yrs BP, covering the Glacial and Last Glacial Maximum. In this case, the presence of periods dominated by a diverse benthos and some aerophilous taxa (*Pinnularia, Luticola, Nitzschia* and *Amphora*), was interpreted as indicative of relatively warm conditions that promoted littoral permafrost melting and increased benthic productivity. Levels dominated by large planktonic diatoms, such as *Cyclostephanos patagonicus*, indicative of fresh and oligotrophic conditions, were associated with cold episodes. Several events of diatom abundance peaks dominated by fragilarioids and other small benthic taxa recorded during the glacial interval were associated with episodic warmings that led to the decline of the permafrost covering the lake shores, consequently promoting the development of benthic taxa. These episodes of warming and increased productivity corresponded well with the Antarctic warm events A1, A2 and A3, supporting the climatic forcing and signaling the magnitude of these events, that had only been described for Antarctica up to that moment [10.55].

10.4 Conclusions

Lakes are good sentinels of climate change, and their sedimentary record store evidence of past climates that constitute the basis of paleoclimatological research in continental settings. In order to successfully recover and interpret those evidences, however, a complete knowledge of the dynamics and internal characteristics of the lake under study is fundamental. Several studies have demonstrated that the impact of climate on limnological variables can vary with the characteristics of the basins, leading in many cases to the impossibility of applying generalizations about the paleoclimatic significance of the signals stored in lake sediments. Therefore, even as climate regulates a series of response variables widely known in the literature, these are not broadly applicable across all lakes, and so, the transfer of known signal/response relationships to a particular system should be evaluated with caution to avoid misinterpretations.

Diatom assemblages have proven to be good indicators of past climatic changes, and their responses to environmental variables related directly and/or indirectly to climate are well documented. Among these, salinity and water level are particularly useful, provided the links between climate and these limnological variables are fully understood on the system of interest. The value of temperature, on the other hand, has been largely questioned, and several evidences suggest that diatoms are not good indicators of water or air temperatures, even as the problems signaled for temperature (such as multiple controlling factors and complex limnological interactions) can also be advocated for other variables such as pH and nutrients. These variables, however, more often present noticeable changes in response to climatic variability that are strong enough to drive shifts in diatom assemblages. The marginal effect of temperature, on the other hand, is usually obscured as major climatic changes only produce small temperature changes (i.e., 1 °C), not large enough to impact diatom physiology and assemblages' composition [10.3]. The exact nature of shifts in diatom assemblages' composition, however, will also be linked to local limnological and ecological conditions, as diatom responses to similar climatic changes can vary among lakes and regions, signaling the importance of local studies to understand the climatic significance of diatom changes over time.

Confounding factors not related to climate can pose critical limitations to diatom-based reconstructions in lakes. Responses of diatoms to climate are complex, and in some cases several intercorrelated factors can obscure the effects of the variable of interest. Among them, the effects of human activities are the main cause of problems in interpreting diatom data from recent periods. In basins located on highly impacted areas, predicting the effects of climate change can be difficult, given the multiple local perturbations. Eutrophication and acid deposition can confound climate-related nutrient and pH signals, being that in many cases it is very difficult, or even impossible, to separate both. Also, land use can confound the impact of climate change on water level. Over longer time periods, other factors non-related to human activities, such as isostatic rebound, groundwater discharge, vegetation changes, and biotic interactions, are also important. In all these cases, diatom studies should rely on other proxies and comparison of multiple records to disentangle the relative impact of these factors through the assessment of independent lines of evidence.

The need for high-quality paleoclimatic data has progressively grown with the recognition of the key role that the fossil record plays in unraveling natural and human-induced influences on climate change. Paleolimnologists are continuously looking for new high-resolution datasets that allow them to understand regional, continental and interhemispheric trends of climate change. Among these, diatoms constitute a useful tool, especially if multi-proxy approaches, including both biological and sedimentological data, are applied. In order to improve our capability of making paleoclimatic inferences from diatoms, however, it becomes imperative to increase our knowledge of the mechanisms that link diatoms, lake chemistry and climate in contemporary settings. To do so, more studies focusing on the links between lake hydrology, chemistry and climatic patterns in modern lakes are needed. Also, increased efforts to comprehend the effect of changes in hydrochemistry and diatom physiology would strengthen the interpretation of changes in past diatom assemblages' composition. Given the strong relationship between local conditions and the responses to climate forcings described in this chapter, the development of regional datasets studying the limnological and climatic forcings driving the distribution of modern diatoms become a key issue to improve the quality of the inferences made. This is particularly important in understudied regions, as those of the Southern Hemisphere, where training sets are scarce, and studies frequently apply data and generalizations made for other regions and systems. Also, expanding the number of studied fossil records in these regions would improve our understanding of past climatic and lacustrine dynamics, thereby allowing new data to be fed to the models needed to predict future climate change.

Acknowledgments

I am very grateful to Dr. Eduardo A. Morales and Dr. Nora I. Maidana for the invitation to contribute to this book. Financial support for my research was granted by the Consejo Nacional de Investigaciones Científicas y técnicas (CONICET, PIP0204/21). I thank Dr. Claudio G. De Francesco for the valuable comments made on earlier versions of the manuscript, and the two anonymous reviewers whose helpful comments greatly improved the manuscript. I want to dedicate this chapter to the memory of Eduardo Morales, a great colleague whose significant contribution to diatom research has been a source of inspiration for my work. I am a member of the Scientific Research Career of the CONICET.

References

[10.1] Adrian, R., O'Reilly, C.M., Zagarese, H., Baines, S.B., Hessen, D.O., Keller, W., Livingstone, D.M., Sommaruga, R., Straile, D., Van Donk, E., Weyhenmeyer, G.A. and Winder, M. (2009) Lakes as sentinels of climate change. *Limnol. Oceanogr.* 54(6), 2283-2297.

[10.2] Althouse, B., Higgins, S. and Zanden, M.J.V. (2014) Benthic and planktonic primary production along a nutrient gradient in Green Bay, Lake Michigan, USA. *Fresh. Sci.* 33(2), 487-498.

[10.3] Anderson, J.N. (2000) Minireview: Diatoms, temperature and climatic change. *Eur. J. Phycol.* 35, 307-314.

[10.4] Bao, R., Hernández, A., Sáez, A., Giralt, S., Prego, R., Pueyo, J. J., Moreno, A. and Valero-Garcés, B. L. (2015) Climatic and lacustrine morphometric controls of diatom paleoproductivity in a tropical Andean lake. *Quat. Sci. Rev.* 129, 96-110.

[10.5] Battarbee, R.W. (2000) Palaeolimnological approaches to climate change, with special regard to the biological record. *Quat. Sci. Rev.* 19, 107-124.

[10.6] Battarbee, R., Jones, V., Flower, R., Cameron, N., Bennion, H., Carvalho, L. and Juggins, S. (2001) Diatoms, In: *Tracking environmental change using lake sediments.* J.P. Smol, H.J. Birks, W. Last, R. Bradley, and K. Alverson, (eds). Springer Netherlands: 155-202.

[10.7] Battarbee, R.W., Charles, D.F., Bigler, C., Cumming, B.F. and Renberg, I. (2010) Diatoms as indicators of surface-water acidity. In: *The diatoms: applications for the environmental and earth sciences.* J.P. Smol and E.F. Stoermer, (eds.). Cambridge University Press, Cambridge: 98-121.

[10.8] Battarbee, R.W., Anderson, N.J., Bennion, H. and Simpson, G. L. (2012) Combining limnological and palaeolimnological data to disentangle the effects of nutrient pollution and climate change on lake ecosystems: problems and potential. *Freshw. Biol.*, 57, 2091-2106.

[10.9] Bennion, H., Carl, D., Sayer, J.T. and Hunter, J.C. (2010) Diatoms as indicators of environmental change in shallow lakes. In: *The diatoms: applications for the environmental and earth sciences.* J.P. Smol and E.F. Stoermer, (eds.). Cambridge University Press, Cambridge: 152-173.

[10.10] Bigler, C. and Hall, R.I. (2003) Diatoms as quantitative indicators of July temperature: a validation attempt at century-scale with meteorological data from northern Sweden. *Palaeogeogr. Palaeoclimatol. Palaeoecol.* 189, 147-160.

[10.11] Birks, H.J.B. (2010) Numerical methods for the analysis of diatom assemblage data. In: *The Diatoms: Applications for the Environmental and Earth Sciences* J.P. Smol and E.F. Stoermer, (eds.). Cambridge University Press, Cambridge: 23-53.

[10.12] Birks, H.J.B., Line, J.M., Juggins, S., Stevenson, A. and Ter Braak, C.J.F. (1990) Diatoms and pH reconstruction. *Philos. Trans. R. Soc. Lond. B Biol. Sci.* 327, 263-278.

[10.13] Burke, K.D., Williams, J.W., Chandler, M.A., Haywood, A.M., Lunt, D.J. and Otto-Bliesner, B.L. (2018) Pliocene and Eocene provide best analogs for near-future climates. *Proc. Natl. Acad. Sci.* 115(52), 13288-13293.

[10.14] Charbit, S., Bouttes, N., Quiquet, A., Bopp, L., Ramstein, G., Dufresne, J.-L. and Cattiaux, J. (2021) From the climates of the past to the climates of the future. In: *Paleoclimatology*. Frontiers in Earth Sciences. G. Ramstein, A. Landais, N. Bouttes, P. Sepulchre and A. Govin, (eds.). Springer, Cham: 443-478.

[10.15] Crespin, J., Sylvestre, F., Alexandre, A., Sonzogni, C., Paillès, C. and Perga, M.E. (2010) Re-examination of the temperature-dependent relationship between δ^{18}Odiatoms and δ^{18}Olake water and implications for paleoclimate inferences. *J. Paleolimnol.* 44(2), 547-557.

[10.16] Cumming, B.F. and Smol, J.P. (1993) Development of diatom-based salinity models for paleoclimatic research from lakes in British Columbia (Canada). *Hydrobiologia* 269/270, 179-196.

[10.17] DeNicola, D.M. (2016) Periphyton responses to temperature at different ecological levels. In: *Algal ecology in freshwater benthic ecosystems.* R.J. Stevenson, M.L. Bothwell and R.L. Lowe, (eds.). Academic Press, San Diego: 149-181.

[10.18] Diaz, M.C. and Hassan, G.S. (2019) Assessing the influence of saline and alkaline environments in the preservation of Pampean diatoms: an experimental approach. *Ameghiniana* 56, 471-492.

[10.19] Douglas, M.S.V. and Smol, J.P. (2010) Freshwater diatoms as indicators of environmental change in the High Arctic. In: *The diatoms: applications for the environmental and earth sciences.* J.P. Smol and E.F. Stoermer, (eds.). Cambridge University Press, Cambridge: 249-266.

[10.20] Dove, P.M. and Crerar, D.A. (1990) Kinetics of quartz dissolution in electrolyte solutions using a hydrothermal mixed flow reactor. *Geochim. Cosmochim. Acta,* 54: 955-969.
[10.21] Fey, M., Korr, C., Maidana, N.I., Carrevedo, M.L., Corbella, H., Dietrich, S., Haberzettl, T., Kuhn, G., Lücke, A., Mayr, C., Ohlendorf, C., Paez, M.M., Quintana, F.A., Schäbitz, F. and Zolitschka, B. (2009) Palaeoenvironmental changes during the last 1600 years inferred from the sediment record of a cirque lake in southern Patagonia (Laguna Las Vizcachas, Argentina). *Palaeogeogr. Palaeoclimatol. Palaeoecol.,* 281(3-4), 363-375.
[10.22] Finkelstein, S.A., Bunbury, J., Gajewski, K., Wolfe, A.P., Adams, J.K. and Devlin, J.E. (2014) Evaluating Diatom-Derived Holocene Ph Reconstructions for Arctic Lakes Using an Expanded 171-Lake Training Set. *J. Quat. Sci.,* 29, 249-60.
[10.23] Fritz, S.C. (2013) Diatom methods. Salinity and climate reconstructions from continental lakes. In: *Encyclopedia of Quaternary Science (Second Edition).* S.A. Elias and C.J. Mock, (eds.). Elsevier, Amsterdam: 507-515.
[10.24] Fritz, S.C., Juggins, S., Battarbee, R.W. and Engstrom, D.R. (1991) Reconstruction of past changes in salinity and climate using a diatom-based transfer function. *Nature* 352, 706-708.
[10.25] Fritz, S.C., Cumming, B.F., Gasse, F. and Laird, K.R. (2010) Diatoms as indicators of hydrologic and climatic change in saline lakes. In: *The diatoms: applications for the environmental and earth sciences.* J.P. Smol and E.F. Stoermer, (eds.). Cambridge University Press, Cambridge: 186-208.
[10.26] Gasse, F., Juggins, S. and Khelifa, L.B. (1995) Diatom-based transfer functions for inferring past hydrochemical characteristics of African lakes. *Palaeogeogr. Palaeoclimatol. Palaeoecol.,* 117, 31-54.
[10.27] Gasse, F., Barker, P., Gell, P.A., Fritz, S.C. and Chalié, F. (1997) Diatom-inferred salinity in paleolakes: an indirect tracer of climate change. *Quat. Sci. Rev.* 16, 547-563.
[10.28] Gomes, D.F., Albuquerque, A.L.S., Torgan, L.C., Turcq, B. and Sifeddine, A. (2014) Assessment of a diatom-based transfer function for the reconstruction of lake-level changes in Boqueirão Lake, Brazilian Nordeste. *Palaeogeogr. Palaeoclimatol. Palaeoecol.,* 415, 105-116.
[10.29] Hall, R.I. and Smol J. P. (2010) Diatoms as indicators of lake eutrophication. In: *The diatoms: applications for the environmental and earth sciences.* J.P. Smol and E.F. Stoermer, (eds.). Cambridge University Press, Cambridge: 122-151.
[10.30] Hassan, G.S. and De Francesco, C.G. (2018) Preservation of *Cyclotella meneghiniana* Kützing (Bacillariophyceae) along a continental salinity gradient: implications for diatom-based paleoenvironmental reconstructions. *Ameghiniana* 55, 263-276.
[10.31] Hassan, G.S., Tietze, E., De Francesco, C.G. and Cristini, P.A. (2011) Problems and potentialities of using diatoms as paleoclimatic indicators

in Central Argentina. In: *Diatoms: Ecology and life cycle.* J.C. Compton, (ed.). Nova Science Publishers, New York: 185-216.

[10.32] Hassan, G.S., De Francesco, C.G. and Díaz, M.C. (2020) Actualistic taphonomy of freshwater diatoms: Implications for the interpretation of the Holocene record in Pampean shallow lakes. In: *Actualistic Taphonomy in South America. Topics in Geobiology 48.* S.A. Martínez, A. Rojas and F. Cabrera, (eds.). Springer Nature Switzerland: 89-109.

[10.33] Hernández, A., Bao, R., Giralt, S., Sáez, A., Leng, M.J., Barker, P.A., Kendrick, C.P. and Sloane, H.J. (2013) Climate, catchment runoff and limnological drivers of carbon and oxygen isotope composition of diatom frustules from the central Andean Altiplano during the Lateglacial and Early Holocene. *Quat. Sci. Rev.* 66, 64-73.

[10.34] Huber, K., Weckström, K., Drescher-Schneider, R., Knoll, J., Schmidt, J. and Schmidt, R. (2010) Climate changes during the last glacial termination inferred from diatom-based temperatures and pollen in a sediment core from Längsee (Austria). *J. Paleolimnol.* 43(1), 131-147.

[10.35] IPCC (Intergovernmental Panel on Climate Change), 2021. Climate change 2021: The physical science basis. Working Group I contribution to the Sixth Assessment Report of the Intergovernmental Panel on Climate Change. V. Masson-Delmotte, P. Zhai, A. Pirani, S.L. Connors, C. Péan, S. Berger, N. Caud, Y. Chen, L. Goldfarb, M.I. Gomis, M. Huang, K. Leitzell, E. Lonnoy, J.B.R. Matthews, T.K. Maycock, T. Waterfield, O. Yelekçi, R. Yu and B. Zhou, (eds.). Cambridge University Press. in press (Available from: www.ipcc.ch).

[10.36] Jeppesen, E., Moss, B., Bennion, H., Carvalho, L., DeMeester, L., Feuchtmayr, H., Friberg, N., Gessner, M., Hefting, M. and Lauridsen, T.L. (2012) Interaction of Climate Change and Eutrophication, In: *Climate Change Impacts on Freshwater Ecosystems*, M. Kernan, R.W. Battarbee, and B. Moss (eds.), Wiley: 119-51.

[10.37] Juggins, S. (2013) Quantitative reconstructions in palaeolimnology: new paradigm or sick science? *Quat. Sci. Rev.* 64(0), 20-32.

[10.38] Juggins, S. and Birks, H.J.B. (2012) Quantitative environmental reconstructions from biological data. In: *Tracking environmental change using lake sediments.* H.J.B. Birks, A.F. Lotter, S. Juggins and J.P. Smol, (eds.), Springer Netherlands: 431-494.

[10.39] Kolbe, R.W. (1927) Zur oekologie, morphologie und systematik del brackwasser-diatomeen. *Pflanzenforschung* 7, 1-146.

[10.40] Kuefner, W., Ossyssek, S., Geist, J. and Raeder, U. (2020a) The silicification value: a novel diatom-based indicator to assess climate change in freshwater habitats. *Diatom Res.* 35, 1-16.

[10.41] Kuefner, W., Hofmann, A. M., Geist, J., & Raeder, U. (2020b) Evaluating climate change impacts on mountain lakes by applying the new silicification value to paleolimnological samples. *Sci. Total Environ.* 715, #136913.

[10.42] Laird, K.R., Kingsbury, M.V. and Cumming, B.F. (2010) Diatom habitats, species diversity and water-depth inference models across surface-sediment transects in Worth Lake, northwest Ontario, Canada. *J. Paleolimnol.* 44(4), 1009-1024.

[10.43] Leavitt, P.R., Fritz, S.C., Anderson, N.J., Baker, P.A., Blenckner, T., Bunting, L., Catalan, J., Conley, D.J., Hobbs, W.O., Jeppesen, E., Korhola, A., McGowan, S., Rühland, K., Rusak, J.A., Simpson, G.L., Solovieva, N. and Werne, J. (2009) Paleolimnological evidence of the effects on lakes of energy and mass transfer from climate and humans. *Limnol. Oceanogr.* 54, 2330-2348

[10.44] Leng, M.J. and Barker, P.A. (2006) A review of the oxygen isotope composition of lacustrine diatom silica for palaeoclimate reconstruction. *Earth-Science Reviews* 75(1), 5-27.

[10.45] Leppäranta, M. (2015) *Freezing of lakes and the evolution of their ice cover.* Springer Science & Business Media.

[10.46] Mackay, A.W., Jones, V.J. and Battarbee, R. (2003) Approaches to Holocene climate reconstruction using diatoms. In: *Global change in the Holocene.* A.W. Mackay, R. Battarbee, H.J.B. Birks and F. Oldfield, (eds). Arnold: 294-309.

[10.47] Massaferro, J., Recasens, C., Larocque-Tobler, I., Zolitschka, B. and Maidana, N.I. (2013) Major Lake level fluctuations and climate changes for the past 16,000 years as reflected by diatoms and chironomids preserved in the sediment of Laguna Potrok Aike, southern Patagonia. *Quat. Sci. Rev.* 71, 167-174.

[10.48] Metcalfe, S.E., Whitney, B.S., Fitzpatrick, K.A., Mayle, F.E., Loader, N.J., Street-Perrott, F.A. and Mann, D.G. (2014) Hydrology and climatology at Laguna La Gaiba, lowland Bolivia: complex responses to climatic forcings over the last 25 000 years. *J. Quat. Sci.* 29(3), 289-300.

[10.49] Moos, M. T., Laird, K. R. and Cumming, B. F. (2009) Climate-related eutrophication of a small boreal lake in northwestern Ontario: a palaeolimnological perspective. *Holocene* 19(3), 359-367.

[10.50] Mourelle, D., del Puerto, L., Perez, L., Bergamino, L. and García-Rodríguez, F. (2020) Late Pleistocene and Late Holocene environment and climate in the campos region of southeastern South America inferred from phytolith, diatom and geochemical data. *Palaeogeogr. Palaeoclimatol. Palaeoecol.* 543, #109586.

[10.51] Neukom, R. and Gergis, J. (2012) Southern Hemisphere high-resolution palaeoclimate records of the last 2000 years. *Holocene* 22(5), 501-524.

[10.52] Pienitz, R., Smol, J.P. and Birks, H.J.B. (1995) Assessment of freshwater diatoms as quantitative indicators of past climatic change in the Yukon and Northwest Territories, Canada. *J. Paleolimnol.* 13(1), 21-49.

[10.53] Psenner, R. and Schmidt, R. (1992) Climate-driven pH control of remote alpine lakes and effects of acid deposition. *Nature* 356, 781-783.

[10.54] Ramstein, G., Landais, A., Bouttes, N., Sepulchre, P., and Govin, A. (eds.). (2021) *Paleoclimatology*. Springer Nature.

[10.55] Recasens, C., Ariztegui, D., Maidana, N. and Zolitschka, B. (2015) Diatoms as indicators of hydrological and climatic changes in Laguna Potrok Aike (Patagonia) since the Late Pleistocene. *Palaeogeogr. Palaeoclimatol. Palaeoecol.*, 417, 309-319.

[10.56] Roubeix, V. and Lancelot, C. (2008) Effect of salinity on growth, cell size and silicification of an euryhaline freshwater diatom: *Cyclotella meneghiniana* Kûtz. *Transit. Waters Bull.* 1, 31-38.

[10.57] Rühland, K., Paterson, A. and Smol, J. (2015) Lake diatom responses to warming: reviewing the evidence. *J. Paleolimnol.* 54(1), 1-35.

[10.58] Ryves, D.B., McGowan, S. and Anderson, N.J. (2002) Development and evaluation of a diatom-conductivity model from lakes in West Greenland. *Freshw. Biol.* 47(5), 995-1014.

[10.59] Ryves, D.B., Battarbee, R., Juggins, S., Fritz, S.C. and Anderson, N.J. (2006) Physical and chemical predictors of diatom dissolution in freshwater and saline lake sediments in North America and West Greenland. *Limnol. Oceanogr.* 51(3), 1355-1368.

[10.60] Schneider, S.H. and Mastrandrea, M.D. (2005) Probabilistic assessment of "dangerous" climate change and emissions pathways. *Proc. Natl. Acad. Sci.* 102(44), 15728-15735.

[10.61] Smol, J.P. (1981) Problems associated with the use of "Species Diversity" in paleolimnological studies. *Quat. Res.* 15, 209-212.

[10.62] Smol, J.P. (1988) Paleoclimate proxy data from freshwater arctic diatoms. *Int. Ver. Theor. Angew. Limnol.* 23(2), 837-844.

[10.63] Smol, J.P., Wolfe, A.P., Birks, H.J.B., Douglas, M.S.V., Jones, V.J., Korhola, A., Pienitz, R., Rühland, K., Sorvari, S., Antoniades, D., Brooks S.J., Fallu, M.E., Hughes, M., Keatley, B.E., Laing, T.E., Michelutti, N., Nazarova, L., Nyman, M., Paterson, A.M., Perren, B., Quinlan, R., Rautio, M., Saulnier-Talbot, E., Siitonen, S., Solovieva, N. and Weckström, J. (2005) Climate-driven regime shifts in the biological communities of arctic lakes. *Proc. Natl. Acad. Sci.* 102(12), 4397-4402.

[10.64] Spaulding, S.A., Van De Vijver, B., Hodgson, D.A., Mcknight, D.M., Verleyen, E. and Stanish, L. (2010) Diatoms as indicators of environmental change in Antarctic and subantarctic freshwaters. In: *The diatoms: applications for the environmental and earth sciences*. J.P. Smol and E.F. Stoermer, (eds.). Cambridge University Press, Cambridge: 267-283.

[10.65] Speyer, S.E. and Brett, C.E. (1986) Trilobite taphonomy and Middle Devonian Taphofacies. *Palaios* 1(3), 312-327.

[10.66] Stone, J.R. and Fritz, S.C. (2004) Three-dimensional modeling of lacustrine diatom habitat areas: Improving paleolimnological interpretation of planktic: benthic ratios. *Limnol. Oceanogr.* 49(5), 1540-1548.

[10.67] Sylvestre, F. (2002) A high-resolution diatom reconstruction between 21,000 and 17,4000 14C yr BP from the southern Bolivian Altiplano (18–23°S). *J. Paleolimnol.* 27, 45-57.

[10.68] Tapia, P.M., Fritz, S.C., Baker, P.A., Seltzer, G.O. and Dunbar, R.B. (2003) A Late Quaternary diatom record of tropical climatic history from Lake Titicaca (Peru and Bolivia). *Palaeogeogr. Palaeoclimatol. Palaeoecol.* 194(1-3), 139-164.

[10.69] Thomson, M.J. (2019) Paleoclimatology. In: *Encyclopedia of Ecology (Second Edition).* B. Fath, (ed.). Elsevier, Oxford: 154-161.

[10.70] Tierney, J.E., Poulsen, C.J., Montañez, I.P., Bhattacharya, T., Feng, R., Ford, H.L., Hönisch, B., Inglis, G.N., Petersen, S.V., Sagoo, N., Tabor, C.R., Thirumalai, K., Zhu, J., Burls, N.J., Foster, G.L., Goddéris, Y., Huber, B.T., Ivany, L.C., Turner, S.K., Lunt, D.J., McElwain, J.C., Mills, B.J.W., Otto-Bliesner, B.L., Ridgwell, A. and Zhang, Y.G. (2020) Past climates inform our future. *Science* 370(6517), #3701.

[10.71] Vélez, M.I., Hooghiemstra, H., Metcalfe, S., Martínez, I. and Mommersteeg, H. (2003) Pollen- and diatom based environmental history since the Last Glacial Maximum from the Andean core Fúquene-7, Colombia. *J. Quat. Sci.* 18(1), 17-30.

[10.72] Williamson, C.E., Saros, J.E., Vincent, W.F. and Smol, J.P. (2009) Lakes and reservoirs as sentinels, integrators, and regulators of climate change. *Limnol. Oceanogr.* 54, 2273-2282.

[10.73] Wolin, J.A. and Stone, J.R. (2010) Diatoms as indicators of water-level change in freshwater lakes. In: *The diatoms: applications for the environmental and earth sciences.* J.P. Smol and E.F. Stoermer, (eds.). Cambridge University Press, Cambridge: 174-185.

[10.74] Zimmermann, J., Glöckner, G., Jahn, R., Enke, N. and Gemeinholzer, B. (2015) Metabarcoding vs. morphological identification to assess diatom diversity in environmental studies. *Mol. Ecol. Resour.* 15(3), 526-542.

11

Diatoms in Temporary Rivers: Importance in a Global Climate Change Context

Maria Helena Novais* **and Manuela Morais**†

Water Laboratory, University of Évora, Évora, Portugal

Abstract

In this chapter, we describe the biological aspects of diatoms in temporary rivers, focusing on their ecology and adaptation mechanisms to the harsh environment in these aquatic ecosystems. Predicted alterations in temporary rivers, according to the climate change scenarios, will most probably affect diatom composition and community structure. Given the importance of diatoms in these ecosystems, changes in diatom taxonomical and functional diversity, physiology, or biomass will be reflected in the whole ecosystem. The adaptation mechanisms to the stressors in temporary rivers can contribute to providing diatoms with an adaptive advantage under the predicted climate change extreme events. Furthermore, the use of diatoms as bioindicators, including the dry phase, contributes to the conservation and management of temporary rivers. However, further studies on diatoms in temporary rivers should be carried out, since their ecology is still poorly known due to the lack of exhaustive and thorough studies on their taxonomy, highlighted by the number of new species recently described from a few sampling sites in Southern Portugal.

Keywords: Adaptation mechanisms, bioindication, community dynamics, desiccation stress, thermal stress, UV radiation exposure

11.1 Introduction

Watercourses that are subject to flow cessation include a wide range of hydrological regimes, namely, temporary or intermittent rivers/streams,

**Corresponding author*: hnovais@uevora.pt

†*Corresponding author*: mmorais@uevora.pt

Nora I. Maidana, Magdalena Licursi and Eduardo Morales (eds.) Diatom Ecology: From Molecules to Metacommunities, (361–392) © 2024 Scrivener Publishing LLC

that may dry up some time throughout the year [11.40], and ephemeral streams, that only flow for a short period, usually after rainfall events [11.40] or after snow melt [11.15]. To describe these watercourses a diverse lexicon is applied, such as temporary, intermittent, seasonally intermittent, non-perennial, episodic or ephemeral [11.38]; or freezing intermittent rivers and ephemeral streams (FIRES), which are lotic systems that have the whole water column periodically freezing [11.73]. In this review, the term "temporary rivers" was selected.

The integration of temporary rivers in riverine ecosystems conceptual models has been critically revised by Allen *et al.* [11.4], stating that the 18 models jointly describe the main drivers of biogeochemical and ecological patterns and processes along the following four dimensions: (i) *longitudinally* (upstream-downstream); (ii) *laterally* (channel-riparian-floodplain); (iii) *vertically* (surface water-groundwater); and (iv) *temporally* across local and landscape scales. Nevertheless, drying represents a significant hydrological process that structures river ecosystems, contributing to the specificities of temporary watercourses that are not represented in these models, namely:

(i) their longitudinal discontinuity, with the formation of isolated pools or disconnected flowing reaches during the dry phase, with consequences for materials transport and for the connectivity of the populations of riverine organisms [11.5];
(ii) the unidirectional lateral connectivity during part of the year (only terrestrial to aquatic ecosystems);
(iii) the variability in the vertical connectivity between surface and subsurface water throughout the year, since it can be unidirectional during the drying (downwelling of surface water into the hyporheic zone, the saturated subsurface zone beneath the river channel) and rewetting phases (upwelling of subsurface water into the river channel). This variability in the vertical connectivity affects the nutrient and organic matter dynamics [11.76], the vertical exchanges of gases, with litter rewetting contributing to CO_2 emissions [11.16], and the hyporheic zone can act as a refuge for the microorganisms [11.55];
(iv) changes in flow caused by the drying process should be considered across temporal and spatial scales [11.4]. Thus, these systems are typically characterized by alternation between wet and dry phases, and are archetypal

aquatic-terrestrial ecosystems, including flowing, ponded, and dry habitats [11.69], as presented in Figures 11.1 and 11.2.

Temporary rivers are considered the most widespread type of flowing waters, dominating arid, semiarid, Mediterranean, and dry-subhumid regions, together covering almost half of the land surface [11.15]. It is estimated [11.49] that respectively 69, 56, 49, 42 and 34% of stream orders 1–5 are ephemeral in COSCATs (coastal segmentation and related catchment) below latitude 60° N, therefore, these ecosystems are important even in temperate and humid areas, since headwaters represent more than 70% of the channel length of most river networks [11.17]. Recent estimations indicate that 51–60% of the global river length is non-perennial [11.42], and in Southern Portugal represents >70% of the river network (Agência Portuguesa do Ambiente, I.P., A.R.H. Alentejo, unpublished data).

Furthermore, temporary rivers are increasing annually, with more of the land surface drying, due to the reduction and seasonal shifts in precipitation and runoff [11.15]. Temporary rivers are among the aquatic ecosystems more susceptible to being affected by climate change [11.19]. This intermittence increase (both in duration, frequency, and extension, in

Figure 11.1 Example of dry and ponded habitats present in temporary rivers. Lampreia stream (left) and Cuncos stream (right), both in Guadiana Basin, southern Portugal.

Figure 11.2 Temporal dynamics of a temporary river with the presence of wet and dry phases. Valverde stream (Sado Basin), southern Portugal.

terms of stretches/areas subject to drying) triggered by climatic drying is aggravated by human activities, such as water abstraction and diversion, with the Anthropocenic intermittence [11.15] creating artificially intermittent reaches.

The importance of naturally temporary rivers for biodiversity support, ecosystem processes and for goods and services provision is already acknowledged [11.1] [11.15] [11.67]. Despite this recent recognition of their value and global increasing spatial and temporal extent, temporary rivers are still largely omitted from legislation and policies, even though the Mediterranean GIG river types recognized by the WFD [11.20] already included "small, lowland, temporary" rivers. Consequently, these rivers are still typically poorly represented in biomonitoring programs, jeopardizing their assessment, and impairing the adoption of adequate management measures to prevent their health degradation [11.66] [11.68]. Temporary rivers are highly vulnerable to multiple anthropogenic pressures, namely agriculture intensification, livestock farming, hydromorphological alteration, and urban wastewater discharge, and these pressures will likely increase under global change [11.2] [11.13] [11.50].

Research on these ecosystems has increased over the past decades, as depicted in a search on Scopus in January 2023 (including data from 1885 until December 2022), where a total of 3,072 manuscripts were published with the terms 'temporary river*' OR 'intermittent stream*' OR 'temporary stream*' OR 'intermittent river*' OR 'ephemeral stream' OR 'non-perennial river*' OR 'non-perennial stream*'. The first record of a published paper on this subject was entitled "Intermittent Streams in Berkshire," written by Rupert Jones in 1885 and published in *Geological Magazine*. During the following decades, almost no interest was devoted to this subject, until the 60s, when one or two papers started to be published each year, increasing in the 70s and onward, until surpassing the threshold of more than 100 publications per year in 2011. Since this date, more than 100 papers were published every year, with a maximum of 193 publications in 2020, as seen in Figure 11.3.

From these, when the search term "alga*" is included, 95 publications are retrieved, from which only 56 are related to diatoms. The first published record on diatoms in temporary rivers is from 1985, and until 2007 only one paper was published per year. In 2008, two papers were published, with this number increasing to four in 2014 and reaching a maximum of seven in 2020 and 2021, revealing a slight increase in interest in this subject, as seen in Figure 11.3. These studies include taxonomical approaches, with the description of three new taxa from temporary rivers (*Punctastriata obstinata* E.Morales, M.L.García & Novais; *Planothidium*

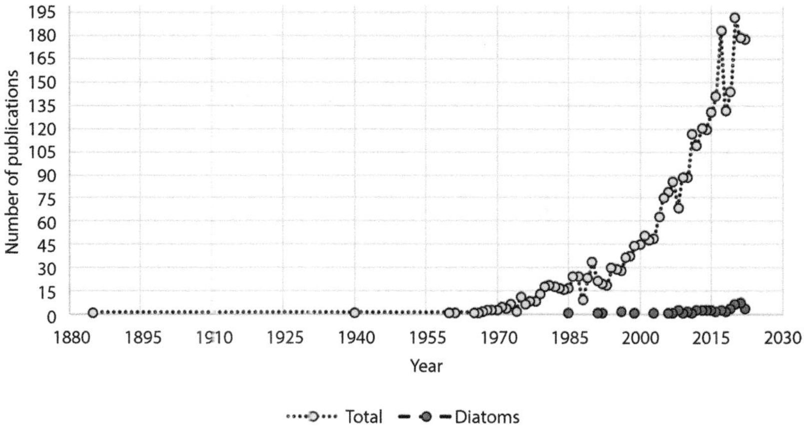

Figure 11.3 Total number of publications on temporary rivers (light grey) and including diatoms (dark grey).

audax E.Morales, Novais & M.L.García; *Fragilaria odeloucaensis* E.Morales, Novais, C.E.Wetzel, Ector & Morais); community dynamics; monitoring and ecological status assessment, detailed in the following sections. The great majority of the studies are focused on drying rivers, whilst diatom studies in intermittent freezing streams are rare, e.g., [11.64] [11.65].

When comparing these results with those for invertebrates, the latter were present in 369 publications (from 1978 until the end of 2022), with this discrepancy revealing there is still a long way to go in diatom research in these ecosystems. Also, Leigh *et al.* [11.38] reviewed the literature on Intermittent Rivers (IR), highlighting the increasing characterization of these systems, considering theoretical aspects, assessment, protection, and habitat restoration. Nevertheless, in their review, only invertebrates and fish were considered among the biota of IR, along with biogeochemistry and hydrological and ecological assessment.

The few publications on algae, and particularly diatoms, do not reflect their essential role in the functioning of temporary rivers as the basis of the trophic network along with the other primary producers (e.g., cyanobacteria, mosses, and macrophytes in the river channel and the riparian vegetation in the margins). These organisms produce organic matter using upstream and lateral inorganic nutrients and carbon dioxide in the photosynthesis process with solar energy [11.56]. In riverine systems, primary producers differ, with the riparian vegetation dominating the terrestrial-aquatic ecotone, whilst the aquatic environment is colonized by macrophytes and algae [11.56].

The structure of algal assemblages is commonly determined by the type of substratum, the water chemistry, light availability, water temperature, and flow [11.72]. In temporary rivers, these environmental factors are coupled with the challenges associated with the hydrological fluctuations, including the long dry periods and the hydric and thermal stress [11.56].

In this chapter, we review 1) the biological aspects of diatoms in temporary rivers, focusing on their adaptation mechanisms to the fluctuating hydrology and associated stressors, such as flow reduction/lentification, increased temperature, desiccation, and UV radiation exposure; 2) diatom community dynamics and the importance of temporary rivers for endangered species conservation; and 3) diatom as bioindicators in temporary rivers.

11.2 Diatoms Adaptation Mechanisms in Temporary Rivers

Temporary rivers represent a harsh and rigorous environment for diatoms, due to the hydric stress related to drying and thermal stress, water flow fluctuations, atmospheric exposure, and desiccation. Globally, water stress directly affects cell structure, biomass, and algal diversity, since water is essential for phototrophs because it is involved in the cells' osmotic equilibrium, in photosynthesis as a reducing agent and in sugar synthesis, and in other metabolic processes [11.56]. Furthermore, vegetative cells of benthic freshwater diatoms are known to be highly sensitive to desiccation, freezing, and abrupt heating, whilst vegetative cells of terrestrial diatoms are more tolerant to temperature extremes [11.62].

Diatoms have developed adaptation mechanisms (either life cycle, morphological or physiological traits) related to the capacity to withstand these harsh conditions and persist in the stream (by staying in, or colonizing potential refuges as pools, leaf litter packs, or subsuperficial sediments) and recover and spread once the flow resumes [11.55]. However, little evidence of diatom adaptation mechanisms is yet available from temporary rivers, as seen in Table 11.1. Thus, in this chapter, information about the possible adaptation mechanisms to the stressors diatoms face in temporary rivers is compiled but includes data available from other ecosystems.

Morphological and physiological changes in freshwater diatoms exposed to adverse conditions (e.g., temperature, light, desiccation, nutrients, pH, or salinity) have been reported [11.41], namely, the formation of resting stages, defined by Evans [11.26] as cells or group of cells that stop growing and have their metabolic processes at a minimum. Two types have been described [11.62]:

(i) resting spores (hypnospores), that differ in morphology and physiology from vegetative cells, with thicker frustules, often a rounder shape, and less elaborated patterns at the cell surface [11.41];
(ii) resting cells, with cytoplasmatic and physiological alterations but with similar morphology to the vegetative cells, a fact that can contribute to their being overlooked.

Occurrence of spores has been more common in marine centric diatoms, with a few freshwater species exceptions, mainly belonging to the genera *Urosolenia* and *Acanthoceras* [11.22], *Aulacoseira* and inland *Chaetoceros* [11.41]; and being rare amongst pennate diatoms, with few exceptions, such as *Craticula cuspidata* (Kützing) Mann, whose resting-spore formation is detailed in [11.58]. Resting cells are more often formed in freshwater and pennate diatoms [11.41] and their existence explains the persistence of viable cells where spores are not found (for instance, a sample collected in Southern Portugal from cobbles that were dry for a long time, at the end of the summer, revealed a high proportion of live diatoms after rewetting in the laboratory). In diatoms, resting stages are formed vegetatively, e.g., [11.21], with a few exceptions of sexual spore formation inside the auxospores in a few marine diatoms, e.g., *Leptocylindrus danicus* Cleve, *L. minimus* Gran [11.41].

Resting stages are reported to withstand unfavorable conditions for years, decades, and even millennia and still be alive, e.g., (i) *Aulacoseira granulata* (Ehrenberg) Simonsen resting cells present in anoxic sediments for twenty years were still able to germinate [11.60]; (ii) ~6600 cal. year BP old *Chaetoceros muelleri* var. *subsalsum* (Lemmermann) Johansen & Rushforth resting spores in hypoxic sediments were still viable (even though the germination rate decreased with increasing age, from three hours to 2–3 days) [11.57]. According to McQuoid and Hobson [11.41], cell survival in the resting stage depends on several important factors, such as light, temperature, and time. They listed 16 non-marine taxa as producing resting spores and 17 producing resting cells:

(i) Resting spores producing taxa: *Acanthoceras zachariasii* (Brun) Simonsen, *Aulacoseira bellicosa* (Héribaud) Simonsen, *A. italica* (Ehr.) Simonsen, *A. skvortzowii* Edlund, Stoermer & Taylor, *Craticula cuspidata* (Kützing)

Mann, *Diatoma anceps* (Ehrenberg) Kirchner, *Eunotia faba* (Ehrenberg) Grunow, *E. pectinalis* (Kützing) Rabenhorst, *E. soleirolii* (Kützing) Rabenhorst, *Melosira charcotii* Peragallo, *M. laevis* (Ehrenberg) Ralfs, *M. semilaevis* Grunow, *M. turgida* Ehrlich, *Meridion circulare* (Greville) C.Agardh, *Urosolenia eriensis* (H.L. Smith) Round & Crawford, and *U. longiseta* (Zacharias) Edlund & Stoermer;

(ii) Resting cells producing taxa: *Actinocyclus normanii* (Gregory ex Greville) Hustedt, *Asterionella formosa* Hassall, *Aulacoseira granulata* (Ehrenberg) Simonsen, *A. islandica* (O.Müller) Simonsen, *A. subarctica* (O.Müller) Haworth, *Diatoma tenuis* C.Agardh, *Fragilaria capucina* Desmazières, *Staurosira construens* Ehrenberg, *Fragilaria crotonensis* Kitton, *F. intermedia* Grunow, *F. pinnata* Ehrenberg, *Stephanodiscus alpinus* Hustedt, *S. binderanus* (Kützing) Krieger, *S. medius* Håkansson, *S. niagarae* Ehrenberg, *Tabellaria fenestrata* (Lyngbye) Kützing, and *T. flocculosa* (Roth) Kützing.

Physiological distinctions between resting stages and vegetative cells have been observed and are reported in [11.41], namely, (i) the higher rates of cellular carbon to nitrogen; (ii) higher carbon and chlorophyll content per cell when in the dark; (iii) more condensed organelles; (iv) more or larger vesicles of storage products; (v) lower sugar phosphate content; (vi) decrease in the respiration rate (to less than 20% when compared with actively growing cells); (vii) reduction in the photosynthesis rate (less than 4% to that in vegetative cells), related to the chloroplasts size reduction, loss, or color change (from green to yellow and orange-red), observed in diatom cells during the non-flow phase of temporary rivers by Timoner *et al.* [11.72]. An accumulation of oil in the cells of *Pinnularia viridis* (Nitzsch) Ehrenberg was also observed by Evans [11.26] in experimental drying. According to his study, *Pinnularia viridis*, *P. major* (Kützing) Rabenhorst, *Cymbella lanceolata* C.Agardh, and *Nitzschia linearis* W.Smith were species mainly found in the water, but able to survive some exposure and drying, while *Nitzschia palea* (Kützing) W.Smith and *Stauroneis phoenicenteron* (Nitzsch) Ehrenberg were able to survive prolonged and severe drought.

Table 11.1 summarizes the adaptation mechanisms to the stressors in temporary rivers.

Table 11.1 Diatoms adaptation mechanisms (in lines) to the main stressors (in columns) in temporary rivers.

	Water flow fluctuations	Flow reduction/ lentification	Desiccation/ drying	Increased temperature	UV radiation exposure
Alterations in the species composition at the community level	●	●	●	●	○
Mobile and mucilage life forms	•	●	●	●	●
Low-profile guild	●	●	○	○	○
Resting stages formation	○	○	●	○	○
Accumulation of oil in the cells	○	○	•	○	○
Decrease in the respiration and photosynthesis rates	○	○	●	○	○
Chloroplasts size reduction and loss or color change	○	○	●	○	○
Increase in extracellular substances production	○	○	•	○	○
Migration to deeper sediments	○	○	●	●	●
Heat shock protein production / Decrease in constitutive protein production	○	○	●	●	●
Changes in fatty acids composition	○	○	●	•	○
Increase in polyamine content	○	○	●	•	○
Compatible solutes increase	○	○	●	•	○
Screening agents' synthesis	○	○	○	○	○

(*Continued*)

Table 11.1 Diatoms adaptation mechanisms (in lines) to the main stressors (in columns) in temporary rivers. (*Continued*)

	Water flow fluctuations	Flow reduction/ lentification	Desiccation/ drying	Increased temperature	UV radiation exposure
DNA repair mechanisms	○	○	○	○	○
Detoxifying mechanisms	○	○	○	○	○
Frustule optical properties	○	○	○	○	○

Not reported ○ Punctually reported ● Reported ⬤

11.3 Flow Reduction/Lentification

In temporary rivers, one of the environmental factors most challenging for biological communities is flow reduction, which can be the flow interruption or even the complete drying of the channel, according to the Aquatic Regime (AR) of the temporary rivers, e.g., temporary pools, temporary drying or episodic, following the updated classification proposed in [11.30], that describes the long-term global variability of the hydrological conditions.

In temporary rivers, the duration and predictability of flow and no-flow periods are major factors shaping biological communities [11.30].

Considering the biological communities' dependence on the changes of the aquatic habitats over time, Gallart *et al.* [11.30] developed a methodology to analyze the aquatic regimes based on aquatic states (AS), that relate to the sets of mesohabitats occurring in a reach at a determined moment, depending on the hydrological conditions (presented in Figure 11.4):

(i) Hyperrheic: High water (flood) that originates the movement of stream bed alluvium and drift of most of the aquatic fauna;
(ii) Eurheic: Optimal hydraulic connectivity and widest range of discharges due to the succession of riffles and pools, presence and connectivity of all mesohabitats;
(iii) Oligorheic: Low water discharge, but enough to connect most pools due to water rivulets;
(iv) Arheic: Surface discharge close to 0, but some disconnected pools are still present in the stream bed;

(v) Hyporheic: Absence of surface water in most of the stream bed, but the alluvium may still be close to saturation;
(vi) Edaphic: No surface water in the entire stream bed, and the dryness of the alluvium does not allow active hyporheic life.

The importance of these aquatic states for the ecological status assessment is also discussed in [11.30], based on macroinvertebrate communities.

Furthermore, the hypothesis that diatom communities also respond to the AS was validated by Novais *et al.* [11.45], based on a one-year study of permanent (as control) and temporary rivers in 2017 in Southern Portugal. 2017 was a very dry year, characterized by severe droughts, with low accumulated precipitation even in winter and resuming only in March 2018. Therefore, the arheic state was distinguished between summer and winter, with temperature being the main environmental factor separating the two types. Diatom communities not only differed between AS (arheic and eurheic), but also reflected the mesohabitats (flowing water and pools). Indicator species of the AS (eurheic in spring 2017, eurheic during the remaining seasons, arheic in summer, arheic in the remaining seasons) were also retrieved and are as follows:

- Arheic (except Summer 2017): *Craticula cuspidata* (Kützing) D.G.Mann, *Fallacia pygmaea* (Kützing) Stickle & D.G.Mann, *Navicula erifuga* Lange-Bertalot, *Nitzschia intermedia* Hantzsch ex Cleve & Grunow, *Nitzschia supralitorea* Lange-Bertalot, *Planothidium pericavum* (J.R.Carter) Lange-Bertalot, *Sellaphora pupula* (Kützing) Mereschkowksy, and *Surirella angusta* Kützing;
- Arheic in Summer 2017: *Achnanthidium eutrophilum* (Lange-Bertalot) Lange-Bertalot, *Cocconeis placentula*

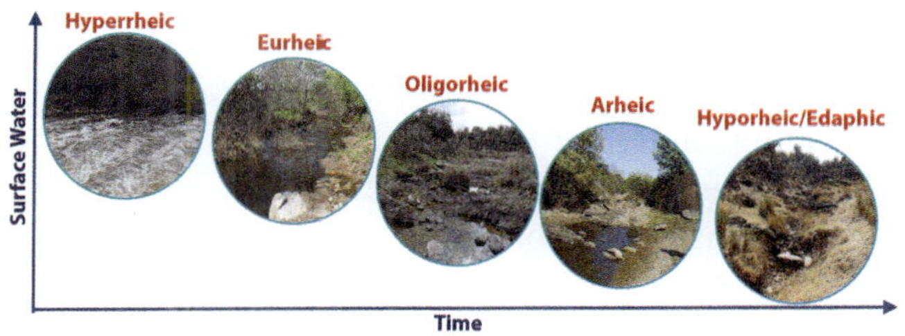

Figure 11.4 Aquatic States (AS), depending on the hydrological conditions (based on [11.30]).

Ehrenberg, *Eolimna minima* (Grunow) Lange-Bertalot, *Epithemia turgida* (Ehrenberg) Kützing, *Karayevia clevei* (Grunow) Bukhtiyarova, *Nitzschia amphibia* Grunow, *N. filiformis* var. *conferta* (Richter) Lange-Bertalot, *N. valdestriata* Aleem & Hustedt, *Pseudofallacia tenera* (Hustedt) Liu, Kociolek & Wang, *Pseudostaurosira brevistriata* (Grunow) D.M.Williams & Round, *Pseudostaurosiropsis* sp. 1, and *Staurosirella pinnata* (Ehrenberg) D.M.Williams & Round;

- Eurheic: *Denticula subtilis* Grunow, *Encyonema minutiforme* Krammer, *Eunotia implicata* Nörpel, Lange-Bertalot & Alles, *E. minor* (Kützing) Grunow, *E. pectinalis* (Kützing) Rabenhorst, *Fragilaria* aff. *capitellata* (Grunow) J.B.Petersen, and *Nitzschia gracilis* Hantzsch;
- Eurheic in Spring 2017: Two *Fragilaria* species with affinity to the pectinalis group, *Gomphonema pumilum* var. *rigidum* E.Reichardt & Lange-Bertalot, *G. truncatum* Ehrenberg, *Melosira varians* C.Agardh, *Planothidium* cf. *frequentissimum* (Lange-Bertalot) Lange-Bertalot, and *Ulnaria acus* (Kützing) Aboal.

These assemblage shifts were also reported by Sabater *et al.* [11.56], who stated that algal assemblages during the flowing phase were similar to permanent streams. With intermittency, a shift in the communities is observed, with lotic (benthic) algae gradually being replaced by lentic communities, with a dominance of the motile form in the pools, related to the nutrient increase and siltation [11.45]. When complete drying does not occur, these pools play an important role in the recolonization process once the flow resumes [11.51].

11.4 Desiccation Stress

Water supports vital biophysical and biochemical processes and influences key cellular functions, such as protein folding and stability, enzyme catalysis, DNA packaging, molecular recognition, cellular tolerance to freezing, intracellular signaling and transport processes; also, biomolecules such as nucleic acids, proteins or membranes are normally in close contact with water solutions [11.43]. Furthermore, alterations in cell organelles' morphology and location have been observed, as water leaves the cell during the desiccation process, e.g., the shrinking of the protoplasmic components and distortion of chloroplasts and pyrenoids of *Stauroneis anceps*

Ehrenberg cells, with dry cells' organelles positioned at the ends of the frustule, with the entire protoplast at one end or divided and occupying both ends of the frustule [11.33].

During the dry phase, biomolecules face air drying/dehydration, which is the process involving the removal of both freezable and non-freezable fractions of water, with extreme dehydration often causing protein denaturation (a process that may be irreversible); destabilization or disruption of membranes [11.14]; and the accumulation of reactive oxygen species [11.61]. Tolerance to desiccation has been mainly observed in resting cells of terrestrial diatoms by Souffreau *et al.* [11.62], with some mechanisms already identified at the cellular level that may contribute to achieving it: (i) the addition of a solute (disaccharides) can contribute to proteins' stabilization, and therefore dehydration to much lower water content. In brief, as dehydration progresses, the solute concentration increases to the point where it penetrates the protein hydration shell (one or two layers of water molecules that are in close interaction with the surface of the proteins) [11.43], allowing the preservation of the conformational state of the dry protein [11.14]; (ii) also, sugars (namely the disaccharide trehalose) are effective in the stabilization of phospholipid bilayers under severe dehydration, allowing dry phospholipids to be in liquid crystalline phase at temperatures they would normally be in the gel phase, avoiding the gel to the liquid crystalline phase transition during rehydration [11.14] [11.77]. Furthermore, according to the principles of the excluded solute theory, all compatible solutes (low molecular mass organic solutes) would be expected to provide some protection to cells and intracellular macromolecules, against the reduced intracellular water potential and the resulting increase in intracellular ionic strength caused by the water loss [11.77]. However, under a severe desiccation state, this mechanism is not efficient, since ca. 0.3–0.4 g H_2O protein^{-1} is needed to maintain even a single monolayer of bound water around the proteins [11.77].

Diatom response in temporary rivers to drying includes the prevalence of the mobile and mucilage-forming life forms [11.56]. The first trait includes species that are able to move (mobile) and are therefore capable of selecting the most suitable habitat [11.46] and migrating, for instance, to deeper sediments where moisture can remain [11.27] [11.55], or to exploit refugia (as pools). Mucilage-forming species (e.g., *Cymbella* or *Gomphonema* species) form mucilaginous stalks or tubes that can protect the cells. The increase in the production of extracellular substances in stress situations is known to occur when there is an inhibition of cell division but photoassimilation continues [11.47].

These adaptations, either as resting stages or certain life-forms prevalence, allow diatom communities to persist in dry stream beds only when the drying process is recurrent and slow [11.56], resulting from a long evolutionary period during which the multihabitat filter process selects diatoms with a set of traits that allows their adaptation [11.23], as in Mediterranean temporary rivers. This acclimation process has been experimentally tested by Souffreau *et al.* [11.62] and Hostetter and Hoshaw [11.33]. The first authors acknowledging that *Stauroneis anceps* survival is increased by slowing drying; and the latter proving that gradual heating before desiccation had positive results only on the survival of resting cells of terrestrial diatoms, that have protection mechanisms to survive in soils and temporary water bodies, being regularly exposed to harsh changes in humidity and temperature.

When diatom communities are subjected to rapid alterations in natural flow conditions, i.e., magnitude, frequency, duration, timing, and predictability [11.23], community recovery depends on their capacity to remain in drought refuges (e.g., permanent pools, moss patches, leaf packs, seeps and woody debris, dry biofilm on stones, dry sediments and leaf litter), from where regrowth or recolonization may occur [11.51].

Adaptations to desiccation play a major role in shaping diatom communities in view of the increasing magnitude, frequency, occurrence and duration of river drying, driven by climate change and water abstraction [11.4] [11.9] [11.17] [11.19]. Furthermore, predicted temperature increase will lead to an increase in evapotranspiration, pushing systems closer or beyond the balance where water losses to the atmosphere exceed inputs. The probability of the river drying is also expected to increase due to the higher probability of dry periods occurrence, i.e., seasonal or multi-year droughts [11.4].

11.5 Thermal Stress

In temporary rivers, temperature increase occurs during the lentification process, as observed in southern Portugal, where a maximum of 33.1 °C was measured in a pool at the end of the summer (Ameira stream, 22nd September 2017), with the median for all the summer pools (11 sites) being 22.4 °C (Interquartile range: 18.35–28.15 °C), whilst the flowing streams (two sites) presented lower temperatures (median: 14.60 °C; interquartile range: 13.95–15.25 °C) as depicted in [11.45]. Thus, diatoms are also exposed to thermal stress during the dry phase, an important issue considering that temperature is a relevant environmental factor that regulates algal growth and species distribution, either seasonally, altitudinally or latitudinally [11.11] [11.39] [11.71].

Diatom specific growth in response to temperature depends on both the ranges and temperature-specific dependency [11.71], which is more evident at low (<10 °C) and high temperatures (>20 °C), but can be modified by light flux density and photoperiod [11.11].

Species growth rate is known to increase up to a temperature maximum, followed by an abrupt decrease until reaching the upper limit of the growth range; therefore, the maximum rate occurs near the upper temperature range [11.71]. Hence, according to the optimum conditions (with a higher relative growth rate), growth tendentially occurs at the temperature at which each species predominates in the field [11.71]. In the absence of data from temporary rivers, a study with planktonic diatoms in temperate climate is presented as an example [11.71]. In this case, temperature dependency was higher in high-temperature species and lethal heating was 5 °C above the higher limit of the growth range of the species, with the highest value of 40 °C for *Chaetoceros pseudocurvisetus* Mang., *Stephanodiscus hantzschii* Grun. and *Skeletonema costatum* (Grev.) Cleve, allowing most of the species to grow under the annual habitat temperature range. Several freshwater planktonic algae, including diatoms, were not able to withstand temperatures above 30 °C even though they were actively growing at 25 °C, e.g., *Asterionella formosa* Hassall, *Tabellaria flocculosa* var. *asterionelloides* (Grunow) Knudson, *Aulacoseira subarctica* (O.Müller) E.Y.Haworth and *Discostella pseudostelligera* (Hustedt) Houk & Klee; except for *Fragilaria crotonensis* Kitton, which was still actively growing at this temperature, but for which 35 °C was lethal [11.11]. Furthermore, in the experiments carried out by [11.63], abrupt heating up to 40 °C was lethal for 40% of the 34 studied species, whilst more species were able to withstand gradual heating.

The stress response to high temperatures is considered to be a universal and highly conservative mechanism, that is typically characterized by a decrease in the production of constitutive proteins coupled with an increase in stress-inducible proteins (heat shock proteins, HSPs). This response can be observed, for instance, in the diatoms *Chaetoceros muelleri* [11.53], *Nitzschia alba* [11.36], and *Skeletonema costatum* [11.59], with a faster response to temperature increase detected in thermotolerant species (e.g., *Chaetoceros muelleri*) possessing higher amounts of constitutively expressed HSPs.

Additionally, changes in diatom fatty acid composition have been reported in response to temperature increases above the optimal growth temperature level, triggered by a particular temperature and not by the stress duration [11.52].

The exposure of *Skeletonema costatum* to heat stress (25 °C at 40 min) has caused a rapid increase in free polyamine content (putrescine and spermidine titers), which can be explained by their release from the bound form (as a "first aid" against the initial shock), followed by a new synthesis of putrescine after 30 °C at 40 min [11.59]. The polycationic nature of polyamines seems to provide the cells with the ability to regulate the intracellular environment, to have a protective effect on membrane integrity by putrescine binding to phospholipids or the stabilization of other cell molecules by inhibition of degradative enzymes [11.59]. Also, conjugated polyamines are hypothesized to be involved in the cellular response to heat stress (e.g., TCA-soluble bound spermine in *Skeletonema costatum*) [11.59].

Thermotolerance of enzymes can also be enhanced by the presence of compatible solutes, since they are preferentially excluded from the immediate hydration shell of proteins; consequently, their presence in high concentrations would tend to inhibit the thermal denaturation of proteins [11.77].

Diatoms' dependency on temperature and sensitivity to abrupt temperature changes make them a highly sensitive group to climate change, with the predicted temperature increase, alteration in precipitation patterns, and increase in the frequency and duration of extreme events such as droughts and floods. These extreme climate events can compromise the conditions beyond which many species are adapted [11.13] [11.48]. Climate change impacts on freshwater ecosystems have already been detected at a global scale, including shifts in species range and in timing. Also, in a 1.5 °C global warming scenario, species and ecosystems (mainly in polar and already warm areas, where most of the non-perennial watercourses occur) will face temperatures beyond their historical values [11.48]. This can severely alter freshwater diatom distribution, favoring species already adapted to more harsh habitats (e.g., terrestrial diatoms) and even impairing their survival, with consequences for the functioning of temporary rivers.

11.6 UV Radiation Exposure

Solar radiation is the electromagnetic radiation emitted by the sun and comprises a wide range of wavelengths, known as the electromagnetic spectrum, from radio waves through the infrared, visible and ultraviolet to X-rays and gamma rays. However, 99% of the energy of solar radiation is comprised in the wavelength band from 150–4000 nm (near ultraviolet, visible, and near-infrared regions of the solar spectrum).

According to the UCAR COMET Program, ca. 43% of radiant energy from the sun is in the visible part of the spectrum (between 400–700 nm), roughly 49% is infrared (between 700 nm–1 mm), about 7% is from ultra-violet (between 100–400 nm), and less than 1% of solar radiation is emitted as X-rays, gamma rays and radio waves (https://sos.noaa.gov/catalog/datasets/climatebits-solar-radiation/).

Considering that shorter wavelength radiation is more energetic, ultraviolet is the range that raises more concerns, due to the possible damage it may cause in living organisms. It is separated into three ranges: UV-C (100–280 nm), which makes up only 0.5% of all solar radiation, and very little reaches the surface because it is absorbed by ozone; UV-B (280–320 nm) is an energetic, photoactivating band of radiation, only partially absorbed in the stratosphere, that can penetrate the water reaching 20 m depth in the ocean; UV-A (320–400 nm) is not absorbed by ozone, though it can be blocked by cloud cover and can penetrate deeper into the water than UV-B (Fondriest Environmental, 2014, https://www.fondriest.com/environmental-measurements/parameters/weather/solar-radiation/).

Under climate change scenarios, drying increase in temporary rivers, both in length, extension, frequency, and duration, may affect the exposure of diatoms to radiation, including photosynthetically active radiation (PAR) and ultraviolet (UV). Thus, organisms in these exposed areas should possess efficient adaptation mechanisms [11.63].

UV-B irradiance can be hazardous to living beings, as it affects biomolecules directly or by inducing the formation of reactive oxygen species (ROS; as hydrogen peroxide and superoxide radicals) inside the cell. ROS can further affect the cellular structures, block enzymatic reactions, interfere with photosynthesis (through changes to the main photosynthetic enzyme, ribulose bis-phosphate carboxylase), the nutrient uptake, DNA replication and physiological responses such as motility and orientation [11.10] [11.18] [11.24] [11.32] [11.75]. Thus, at the cellular level, the toxic effect of UV radiation can be triggered in two photochemical pathways: (i) direct or primary mechanisms – biomolecules as proteins and nucleic acids have chromophores that absorb in the UV region of the radiation spectrum and are photochemically degraded or transformed under high UV flux, impairing or losing their biological function; (ii) indirect mechanisms – intermediate compounds absorb UV (photosensitizing agents) either inside or outside the cell, and produce reactive oxygen species (ROS). These high-energy oxidants, such as hydrogen peroxide, superoxide or hydroxyl radicals, can further react with other cellular components [11.10]. DNA and RNA are especially sensitive to UV radiation because of their highest absorbance coefficients for short wavelength UV [11.75], making genetic one of the most reported damages, also for diatoms [11.10].

In the absence of specific information about diatoms in temporary rivers, the results of Buma and colleagues using cultured marine diatom species in laboratory experiments are addressed [11.10]. Marine diatom species (*Cyclotella* sp., *Nitzschia closterium* and *Thalassiosira nordenskioldii*) exposed to an increasing range of daily doses of UV-B (280–320 nm) radiation revealed that, at the lowest dose (<2000 J m^{-2} d^{-1}, DNA weighted biologically effective dose (BED), normalized at 300 nm: daily $BED_{DNA\ 300\ nm}$), division rates decreased, volume enlargement occurred and high cellular protein and pigment content levels were detected, including the photoprotective carotenoids (diadinoxanthin + diatoxanthin), or both, in the *Cyclotella* species. The highest UV-B treatments (between 2000 and 3800 J m^{-2} d^{-1} daily $BED_{DNA\ 300\ nm}$) resulted in complete growth inhibition, with minor changes in protein, pigments and cell volume [11.10]. Furthermore, high UV-B radiation levels induced plasmolysis and disorientation of cell organelles in *Cyclotella* sp., whilst lower levels seemed to increase chloroplasts in number and volume [11.10]. Thus, these results support the hypothesis that UV-B radiation causes growth without cell division, due to DNA damage and the arrest in the S or G2 phase of the cell cycle [11.10]. Most microalgae developed strategies to minimize the UV-induced damage. According to Roy [11.54], these include:

(i) Avoidance mechanisms that control their movements towards light or water surface, as is the case of motile diatoms that can migrate to less exposed areas [11.18];
(ii) Screening that reduces the effective UV radiation that penetrates the cell, by synthesizing compounds that absorb the damaging wavelengths (screening agents), either intracellular, as the mycosporine-like amino acids (MAAs), (also reported by Ha *et al.* [11.31] as being involved in the defense strategy in response to UV-induced damage in the diatom *Porosira glacialis*) or extracellular (at the cell walls where they block incoming UV radiation);
(iii) Repair of UV-induced damages to the nuclear material (the most important damages, for which cells have developed repair mechanisms generally involving biosynthetic pathways mediated by stress proteins), proteins and enzymes, pigments and photosynthetic reaction centers, lipids and membranes. In the case of DNA repair, several mechanisms have been identified, such as photoreactivation (light-induced repair), nucleotide and base excision repair and recombinational repair; and the activation of

heat shock proteins for protein damage repair. For the UV damage caused by ROS, detoxifying mechanisms have been developed by means of antioxidant enzymes, lipid-soluble antioxidants in cellular membranes (e.g., carotenoids as diadinoxanthin and beta-carotene), and water-soluble reductants in the cytosol;

(iv) Acclimation, which is the process by which organisms adapt to changes in the environment, either by physiological changes or by alterations in the species composition at the community level.

(v) Recent research also verified the hypothesis that the optical properties of centric diatoms frustule can protect the cell from harmful radiation (UVB and UVC), and three main mechanisms involved in cell protection were identified using *Coscincdiscus wailesii* Gran & Angst and *C.* cf. *radiatus* Ehrenberg as a model [11.3] [11.18]. These are: absorption by silica and trace organic compounds incorporated into the frustule; diffraction (due to the geometry of the valve and refractive index contrast with respect to the surrounding environment); and photoluminescence (mainly due to chemical surface defects of hydrated amorphous porous silica). This hypothesis arose from the fact that diatoms evolved under low concentrations of atmospheric oxygen and ozone and are characterized by high tolerance to UV irradiation, even with low concentrations of screening agents such as MAAs [11.3] [11.18]. Nonetheless, it could be interesting to verify this in pennate diatoms, since the test with *Navicula perminuta* was not conclusive due to its small dimensions and high curvature [11.3]. Thus, this distribution of UV radiation due to their SiO_2 frustules can be an important evolutionary aspect to be considered for the presence and evolution of frustules in diatoms, contributing to a decrease in the rate of UV radiation-induced damage in the cells [11.3]. Furthermore, the review made by Ellegaard *et al.* [11.24] concluded that research into the role of diatom frustules in UV protection shows great potential, both in relation to diatom biology and potential technological applications, in spite of the few number of diatom species analyzed so far.

The development of these strategies is related to the previous light history (e.g., for the development of UV screening) and cell size, for instance, in the production of MAAs [11.54], and a differential response to UV radiation was observed for two diatoms from contrasting niches, planktic (collected offshore) and benthic (collected from intertidal sediments). Benthic species presented higher tolerance to extreme temperature and UV radiation and a better response was observed under elevated temperature [11.78]. Given these environmental-specific responses, it would be interesting to carry out detailed studies on the adaptation mechanisms to UV exposure of diatoms in temporary rivers.

Thus, as an overview, UV exposure net stress reflects the balance between damage, repair and the energetic costs of protection, and can be presented in terms of an increase in energy demand, changes in cell composition, and a decrease in growth and survival rates, as depicted in the scheme in [11.75]. Given that the physiological changes observed in response to the increase in UV-B radiation do not confer full protection to the organisms, and a number of organisms have already reached their UV-B tolerance limits [11.54], the diatom frustule UV protection detailed in [11.24] may confer on them an additional advantage in temporary rivers, more susceptible to drying.

11.7 Community Dynamics in Temporary Rivers

Community response to perturbations depends on the environment where it is developed, with low-stress environments supporting complex communities, but fragile when compared to those subjected to environmental stress, e.g., pollutants exposure, light, current or desiccation, as reported in the diverse studies on diatom communities cited in [11.47].

According to Sabater *et al.* [11.56], flow intermittence is a selective force on algal community composition, usually causing a decrease in alga diversity. Indeed, since a long drying phase can eliminate most of the sensitive species, few species are able to resist drying and there are few pioneer species able to recolonize when flow resumes. Thus, flow intermittency also alters diatom composition throughout the year, with shifts from lotic to species adapted to lentic conditions, even up to aerophilous species during the lentification process and drying up of the riverbed [11.28] [11.37]. Nevertheless, when drying is not complete, beta diversity is higher than expected, since lotic, lentic, and terrestrial habitats co-occur sequentially in a longitudinal direction [11.56].

Furthermore, according to Tornés and Ruhí [11.74], diatom communities in temporary rivers have less species and a lower proportion of specialist taxa than perennial streams. Also, Falasco *et al.* [11.28] detected taxonomic and functional differences between temporary and permanent reaches in Liguria (NW Italy), both at local and regional scales, with the temporary reaches being characterized by less heterogeneous communities, with higher percentages of small, mainly stalked and pioneer taxa belonging to the low-profile guild, which are able to tolerate physical disturbance. These results are in accordance with B-Béres *et al.* [11.8]. In their study of intermittent and permanent lowland rivers in the Carpathian Basin, the authors observed that intermittent rivers were characterized by lower taxonomical diversity, higher Berger-Parker index (a metric to measure dominance, defined by the proportional abundance of the dominant species in the assemblage), and lower functional richness and diversity. Disturbed sections were dominated by small-sized generalists (including pioneers and small low-profile taxa) and aerophilic taxa, the latter considered good indicators of drought, since they are able to survive the drying up of the riverbed and colonize the habitats rapidly once flow resumes. The importance of hydrology-related variables in determining diatom species distribution and community ecological attributes in Cyprus was also acknowledged by Cantonati *et al.* [11.12]. Furthermore, even in large rivers, a continuous decrease in precipitation (drying period) significantly decreased benthic diatoms in the number of taxa, Shannon diversity, and functional richness [11.44], addressing the implications of the longer periods with low precipitation, as predicted in the climate change scenarios.

The importance of temporary rivers for species conservation is another issue that should be highlighted, thus reinforcing the need for the adoption of appropriate conservation and management measures. This is supported by the relevance and frequency of species that fulfil the descriptions of Red List threat categories and that are presented in the diatom Red List for Central Europe (2018) in intermittent streams in Cyprus [11.12]. This is in accordance with the findings of Novais *et al.* [11.45] and Falasco *et al.* [11.29], who highlighted the fact that flow instability, lentification and habitat fragmentation represented the major threats to endangered species, whilst water stability and river habitat heterogeneity favored their presence. These results highlighted the importance of isolated pools as refugia for benthic diatoms during the dry phase, especially under the climate change scenarios, with increasing drought causing the persistence of isolated pools beyond the summer season (e.g., in 2007, southern Portugal was subjected to a severe drought, with isolated pools persisting during the winter until flow resuming only in the next spring).

11.8 Diatoms as Bioindicators in Temporary Rivers

Diatoms have been largely used as bioindicators in the framework of monitoring in accordance with the Water Framework Directive [11.20], being commonly used as proxies for the biological quality element "Phytobenthos" [11.35]. However, for the ecological status assessment, diatom sampling should be carried out in stable flow conditions [11.34], avoiding zones of very slow current [11.25]. The fact that the dry phase is excluded gives an incomplete idea of the habitat quality in temporary rivers [11.70]; thus, complementary methods for ecological status assessment should be designed and tested. Furthermore, an adaptation or development of novel metrics to assess the ecological status in temporary rivers is envisaged, since metrics developed for permanent streams may underestimate their ecological status since sensitivity to intermittency and to environmental degradation often co-vary [11.70]. Diatoms have the potential to fill this lacune, as presented in the preliminary results depicted in [11.70] and further tested by Novais *et al.* [11.45], with the specific pollution sensitive index (SPI) negatively correlated with the percentage of dry riverbeds and presenting a good response over time, independent of the aquatic state. The ecological quality ratio (EQR) for dry biofilm samples collected in summer was comparable with the values from the previous spring, thus dry biofilm can be considered a viable alternative for the ecological status assessment during the dry phase, instead of the isolated pools, whose communities are not considered by Gallart *et al.* [11.30] to be representative of the ecological status of the stream. Also, Barthés *et al.* [11.6] observed that the biological diatom index (BDI) was slightly affected by drying, but these effects had no significant influence on the ecological quality assessment. B-Béres *et al.* [11.7] also validated this hypothesis in a study conducted in small hilly streams in the Pannonian ecoregion of Hungary, with drying only negatively affecting the Rott trophic index (TID), whilst the specific pollution sensitivity index (IPS), Rott saprobic index (SID) and the Hungarian phytobenthos metric (IPSITI) were not impacted.

11.9 Final Remarks

Diatoms have a recognized importance in temporary rivers as the main algal component of biofilms, with any changes in their taxonomical and functional diversity, physiology or biomass being repercussed in the whole ecosystem. Furthermore, adaptation mechanisms of diatoms have already been detected, and contribute to enhance their response to main environmental stressors (e.g., flow reduction, temperature increase, increased exposure to UV radiation and desiccation), and reveal that diatoms can have an adaptative advantage under the predicted climate change extreme conditions. However, some of these strategies were tested with diatoms adapted to other aquatic ecosystems. Therefore, detailed studies should be carried out with diatoms adapted to naturally intermittent rivers to validate their effectiveness in the particular conditions characterizing these ecosystems. Furthermore, diatoms can also play an important role in the management of temporary rivers, since they can be used as bioindicators even during the dry phase. Appropriate management of naturally temporary rivers is a must since, apart from their value and global importance, they also host fragile diatoms (still poorly explored) and include threatened species. Further studies on diatoms in temporary rivers should be addressed since the ecology of most species inhabiting these ecosystems is still poorly studied.

References

[11.1] Acuña, V., Datry, T., Marshall, J., Barceló, D., Dahm, C. N., Ginebreda, A., McGregor, G., Sabater, S., Tockner, K., & Palmer, M. A. (2014). Why should we care about temporary waterways? In *Science* (Vol. 343, Issue 6175, pp. 1080–1081). American Association for the Advancement of Science. https://doi.org/10.1126/science.1246666

[11.2] Acuña, V., Hunter, M., & Ruhí, A. (2017). Managing temporary streams and rivers as unique rather than second-class ecosystems. *Biological Conservation, 211*, 12–19. https://doi.org/10.1016/j.biocon.2016.12.025

[11.3] Aguirre, L. E., Ouyang, L., Elfwing, A., Hedblom, M., Wulff, A., & Inganäs, O. (2018). Diatom frustules protect DNA from ultraviolet light. *Scientific Reports, 8*(1). https://doi.org/10.1038/s41598-018-21810-2

[11.4] Allen, D. C., Datry, T., Boersma, K. S., Bogan, M. T., Boulton, A. J., Bruno, D., Busch, M. H., Costigan, K. H., Dodds, W. K., Fritz, K. M., Godsey, S. E., Jones, J. B., Kaletova, T., Kampf, S. K., Mims, M. C., Neeson, T. M., Olden, J. D., Pastor, A. V., Poff, N. L. R., & Zimmer, M. (2020). River ecosystem conceptual models and non-perennial rivers: A critical review. In *Wiley Interdisciplinary Reviews: Water* (Vol. 7, Issue 5). John Wiley and Sons Inc. https://doi.org/10.1002/wat2.1473

[11.5] Allen, D. C., Kopp, D. A., Costigan, K. H., Datry, T., Hugueny, B., Turner, D. S., Bodner, G. S., & Flood, T. J. (2019). Citizen scientists document longterm streamflow declines in intermittent rivers of the desert southwest, USA. *Freshwater Science, 38*(2), 244–256. https://doi.org/10.1086/701483

[11.6] Barthés, A., Leflaive, J., Coulon, S., Peres, F., Rols, J.-L., Ten-Hage, L., 2015. Impact of drought on diatom communities and the consequences for the use of diatom index values in the river Maureillas (Pyrénées-Orientales, France). *River Res. Appl.,* 31, 993–1002. https://doi.org/10.1002/rra.2793

[11.7] B-Béres V., Kókai Z., Várbiró G., Mustazhapova G., Csabai, Z., Pernecker B., Borics, G., Bácsi, I., Boda, P. (2022). Flow Intermittence Drives the Benthic Algal Composition, Biodiversity and Diatom-Based Quality of Small Hilly Streams in the Pannonian Ecoregion, Hungary, *Front Ecol Evol,* vol. 10, https://doi.org/10.3389/fevo.2022.834548

[11.8] B-Béres, V., Tóthmérész, B., Bácsi, I., Borics, G., Abonyi, A., Tapolczai, K., Rimet, F., Bouchez, Á., Várbíró, G., & Török, P. (2019). Autumn drought drives functional diversity of benthic diatom assemblages of continental intermittent streams. *Advances in Water Resources, 126,* 129–136. https://doi.org/10.1016/j.advwatres.2019.02.010

[11.9] Brooks, R. T. (2009). Potential impacts of global climate change on the hydrology and ecology of ephemeral freshwater systems of the forests of the northeastern United States. *Climatic Change, 95*(3–4), 469–483. https://doi.org/10.1007/s10584-008-9531-9

[11.10] Buma, A. G. J., Zemmelink, H. J., Sjollema, K., & Gieskes, W. W. C. (1996). UVB radiation modifies protein and photosynthetic pigment content,volume and ultrastructure of marine diatoms. *Marine Ecology Progress Series, 142*(1–3), 47–54. https://doi.org/10.3354/meps142047

[11.11] Butterwick, C., Heaney, S. I., & Talling, J. F. (2005). Diversity in the influence of temperature on the growth rates of freshwater algae, and its ecological relevance. *Freshwater Biology, 50*(2), 291–300. https://doi.org/10.1111/j.1365-2427.2004.01317.x

[11.12] Cantonati, M., Kelly, M. G., Demartini, D., Angeli, N., Dörflinger, G., Papatheodoulou, A., & Armanini, D. G. (2020). Overwhelming role of hydrology-related variables and river types in driving diatom species distribution and community assemblage in streams in Cyprus. *Ecological Indicators, 117.* https://doi.org/10.1016/j.ecolind.2020.106690

[11.13] Chiu, M. C., Leigh, C., Mazor, R., Cid, N., & Resh, V. (2017). Anthropogenic Threats to Intermittent Rivers and Ephemeral Streams. In *Intermittent Rivers and Ephemeral Streams: Ecology and Management* (pp. 433–454). Elsevier Inc. https://doi.org/10.1016/B978-0-12-803835-2.00017-6

[11.14] Crowe, J. H., Carpenter, J. F., Crowe, L. M., & Anchordoguy, T. J. (1990). Are Freezing and Dehydration Similar Stress Vectors? A Comparison of Modes of Interaction of Stabilizing Solutes with Biomolecules'. In *CRYOBIOLOGY* (Vol. 27).

[11.15] Datry, T., Boulton, A. J., Bonada, N., Fritz, K., Leigh, C., Sauquet, E., Tockner, K., Hugueny, B., & Dahm, C. N. (2018a). Flow intermittence and ecosystem services in rivers of the Anthropocene. In *Journal of*

Applied Ecology (Vol. 55, Issue 1, pp. 353–364). Blackwell Publishing Ltd. https://doi.org/10.1111/1365-2664.12941

[11.16] Datry, T., Foulquier, A., Corti, R., Von Schiller, D., Tockner, K., Mendoza-Lera, C., Clément, J. C., Gessner, M. O., Moleón, M., Stubbington, R., Gücker, B., Albarinõ, R., Allen, D. C., Altermatt, F., Arce, M. I., Arnon, S., Banas, D., Banegas-Medina, A., Beller, E., Zoppini, A. (2018b). A global analysis of terrestrial plant litter dynamics in non-perennial waterways. *Nature Geoscience, 11*(7), 497–503. https://doi.org/10.1038/s41561-018-0134-4

[11.17] Datry, T., Larned, S. T., & Tockner, K. (2014). Intermittent rivers: A challenge for freshwater ecology. In *BioScience* (Vol. 64, Issue 3, pp. 229–235). Oxford University Press. https://doi.org/10.1093/biosci/bit027

[11.18] De Tommasi, E., Congestri, R., Dardano, P., De Luca, A. C., Managò, S., Rea, I., & De Stefano, M. (2018). UV-shielding and wavelength conversion by centric diatom nanopatterned frustules. *Scientific Reports, 8*(1). https://doi.org/10.1038/s41598-018-34651-w

[11.19] Dhungel, S., Tarboton, D. G., Jin, J., & Hawkins, C. P. (2016). Potential Effects of Climate Change on Ecologically Relevant Streamflow Regimes. *River Research and Applications, 32*(9), 1827–1840. https://doi.org/10.1002/rra.3029

[11.20] Directive 2000/60/EC (2000). Commission Directive 2000/60/EC of the European Parliament and the Council establishing the framework for community action in the field of water policy. *Off. J. Eur. Union* 327.

[11.21] Drebes, G. (1966). On the life history of the marine plankton diatom Stephanopyxis palmeriana. *Helgol. Wiss. Meeresunters., 13*, 101–114.

[11.22] Edlund, M. B., & Stoermer, E. F. (1993). Resting spores of the freshwater diatoms Acanthoceras and Urosolenia. In *Journal of Paleolimnology* (Vol. 9).

[11.23] Elias, C. L., Calapez, A. R., Almeida, S. F. P., & Feio, M. J. (2015). From perennial to temporary streams: An extreme drought as a driving force of freshwater communities' traits. *Marine and Freshwater Research, 66*(5), 469–480. https://doi.org/10.1071/MF13312

[11.24] Ellegaard, M., Lenau, T., Lundholm, N., Maibohm, C., Friis, S. M. M., Rottwitt, K., & Su, Y. (2016). The fascinating diatom frustule—can it play a role for attenuation of UV radiation? In *Journal of Applied Phycology* (Vol. 28, Issue 6, pp. 3295–3306). Springer Netherlands. https://doi.org/10.1007/s10811-016-0893-5

[11.25] European Committee for Standardization (2003). European Standard EN 13946:2003. Water quality - Guidance standard for the routine sampling and pretreatment of benthic diatoms from rivers.

[11.26] Evans, J. H. (1958). The Survival of Freshwater Algae During Dry Periods: Part I. An Investigation of the Algae of Five Small Ponds. In *Source: Journal of Ecology* (Vol. 46, Issue 1). https://www.jstor.org/stable/2256910

[11.27] Evans, J. H. (1959). The Survival of Freshwater Algae During Dry Periods: Part II. Drying Experiments: Part III. Stratification of Algae in Pond Margin Litter and Mud. In *Source: Journal of Ecology* (Vol. 47, Issue 1).

[11.28] Falasco, E., Bona, F., Risso, A. M., & Piano, E. (2021). Hydrological intermittency drives diversity decline and functional homogenization in benthic diatom communities. *Science of the Total Environment, 762.* https://doi.org/10.1016/j.scitotenv.2020.143090

[11.29] Falasco, E., Piano, E., & Bona, F. (2016). Diatom flora in Mediterranean streams: flow intermittency threatens endangered species. *Biodiversity and Conservation, 25*(14), 2965–2986. https://doi.org/10.1007/s10531-016-1213-8

[11.30] Gallart, F., Prat, N., Garca-Roger, E. M., Latron, J., Rieradevall, M., Llorens, P., Barbera, G. G., Brito, D., De Girolamo, A. M., Lo Porto, A., Buffagni, A., Erba, S., Neves, R., Nikolaidis, N. P., Perrin, J. L., Querner, E. P., Quinonero, J. M., Tournoud, M. G., Tzoraki, O., … Froebrich, J. (2012). A novel approach to analysing the regimes of temporary streams in relation to their controls on the composition and structure of aquatic biota. *Hydrology and Earth System Sciences, 16*(9), 3165–3182. https://doi.org/10.5194/hess-16-3165-2012

[11.31] Ha, S. Y., La, H. S., Min, J. O., Chung, K. H., Kang, S. H., & Shin, K. H. (2014). Photoprotective function of mycosporine-like amino acids in a bipolar diatom (Porosira glacialis): Evidence from ultraviolet radiation and stable isotope probing. *Diatom Research, 29*(4), 399–409. https://doi.org/10.1080/0269249X.2014.894945

[11.32] Häder, D. P., Williamson, C. E., Wängberg, S. Å., Rautio, M., Rose, K. C., Gao, K., Helbling, E. W., Sinha, R. P., & Worrest, R. (2015). Effects of UV radiation on aquatic ecosystems and interactions with other environmental factors. In *Photochemical and Photobiological Sciences* (Vol. 14, Issue 1, pp. 108–126). Royal Society of Chemistry. https://doi.org/10.1039/c4pp90035a

[11.33] Hostetter, H. P., & Hoshaw, R. W. (1970). Environmental Factors Affecting Resistance to Desiccation in the Diatom Stauroneis anceps. *American Journal of Botany, 57*(5), 512–518. http://www.jstor.orghttp://www.jstor.org/stable/2441048

[11.34] INAG I.P. (2008). Manual para a avaliação biológica da qualidade da água em sistemas fluviais segundo a Directiva-Quadro da Água. Protocolo de amostragem e análise para o Fitobentos-Diatomáceas.

[11.35] Kelly, M. G., King, L., Jones, R. I., Barker, P. A., & Jamieson, B. J. (2008). Validation of diatoms as proxies for phytobenthos when assessing ecological status in lakes. *Hydrobiologia, 610*(1), 125–129. https://doi.org/10.1007/s10750-008-9427-8

[11.36] Lai, Y.-K., Li, C.-W., Hu, C.-H., & Lee, M.-L. (1988). Quantitative and qualitative analyses of protein synthesis during heat shock in the marine

diatom Nitzschia alba (Bacillariophyceae). *Journal of Phycology, 24,* 509–514. https://doi.org/10.1111/j.1529-8817.1988.tb04255.x
[11.37] Leigh, C. & Datry, T. (2017). Drying as a primary hydrological determinant of biodiversity in river systems: a broad-scale analysis', *Ecography,* vol. 40, no. 4, pp. 487–499. https://doi.org/10.1111/ecog.02230
[11.38] Leigh, C., Boulton, A. J., Courtwright, J. L., Fritz, K., May, C. L., Walker, R. H., & Datry, T. (2015). Ecological research and management of intermittent rivers: an historical review and future directions. *Freshwater Biology, 61*(8), 1181–1199. https://doi.org/10.1111/fwb.12646
[11.39] Lund, J. W. G. (1949). Studies on Asterionella: I. The Origin and Nature of the Cells Producing Seasonal Maxima. *Source: Journal of Ecology, 37*(2), 389–419.
[11.40] Magand, C., Helena Alves, M., Calleja, E., Datry, T., Dörflinger, G., & England, J. (2020). *Intermittent Rivers and Ephemeral streams: What water managers need to know.* https://doi.org/10.5281/zenodo.3888474
[11.41] McQuoid, M. R., & Hobson, L. A. (1996). Review diatom resting stages. In *J. Phycol* (Vol. 32).
[11.42] Messager, M. L., Lehner, B., Cockburn, C., Lamouroux, N., Pella, H., Snelder, T., Tockner, K., Trautmann, T., Watt, C., & Datry, T. (2021). Global prevalence of non-perennial rivers and streams. *Nature, 594*(7863), 391–397. https://doi.org/10.1038/s41586-021-03565-5
[11.43] Morón, M. C. (2021). Protein hydration shell formation: Dynamics of water in biological systems exhibiting nanoscopic cavities. *Journal of Molecular Liquids, 337.* https://doi.org/10.1016/j.molliq.2021.116584
[11.44] Nemes-Kókai, Z., Kovács, K., Borics, G., Mayer, R., Novák, Z., Robotka, A. G., József, J., Érczes, K., Lukács, A., & B-Béres, V. (2023). Continuous precipitation loss induced more pronounced compositional and diversity changes in the lotic phytobenthos than one-off drought events. *Ecol. Indic.,* 148, 110051. https://doi.org/10.1016/j.ecolind.2023.110051
[11.45] Novais, M. H., Morales, E. A., Penha, A. M., Potes, M., Bouchez, A., Barthès, A., Costa, M. J., Salgado, R., Santos, J., & Morais, M. (2020). Benthic diatom community dynamics in Mediterranean intermittent streams: Effects of water availability and their potential as indicators of dry-phase ecological status. *Science of the Total Environment,* 719. https://doi.org/10.1016/j.scitotenv.2020.137462
[11.46] Passy, S.I. (2007). Diatom ecological guilds display distinct and predictable behavior along nutrient and disturbance gradients in running waters. *Aquat. Bot.* 86, 171–178. https://doi.org/10.1016/j.aquabot.2006.09.018
[11.47] Peterson, C. G. (1987). Influences of Flow Regime on Development and Desiccation Response of Lotic Diatom communities *Ecology,* 68 (4), 946–954. https://doi.org/10.2307/1938366
[11.48] Pörtner, H.-O., D.C. Roberts, H. Adams, I. Adelekan, C. Adler, R. Adrian, P. Aldunce, E. Ali, R. Ara Begum, B. Bednar-Friedl, R. Bezner Kerr, R. Biesbroek, J. Birkmann, K. Bowen, M.A. Caretta, J. Carnicer, E.

Castellanos, T.S. Cheong, W. Chow, G. Cissé, S. Clayton, A. Constable, S.R. Cooley, M.J. Costello, M. Craig, W. Cramer, R. Dawson, D. Dodman, J. Efitre, M. Garschagen, E.A. Gilmore, B.C. Glavovic, D. Gutzler, M. Haasnoot, S. Harper, T. Hasegawa, B. Hayward, J.A. Hicke, Y. Hirabayashi, C. Huang, K. Kalaba, W. Kiessling, A. Kitoh, R. Lasco, J. Lawrence, M.F. Lemos, R. Lempert, C. Lennard, D. Ley, T. Lissner, Q. Liu, E. Liwenga, S. Lluch-Cota, S. Löschke, S. Lucatello, Y. Luo, B. Mackey, K. Mintenbeck, A. Mirzabaev, V. Möller, M. Moncassim Vale, M.D. Morecroft, L. Mortsch, A. Mukherji, T. Mustonen, M. Mycoo, J. Nalau, M. New, A. Okem, J.P. Ometto, B. O'Neill, R. Pandey, C. Parmesan, M. Pelling, P.F. Pinho, J. Pinnegar, E.S. Poloczanska, A. Prakash, B. Preston, M.-F. Racault, D. Reckien, A. Revi, S.K. Rose, E.L.F. Schipper, D.N. Schmidt, D. Schoeman, R. Shaw, N.P. Simpson, C. Singh, W. Solecki, L. Stringer, E. Totin, C.H. Trisos, Y. Trisurat, M. van Aalst, D. Viner, M. Wairiu, R. Warren, P. Wester, D. Wrathall, and Z. Zaiton Ibrahim, 2022: Technical Summary. [H.-O. Pörtner, D.C. Roberts, E.S. Poloczanska, K. Mintenbeck, M. Tignor, A. Alegría, M. Craig, S. Langsdorf, S. Löschke, V. Möller, A. Okem (eds.)]. In: Climate Change 2022: Impacts, Adaptation and Vulnerability. Contribution of Working Group II to the Sixth Assessment Report of the Intergovernmental Panel on Climate Change [H.-O. Pörtner, D.C. Roberts, M. Tignor, E.S. Poloczanska, K. Mintenbeck, A. Alegría, M. Craig, S. Langsdorf, S. Löschke, V. Möller, A. Okem, B. Rama (eds.)]. Cambridge University Press, Cambridge, UK and New York, NY, USA, pp. 37–118, https://doi:10.1017/9781009325844.002

[11.49] Raymond, P. A., Hartmann, J., Lauerwald, R., Sobek, S., McDonald, C., Hoover, M., Butman, D., Striegl, R., Mayorga, E., Humborg, C., Kortelainen, P., Dürr, H., Meybeck, M., Ciais, P., & Guth, P. (2013). Global carbon dioxide emissions from inland waters. *Nature, 503*(7476), 355–359. https://doi.org/10.1038/nature12760

[11.50] Reid, A. J., Carlson, A. K., Creed, I. F., Eliason, E. J., Gell, P. A., Johnson, P. T. J., Kidd, K. A., MacCormack, T. J., Olden, J. D., Ormerod, S. J., Smol, J. P., Taylor, W. W., Tockner, K., Vermaire, J. C., Dudgeon, D., & Cooke, S. J. (2019). Emerging threats and persistent conservation challenges for freshwater biodiversity. *Biological Reviews, 94*(3), 849–873. https://doi.org/10.1111/brv.12480

[11.51] Robson, B. J., Matthews, T. G., Lind, P. R., & Thomas, N. A. (2008). Pathways for algal recolonization in seasonally-flowing streams. *Freshwater Biology, 53*(12), 2385–2401. https://doi.org/10.1111/j.1365-2427.2008.02061.x

[11.52] Rousch, J. M., Bingham, S. E., & Sommerfeld, M. R. (2003). Changes in fatty acid profiles of thermo-intolerant and thermo-tolerant marine diatoms during temperature stress. *Journal of Experimental Marine Biology and Ecology, 295*(2), 145–156. https://doi.org/10.1016/ S0022-0981(03)00293-4

[11.53] Rousch, J. M., Bingham, S. E., & Sommerfeld, M. R. (2004). Protein expression during heat stress in thermo-intolerant and thermo-tolerant diatoms. *Journal of Experimental Marine Biology and Ecology, 306*(2), 231–243. https://doi.org/10.1016/j.jembe.2004.01.009

[11.54] Roy, S. (2009). Strategies for the minimisation of UV-induced damage. In *The Effects of UV Radiation in the Marine Environment* (pp. 177–205). Cambridge University Press. https://doi.org/10.1017/cbo9780511535444.008

[11.55] Sabater, S., Timoner, X., Borrego, C., & Acuña, V. (2016). Stream biofilm responses to flow intermittency: From cells to ecosystems. In *Frontiers in Environmental Science* (Vol. 4, Issue MAR). Frontiers Media S.A. https://doi.org/10.3389/fenvs.2016.00014

[11.56] Sabater, S., Timoner, X., Bornette, G., De Wilde, M., Stromberg, J. C., & Stella, J. C. (2017). The Biota of Intermittent Rivers and Ephemeral Streams: Algae and Vascular Plants. In *Intermittent Rivers and Ephemeral Streams: Ecology and Management* (pp. 189–216). Elsevier Inc. https://doi.org/10.1016/B978-0-12-803835-2.00016-4

[11.57] Sanyal, A., Larsson, J., van Wirdum, F., Andrén, T., Moros, M., Lönn, M., & Andrén, E. (2022). Not dead yet: Diatom resting spores can survive in nature for several millennia. *American Journal of Botany, 109*(1), 67–82. https://doi.org/10.1002/ajb2.1780

[11.58] Schmid A. (2009). Induction of resting-spores in the pennate diatom Navicula (Craticula) cuspidata by uncoupling of the cell and plastid cycles. Nova Hedwigia. [Online] 85–101.

[11.59] Scoccianti V., Penna A., Penna N., & Magnani M. (1995). Effect of heat stress on polyamine content and protein pattern in Skeletonema costatum. *Marine Biology, 121*, 549–554.

[11.60] Sicko-Goad, L., Stoermer, E. F., & Fahnensliel, G. (1986). Rejuvenation of Melosira granulata (Bacillariophyceae) resting cells from the anoxic sediments of Douglas Lake, Michigan. I. Light Microscopy and 14C uptake. In *J Phycol* (Vol. 22).

[11.61] Smirnoff, N. (1993). The role of active oxygen in the response of plants to water deficit and desiccation. *New Phytologist, 125*, 27–58.

[11.62] Souffreau, C., Vanormelingen, P., Sabbe, K., & Vyverman, W. (2013). Tolerance of resting cells of freshwater and terrestrial benthic diatoms to experimental desiccation and freezing is habitat-dependent. *Phycologia, 52*(3), 246–255. https://doi.org/10.2216/12-087.

[11.63] Souffreau, C., Vanormelingen, P., Verleyen, E., Sabbe, K., & Vyverman, W. (2010). Tolerance of benthic diatoms from temperate aquatic and terrestrial habitats to experimental desiccation and temperature stress. *Phycologia, 49*(4), 309–324. https://doi.org/10.2216/09-30.1

[11.64] Stanish, L. F., Kohler, T. J., Esposito, R. M. M., Simmons, B. L., Nielsen, U. N., Wall, D. H., Nemergut, D. R., & McKnight, D. M. (2012). Extreme streams: Flow intermittency as a control on diatom communities in meltwater streams in the McMurdo Dry Valleys, Antarctica. *Canadian Journal of Fisheries and Aquatic Sciences, 69*(8), 1405–1419. https://doi.org/10.1139/F2012-022

[11.65] Stanish, L. F., O'Neill, S. P., Gonzalez, A., Legg, T. M., Knelman, J., Mcknight, D. M., Spaulding, S., & Nemergut, D. R. (2013). Bacteria and diatom co-occurrence patterns in microbial mats from polar desert streams. *Environmental Microbiology, 15*(4), 1115–1131. https://doi.org/10.1111/j.1462-2920.2012.02872.x

[11.66] Steward, A. L., Negus, P., Marshall, J. C., Clifford, S. E., & Dent, C. (2018). Assessing the ecological health of rivers when they are dry. *Ecological Indicators, 85*, 537–547. https://doi.org/10.1016/j.ecolind.2017.10.053

[11.67] Stubbington, R., Acreman, M., Acuña, V., Boon, P. J., Boulton, A. J., England, J., Gilvear, D., Sykes, T., & Wood, P. J. (2020). Ecosystem services of temporary streams differ between wet and dry phases in regions with contrasting climates and economies. *People and Nature, 2*(3), 660–677. https://doi.org/10.1002/pan3.10113

[11.68] Stubbington, R., Chadd, R., Cid, N., Csabai, Z., Miliša, M., Morais, M., Munné, A., Pařil, P., Pešić, V., Tziortzis, I., Verdonschot, R. C. M., & Datry, T. (2018). Biomonitoring of intermittent rivers and ephemeral streams in Europe: Current practice and priorities to enhance ecological status assessments. In *Science of the Total Environment* (Vol. 618, pp. 1096–1113). Elsevier B.V. https://doi.org/10.1016/j.scitotenv.2017.09.137

[11.69] Stubbington, R., England, J., Wood, P. J., & Sefton, C. E. M. (2017). Temporary streams in temperate zones: recognizing, monitoring and restoring transitional aquatic-terrestrial ecosystems. *Wiley Interdisciplinary Reviews: Water, 4*(4). https://doi.org/10.1002/WAT2.1223

[11.70] Stubbington, R., Paillex, A., England, J., Barthès, A., Bouchez, A., Rimet, F., Sánchez-Montoya, M. M., Westwood, C. G., & Datry, T. (2019). A comparison of biotic groups as dry-phase indicators of ecological quality in intermittent rivers and ephemeral streams. *Ecological Indicators, 97*, 165–174. https://doi.org/10.1016/j.ecolind.2018.09.061

[11.71] Suzuki Y., & Takahashi, M. (1995). Growth responses of several diatom species isolated from various environments to temperature. *J. Phycol., 31*, 880–888.

[11.72] Timoner, X., Buchaca, T., Acuña, V., & Sabater, S. (2014). Photosynthetic pigment changes and adaptations in biofilms in response to flow intermittency. *Aquatic Sciences, 76*(4), 565–578. https://doi.org/10.1007/s00027-014-0355-6

[11.73] Tolonen, K. E., Picazo, F., Vilmi, A., Datry, T., Stubbington, R., Pařil, P., Perez Rocha, M., & Heino, J. (2019). Parallels and contrasts between intermittently freezing and drying streams: From individual adaptations to biodiversity variation. In *Freshwater Biology* (Vol. 64, Issue 10, pp. 1679–1691). Blackwell Publishing Ltd. https://doi.org/10.1111/fwb.13373

[11.74] Tornés, E., & Ruhí, A. (2013). Flow intermittency decreases nestedness and specialisation of diatom communities in Mediterranean rivers. *Freshwater Biology, 58*(12), 2555–2566. https://doi.org/10.1111/fwb.12232

[11.75] Vincent, W. F., & Neale, P. J. (2009). Mechanisms of UV damage to aquatic organisms. In *The Effects of UV Radiation in the Marine Environment* (pp. 149–176). Cambridge University Press. https://doi.org/10.1017/cbo9780511535444.007

[11.76] von Schiller, D., Bernal, S., Dahm, C. N., & Martí, E. (2017). Nutrient and Organic Matter Dynamics in Intermittent Rivers and Ephemeral Streams. In *Intermittent Rivers and Ephemeral Streams: Ecology and Management* (pp. 135–160). Elsevier Inc. https://doi.org/10.1016/B978-0-12-803835-2.00006-1

[11.77] Welsh, D. T. (2000). Ecological significance of compatible solute accumulation by micro-organisms: from single cells to global climate. *FEMS Microbiology Reviews, 24*(3), 263–290. https://doi.org/10.1111/j.1574-6976.2000.tb00542.x

[11.78] Wu, Y., Yue, F., Xu, J., & Beardall, J. (2017). Differential photosynthetic responses of marine planktonic and benthic diatoms to ultraviolet radiation under various temperature regimes. *Biogeosciences, 14*(22), 5029–5037. https://doi.org/10.5194/bg-14-5029-2017

12

The Ecology of Diatoms in Peatlands: Communities from Tierra del Fuego Peat Bogs as a Study Case

Valeria Casa[1]*, María V. Quiroga[2], Bart Van de Vijver[3,4] and Gabriela Mataloni[1,2]

[1]*Environmental Research and Engineering Institute (IIIA) UNSAM-CONICET, San Martín, Argentina*
[2]*Chascomús Technological Institute (CONICET-UNSAM), School of Bio and Nanotechnologies, Buenos Aires, Argentina*
[3]*Meise Botanic Garden, Research Department, Meise, Belgium*
[4]*University of Antwerp, Department of Biology, ECOSPHERE, Antwerp, Belgium*

Abstract

Wetlands are valued for their biodiversity, water reservoir function, and contributions to human welfare, they provide 40% of Nature's benefits to humanity but face threats from land use changes, pollution, and overexploitation. Peatlands, a distinct wetland type that covers 3% of the global land surface, host a remarkable diversity of diatom species. This chapter first explores peatland characteristics and diatom communities globally, highlighting key genera, ecological aspects, and adaptations. Physical-chemical factors, especially pH, significantly affect diatom

**Corresponding author*: vcasa@unsam.edu.ar

Valeria Casa: https://www.researchgate.net/profile/Valeria-Casa, https://iiia.conicet.gov.ar/casa-valeria/, https://www.conicet.gov.ar/new_scp/detalle.php?keywords=casa%2Bvaleria&id=51548&datos_academicos=yes

María V. Quiroga: https://orcid.org/0000-0002-6223-334X, https://www.conicet.gov.ar/new_scp/detalle.php?id=33283&keywords=maria%2Bvictoria%2Bquiroga&datos_academicos=yes

Bart Van de Vijver: https://www.plantentuinmeise.be/nl/pQv6OuV/bart-van-de-vijver, https://www.researchgate.net/profile/Bart-Van-De-Vijver

Gabriela Mataloni: https://www.researchgate.net/profile/Gabriela-Mataloni, https://www.conicet.gov.ar/new_scp/detalle.php?id=22176&keywords=mataloni&datos_academicos=yes

Nora I. Maidana, Magdalena Licursi and Eduardo Morales (eds.) Diatom Ecology: From Molecules to Metacommunities, (393–420) © 2024 Scrivener Publishing LLC

distribution, decreasing species richness in mineral-poor sites. Landscape-scale variables, like water-table depth and microtopography, create diverse niches for diatom. The second part focuses on a case study of two peat bogs in Tierra del Fuego, Argentina: Rancho Hambre and Valle de Andorra. We compare their diatom communities to those in peripheral limnetic environments. Despite being 50 km apart, both peat bogs exhibited remarkably similar species compositions and unique diatom assemblages distinct from their surroundings. Environmental filters, such as pH, total hardness, total phosphorus, and conductivity, were responsible for the observed species turnover. Ponds proved to be the most species-rich habitats, while the *Sphagnum magellanicum* matrix hosted relatively few species. Among pond ecosystems, benthic samples displayed the highest species richness and unique species.

***Keywords*:** Bacillariophyceae, moss, ponds, *Sphagnum*, diatom communities, wetlands, biodiversity

12.1 Introduction

12.1.1 Wetlands and Peatlands: General Characteristics and Global Significance

In the last decades, there has been an increasing interest in studies in wetland ecology, mainly because of their unique biodiversity, their primordial role as water reservoirs and their contributions to human welfare (food/materials suppliers). Wetlands provide approximately 40% of all Nature's contribution to mankind [12.26]. Nevertheless, they also rank amongst the most vulnerable environments worldwide. The main threats they face include dramatic changes in land use, mainly due to increasing urbanization and uncontrolled expansion of agriculture/livestock lands, pollution and overexploitation of their resources [12.33] [12.57]. As discussed in Quintana and Mataloni [12.99], the definition of wetlands is broad and has varied with time and authors. Nonetheless, wetlands can be summarized as a diverse set of ecosystems characterized by varying amounts of standing or subsuperficial water, anoxic soils, and plant and animal species adapted to these environmental conditions. The hydrological cycle defines the structural and functional characteristics of these environments.

One-third of all wetlands worldwide are peatlands, an ecosystem type present in at least 180 countries and representing 3% of the total land surface [12.3] [12.94]. Peatlands are established in areas where water can accumulate due to a positive precipitation/evaporation balance, which explains their predominance in cold (boreal and sub-Arctic) and humid (oceanic and humid tropical) regions [12.33]. Thus, climate type is essential for the formation of peatlands, as it defines the temperatures and precipitation of the region. As this relates to the geographical distribution of continents, climate type explains the higher frequency of peatlands in the Northern compared to the Southern Hemisphere. Under such particular environmental conditions, the rate of primary production is much higher than the decomposition rate, causing an accumulation of dead organic matter (peat). This allows for a long-term net carbon storage [12.7] [12.83]. Therefore, from a climate regulation point of view, peatlands play a crucial role on a global scale, as they are involved in the worldwide greenhouse gas balance [12.56]. Nevertheless, these balances heavily depend on the ecosystem's health. Degraded peatland stops sequestering CO_2 from the atmosphere and the accumulated peat begins to decompose aerobically, releasing massive amounts of carbon into the system. Large amounts of historically sequestered carbon can be released because of degradation through drainage, excavation or fertilization [12.7]. For example, Sacramento–San Joaquin Delta in California is an example of a peatland undergoing rapid environmental transformation, where disturbances have led to irreversible changes in the carbon cycle, driven primarily by processes such as drainage, conversion for agricultural purposes, and subsequent significant rates of land subsidence [12.1] [12.52].

Peatlands also have an important hydrological role. At a global scale, they are natural freshwater reservoirs (10% of freshwater). Due to their exceptional water-holding capacity, they are able to buffer water supply through long droughts but also during heavy rainfall or flooding. These wetland ecosystems possess the ability to serve as extensive natural filters, effectively retaining sediments and sequestering contaminants [12.24] [12.79].

12.1.2 Peatland Classification and Environmental Characterization

A peatland area encompassing potentially peat-forming vegetation is commonly termed mire. Mires can be classified into two types based on their water source [12.75]: minerotrophic fens are flat or depressed systems fed by groundwater or surface water flows, while ombrotrophic bogs are isolated from groundwater influence, and thus water and nutrient supply exclusively depend on direct precipitation [12.107]. Peat accumulation in ombrotrophic bogs results in the development of a domed-shaped morphology (known as raised bogs) [12.105]. Peat-forming vegetation starts to develop at the edges of water bodies and gradually the entire basin gets filled with peat, eventually resulting in the transformation of the lentic habitat into a more terrestrial system. Once the mire basin is entirely filled with peat, *Sphagnum* mosses can generate small hummocks above the groundwater table and grow eventually into elevated domes isolated from groundwater influence.

In general, mires with different hydrological regimes display distinct pH and nutrient conditions. The transition from minerotrophic fens to ombrotrophic bogs causes a progressive decrease in pH, from circumneutral to acidic conditions, consequently decreasing nutrient availability in dystrophic bogs [12.33]. *Sphagnum* mosses play a key role in this environmental change as they efficiently remove cations, liberating protons [12.23], thus acidifying ombrotrophic mires.

The complex topography of peatland landscapes generally results in the development of a mosaic of different environments [12.47]. This environmental diversity can be observed at several spatial scales. For example, a gradual decrease in pH and increase in conductivity was observed over short distances along a terrestrialization gradient from open water of ponds to the drier surrounding mosses [12.14] [12.25]. Furthermore, the microtopography-determined hydrological network increases habitat complexity even in small areas, with ponds only meters apart displaying very different chemical features [12.45].

From a biodiversity point of view, peatlands are regarded as important hotspots because they support a high number of specialized organisms that can cope with many inimical conditions such as acidity and dystrophy [12.43] [12.54] [12.55] [12.100] [12.107]. Nutrient-poor peat bogs are essential ecosystems maintaining the biodiversity of microalgae sensitive to eutrophication [12.27].

12.2 Singularity of Peatland Diatoms

12.2.1 Main Genera and Species in Peatlands

Among autotrophs, diatoms (Bacillariophyceae) and desmids (Desmidiaceae) represent some of the richest taxonomic groups, with a high number of taxa restricted to these habitats [12.14] [12.19] [12.27] [12.69] [12.82] [12.90] [12.91] [12.92].

Peatland diatom communities are dominated by a handful of genera, among which *Eunotia* and *Pinnularia* are the most species-rich [12.19]

[12.33] [12.95] [12.97] [12.98]. Most *Eunotia* species have narrow ecological preferences and are typically observed in oligotrophic waters, often rich in humic acids and strongly associated with *Sphagnum* spp. in peat bogs or related habitats [12.73]. *Pinnularia*, in turn, is a large genus containing more than 1800 species that can be found in a variety of habitats, many of which are indicative of acidic and oligotrophic freshwaters [12.65]. Several other genera, such as *Frustulia, Neidium, Brachysira, Kobayasiella* and *Encyonema*, are also frequently recorded in oligotrophic, acidic habitats and therefore represent an important part of peatland diatom communities [12.64] [12.69] [12.71] [12.72].

Some of the most frequently cited species in peatlands include *Eunotia sphagnicola* Van de Vijver, A.Mertens & Lange-Bert.(often erroneously reported as *Eunotia paludosa* Grunow, [12.115]), *Eunotia seminulum* Nörpel-Schempp & Lange-Bert., *Eunotia exigua* (Bréb. *ex* Kütz.) Rabenh., *Pinnularia borealis* Ehrenb., *Pinnularia microstauron* (Ehrenb.) Cleve, *Pinnularia viridis* (Nitzsch) Ehrenb., *Tabellaria flocculosa* (Roth) Kütz., *Achnanthidium minutissimum* (Kütz.) Czarn., *Frustulia saxonica* Rabenh. and others. Many of these species are considered cosmopolitan, with a wide distribution on a global scale. However, it is essential to note that diatom diversity and possible local endemisms, especially outside Europe and North America, have long been underestimated, primarily due to force-fitting [12.112]. In turn, this could have led to overestimate the distribution area of iconic species by misidentifying similar locally distributed taxa. For instance, a recent study on the type material of *Eunotia paludosa* proved our current concept of this species wrong. The analysis resulted in the description of at least three different species, with similar but distinguishable morphologies, and redefined geographical distributions and ecological preferences [12.115]. Also, the combination of morphometric and molecular studies in the *Pinnularia borealis* complex has revealed truly cryptic species [12.96].

Flaws in the identification of diatom species could impair their usefulness as environmental indicators in paleoenvironmental reconstructions or in biogeographical studies. In this context, there is a pressing need for the detailed study of peatland diatoms in order to refine taxonomic and ecological information about well-known, but also new, potentially endemic taxa. The analysis of type material and other historic material is crucial to establish both the correct taxonomic identity of the species and their ecological preferences. The recent revision of the European *Brachysira* flora is a clear example of how such studies might be undertaken [12.115] [12.122] [12.123].

12.2.2 Physical, Chemical and Landscape-Scale Drivers of Diatom Communities

Amongst the physical and chemical variables of freshwater environments, pH is the main driver of diatom distribution, since most species have a relatively narrow pH range [12.5] [12.28] [12.113]. DeNicola [12.28] reviewed diatoms from highly acidic environments and concluded that only a few taxa are capable of maintaining a viable population below pH 3.5. However, most aquatic environments dominated by *Sphagnum* rarely have lower pH values [12.53], which allows them to support a rich diatom flora. A high influence of pH and moisture on diatom communities was observed in many peatlands or mossy environments in Europe [12.33] [12.91] [12.97], North America [12.50], South America [12.82], the sub-Antarctic region [12.119] and Oceania [12.61]. The diversity of diatoms in peatlands usually increases with local pH values [12.90].

For example, in 1982, Kingston conducted a study of Minnesota peat bogs, shedding light on the characteristic diatom assemblages and their correlation with the trophic status of these ecosystems. This study identified several diatom species as trophic indicators in peatland ecosystems, significantly advancing our understanding of these environments. More recently, a study on diatom communities along a mineral gradient in fens in the Czech Republic and Slovakia demonstrated that water chemistry substantially influenced the spring fen biota [12.33]. These authors found that species richness decreased significantly from mineral-rich to mineral-poor sites, and the structure and composition of diatom assemblages were ruled by a strong environmental gradient of pH, calcium, and conductivity.

At a landscape scale, peatlands host a great diversity of habitats and microhabitats. In ombrotrophic peat bogs, the moss matrix frequently encompasses water bodies ranging from numerous small ponds to a few shallow lakes with contrasting community compositions. Also, the microtopography of peatland surfaces has key implications for hydrologic conditions, generating small-scale variations in environmental features that translate into more ecological niches wherein a greater diversity of microorganisms can be established. In the southernmost area of South America, Mataloni [12.82] found a high species turnover in progressively impoverished microalgal communities along transects from the open water of ponds to the water among *Sphagnum* mosses ca. 3 m away from the pond edge, in accordance with constantly decreasing pH and humidity and increasing conductivity.

Moisture of the sampling site strongly depends on water table depth and is influenced by its position on the mire surface. Poulíčková *et al.* [12.97] studied three spring fens from Czechia and Slovakia. They demonstrated that diatom abundance, species richness and diversity decrease along with decreasing moisture in the microsite. *Eunotia paludosa* was a typical species of dry hummock tops, accompanied by *Pinnularia rupestris* Hantzsch and *Achnanthes lanceolata* (Bréb. ex Kütz.) Grunow. Later, Chen *et al.* [12.20] studied diatom communities from different microhabitats within three montane peatlands in central China. They found that the desiccation-tolerant genera *Luticola* and *Hantzschia*, as well as *Eunotia paludosa*, *Hantzschia amphioxys* (Ehrenb.) Grunow, *Pinnularia borealis* and *Luticola mutica* (Kütz.) Mann were good indicators for hummocks microhabitats, characterized by moisture-limited and oxidizing conditions. Meanwhile, *Eunotia minor* (Kütz.) Grunow and *Eunotia seminulum* preferred the wetter hollows. In addition, other species, such as *Aulacoseira ambigua* (Grunow) Simonsen or *Neidium ampliatum* (Ehrenb.) Krammer, were good indicators for ditch edges, a hydrologically unstable microhabitat with highly variable environmental conditions wherein diatoms with different life strategies coexisted.

Historically, as for freshwater ecosystems, the planktonic community of water bodies has received primary attention in most peat bog studies; however, other communities such as periphyton or benthos largely contribute to diversity, especially in shallow ponds from peatlands. For example, at the local scale, the community structure of diatoms differed between *Sphagnum* periphyton and epipelon of peatland pools from Czechia [12.88]. Also, the identity of the moss species forming the peat bog is a source of environmental variation. The morphology of the cauloids and phyllids and the cation exchange with the environment can differ between moss species, thus influencing the growth and development of epiphytic algae [12.29] [12.97].

In sum, in agreement with Gaiser and Rühland [12.41] and Simkhada and Jüttner [12.109], detailed sampling of the different habitats, microhabitats or substrates within wetlands is paramount, since these systems host a great environmental heterogeneity which is essential to sustain their biodiversity. Therefore, when surveying diatom diversity of such ecosystems, we strongly recommend sampling the largest possible variety of environments across different space scales to include even those species restricted to a particular habitat type.

12.3 Diatoms in Tierra del Fuego Peat Bogs

12.3.1 Prior Knowledge

As mentioned previously, the world's largest peatland areas are situated in the Northern Hemisphere (Europe, Asia and North America). Also, given the higher concentration of diatom scientists on these continents, there is a high number of studies dealing with diatoms of these systems [12.34] [12.69] [12.80] [12.88] [12.91] [12.111], while those on diatoms from Southern Hemisphere peat bogs are much more scarce [12.14] [12.60]. Due to the high rate of wetlands loss on a global scale, Gaiser and Rühland [12.41] call to double efforts to increase our existing knowledge of their flora. This is especially true for high-latitude wetlands, wherein the effect of climate change is so remarkable [12.41], and particularly for good climatic indicators such as diatoms. The next section will therefore focus on the peat bogs of Tierra del Fuego in southern Patagonia.

The province of Tierra del Fuego, Antarctica and South Atlantic Islands encompasses 95% of all peat bogs in Argentina [12.103], representing almost 12.5% of the surface of the Isla Grande de Tierra del Fuego [12.56]. Most of these are found in Mitre Peninsula, an area dominated by *Sphagnum magellanicum* peat bogs mostly inaccessible by land. Fuegian wetlands are highly at risk due to increasing human activities on a local scale but also following deterioration due to climate change.

Diatom studies records from Tierra del Fuego (Argentina) are scarce, with those focused on peatland diatoms even fewer. The oldest study was performed by Cleve [12.21] and dealt with marine and freshwater diatoms from estuaries around Río Grande (east coast of Tierra del Fuego), and Isla Desolación (west coast). In turn, Frenguelli carried out very detailed studies on lakes and streams [12.35] [12.36] [12.37] [12.38] and a core from Río de la Misión, Río Grande peat bog [12.40] [12.39]. Further historical surveys, often restricted to a handful of samples, were performed by Cleve-Euler [12.22] and Krasske [12.66], both reporting on freshwater diatoms from *Sphagnum* peat bogs. After an almost five-decade pause, Mataloni and Tell [12.85] and Mataloni [12.82] resumed floristic and ecological diatom research in Tierra del Fuego by identifying and illustrating 113 diatom taxa from different peat bogs. Later, in an impressive taxonomic work, Rumrich *et al.* [12.106] discussed and illustrated the diatom flora of the Andes from Venezuela to Tierra del Fuego. Although only part of this floristic study is devoted to Tierra del Fuego, the book is still considered a cardinal taxonomic work for those working on its diatom flora. Over the past 20 years, most research in the region has focused on specific diatom groups such as Achnanthaceae [12.78], Epithemiaceae and Surirellaceae [12.46] [12.48], Orthoseirales [12.49] and the invasive species *Didymosphenia geminata* (Lyngbye) Schmidt [12.45]. In addition to studies dealing with modern diatoms, several studies applied diatoms for paleoenvironmental reconstructions, [12.8] [12.12] [12.31] [12.32] [12.74] [12.87] [12.126], among others.

Our own recent research dealt with diatom communities from Tierra del Fuego, as it belongs to the southernmost extensive peatland area in the world. We selected two dome-shaped peat bogs—Rancho Hambre and Valle de Andorra—composed of several shallow ponds in a *Sphagnum magellanicum* matrix. To survey their diatom assemblages and their relationship with environmental parameters at different scales, diatom communities were compared to those of peripheral limnetic environments (moats, rivers, beaver dams), within peat bog environments (ponds vs. moss matrix), and within pond habitats (sediments vs. periphyton).

In this study, we hypothesized that the ombrotrophic characteristics of both systems impose a robust environmental filter, structuring the diatom communities and differentiating them from the surrounding aquatic environments. Furthermore, we postulated that such environmental filters are common to peatlands with similar features within the same region. Thus, we predicted that diatom assemblages would be more similar between peat bogs than those of surrounding environments with different characteristics.

12.3.2 Study Area, Sampling and Statistical Analysis

Valle de Andorra peat bog (54° 45'S, 68° 20'W) is located 6 km away from Ushuaia City (Tierra del Fuego, Argentina) and belongs in the Ramsar-protected site "Vinciguerra Glacier and Associated Peatlands" [12.104]. On the other hand, Rancho Hambre peat bog (54° 47'S, 68° 19'W) is located 50 km east from Valle de Andorra and is included in the Valle Tierra Mayor Natural Reserve [12.108]. Both systems were sampled in February 2014 and November 2016. The central area of each of these *Sphagnum*-dominated peat bogs hosts several shallow (less than 2 m deep) ponds. In each peat bog, nine ponds and four points in the *Sphagnum* matrix were sampled across each valley to obtain a representative picture of these systems (Figure 12.1).

In each pond, three samples were taken: one from the plankton, one from the benthos inhabiting the upper sediment layer and one periphyton sample from submerged mosses, including *Sphagnum magellanicum* Bridel, *S. fimbriatum* Wilson and *Sanionia uncinata* (Hedw.) Loeske. In 2016, three peripheral environments were also sampled at each location

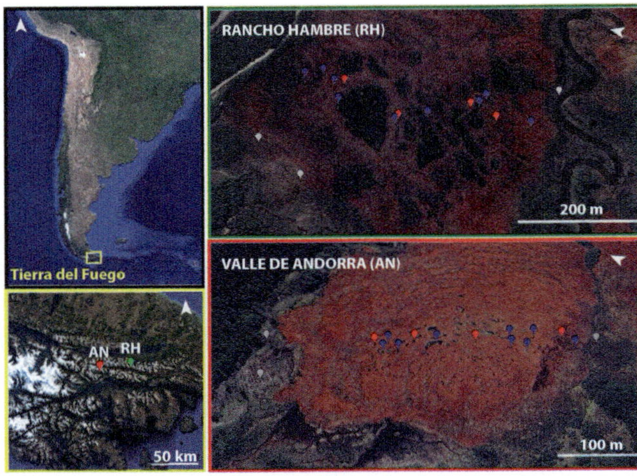

Figure 12.1 Location of the study area and spatial distribution of the sampled points. Markers: *Sphagnum* matrix sites (in red); ponds (in blue); peripheral environments (in gray). The maps were generated using Google Earth Pro.

for comparison with those within the peat bog. These included a natural moat running along the edge of the peat dome (Moat), the river running along each peat bog (River), and a beaver dam built on its margin (Beaver).

In this study, 181 samples were collected from both peatbogs and their surrounding environments during two sampling events. Out of these, 116 samples containing sufficient diatom valves were analyzed. Sixty of them pertained to Valle de Andorra (8 from the *Sphagnum* matrix, 49 from the different water body communities, and 3 from peripheral environment sediments), while the other 56 concerned Rancho Hambre (8 from *Sphagnum* matrix, 45 from water body communities, and 3 from peripheral environment sediments).

A thorough physical and chemical characterization of the waters at each site included measurements of temperature, pH, conductivity, total hardness, and ammonium, phosphate, total nitrogen, total phosphorus, chlorophyll-*a* concentration and dissolved organic carbon (DOC) concentrations. The corresponding sampling, conservation and measurement methods are described in Casa [12.14].

Samples for light microscopy observation were prepared according to van der Werff [12.124]. Permanent slides were prepared and analyzed as described in Casa *et al.* [12.16]. Diatom populations of each sample were photographed, measured and identified using the currently available literature.

Differences in species composition between groups of samples were visualized through Venn diagrams. Beta diversity was calculated using the Sørensen dissimilarity index (βsor) [12.77] [12.110]. Lower values indicate higher similarities between the compared communities. Then, we calculated the two components of Beta diversity—turnover (βturn) and nestedness (βnest)—as they reflect the two primary mechanisms underlying the overall change in species identities across the landscape: the replacement of species or their loss, respectively [12.10].

The relationship between diatom composition and environmental variables was studied using a distance-based redundancy analysis (dbRDA) with Jaccard distance matrix as the response variable. Significantly explanatory environmental features were identified with a stepwise model selection approach. The contribution of each significant feature in shaping the diatom composition was quantified with a permutational multivariate analysis of variance (PERMANOVA) [12.2]. Statistical analyses were performed using *vegan* [12.93], *betapart* [12.4] and *venn* [12.30] packages in R [12.102].

12.3.3 Changes in Diatom Composition Between Peat Bogs and Their Surrounding Environments

A total of 171 diatom taxa belonging to 52 genera were identified in all samples. Almost 60% of these taxa were present in both Rancho Hambre and Valle de Andorra (Figure 12.2.a). The most remarkable difference in composition was due to the species turnover between localities (βsor=0.25; %βturn=68; %βnest=32). Sørensen dissimilarity indices and their partitions also showed that the diatom flora of each peat bog differed from that of the peripheral environments (Figure 12.2.b). This was mainly due to turnover, particularly in Rancho Hambre ($βsor_{AN}$=0.6; $\%βturn_{AN}$=65; $\%βnest_{AN}$=35; $βsor_{RH}$=0.54; $\%βturn_{RH}$=81; $\%βnest_{RH}$=19). Accordingly, samples from the peripheral environments of both localities grouped together in the dbRDA ordination plot (Figure 12.3), indicating that the composition of their communities clearly differed from those of the peat bogs. While the PERMANOVA analysis demonstrated that variations in diatom composition across all sites were significantly associated with pH ($R^2 = 8.7$, $P = 0.001$), total hardness ($R^2 = 3.4$, $P = 0.016$), total phosphorus ($R^2 = 3.3$, $P = 0.015$), and conductivity ($R^2 = 2.9$, $P = 0.029$), the relatively low R^2 values, combined with the limited proportion of community composition variance (14%) explained by these chemical factors in the dbRDA (as shown in Figure 12.3), suggest that they only contribute partially to shaping diatom composition. Instead, it is plausible that unmeasured factors more closely linked to water availability, such as water table depth, or the nature of the lakebed substrate, could exert a more pronounced influence on these communities. This observation aligns with findings from similar ecosystems in Canada [12.51], Spain [12.11] and Slovenia [12.62]. Hence, based on our experience, we recommend considering these parameters when conducting studies in similar environments.

The number of diatom species found in our studied environments is similar to that reported by Frenguelli for lakes of the region, whose richness varied between 46, 47 and 90 species [12.35] [12.36] [12.37] [12.38]. However, the diatom flora of these environments is quite different, with only a few common taxa, such as *Neidium magellanicum* Cleve (as *Navicula magellanicum* in Frenguelli) and some widespread species of *Pinnularia*. Conversely, the diversity recorded in paleoenvironmental reconstructions involving diatoms from Tierra del Fuego revealed significantly lower richness, encompassing only a few dozen species per locality [12.8] [12.31] [12.32].

The highest taxonomic richness in the entire study was observed in Valle de Andorra's beaver dam (AN BEAVER), with 108 species, followed by the connecting Arroyo Grande River (68 species). Both sites account for the higher total richness of Valle de Andorra as compared to Rancho Hambre (Figure 12.2.a). Beaver dams are known to affect the hydrological dynamics of rivers, creating *de novo* lentic environments in sections of lotic ones. These precipitate and retain large amounts of particles, allowing for the development of different diatom floras [12.42] [12.89]. Unlike the RH beaver dam, AN BEAVER was actively occupied during the study period, and increased its surface between 2014 and 2016. It is therefore possible that, apart from acting as a diatom trap concentrating species from the Arroyo Grande River, AN BEAVER also acquired additional species transported by beaver from other areas inside their home range.

12.3.4 Diatom Communities Inside the Peat Bogs

A total of 104 taxa belonging to 36 genera were observed in the peat bog samples. *Eunotia* was the most species-rich genus with 23 species, followed by *Pinnularia* (12), *Nitzschia* (6), *Frustulia* (5), *Gomphonema* (5) and *Neidium* (4). Other genera such as *Fragilaria*, *Brachysira*, *Kobayasiella* and *Encyonema* only had very low species numbers. On a genus level, the observed diatom flora was very similar to that of peatlands from different regions of the world [12.20] [12.67] [12.69] [12.70]. These taxa are usually associated with acid, dystrophic environments with low concentration

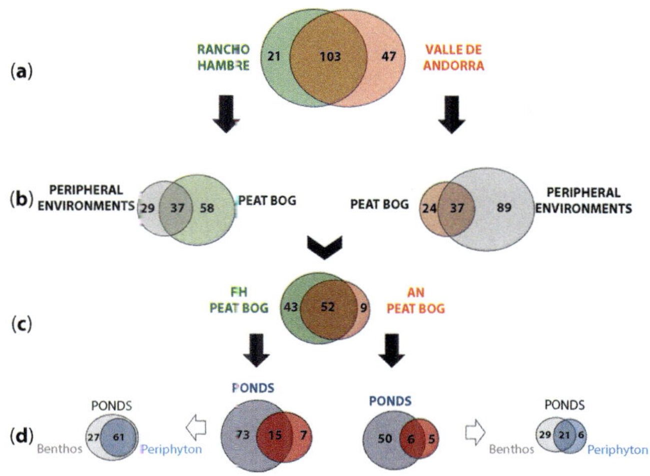

Figure 12.2 Venn diagrams representing a general comparison of the diatom flora between the two localities (a), breakdown of local species composition among each peat bog and its peripheral environments (b), comparison of diatom flora between peat bogs (c), and comparisons between the two landscape components of each peat bog (ponds vs. *Sphagnum* matrix), and among the benthic and periphytic communities of the ponds (d). RH=Rancho Hambre; AN=Valle de Andorra; SM=*Sphagnum magellanicum* matrix.

of nutrients and often related to mosses [12.28] [12.60] [12.114] [12.117] [12.125]. Also in this study, community composition both in ponds and in the *Sphagnum* matrix was related to pH, conductivity and total hardness values (Figure 12.3).

A large proportion of the observed diatom taxa could not be identified using the available up-to-date literature. Detailed analysis using both light and scanning electron microscopy resulted in the description of several new species for the genera *Aulacoseira* [12.17], *Stauroneis* [12.15], *Frustulia* [12.16] and *Distrionella* [12.18]. Detailed analysis of the two richest genera, *Eunotia* and *Pinnularia*, also revealed a high number of unknown species that are currently under description [12.14]. Moreover, some of the Eunotia species observed were recently described from the nearby Islas Malvinas/Falkland Islands [12.59]. As in our study area, many species have been newly discovered in the Antarctic and sub-Antarctic regions in recent years, highlighting their remarkable levels of endemism and high richness [12.59] [12.63] [12.118] [12.120] [12.127].

The most commonly reported diatom taxa in the samples include three unknown species of *Eunotia*, *Eunotia tenuivalva* Simonsen, *Chamaepinnularia* aff. *begeri* (Krasske) Lange-Bert., *Frustulia australocrassinervia* Casa, Mataloni & Van de Vijver, *Tabellaria flocculosa* (Roth) Kützing, *Kobayasiella*

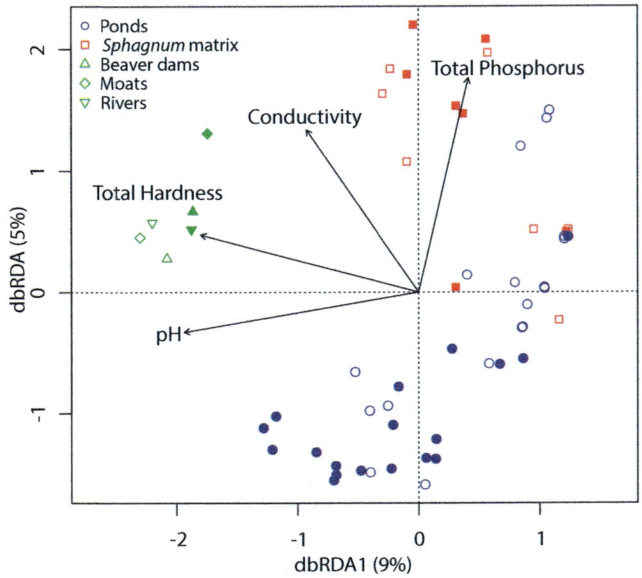

Figure 12.3 Distance-based redundancy analyses (dbRDA) based on Jaccard distance matrix. The environmental features that significantly explained variability in diatom composition were identified with a stepwise model selection approach. For simplicity, the benthic and periphytic communities were grouped at each pond. Filled markers represent Rancho Hambre samples, empty ones Valle de Andorra samples.

micropunctata (Germain) Lange-Bertalot, the *Pinnularia borealis* complex, *Pinnularia viridiformis* Krammer, *Frustulia delicatula* Casa, Mataloni & Van de Vijver, an unknown species of *Pinnularia* and *Encyonema neogracile* Krammer.

Despite being 50 km apart and located in different mountain valleys, the composition of both peat bogs was highly similar (βsor=0.33; %βturn=44; %βnest=56) and the main cause for dissimilarity between them was nesting. The Rancho Hambre peat bog had a richer community than that of Valle de Andorra, which was almost a subset of the former (Figure 12.2.c). This may relate to the higher environmental heterogeneity of Rancho Hambre, as we could observe clear differences in size and trophic conditions amongst different ponds, linked to the rugged microtopography of its matrix [12.44]. Ponds in the Valle de Andorra peat bog were smaller and more homogeneous in size than those of Rancho Hambre. This may not only reduce the supply of available niches in Andorra but also induce a greater influence of the *Sphagnum* matrix on its landscape dynamics. Indeed, samples taken from the *Sphagnum* matrix (SM) had the lowest species richness (RHSM=22; ANSM=11) (Figure 12.2.d) and were dominated by ubiquitous *Eunotia* sp1 (present in every sample), followed by typically aerophilic limno-terrestrial taxa such as *Hantzschia amphioxys* and members of the *Pinnularia borealis* complex. *Sphagnum* sites also showed higher concentrations of ammonium, chlorophyll-*a* and dissolved organic carbon compared to ponds (Table 12.1), the latter most probably related to the partial decomposition of plant material. Remarkably, most diatom communities inhabiting the *Sphagnum* matrix were subjected to the lowest pH combined with the highest conductivity values and total

Table 12.1 Mean and standard deviation of main chemical parameters.

	Andorra			Rancho Hambre		
	Ponds	*Sphagnum* matrix	Peripheral environments	Ponds	*Sphagnum* matrix	Peripheral environments
Temperature (°C)	14.9 ± 2.4	14.1 ± 2.0	12.2 ± 2.7	15.7 ± 4.4	13.8 ± 3.6	13.1 ± 12.8
pH	4.2 ± 0.3	3.9 ± 0.3	6.5 ± 0.2	4.7 ± 0.6	4.1 ± 0.3	6.7 ± 0.5
Conductivity (μS cm^{-1})	34.0 ± 9.5	93.4 ± 36.3	80.7 ± 10.5	34.6 ± 9.0	77.8 ± 12.9	260.7 ± 1512.1
Ammonium (μg L^{-1})	16.1 ± 14.2	75.0 ± 40.9	13.3 ± 12.5	36.1 ± 36.7	67.5 ± 47.1	16.7 ± 17.0
Total Nitrogen (μg L^{-1})	6422.8 ± 3276.1	3052.5 ± 1882.7	1906.7 ± 288.7	3776.7 ± 2918.6	26812.3 ± 1948.4	4876.7 ± 2122.2
Dissolved Reactive Phosphorus (μg L^{-1})	108.3 ± 53.4	105.0 ± 59.8	136.7 ± 60.2	102.2 ± 45.3	1112.3 ± 72.4	73.3 ± 18.9
Total Phosphorus (μg L^{-1})	306.2 ± 163.5	698.5 ± 363.1	315.3 ± 174.5	227.3 ± 96.7	834.6 ± 518.0	165.0 ± 53.9
Chlorophyll *a* (μg L^{-1})	4.1 ± 2.6	212.2 ± 6.2	2.8 ± 2.2	3.5 ± 2.7	17.4 ± 13.4	5.0 ± 3.6
Total Hardness (mg equiv. $CaCO_3$ L^{-1})	14.3 ± 10.3	16.3 ± 9.5	67.9 ± 20.8	16.0 ± 6.2	212.7 ± 8.9	62.3 ± 2.7
Dissolved Organic Carbon (mg L^{-1})	12.1 ± 3.3	20.1 ± 4.6	13.7 ± 15.1	14.3 ± 5.3	312.3 ± 8.6	8.1 ± 4.9

phosphorus concentrations (Figure 12.3). This suggests that the *Sphagnum* matrix exerts a strong environmental filtering, which only few diatom species manage to survive.

Our results also show that remoteness did not limit the dispersal of most diatom species between these mountain valleys. On this basis, we conclude that peat bogs with similar environmental characteristics in the region would host similar diatom communities, since they would be driven by the same selection agents, as previously shown for phytoplankton communities [12.82].

Almost 87.5% of all species could be classified as raphid diatoms. Only one centric species, *Aulacoseira frenguelliana* Casa & Van de Vijver, and twelve araphid taxa were observed. The latter include *Tabellaria flocculosa*, *Staurosira* cf. *venter* (Ehrenb.) Cleve & Möller and *Distrionella coxiana* Casa & Van de Vijver. A similar distribution between the main diatom groups is commonly observed in other peatlands [12.33] [12.68] owing to the fact that the evolution of the raphe system allowed diatoms to thrive in a broader range of substrates, particularly bottom sediments. Colonization by benthic diatoms depends mainly on the availability of a suitable substrate and light penetration [12.6]. In these shallow peat bog ponds, wherein light incidence is not a limiting factor [12.44], the benthic community was by far the richest (88.5% of all species), and included most periphyton species (Figure 12.2.d), thus becoming the most representative community of the studied wetlands.

The low number of diatom valves observed in the planktonic assemblage aligns with prior investigations of microalgae within the same ecosystems [12.81] [12.82] [12.84] [12.85]. In a comprehensive two-year study of five ponds in Rancho Hambre peat bog [12.84], diatoms accounted for 21% of the planktonic community richness, but did not exceed 3% in any sample. This pattern persisted across six other peat bogs in Tierra del Fuego, as evidenced by Mataloni in 1999 [12.82]. In this case, although diatoms accounted for 38% of all 194 species in the qualitative analysis, only 7 of them had a relative abundance exceeding 5%. The examination of these taxa, belonging to the genera *Cymbella*, *Eunotia*, and *Frustulia*, suggests that they may be incidental members of the plankton.

Considering these findings, it can be deduced that diatoms do not establish a robust presence within the planktonic community of Tierra del Fuego peat bogs. A possible explanation for this could be their lower competitive ability as compared to other algal groups. In Tierra del Fuego peat bogs, diatoms appeared to be outcompeted by flagellated phytoplankton, notably mixotrophic Chrysophyceae [12.84] [12.101]. A similar trend was also observed in aquatic environments of the Antarctic and sub-Antarctic regions, where oligotrophic ponds and lakes showed a sparse diatom-phytoplankton community, but also supported a well-developed phytobenthos [12.58] [12.76] [12.116] [12.121], often also dominated by small flagellated Chrysophyceae [12.58]. The exact cause for such planktonic diatom paucity remains to be understood, and it does not appear to relate to low nutrient concentrations, as neither eutrophic or hypertrophic waterbodies are dominated by diatoms but flagellated Volvocales [12.86].

Furthermore, none of the grazers appear to limit the plank diatom abundance, as a detailed analysis of the plankton common to RH showed that these accounted for shifts in the dominance of different nanoplanktonic groups, yet diatoms remained very sparse in ponds with different features over the entire growth season [12.101].

12.4 Closing Remarks and Future Perspectives

Although studies of peatland diatoms have been carried out worldwide, much remains to be learned about this peculiar flora. Peatlands are biodiversity hotspots; in particular, the acidic, dystrophic, and nutrient-poor environments within peat bogs allow the establishment of diverse and unique diatom communities very different from those of neighboring aquatic environments. Local environmental conditions are essential in structuring diatom assemblages, mainly through pH and moisture availability. These parameters are closely related to the microtopography and habitat heterogeneity of each peatland system. In particular, this study case showed the extreme importance of thoroughly sampling a wide range of habitats and microhabitats in order to capture the largest possible diatom diversity and refine the knowledge on their ecological preferences.

Obtaining comprehensive information on the flora and environmental preferences of each diatom taxon or species group is crucial in using diatoms as bioindicators. While diatoms proved to be powerful proxies in reconstructing past changes in paleoenvironmental studies [12.8] [12.9] [12.13] [12.41], they can play an even more remarkable role as bioindicators in present-day ecological and conservation studies [12.100]. The more we understand present-day diatoms, the more robust they become as tools for assessing the ecological integrity of these singular environments in a rapidly changing world.

References

[12.1] Armentano, T.V. (1980) Drainage of organic soils as a factor in the world carbon cycle. *Bioscience*, 30(12), 825-830.

[12.2] Anderson, M.J. (2001) A new method for non-parametric multivariate analysis of variance. *Austral Ecology* 26, 32–46.

[12.3] Bain, C.G., Bonn, A., Stoneman, R., Chapman, S., Coupar, A., Evans, M., Gearey, B., Howat, M., Joosten, H., Keenleyside, C., Labadz, J., Lindsay, R., Littlewood, N., Lunt, P., Miller, C.J., Moxey, A., Orr, H., Reed, M., Smith, P., Swales, V., Thompson, D.B.A., Thompson, P.S., Van de Noort, R., Wilson, J.D. and Worrall, F. (2011) IUCN UK *Commission of Inquiry on Peatlands*. IUCN UK Peatland Programme, Edinburgh.

[12.4] Baselga, A., Orme, D., Villeger, S., De Bortoli, J., Leprieur F. and Logez, M. (2022) betapart: Partitioning Beta Diversity into Turnover and Nestedness Components. R package version 12.5.6.

[12.5] Battarbee, R.W., Smol J. P., and Meriläinen J. (1986) Diatoms as indicators of pH: An historical review. In: *Diatoms and Lake Acidity*. J.P. Smol, Battarbee, R.W., Davis R.B. and J. Meriläinen, (eds) Dr W. Junk Publishers, Dordrecht: 5-14.

[12.6] Battarbee, R.W., Charles, D.F., Bigler, C., Cumming, B.F. and Renberg, I. (2010) Diatoms as indicators of surface-water acidity. In: *The Diatoms: Applications for the Enviromental and Earth Sciences*. J. P. Smol, and E.F. Stoermer, (eds) Cambridge University Press, New Work: 98-121.

[12.7] Bobbink, R., Whigham, D.F., Beltman, B., and Verhoeven, J.T. (2006) Wetland functioning in relation to biodiversity conservation and restoration. In: *Wetlands: Functioning, biodiversity conservation, and restoration*. Bobbink, R., Beltman, B., Verhoeven, J.T.A., Whigham, D.F. (eds) Springer, Berlin, Heidelberg: 1-12.

[12.8] Borromei, A.M., Coronato, A., Franzén, L.G., Ponce, J.F., Sáez, J.A.L., Maidana, N., Rabassa, J. and Candel, M.S. (2010) Multiproxy record of Holocene paleoenvironmental change, Tierra del Fuego, Argentina. *Palaeogeography, Palaeoclimatology, Palaeoecology* 286(1-2), 1-16.

[12.9] Brugam, R.B. and Swain, P. (2000) Diatom indicators of peatland development at Pogonia Bog Pond, Minnesota, USA. *The Holocene* 10(4), 453-464.

[12.10] Calderón-Patrón J.M., Moreno C.E. and Zuria I. (2012). La diversidad beta: medio siglo de avances. *Revista mexicana de biodiversidad* 83, 879-8912.

[12.11] Cambra, J. (2015) Micro-scale distribution of algae in a Pyrenean peatbog, Spain. *Hidrobiológica* 25(2), 213-222.

[12.12] Candel, M.S., Díaz, P.E., Borromei, A.M., Fernández, M., Montes, A. and Santiago, F.C. (2020) Multiproxy analysis of a Lateglacial-Holocene sedimentary section in the Fuegian steppe (northern Tierra del Fuego, Argentina): Implications for coastal landscape evolution in relation to climatic variability and sea-level fluctuations. *Palaeogeography, Palaeoclimatology, Palaeoecology* 557, #1099412.

[12.13] Carballeira, R., and Pontevedra-Pombal, X. (2020) Diatoms in paleoenvironmental studies of peatlands. *Quaternary* 3(2), 10.

[12.14] Casa V. (2020). Diatomeas de turberas de Tierra del Fuego: diversidad, ecología y evaluación de su uso como indicadores ambientales. (*Doctoral Thesis,* Universidad de Buenos Aires. Facultad de Ciencias Exactas y Naturales). Retrived https://hdl.handle.net/20.500.12110/tesis_n6775_Casa

[12.15] Casa V., Mataloni G. and Van de Vijver B. (2017a) *Stauroneis fuegiana*, a new *Stauroneis* species (Bacillariophyta) from Tierra del Fuego, southern Argentina. *Phytotaxa* 311, 85–92. https://doi.org/10.11646/phytotaxa.3112.12.8

[12.16] Casa V., Mataloni G. and Van de Vijver B. (2018) Six new *Frustulia* species (Bacillariophyta) in Tierra del Fuego peatbogs, Patagonia, Argentina. *Fottea* 18, 55–712. https://doi.org/10.5507/fot.2017.016

[12.17] Casa, V., López Bedogni, G. and Van de Vijver, B. (2017b) A new *Aulacoseira* species (Bacillariophyta) from Tierra del Fuego (Argentina) and comparison with the type material of *Melosira laevis* var. *fuegiana* Frenguelli. *Diatom research* 32(4), 409-416.

[12.18] Casa, V., Mataloni, G. and Van de Vijver, B. (2019) *Distrionella coxiana*, a new fragilarioid diatom species (Bacillariophyceae) from Tierra del Fuego, Patagonia, Argentina. *Plant Ecology and Evolution* 152(2), 385-3912.

[12.19] Chen, X., Bu, Z., Stevenson, M. A., Cao, Y., Zeng, L. and Qin, B. (2016) Variations in diatom communities at genus and species levels in peatlands (central China) linked to microhabitats and environmental factors. *Science of the total environment* 568, 137-146.

[12.20] Chen, X., McGowan, S., Bu, Z. J., Huang, X. Y., Bai, X., Zhang, Y. K., Peng, J. and Li, J. L. (2022) Diatom-inferred microtopography formation in peatlands. *Earth Surface Processes and Landforms* 47(2), 672-687.

[12.21] Cleve, P.T. (1900): Report on the Diatoms of the Magellan Territories. In: *Svenska Expeditionen till Magellanslanderna (under Leitung Otto Nordenskjöld)*, Stockholm 3, 273–282.

[12.22] Cleve–Euler, A. (1948): Susswasserdiatomeen aus dem Feuerland. – *Acta Geographica* (Societas Geographica Fennica) 10,1–61.

[12.23] Clymo, R.S. (1964) The origin of acidity in Sphagnum bogs. *The Bryologist* 67, 427–4312.

[12.24] Clymo, R.S. and Hayward, P.M. (1982) The ecology of *Sphagnum*. In: *Bryophyte Ecology.* A.J.E. Smith (ed.), Chapman & Hall, London-New York: 229-290.

[12.25] Coesel, P.F.M. (1986) Structure and dynamics of desmid communities in hydrosere vegetation in a mesotrophic quivering bog. *Beih. zur Nova Hedwigia* 56, 119-143.

[12.26] Costanza, R., d'Arge, R., De Groot, R., Farber, S., Grasso, M., Hannon, B., Limburg, K., Naeem, S., O'Neill, R.V., Paruelo, J., Raskin, R.G., Sutton, P., Van Den Belt, M., (1997). The value of the world's ecosystem services and natural capital. *Nature* 387(6630): 253-260.

[12.27] Czerwik-Marcinkowska, J., Wojciechowska, A., Mrozińska, T. and Wojtal, A. (2019) Algal diversity and community composition of peat bogs in Poland (Central Europe). *Phytocoenologia* 49(3), 249-262.

[12.28] DeNicola, D.M. (2000) A review of diatoms found in highly acidic environments. *Hydrobiologia* 433(1-3), 111-122.

[12.29] During, H.J. and Van Tooren, B.F. (1990) Bryophyte interactions with other plants. *Botanical Journal of the Linnean Society* 104(1-3), 79-98.

[12.30] Dusa, A. (2022) venn: Draw Venn Diagrams. R package version 12.112.

[12.31] Espinosa, M.A. (2008) Diatoms from Patagonia and Tierra del Fuego. Developments in *Quaternary Sciences* 11, 383-392.

[12.32] Fernández, M., Maidana, N., y Rabassa, J. (2012) Palaeoenvironmental conditions during the Middle Holocene at Isla de los Estados (Staaten Island, Tierra del Fuego, 54 S, Argentina) and their influence on the possibilities for human exploration. *Quaternary international* 256, 78-87.

[12.33] Finlayson C.M. and Milton G.R. (2016) Peatlands. In: *The Wetland Book*. C. Finlayson, G.Milton, R. Prentice, and N. Davidson, (eds) Springer, Dordrecht. https://doi.org/10.1007/978-94-007-6173-5_202-1

[12.34] Fránková, M., Bojková, J., Poulíčková, A. and Hájek, M. (2009) The structure and species richness of the diatom assemblages of the Western Carpathian spring fens along the gradient of mineral richness. *Fottea* 9(2), 355-368.

[12.35] Frenguelli, J. (1923) Resultados de la Primera Expedición a Tierra del Fuego (1921). Diatomeas de Tierra del Fuego. *Anales de la Sociedad Científica Argentina* 96:225–263.

[12.36] Frenguelli, J. (1924a) Resultados de la Primera Expedición a Tierra del Fuego (1921). Diatomeas de Tierra del Fuego. *Anales de la Sociedad Científica Argentina* 97:87–118.

[12.37] Frenguelli, J. (1924b) Resultados de la Primera Expedición a Tierra del Fuego (1921). Diatomeas de Tierra del Fuego. *Anales de la Sociedad Científica Argentina* 97:231–266.

[12.38] Frenguelli, J. (1924c) Resultados de la Primera Expedición a Tierra del Fuego (1921). Diatomeas de Tierra del Fuego. *Anales de la Sociedad Científica Argentina* 98: 5–63.

[12.39] Frenguelli, J. (1951) Análisis microscópico de las muestras de la turbera del Río de la Misión, Río Grande, Tierra del Fuego. *Annales Academiae Scieniarum Fennicae* (s. A), Geologica–Geographica 26, 1-60.

[12.40] Frenguelli, J. (1953) Análisis microscópico de una segunda serie de muestras de la turbera del Rio de la Misión, Rio Grande, Tierra del Fuego extraídas por el Dr. Väinö *Auer. Annales Academiae Scieniarum Fennicae* (s. A) 3, Geologica–Geographica 34: 1–42.

[12.41] Gaiser, E., and Rühland, K. (2009) Diatoms as indicators of environmental change in wetlands and peatlands. In: *The diatoms: applications for the environmental and earth sciences*, (eds) J. P. Smol, and E. F. Stoermer, Cambridge, England: Cambridge University Press (Ed. 2), 473-496.

[12.42] García, V.J. and Rodríguez, P.L. (2018) Efecto del castor en el metabolismo del perifiton y en variables limnológicas de ríos y arroyos fueguinos. *Ecología Austral* 28(3), 593-605.

[12.43] Gilbert, D. and Mitchell, E.A.D. (2006) Microbial diversity in *Sphagnum* peatlands. In: I. Martini, A. Martínez Cortizas, W. Chesworth (eds.) *Peatlands: Evolution and Records of Environmental and Climatic Changes.* Elsevier, Amsterdam: 287–318.

[12.44] González Garraza G, Mataloni G, Iturraspe R *et al* (2012) The limnological character of bog pools in relation to meteorological and hydrological features. *Mires Peat* 10, #7.

[12.45] González Garraza, G., and Rodríguez, P. (2022) Updated distribution of the diatom *Didymosphenia geminata* (Lyngbye) Schmidt in Tierra del Fuego, Patagonia. *Diatom Research* 37(1), 71-79.

[12.46] Gorritti, G., Sala, S. E. and Guerrero, J. M. (2000). Sobre la validez taxonómica de *Epithemia zebra var. elongata* (Epithemiaceae, Bacillariophyceae). *Darwiniana* 38(3-4), 285-289.

[12.47] Grootjans A., Iturraspe R., Lanting A., Fritz C. and Joosten H. (2010) Ecohydrological features of some contrasting mires in Tierra del Fuego, Argentina. *Mires and Peat* 6, #012.

[12.48] Guerrero, J.M., Sala, S.E., and Gorriti, G. (2001) Epithemiaceae y Surirellaceae (Bacillariophyceae) de Tierra del Fuego, Argentina. *Boletín de la Sociedad Argentina de Botánica* 36(1-2), 29-45.

[12.49] Guerrero, J.M., Vouilloud, A.A., Sala, S.E., Kociolek, J.P., and Van de Vijver, B. (2018) New species and a new genus of the Orthoseirales from Patagonia, Argentina, with comments on systematic affinities within the Order. *Phytotaxa* 345(2), 119-132.

[12.50] Hargan, K.E. (2014) Diatoms as indicators of environmental and climatic change in peatlands and lakes located across the Boreal Shield and Hudson Bay Lowlands of Canada (*Ph. D. Thesis,* Queen's University, Canada).

[12.51] Hargan, K.E., Rühland, K.M., Paterson, A.M., Finkelstein, S.A., Holmquist, J.R., MacDonald, G.M., Keller, W. and Smol, J.P. (2014) The influence of water-table depth and pH on the spatial distribution of diatom species in peatlands of the Boreal Shield and Hudson Plains, Canada. *Botany* 93(2), 57-74.

[12.52] Hatala, J.A., Detto, M., Sonnentag, O., Deverel, S.J., Verfaillie, J. and Baldocchi, D.D. (2012) Greenhouse gas (CO2, CH4, H2O) fluxes from drained and flooded agricultural peatlands in the Sacramento-San Joaquin Delta. *Agriculture, Ecosystems & Environment* 150, 1-18.

[12.53] Hutchinson, G. E., 1957. *A Treatise on Limnology*. Geography, Physics and Chemistry. John Wiley and Sons Inc, N.Y.

[12.54] IPBES (2019a) Chapter 2.3. Status and Trends - Nature's Contributions to People (NCP). In:. S. Díaz, J. Settele, E. S. Brondízio, *et al.*, (eds.) *IPBES Global Assessment Report on Biodiversity and Ecosystem Services* IPBES secretariat, Bonn, Germany: 56.

[12.55] IPBES (2019b) Summary for policy makers; Intergovernmental Science-Policy Platform on Biodiversity and Ecosystem Services. In: S. Díaz, J. Settele, E. S. Brondízio, *et al.*, (eds.) *IPBES Global Assessment Report on Biodiversity and Ecosystem Services* IPBES secretariat, Bonn, Germany: 56.

[12.56] Iturraspe, R. (2010) *Las Turberas de Tierra del Fuego y el Cambio Climático Global*. Fundacion Humedales / Wetlands International, Buenos Aires.

[12.57] Iturraspe, R. (2016). Patagonian Peatlands (Argentina and Chile). In: *The Wetland Book*. C. Finlayson, G. Milton, R. Prentice and N. Davidson, (eds) Springer, Dordrecht: 1-10. https://doi.org/10.1007/978-94-007-6173-5_230-1

[12.58] Izaguirre, I., Vinocur, A., Mataloni, G. and Pose, M. (1998) Phytoplankton communities in relation to trophic status in lakes from Hope Bay (Antarctic Peninsula). *Hydrobiologia* 369(0), 73-87.

[12.59] Jüttner, I., Van de Vijver, B., Williams, D. M., Lange-Bertalot, H. and Ector, L. (2018) The genus Eunotia (Bacillariophyta) in the Falkland Islands and species-area relationships in sub-Antarctic islands. *Diatom Research* 33(4), 413-452.

[12.60] Kilroy, C. (2007) Diatom communities in New Zealand subalpine mire pools: distribution, ecology and taxonomy of endemic and cosmopolitan taxa. (*Ph.D. Thesis*, University of Canterbury, Christchurch)

[12.61] Kilroy, C., Biggs, B.J., Vyverman, W. and Broady, P.A. (2006) Benthic diatom communities in subalpine pools in New Zealand: relationships to environmental variables. *Hydrobiologia* 561(1), 95-110.

[12.62] Klemenčič, A.K., Smolar-Žvanut, N., Istenič, D. and Griessler-Bulc, T. (2010) Algal community patterns in Slovenian bogs along environmental gradients. *Biologia* 65, 422-437.

[12.63] Kociolek, J. P., Kopalová, K., Hamsher, S. E., Kohler, T. J., Van de Vijver, B., Convey, P., y McKnight, D. M. (2017) Freshwater diatom biogeography and the genus *Luticola*: an extreme case of endemism in Antarctica. *Polar Biology* 40(6), 1185-1196.

[12.64] Krammer, K. (1997) Die cymbelloiden Diatomeen. Teil 2. *Encyonema* part., *Encyonopsis* and *Cymbellopsis*. *Bibliotheca diatomologica* 37, 1-469.

[12.65] Krammer, K. (2000). The genus *Pinnularia*. In: *Diatoms of Europe, Diatoms of the European Inland waters and comparable habitats*. H. Lange-Bertalot, (ed.) A.R.G. Gantner Verlag K.G.

[12.66] Krasske, G. (1949) Subfossile diatomeen aus den mooren Patagoniens und Feuerlands. *Annales Academiae Scientiarum fennicae IV. Biologica* 14, 1-92.

[12.67] Kulikovskiy, M.S. (2008) The species composition and distribution of diatoms in sphagnum bogs of European Russia: Ecosystems of the Volga Upland. *Inland Water Biology* 1(4), 347-355.

[12.68] Kulikovskiy, M. S. (2009) The species composition and distribution of diatom algae in sphagnum bogs in European Russia: The Polistovo-Lovatskii land tract. *Inland Water Biology* 2(2), 135-143.

[12.69] Kulikovskiy, M.S.; Lange-Bertalot, H.; Witkowski, A.; Dorofeyuk, N.I.; Genkal, S.I. (2010) *Diatom Assemblages from Sphagnum Bogs of the World. I. Nur Bog in Northern Mongolia.* Bibliotheca Diatomologica 55. Schweizerbart Science Publishers, Stuttgart, Germany.

[12.70] Küttim, L., Küttim, M., Puusepp, L., and Sugita, S. (2017). The effects of ecotope, microtopography and environmental variables on diatom assemblages in hemiboreal bogs in Northern Europe. *Hydrobiologia* 792(1), 137-149.

[12.71] Lange-Bertalot, H. and Moser, G. (1994) *Brachysira Monographie der Gattung. Wichtige Indikator–Species für das Gewässer Monitoring und Naviculadicta nov. gen. Ein Lösungsvorschlag zu dem Problem Navicula sensu lato ohne Navicula sensu stricto.* Bibliotheca Diatomologica 29. J. Cramer (ed), Berlin–Stuttgart.

[12.72] Lange-Bertalot, H. (2001) *Navicula* sensu stricto, 10 genera separated from *Navicula* sensu lato, *Frustulia. Diatoms of Europe: diatoms of the European inland waters and comparable habitats, 2.* A.R.G. Gantner Verlag K.G, Ruggell.

[12.73] Lange-Bertalot, H., Bąk, M., Witkowski, A. and Tagliaventi, N. (2011) *Diatoms of Europe, vol. 6: Eunotia and some related taxa. Diatoms of the European inland water and comparable habitats.* ARG Gantner Verlag KG, Ruggell.

[12.74] Laprida, C., Orgeira, M.J., Fernández, M., Tófalo, R., Mercau, J.R., Silvestri, G.E., Berman, A.L., García Chapori, N., Plastani, M.S. and Alonso, S. (2021) The role of Southern Hemispheric Westerlies for Holocene hydroclimatic changes in the steppe of Tierra del Fuego (Argentina). *Quaternary International* 571, 11-25.

[12.75] Lindsay, R. (2016) Peatland (Mire Types): Based on Origin and Behavior of Water, Peat Genesis, Landscape Position, and Climate. In: *The Wetland Book.* C. Finlayson, G. Milton, R. Prentice and N. Davidson, (eds) Springer, Dordrecht. https://doi.org/10.1007/978-94-007-6173-5_279-1

[12.76] Llames, M.E. and Vinocur, A. (2007) Phytoplankton structure and dynamics in a volcanic lake in Deception Island (South Shetland Islands, Antarctica). *Polar Biology* 30(7), 849-857.

[12.77] Magurran, A.E. (2004) *Measuring biological diversity.* Blackwell Science, Malden, Massachusetts, USA.

[12.78] Maidana, N.I. (2000) Achnanthaceae (Bacillariophyceae) continentales de Tierra del Fuego, Argentina. I. *Boletin de la Sociedad Argentina Botanica* 35, 49-612.

[12.79] Martínez Cortizas, A., Pontevedra Pombal, X., Novoa Muñoz, J.C., Rodríguez Fernández, R. and López-Sáez, J.A. (2009) Turberas ácidas de esfagnos. In: *Bases ecológicas preliminares para la conservación de los tipos de hábitat de interés comunitario en España*. VV. AA. (eds.) Ministerio de Medio Ambiente y Medio Rural y Marino, Madrid: 1-64.

[12.80] Mašić, E. (2020) Patterns of distribution of diatom assemblages in peatlands ecosystem on Vranica Mountain (Bosnia and Herzegovina). *Works of the Faculty of Agriculture and Food Sciences*, University of Sarajevo, 65(70)

[12.81] Mataloni, G. (1994). Estudios florísticos y ecológicos sobre las algas de agua dulce de turberas de Tierra del Fuego. (*Ph. D. Thesis,* Universidad de Buenos Aires. Facultad de Ciencias Exactas y Naturales).

[12.82] Mataloni, G. (1999) Ecological studies on algal communities from Tierra del Fuego peat bogs. *Hydrobiologia* 391,157-1712.

[12.83] Mataloni, G. (2016) Diversity patterns of plankton communities in Tierra del Fuego peat bog pools as sentinels of climate change. *Biodiversity* 17(1-2), 26-33.

[12.84] Mataloni, G., González Garraza, G. and Vinocur, A. (2015) Landscape-driven environmental variability largely determines abiotic characteristics and phytoplankton patterns in peat bog pools (Tierra del Fuego, Argentina). *Hydrobiologia* 751, 105-125.

[12.85] Mataloni, G. and Tell, G. (1996) Comparative analysis of the phytoplankton communities of a peat bog from Tierra del Fuego (Argentina). *Hydrobiologia* 325(2), 101-112.

[12.86] Mataloni, G., Tesolín, G., Sacullo and Tell, G. (2000) Factors regulating summer phytoplankton in a highly eutrophic Antarctic Lake. Hydrobiologia 432, 65–72.

[12.87] Moy, C.M., Dunbar, R.B., Guilderson, T.P., Waldmann, N., Mucciarone, D.A., Recasens, C., Austin, J A. and Anselmetti, F.S. (2011) A geochemical and sedimentary record of high southern latitude Holocene climate evolution from Lago Fagnano, Tierra del Fuego. *Earth and Planetary Sciences Letters* 302(1-2), 1-13.

[12.88] Mutinová, P.T., Kulichová, J. and Ugland, K.I. (2017) Temporal and spatial dynamics of diatom (Bacillariophyceae) communities in a peatland area. *Cryptogamie, Algologie*, 38(3), 253-266.

[12.89] Naiman, R.J., Melillo, J.M. and Hobbie, J.E. (1986) Ecosystem alteration of boreal forest streams by beaver (Castor canadensis). *Ecology* 67, 1254-1269.

[12.90] Negro, A.I., De Hoyos, C. and Aldasoro, J.J. (2003) Diatom and desmid relationships with the environment in mountain lakes and mires of NW spain. *Hydrobiologia* 505, 1-13.

[12.91] Neustupa, J., Veselá, J. and Šťastný, J. (2013) Differential cell size structure of desmids and diatoms in the phytobenthos of peatlands. *Hydrobiologia* 709(1), 159-1712.

[12.92] Noga, T. and Poradowska, A. (2022) The Occurrence of Diatoms from Genus *Kobayasiella* in Southern Poland. Journal of Ecological Engineering 23(4), 197-200.

[12.93] Oksanen, J., Simpson, G., Blanchet, F., Kindt, R., Legendre, P., Minchin, P., O'Hara, R., Solymos, P., Stevens, M., Szoecs, E., Wagner, H., Barbour, M., Bedward, M., Bolker, B., Borcard, D., Carvalho, G., Chirico, M., De Caceres, M., Durand, S., Evangelista, H., FitzJohn, R., Friendly, M., Furneaux, B., Hannigan, G., Hill, M., Lahti, L., McGlinn, D., Ouellette, M., Ribeiro Cunha, E., Smith, T., Stier, A., Ter Braak and C., Weedon, J. (2022) vegan: Community Ecology Package. R package version 2.6-4.

[12.94] Parish, F., Sirin, A.A., Charman, D., Joosten, H., Minaeva, T.Y. and Silvius, M. (2008) *Assessment on peatlands, biodiversity and climate change*. Main Report. Global Environment Centre, Kuala Lumpur and Wetlands International, Wageningen.

[12.95] Pienkowski, T.P. and Wujek, D.E. (1988) The Diatom Flora of the Red Lake Peatland, Minnesota. *Journal of the Minnesota Academy of Science* 53(3), 7-13.

[12.96] Pinseel, E., Kulichová, J., Scharfen, V., Urbánková, P., Van de Vijver, B. and Vyverman, W. (2019) Extensive cryptic diversity in the terrestrial diatom *Pinnularia borealis* (Bacillariophyceae). *Protist* 170(2), 121-140.

[12.97] Poulíčková, A., Bergová, K., Hnilica, R. and Neustupa, J. (2013) Epibryic diatoms from ombrotrophic mires: diversity, gradients and indicating options. *Nova Hedwigia* 96(3-4), 351-365.

[12.98] Poulíčková, A., Hájková, P., Krenková, P. and Hájek, M. (2004) Distribution of diatoms and bryophytes on linear transects through spring fens. *Nova Hedwigia* 78 (3-4), 411-424.

[12.99] Quintana, R. D .and Mataloni, G. (2022) Are Wetlands Freshwaters? In: *Freshwaters and Wetlands of Patagonia. Ecosystems and socioecological aspects*. G. Mataloni, R. D. Quintana (eds.) Springer, Cham: 19-29.

[12.100] Quiroga, M.V., Casa, V., Garcia, P.E., Kuppers, G. and Mataloni, G. (2022) Diversity Patterns Across Aquatic Communities from Peat Bogs in Changing Environmental Scenarios. In: *Freshwaters and Wetlands of Patagonia. Ecosystems and socioecological aspects*. G. Mataloni, R. D. Quintana (eds.) Springer, Cham: 117-135.

[12.101] Quiroga, M.V., Unrein, F., González Garraza, G., Küppers, G., Lombardo, R., Marinone, M.C., Menu Marque, S., Vinocur, A. and Mataloni, G. (2013) The plankton communities from peat bog pools: structure, temporal variation and environmental factors. *Journal of Plankton Research* 35(6), 1234-1253.

[12.102] R Core Team (2021) R: A language and environment for statistical computing. R Foundation for Statistical Computing, Vienna, Austria.

[12.103] Rabassa J., Coronato, A. and Roig, C. (1996) The peat-bogs of Tierra del Fuego, Argentina. In: *Global Peat Resources.* E. Lappalainen (ed) International Peat Society Publisher, Jyskä, Finland: 261-266.
[12.104] Ramsar (2009). Ramsar Sites Information Service. https://rsis.ramsar.org/ris/1886?language=en.
[12.105] Roig, C. and Roig, F.A. (2004) Consideraciones generales. In: *Los turbales de la Patagonia: bases para su inventario y la conservación de su biodiversidad.* D.E. Blanco and V.M. de la Balze (eds) Wetlands International, Buenos Aires, Argentina: 5-212.
[12.106] Rumrich, U., Lange-Bertalot, H. and Rumrich, M. (2000) *Diatoms of the Andes: from Venezuela to Patagonia/Tierra del Fuego. Iconographia Diatomologica* 9, 1-649.
[12.107] Rydin, H. and Jeglum, J. (2006) *The biology of peatlands. Biology of habitats.* Oxford University Press, lugar.
[12.108] SIB (2022). Sistema de Información de Biodiversidad. https://sib.gob.ar/area/TIERRA%20DEL%20FUEGO*TM*Valle%20Tierra%20Mayor.
[12.109] Simkhada, B. and Jüttner, I. (2006) Diatoms in ponds and small lakes of the Kathmandu Valley, Nepal–relationships with chemical and habitat characteristics. *Archiv für Hydrobiologie* 166(1), 41-65.
[12.110] Sørensen, T.A. (1948) A method of stablishing groups of equal amplitude in plant sociology based on similarity of species content, and its application to analyses of the vegetation on Danish commons. *Kongelige Danske Videnskabernes Selskabs Biologiske Skrifter* 5, 1-34.
[12.111] Stanković, I., Udovič, M.G., Šušnjara, M., Alegro, A. and Levkov, Z. (2022) Diversity and ecology of benthic diatoms in a mire with narrow pH gradient. *Fottea* (1), 39-55.
[12.112] Tyler, P.A. (1996) Endemism in freshwater algae. *Hydrobiologia* 336: 127–135.
[12.113] Van Dam, H., Mertens, A. and Sinkeldam, J. (1994) A coded checklist and ecological indicator values of freshwater diatoms from the Netherlands. *Netherland Journal of Aquatic Ecology* 28(1), 117-133.
[12.114] Van de Vijver, B. and Gremmen, N. (2006) Three new moss–inhabiting diatom species from sub–Antarctic Marion Island. *Diatom Research* 21: 427-439.
[12.115] Van de Vijver, B., Lange-Bertalot, H., Goeyers, C., Mertens, A., Schuster, T. M. and Ector, L. (2022) The identity of *Eunotia paludosa* Grunow 1862 (Eunotiaceae, Bacillariophyta), a revision, and the description of three new species of *Eunotia* Ehrenberg. *Phytotaxa* 545(3), 261-277.
[12.116] Van de Vijver, B. and Beyens, L. (1999) Biogeography and ecology of freshwater diatoms in Subantarctica: a review. *Journal of biogeography* 26(5), 993-1000.
[12.117] Van de Vijver, B., Beyens, L. and Lange-Bertalot, H. (2004) *The genus Stauroneis in the Arctic and (Sub-)Antarctic Regions.* In: Bibliotheca Diatomologica, 512. Gebrüder Borntraeger, Stuttgart.

[12.118] Van De Vijver, B., Gremmen, N. J. M. and Beyens, L. (2005) The genus *Stauroneis* (Bacillariophyceae) in the Antarctic Region. *Journal of Biogeography* 32, 1791-1798.

[12.119] Van de Vijver, B., Gremmen, N. and Smith, V. (2008) Diatom communities from the sub-Antarctic Prince Edward Islands: diversity and distribution patterns. *Polar biology* 31(7), 795-808.

[12.120] Van de Vijver, B., Kopalová, K., Zidarova, R. and Kociolek, J.P. (2016) Two new *Gomphonema* species (Bacillariophyta) from the Maritime Antarctic Region. *Phytotaxa* 269(3), 209-220.

[12.121] Van de Vijver, B., Ledeganck, P. and Beyens, L. (2001) Habitat preferences in freshwater diatom communities from sub-Antarctic Iles Kerguelen. *Antarctic Science* 13(1), 28-36.

[12.122] Van de Vijver, B., Schuster, T.M., Kusber, W.H., Hamilton, P.B., Wetzel, C.E. and Ector, L. (2021a) Revision of European *Brachysira* species (Brachysiraceae, Bacillariophyta): I. The *Brachysira microcephala-B. neoexilis* enigma. *Botany Letters* 168(4), 467-484.

[12.123] Van de Vijver, B., Schuster, T.M., Kusber, W.H., Williams, D.M., Wetzel, C.E. and Ector, L. (2021b) Revision of European *Brachysira* species (Brachysiraceae, Bacillariophyta): II. The *Brachysira styriaca* and *B. zellensis* group. *Botany Letters* 168(4), 485-502.

[12.124] van der Werff, A. (1955) A new method for cleaning and concentrating diatoms and other organisms. *Verhandlungen der Internationalen Vereinigung für theoretische und angewandte Limnologie* 12, 276–277.

[12.125] Vinšová, P. (2016) Taxonomy, diversity and ecology of freshwater diatom communities (Bacillariophyta) from moss habitats of Gough Island (southern Atlantic Ocean). (*Thesis,* Univerzita Karlova, Přírodovědecká fakulta, Katedra ekologie, Praha).

[12.126] Waldmann, N., Ariztegui, D., Anselmetti, F.S., Austin, Jr J.A., Moy, C.M., Stern, C., Recasens, C. and Dunbar, R.B. (2010) Holocene climatic fluctuations and positioning of the Southern Hemisphere westerlies in Tierra del Fuego (54° S), Patagonia. Journal of *Quaternary Science* 25, 1063-1075.

[12.127] Williams, D.M. and Reid, G. (2006) Diatom biogeography and water babies: the search for meaning among the protists. *Diatom Research* 21 (2), 457-462.

Index

18S rDNA (SSU), 126–138, 147, 148

Abiotic factors, 142–145
Abundance hypothesis, 92, 96–97, 102
Abundance-occupancy relationship (AOR), 41–43
Adaptation mechanisms, 366, 368, 369, 370, 377, 380, 382
Aerophilous, 380
Alpha diversity, 91, 97, 205
Analysis of covariance (ANCOVA), 41
Anthropocenic intermittence, 364
Anthropogenic, 295–298, 302
Aquatic diatom, 288, 290, 292–294, 298, 302
Aquatic regime (AR), 370
Aquatic states, 370, 371
Arheic, 370, 371
Art, 26
Artificial, 203, 207
Assessment, 15
Asynchronous sexuality, 72
Aulacoseira life cycle, 68–71
Autecological, 296–297, 303
Automixis, 73
Auxospore, 63–64
Auxospores, 367
Auxosporulation, 59–76

Bacillariophyta, 41
Benthic, 88, 90–91, 93 95, 98–99, 101–102, 103–104
Benthic ecology, 175
Berger-Parker index, 381
Beta diversity, 91, 97, 101, 104, 205, 380, 403
Biodiversity, 86–87, 93–95, 106, 396, 399, 409
Biogeographical challenges, 125
Biogeography, 123, 124
Bioindicator, 207, 213, 214, 215, 281, 283, 297, 303
Bioindicators, 176, 361, 366, 382, 383
Biological condition gradient, 16
Biological diatom index (BDI), 382
Biomonitoring, 281, 294, 297, 300–301, 303
Biomonitoring (water quality assessment), 99
Biotic effects on auxosporulation, 62
Biotic integrity, 19
Biotic interactions, 86–88, 98, 143, 145, 149
Biovolume, 41, 44
Brachysira, 397, 404
Brownification, 94–95
Buried slide technique, 288

Calibration sets, 329, 340
Carotenoids, 378, 379
Chromophores, 377
Citizen science, 26
Climate, 293, 302–303
Climate change, 330, 331, 332, 334, 335, 337, 341, 342, 343, 346, 347, 349, 351, 352
Climatic shifts, 330
Cold-water diatoms, 242

Community assembly, 93, 100, 102–103, 105–106, 124, 143
Community dynamics, 365, 366, 380
Conceptual ecological model, 11
Conductivity, 329, 332, 335, 336, 340, 345, 348, 396, 403, 405
Conservation, 203, 204, 207, 210, 211, 212, 213, 214, 215, 216, 217, 218, 219, 220
Costs of sex, 70, 72
Cryptic, 281, 287, 301
Cryptic species, 93
Culture, 285, 288–289
Cyanobacteria, 283, 288, 294

Desiccation stress, 361, 372
Diat.barcode, 147
Diatom art, 27
Diatom distributions
- *Achnanthidium munitissimum,* 126, 148
- aeolian diatoms, 140
- *Aneumastus* spp., 127
- *Asterionella formosa,* 127
- *Aulacoseira* spp., 127
- *Chaetoceros curvisetus,* 128, 147, 148
- *Chaetoceros socialis,* 128, 144
- *Chaetoceros* spp., 129, 144
- *Cylindrotheca closterium,* 129, 130
- *Didymosphenia geminata,* 130
- *Ditylum brightwellii,* 131
- *Eunotia bilunaris,* 131
- *Fistulifera saprophila,* 126, 148
- *Fragilaria* spp., 132
- *Fragilariopsis kerguelensis,* 132
- *Frustulia* spp., 132, 133
- *Gomphonema parvulum,* 133, 145, 148
- *Gomphonema truncatum,* 148
- *Hantzschia amphioxys,* 133
- *Leptocylindraceae,* 134
- marine planktonic diatoms, 141
- *Navicula cryptocephala,* 134
- *Navicula phyllepta,* 130, 134, 144
- *Navicula radiosa,* 148
- *Navicula tripunctata,* 148
- *Nitzschia anatoliensis,* 134
- *Nitzschia inconspicua,* 126, 135, 144, 148
- *Nitzschia palea,* 135, 146, 148
- *Nitzschia soratensis,* 126, 135
- *Pinnularia borealis,* 133, 135, 145, 148
- *Pinnularia gibba,* 136
- *Pinnularia* spp., 148
- *Pinnularia subgibba,* 136
- *Pseudo-nitzschia arctica,* 136
- *Pseudo-nitzschia pungens,* 136, 137
- *Pseudo-nitzschia* spp., 137, 144
- *Sellaphora capitata,* 131, 137
- *Sellaphora* spp., 144
- *Seminavis robusta,* 130, 138
- *Skeletonema marinoi,* 138
- *Skeletonema* spp., 139, 144
- *Tetracyclus* spp., 124
- *Thalassiosira pseudonana,* 140
- *Thalassiosira rotula,* 139
- *Thalassiosira* spp., 140
- *Ulnaria* spp., 132

Diatom identification, 125, 143, 147
Discovery of auxosporulation, 60
Dispersal, 125, 142–146
Dispersal limitation, 89, 92, 99, 101
Dissolution, 333, 334, 336, 337
Distribution patterns, 125, 142, 146, 148
Disturbance, 237
Drainage systems, 42
Drought refuges, 374
Dry phase, 361–363, 381, 382, 383
Duero River basin, 41, 43–44

Ecological condition, 22
Ecological drift, 86, 89, 100, 106, 143
Ecological guild, 90, 98
Ecological guilds, 41, 44

Ecological niche, 143, 144
Ecological preferences, 166, 176–177, 182
Ecological quality ratio (EQR), 382
Ecoregions, 42
Ecosystem recovery, 4
Ecosystem rehabilitation, 4
Ecosystem restoration, 3
Edaphic, 371
Encyonema, 404, 406
Endangered species, 366, 381
Endemism, 397, 405
Environmental conditions, 174–175, 179–180
Environmental filtering, 86, 87, 91–92, 96–97, 100–101, 103–106
Environmental stressor, 14
Environmental variables, 329, 331, 332, 334, 335, 341, 351
Ephemeral streams, 362
Epilithon, 74
Epipelon, 65, 74–75
Epiphytic diatoms, 41, 43
Epiphyton, 74
Epipsammon, 74–75
Eunotia, 396–397, 399, 404–405, 407–408
Eurheic, 370–372
Eurheic condition, 237
Eutrophication, 204, 205, 207, 214, 216, 217, 346, 352
Experimental studies, 175, 178–180

Facilitation, 237
Field conductivimeter, 44
Flagship species, 125
Forensic science, 281, 301, 303
Freezing intermittent rivers, 362
Frustulia, 397, 404–406
Functional biogeography, 106
Functional group, 237
Fundamental niche, 237

Gamma diversity, 205
Geitler, L., 67, 69–70
Generalists, 96, 101
Gomphonella olivacea auxosporulation, 68
Grazing effects on size spectra, 73
Growth forms, 237
Guild, 237

Haplotype, 142–146, 148–150
HDI, 209, 211, 212
Heat shock proteins, 375, 379
Heterogeneity, 205, 214, 218
Heterovalvy, 171–172
High profile, 90, 97–98, 99, 101, 105
High-throughput sequencing (HTS), 146, 148
Historical, 205, 206, 211, 218
Hungarian phytobenthos metric (IPSITI), 382
Hydric stress, 366
Hydrological regime, 394–396, 398, 404
Hydrological tracer, 281, 298–299
Hyperrheic, 370, 371
Hypnospores, 367
Hyporheic, 371
Hyporheic zone, 362

Ice cover, 330, 331, 332, 348, 349, 350
Indicator criteria, 5
Indicator organisms, 5
Indicator value, 41
Induction of sexuality, 61, 62, 65, 69–70
Intermittent rivers (IR)/streams, 361, 362, 364, 365, 381, 383
Ion chromatography, 44

Jewson, D.H., 68–72

Kobayasiella, 397, 404–405

Lake Baikal, 68–69
Lentification, 366, 369, 370, 374, 380, 381
Life cycle, 60–62
Life forms, 373, 374
Life history strategies, 71–72
Limnological variables, 43–44
Lineage, 142–146, 148–150
Low profile, 90, 97–98, 99, 101
Low-profile guild, 369, 381

Management, 203, 207, 211, 215, 216, 217, 219
Mass effects, 86, 89, 100, 106
Mediterranean GIG river types, 364
Mesoscale level, 43
Metabarcoding, 93, 106, 146–149
Metacommunity, 87–89, 93–94, 98, 102–104, 106, 142–143, 254
Metrics, 18
Microtopography, 396, 398, 406, 409
Minerotrophicfens, 395–396
Mobile life form, 369, 373
Moisture, 398–399
Molecular markers of auxosporulation, 76
Molecular studies, 165, 168–170, 180–181
Molecular techniques, 281, 285, 288, 301, 303
Morphological variability, 166, 170–173, 176, 182–183
Morphospecies, 143–145, 148, 150
Motile, 91, 97–99, 101, 105
Mucilage life form, 369, 373
Multivariate techniques, 331
Mutations, 143
Mycosporine-like amino acids (MAAs), 378

Natural, 203, 204, 205, 206, 211, 219, 220
Neidium, 397, 399, 404
Nestedness, 237, 403
Niche breadth, 42–43, 92, 96, 98
Niche characteristics, 41–42, 44
Niche position, 42, 92, 96–97
Nitrogen-fixing guild, 241
Non-urban, 205, 207, 208, 210, 211, 213, 216, 217
Novel, 204, 205, 206, 208, 211, 213, 214, 215, 216
Nutrients, 329, 330, 331, 332, 333, 340, 345, 346, 351, 395–396, 405, 409

Observer bias, 69–70
Oligorheic, 370
Oligotrophic, 21
Ombrotrophicbogs, 395–396, 398, 401

P. borealis complex, 285, 293, 301
Paleoclimatology, 330, 331
Paleolimnology, 12, 176
Parasitism effects on size spectra, 73
Participatory science, 26
Peatland, 393–402, 404, 408–409
Performance measure, 11
Pfitzer, E., 61, 64, 66
pH, 329, 331, 332, 333, 340, 341, 343, 344, 345, 348, 351, 352, 393–394, 396–398, 402–403, 405, 407, 409
Phenology, 60, 66–69
Phenotypic plasticity, 171–173, 181
Photosensitizing agents, 377
Phytobenthos, 382
Phytoplankton, 174, 178
Pinnularia, 396–397, 399, 404–407
Plankton life cycles, 74
Planktonic, 89, 91, 96, 98–99, 101, 104–105
Planktonic diatoms, 41, 43, 375
Pollution, 296–298, 303
Polyamine, 369, 376
Preservation, 334, 336, 337
Priority effect, 146
Priority effects, 86, 90, 102–103
Protein denaturation, 373
Protein hydration shell, 373

Proximate factors, 143
Public communication, 25
Public engagement, 25

Quantitative reconstructions, 335

Ramsar site, 401
Rarity, 149
"Rate of life" model, 42
rbcL, 126–138, 147, 148
Reactive oxygen species, 373, 377
Reference condition, 14
Reference sequence database, 146, 147
Relative abundance, 177–178
Resilience, 203, 205, 206
Resource, 237
Resource use hypothesis, 92, 96
Resting spores, 367
Resting stages, 366–369, 374
Restoration goals, 10
Richness, 398–399, 404–405, 407–408
Rott trophic index (TID), 382

Scenarios, 23
Scientific expeditions, 124, 147
Screening agents, 369, 379
Seasonality of auxosporulation, 60, 66–69
Sedimentary record, 332, 351
Selection, 143
Sexual reproduction, 60, 73
Sexualization, 61, 62, 65, 69–70
Shannon diversity, 381
Shape analysis, 170–171
Size spectra, 67, 76
Snow-melt effects on auxosporulation, 68
Soil quality, 281, 283, 297, 301, 303
Spatial scaling, 9
Specialists, 96, 99, 101
Species complex, 125, 126, 128, 129, 133–136, 142–145, 148–150
Species limits, 168–170, 173
Species turnover, 394, 398, 403
Specific pollution sensitive index (SPI), 382
Spectrophotometry, 44
Sphagnum, 394, 396–402, 405–408
Spring peak of auxosporulation, 67
Stephanodiscus life cycle, 68–69, 71
Stress, 237
Survey, 15
Synchronous sexuality, 71

Taphofacies analysis, 337
Taphonomy, 337
Target setting, 12
Taxonomic concepts, 165–167, 169–170, 173
Taxonomical diversity, 381
Temperature, 329, 331, 332, 333, 334, 340, 342, 343, 345, 347, 348, 351
Temporal scaling, 9
Terrestrial diatoms, 366, 373, 374, 376
Terrestrialization gradient, 396
Thermal stress, 361, 366, 374
Thermotolerant, 375
Threshold indicator taxa analysis, 20
Thwaites, G.H., 60
Tierra del Fuego, 393, 400–402, 404, 408
Time-stability hypothesis, 102
Trait, 95, 99
Transfer functions, 329, 331, 335, 336, 339, 341, 344, 345, 348
Type method, 165, 167–168

Ubiquitous dispersal, 125, 142, 148
Undersampling, 149

Vegetative cell enlargement, 73
Vicariance, 143

Warm-water diatoms, 242
Water framework directive, 382
Water level, 329, 330, 331, 332, 337, 338, 344, 346, 347, 351, 352
Wetlands, 393–395, 399–400, 408

Also of Interest

Check out these published and forthcoming titles in "Diatoms: Biology and Applications" series from Scrivener Publishing

Diatom Cultivation for Biofuel, Food, and High-Value Products
Edited by Vandana Vinayak and Richard Gordon
Forthcoming 2024. ISBN 978-1-394-17448-5

Diatom Ecology
Edited by Nora I. Maidana, Magdalena Licursi and Eduardo Morales
Published 2024. ISBN 978-1-394-17445-4

Diatom Photosynthesis: From High-Value Molecules to Primary Production
Edited by Johannes Wilhelm Goessling, João Serôdio and Johann Lavaud
Published 2024. ISBN 978-1-119-84208-8

Mathematical Macroevolution in Diatom Research
By Janice L. Pappas
Published 2023. ISBN 978-1-119-75043-7

The Mathematical Biology of Diatoms
Edited by Janice L. Pappas
Published 2023. ISBN 978-1-119-74985-1

Diatom Microscopy
Edited by Nirmal Mazumder and Richard Gordon
Published 2022. ISBN 9781119711537

Diatoms Morphogenesis
Edited by Vadim V. Annenkov, Joseph Seckbach and Richard Gordon
Published 2021. ISBN 978-1-119-48795-1

Diatoms Motility
Edited by Stanley A. Cohn, Kalina M. Manoylov and Richard Gordon
Published 2021. ISBN 978-1-119-52635-3

Diatoms
Fundamentals and Applications
Edited by Joseph Seckbach and Richard Gordon
Published 2019. ISBN 978-1-119-37021-5

Printed and bound by CPI Group (UK) Ltd, Croydon, CR0 4YY
17/09/2026
14964278-0001